LONDON MATHEMATICAL SOCIETY LECTURE NOTE SERIES

Managing Editor: Professor Endre Süli, Mathematical Institute, University of Oxford, Woodstock Road, Oxford OX2 6GG, United Kingdom

The titles below are available from booksellers, or from Cambridge University Press at www.cambridge.org/mathematics

366 Highly oscillatory problems, B. ENGQUIST, A. FOKAS, E. HAIRER & A. ISERLES (eds)
367 Random matrices: High dimensional phenomena, G. BLOWER
368 Geometry of Riemann surfaces, F.P. GARDINER, G. GONZÁLEZ-DIEZ & C. KOUROUNIOTIS (eds)
369 Epidemics and rumours in complex networks, M. DRAIEF & L. MASSOULIÉ
370 Theory of p-adic distributions, S. ALBEVERIO, A.YU. KHRENNIKOV & V.M. SHELKOVICH
371 Conformal fractals, F. PRZYTYCKI & M. URBAŃSKI
372 Moonshine: The first quarter century and beyond, J. LEPOWSKY, J. MCKAY & M.P. TUITE (eds)
373 Smoothness, regularity and complete intersection, J. MAJADAS & A. G. RODICIO
374 Geometric analysis of hyperbolic differential equations: An introduction, S. ALINHAC
375 Triangulated categories, T. HOLM, P. JØRGENSEN & R. ROUQUIER (eds)
376 Permutation patterns, S. LINTON, N. RUŠKUC & V. VATTER (eds)
377 An introduction to Galois cohomology and its applications, G. BERHUY
378 Probability and mathematical genetics, N. H. BINGHAM & C. M. GOLDIE (eds)
379 Finite and algorithmic model theory, J. ESPARZA, C. MICHAUX & C. STEINHORN (eds)
380 Real and complex singularities, M. MANOEL, M.C. ROMERO FUSTER & C.T.C WALL (eds)
381 Symmetries and integrability of difference equations, D. LEVI, P. OLVER, Z. THOMOVA & P. WINTERNITZ (eds)
382 Forcing with random variables and proof complexity, J. KRAJÍČEK
383 Motivic integration and its interactions with model theory and non-Archimedean geometry I, R. CLUCKERS, J. NICAISE & J. SEBAG (eds)
384 Motivic integration and its interactions with model theory and non-Archimedean geometry II, R. CLUCKERS, J. NICAISE & J. SEBAG (eds)
385 Entropy of hidden Markov processes and connections to dynamical systems, B. MARCUS, K. PETERSEN & T. WEISSMAN (eds)
386 Independence-friendly logic, A.L. MANN, G. SANDU & M. SEVENSTER
387 Groups St Andrews 2009 in Bath I, C.M. CAMPBELL *et al* (eds)
388 Groups St Andrews 2009 in Bath II, C.M. CAMPBELL *et al* (eds)
389 Random fields on the sphere, D. MARINUCCI & G. PECCATI
390 Localization in periodic potentials, D.E. PELINOVSKY
391 Fusion systems in algebra and topology, M. ASCHBACHER, R. KESSAR & B. OLIVER
392 Surveys in combinatorics 2011, R. CHAPMAN (ed)
393 Non-abelian fundamental groups and Iwasawa theory, J. COATES *et al* (eds)
394 Variational problems in differential geometry, R. BIELAWSKI, K. HOUSTON & M. SPEIGHT (eds)
395 How groups grow, A. MANN
396 Arithmetic differential operators over the p-adic integers, C.C. RALPH & S.R. SIMANCA
397 Hyperbolic geometry and applications in quantum chaos and cosmology, J. BOLTE & F. STEINER (eds)
398 Mathematical models in contact mechanics, M. SOFONEA & A. MATEI
399 Circuit double cover of graphs, C.-Q. ZHANG
400 Dense sphere packings: a blueprint for formal proofs, T. HALES
401 A double Hall algebra approach to affine quantum Schur–Weyl theory, B. DENG, J. DU & Q. FU
402 Mathematical aspects of fluid mechanics, J.C. ROBINSON, J.L. RODRIGO & W. SADOWSKI (eds)
403 Foundations of computational mathematics, Budapest 2011, F. CUCKER, T. KRICK, A. PINKUS & A. SZANTO (eds)
404 Operator methods for boundary value problems, S. HASSI, H.S.V. DE SNOO & F.H. SZAFRANIEC (eds)
405 Torsors, étale homotopy and applications to rational points, A.N. SKOROBOGATOV (ed)
406 Appalachian set theory, J. CUMMINGS & E. SCHIMMERLING (eds)
407 The maximal subgroups of the low-dimensional finite classical groups, J.N. BRAY, D.F. HOLT & C.M. RONEY-DOUGAL
408 Complexity science: the Warwick master's course, R. BALL, V. KOLOKOLTSOV & R.S. MACKAY (eds)
409 Surveys in combinatorics 2013, S.R. BLACKBURN, S. GERKE & M. WILDON (eds)
410 Representation theory and harmonic analysis of wreath products of finite groups, T. CECCHERINI-SILBERSTEIN, F. SCARABOTTI & F. TOLLI
411 Moduli spaces, L. BRAMBILA-PAZ, O. GARCÍA-PRADA, P. NEWSTEAD & R.P. THOMAS (eds)
412 Automorphisms and equivalence relations in topological dynamics, D.B. ELLIS & R. ELLIS
413 Optimal transportation, Y. OLLIVIER, H. PAJOT & C. VILLANI (eds)
414 Automorphic forms and Galois representations I, F. DIAMOND, P.L. KASSAEI & M. KIM (eds)
415 Automorphic forms and Galois representations II, F. DIAMOND, P.L. KASSAEI & M. KIM (eds)
416 Reversibility in dynamics and group theory, A.G. O'FARRELL & I. SHORT
417 Recent advances in algebraic geometry, C.D. HACON, M. MUSTAŢĂ & M. POPA (eds)
418 The Bloch–Kato conjecture for the Riemann zeta function, J. COATES, A. RAGHURAM, A. SAIKIA & R. SUJATHA (eds)

419 The Cauchy problem for non-Lipschitz semi-linear parabolic partial differential equations, J.C. MEYER & D.J. NEEDHAM
420 Arithmetic and geometry, L. DIEULEFAIT *et al* (eds)
421 O-minimality and Diophantine geometry, G.O. JONES & A.J. WILKIE (eds)
422 Groups St Andrews 2013, C.M. CAMPBELL *et al* (eds)
423 Inequalities for graph eigenvalues, Z. STANIĆ
424 Surveys in combinatorics 2015, A. CZUMAJ *et al* (eds)
425 Geometry, topology and dynamics in negative curvature, C.S. ARAVINDA, F.T. FARRELL & J.-F. LAFONT (eds)
426 Lectures on the theory of water waves, T. BRIDGES, M. GROVES & D. NICHOLLS (eds)
427 Recent advances in Hodge theory, M. KERR & G. PEARLSTEIN (eds)
428 Geometry in a Fréchet context, C.T.J. DODSON, G. GALANIS & E. VASSILIOU
429 Sheaves and functions modulo p, L. TAELMAN
430 Recent progress in the theory of the Euler and Navier–Stokes equations, J.C. ROBINSON, J.L. RODRIGO, W. SADOWSKI & A. VIDAL-LÓPEZ (eds)
431 Harmonic and subharmonic function theory on the real hyperbolic ball, M. STOLL
432 Topics in graph automorphisms and reconstruction (2nd Edition), J. LAURI & R. SCAPELLATO
433 Regular and irregular holonomic D-modules, M. KASHIWARA & P. SCHAPIRA
434 Analytic semigroups and semilinear initial boundary value problems (2nd Edition), K. TAIRA
435 Graded rings and graded Grothendieck groups, R. HAZRAT
436 Groups, graphs and random walks, T. CECCHERINI-SILBERSTEIN, M. SALVATORI & E. SAVA-HUSS (eds)
437 Dynamics and analytic number theory, D. BADZIAHIN, A. GORODNIK & N. PEYERIMHOFF (eds)
438 Random walks and heat kernels on graphs, M.T. BARLOW
439 Evolution equations, K. AMMARI & S. GERBI (eds)
440 Surveys in combinatorics 2017, A. CLAESSON *et al* (eds)
441 Polynomials and the mod 2 Steenrod algebra I, G. WALKER & R.M.W. WOOD
442 Polynomials and the mod 2 Steenrod algebra II, G. WALKER & R.M.W. WOOD
443 Asymptotic analysis in general relativity, T. DAUDÉ, D. HÄFNER & J.-P. NICOLAS (eds)
444 Geometric and cohomological group theory, P.H. KROPHOLLER, I.J. LEARY, C. MARTÍNEZ-PÉREZ & B.E.A. NUCINKIS (eds)
445 Introduction to hidden semi-Markov models, J. VAN DER HOEK & R.J. ELLIOTT
446 Advances in two-dimensional homotopy and combinatorial group theory, W. METZLER & S. ROSEBROCK (eds)
447 New directions in locally compact groups, P.-E. CAPRACE & N. MONOD (eds)
448 Synthetic differential topology, M.C. BUNGE, F. GAGO & A.M. SAN LUIS
449 Permutation groups and cartesian decompositions, C.E. PRAEGER & C. SCHNEIDER
450 Partial differential equations arising from physics and geometry, M. BEN AYED *et al* (eds)
451 Topological methods in group theory, N. BROADDUS, M. DAVIS, J.-F. LAFONT & I. ORTIZ (eds)
452 Partial differential equations in fluid mechanics, C.L. FEFFERMAN, J.C. ROBINSON & J.L. RODRIGO (eds)
453 Stochastic stability of differential equations in abstract spaces, K. LIU
454 Beyond hyperbolicity, M. HAGEN, R. WEBB & H. WILTON (eds)
455 Groups St Andrews 2017 in Birmingham, C.M. CAMPBELL *et al* (eds)
456 Surveys in combinatorics 2019, A. LO, R. MYCROFT, G. PERARNAU & A. TREGLOWN (eds)
457 Shimura varieties, T. HAINES & M. HARRIS (eds)
458 Integrable systems and algebraic geometry I, R. DONAGI & T. SHASKA (eds)
459 Integrable systems and algebraic geometry II, R. DONAGI & T. SHASKA (eds)
460 Wigner-type theorems for Hilbert Grassmannians, M. PANKOV
461 Analysis and geometry on graphs and manifolds, M. KELLER, D. LENZ & R.K. WOJCIECHOWSKI
462 Zeta and L-functions of varieties and motives, B. KAHN
463 Differential geometry in the large, O. DEARRICOTT *et al* (eds)
464 Lectures on orthogonal polynomials and special functions, H.S. COHL & M.E.H. ISMAIL (eds)
465 Constrained Willmore surfaces, Á.C. QUINTINO
466 Invariance of modules under automorphisms of their envelopes and covers, A.K. SRIVASTAVA, A. TUGANBAEV & P.A. GUIL ASENSIO
467 The genesis of the Langlands program, J. MUELLER & F. SHAHIDI
468 (Co)end calculus, F. LOREGIAN
469 Computational cryptography, J.W. BOS & M. STAM (eds)
470 Surveys in combinatorics 2021, K.K. DABROWSKI *et al* (eds)
471 Matrix analysis and entrywise positivity preservers, A. KHARE
472 Facets of algebraic geometry I, P. ALUFFI *et al* (eds)
473 Facets of algebraic geometry II, P. ALUFFI *et al* (eds)
474 Equivariant topology and derived algebra, S. BALCHIN, D. BARNES, M. KĘDZIOREK & M. SZYMIK (eds)
475 Effective results and methods for Diophantine equations over finitely generated domains, J.-H. EVERTSE & K. GYŐRY

London Mathematical Society Lecture Note Series: 473

Facets of Algebraic Geometry

A Collection in Honor of William Fulton's 80th Birthday

VOLUME 2

Edited by

PAOLO ALUFFI
Florida State University

DAVID ANDERSON
Ohio State University

MILENA HERING
University of Edinburgh

MIRCEA MUSTAŢĂ
University of Michigan, Ann Arbor

SAM PAYNE
University of Texas, Austin

Shaftesbury Road, Cambridge CB2 8EA, United Kingdom

One Liberty Plaza, 20th Floor, New York, NY 10006, USA

477 Williamstown Road, Port Melbourne, VIC 3207, Australia

314–321, 3rd Floor, Plot 3, Splendor Forum, Jasola District Centre, New Delhi – 110025, India

103 Penang Road, #05–06/07, Visioncrest Commercial, Singapore 238467

Cambridge University Press is part of Cambridge University Press & Assessment, a department of the University of Cambridge.

We share the University's mission to contribute to society through the pursuit of education, learning and research at the highest international levels of excellence.

www.cambridge.org
Information on this title: www.cambridge.org/9781108792516

DOI: 10.1017/9781108877855

First published 2022

A catalogue record for this publication is available from the British Library

Library of Congress Cataloging-in-Publication data
Names: Aluffi, Paolo, 1960- editor.
Title: Facets of algebraic geometry : a collection in honor of William Fulton's 80th birthday / edited by Paolo Aluffi, Florida State University, David Anderson, Ohio State University, Milena Hering, University of Edinburgh, Mircea Mustaţă, University of Michigan, Ann Arbor, Sam Payne, University of Texas, Austin.
Description: Cambridge, United Kingdom ; New York, NY : Cambridge University Press, 2022. | Series: London mathematical society lecture note series ; 472, 473 | Includes bibliographical references.
Identifiers: LCCN 2021063035 (print) | LCCN 2021063036 (ebook) | ISBN 9781108870061 (2 volume set ; paperback) | ISBN 9781108792509 (volume 1 ; paperback) | ISBN 9781108792516 (volume 2 ; paperback) | ISBN 9781108877831 (volume 1 ; epub) | ISBN 9781108877855 (volume 2 ; epub)
Subjects: LCSH: Geometry, Algebraic. | Topology. | BISAC: MATHEMATICS / Topology
Classification: LCC QA564 .F33 2022 (print) | LCC QA564 (ebook) | DDC 516.3/5–dc23/eng/20220126
LC record available at https://lccn.loc.gov/2021063035
LC ebook record available at https://lccn.loc.gov/2021063036

ISBN – 2 Volume Set 978-1-108-87006-1 Paperback
ISBN – Volume 1 978-1-108-79250-9 Paperback
ISBN – Volume 2 978-1-108-79251-6 Paperback

We dedicate these volumes with gratitude and admiration to William Fulton, on the occasion of his 80th birthday. We hope that the articles collected here, written by colleagues, friends, and students, may reflect the breadth of Bill's impact on mathematics, along with the depth of his enduring influence on all of us who have had the great good fortune to learn from and be inspired by him.

Contents of Volume 2

Contents of Volume 1

Contributors (Volume 2)

Milena Hering
University of Edinburgh

Philipp Jell
University of Regensburg

Allen Knutson
Cornell University, New York

Khazhgali Kozhasov
Technical University of Braunschweig

Thomas Lam
University of Michigan

Seung Jin Lee
Seoul National University

Joshua F. Lieber
California Institute of Technology

Yuri I. Manin
Max Planck Institute for Mathematics

Matilde Marcolli
California Institute of Technology

Mateusz Michałek
University of Konstanz

Benjamin Nill
Otto-von-Guericke-Universität Magdeburg

John Christian Ottem
University of Oslo

Ragni Piene
University of Oslo

Sutipoj Promtapan
University of North Carolina, Chapel Hill

Richárd Rimányi
University of North Carolina, Chapel Hill

Colleen Robichaux
University of Illinois at Urbana-Champaign

Mark Shimozono
Virginia Tech

Frank Sottile
Texas A & M University

Bernd Sturmfels
Max Planck Institute for Mathematics in the Sciences, Leipzig

Hendrik Süß
University of Manchester

Richard P. Thomas
Imperial College London

Robert Williams
Sam Houston State University

Harshit Yadav
Rice University, Houston

Li Ying
Vanderbilt University, Tennessee

Alexander Yong
University of Illinois at Urbana-Champaign

Contributors (Volume 1)

Paolo Aluffi,
Florida State University

Asher Auel,
Dartmouth College, New Hampshire

Michel Brion,
Institut Fourier, Université Grenoble Alpes

Melody Chan,
Brown University, Rhode Island

Alessio Corti,
Imperial College London

Chiara Damiolini, ,
Rutgers University, New Jersey

Dan Edidin,
University of Missouri, Columbia

Lawrence Ein,
University of Illinois, Chicago

Gavril Farkas,
Humboldt-Universität zu Berlin

Matej Filip,
University of Ljubljana

Søren Galatius,
University of Copenhagen

Noah Giansiracusa,
Bentley University, Massachusetts

Angela Gibney,
Rutgers University, New Jersey

Eduardo González,
University of Massachusetts, Boston

William Graham,
University of Georgia

Johan P. Hansen,
Aarhus Universitet, Denmark

Robert Lazarsfeld,
Stony Brook University, New York

Leonardo C. Mihalcea,
Virginia Tech University

Sam Payne,
University of Texas, Austin

Andrea Petracci,
Università di Bologna

Ryan Richey,
University of Missouri, Columbia

Jörg Schürmann,
Universität Münster

Changjian Su,
University of Toronto

Nicola Tarasca,
Virginia Commonwealth University

Chris T. Woodward,
Rutgers University, New Jersey

Xian Wu,
Jagiellonian University, Krakow

14

Stability of Tangent Bundles on Smooth Toric Picard-rank-2 Varieties and Surfaces

Milena Hering[a], Benjamin Nill[b] and Hendrik Süß[c]

To Bill Fulton on the occasion of his 80th birthday.

Abstract. We give a combinatorial criterion for the tangent bundle on a smooth toric variety to be stable with respect to a given polarisation in terms of the corresponding lattice polytope. Furthermore, we show that for a smooth toric surface X and a smooth toric variety of Picard rank 2, there exists an ample line bundle with respect to which the tangent bundle is stable if and only if it is an iterated blow-up of projective space.

1 Introduction

Let X be a smooth toric variety of dimension n over a field of characteristic 0, with tangent bundle $\mathcal{T}_X$. Let $\mathcal{O}(D)$ be an ample line bundle. Recall that the slope of a torsion-free sheaf $\mathcal{E}$ on a normal projective variety X with respect to a nef line bundle $\mathcal{O}(D)$ is defined to be

$$\mu(\mathcal{E}) = \frac{c_1(\mathcal{E}) \cdot D^{n-1}}{\operatorname{rank}(\mathcal{E})},$$

and that $\mathcal{E}$ is *stable* (resp. *semistable*) with respect to $\mathcal{O}(D)$ if for any subsheaf $\mathcal{F}$ of $\mathcal{E}$ of smaller rank, we have $\mu(\mathcal{F}) < \mu(\mathcal{E})$ (resp. $\mu(\mathcal{F}) \leq \mu(\mathcal{E})$). A direct sum of stable sheaves with identical slope is called *polystable*. A situation of particular interest is when X is Fano, $\mathcal{E} = \mathcal{T}_X$ is the tangent bundle, and $D = -K_X$ the anticanonical divisor, in particular, since the existence of a Kähler–Einstein metric on a Fano variety implies that the tangent bundle is polystable with respect to the anticanonical polarisation, see Section 1.1 for more details.

[a] University of Edinburgh
[b] Otto-von-Guericke-Universität Magdeburg
[c] The University of Manchester

The main question we discuss in this article is when toric varieties admit a polarisation $\mathcal{O}(D)$ such that the tangent bundle $\mathcal{T}_X$ is stable with respect to $\mathcal{O}(D)$. This question has been studied in [29] and recently also by Biswas, Dey, Genc, and Poddar in [2]. Note that it is well-known that the tangent bundle on projective space is stable with respect to $\mathcal{O}_{\mathbb{P}^n}(1)$.

Theorem 1.1 *Let X be a smooth toric surface or a smooth toric variety of Picard rank 2. Then there exists an ample line bundle $\mathcal{L}$ on X such that T_X is stable with respect to $\mathcal{L}$ if and only if it is an iterated blow-up of projective space.*

For more precise statements, see Theorems 1.3 and 1.4. Theorem 1.4 and a more detailed discussion of the Fano case has been independently obtained by Dasgupta, Dey, and Khan [11]. While for smooth toric varieties of Picard rank 3 it is an open question whether Theorem 1.1 holds, there exists a toric Fano 3-fold of Picard rank 4 whose tangent bundle is stable with respect to the anticanonical polarisation, but that does not admit a morphism to $\mathbb{P}^3$, see Example 5.1.

We deduce the following criterion for the tangent bundle $\mathcal{T}_X$ on a toric variety X to be stable with respect to a given polarisation $\mathcal{O}(D)$ from well-known descriptions of stability conditions in terms of the Klyachko filtrations associated to the tangent bundle (see, for example, [20, 21, 23]). Now fix a fan Σ corresponding to X. Let P_D be the lattice polytope associated to D. For each ray ρ in Σ, let P_D^ρ denote the facet corresponding to ρ, and let v_ρ denote the primitive vector generating ρ.

Proposition 1.2 *The tangent bundle on a smooth projective toric variety X of dimension n is (semi)-stable with respect to an ample line bundle $\mathcal{O}(D)$ on X if and only if for every proper subspace $F \subsetneq N \otimes k$ that is generated by primitive vectors v_ρ generating rays in the fan Σ, the following inequality holds:*

$$\frac{1}{\dim F} \sum_{v_\rho \in F} \operatorname{vol}(P_D^\rho) \overset{(\leq)}{<} \frac{1}{n} \sum_{\rho} \operatorname{vol}(P_D^\rho) = \frac{1}{n} \operatorname{vol} \partial P_D. \tag{1}$$

Here, $\operatorname{vol}(P^\rho)$ denotes the lattice volume inside the affine span of P^ρ with respect to the lattice $\operatorname{span}(P^\rho) \cap M$.

We now present our results with more details. Let $\operatorname{Amp}(X) \subset N^1(X)_{\mathbb{R}}$ denote the ample cone of X. It is convenient to define

$$\operatorname{Stab}(\mathcal{T}_X) = \{D \in \operatorname{Amp}(X) \mid \mathcal{T}_X \text{ is stable with respect to } \mathcal{O}(D)\}, \text{ and}$$
$$\operatorname{sStab}(\mathcal{T}_X) = \{D \in \operatorname{Amp}(X) \mid \mathcal{T}_X \text{ is semistable with respect to } \mathcal{O}(D)\}.$$

Using results from [14] one can show that if for a $\mathbb{Q}$-factorial variety $\mathrm{Stab}(\mathcal{T}_X)$ is non-empty, then for any birational morphism $X' \to X$, $\mathrm{Stab}(\mathcal{T}_{X'}) \neq \emptyset$, see 2.7. In particular, since the tangent bundle to $\mathbb{P}^n$ is stable with respect to the anticanonical polarisation, any iterated blow-up of projective space has $\mathrm{Stab}(\mathcal{T}_X) \neq \emptyset$.

Recall that every smooth toric surface is either a successive toric blow-up of $\mathbb{P}^2$ or of a Hirzebruch surface $\mathbb{F}_a$. In Lemma 3.2, we characterise the fans of smooth toric surfaces that are not a blow-up of $\mathbb{P}^2$ and use this to prove the following theorem.

Theorem 1.3 *Let X be a smooth toric surface. Then*

1 $\mathrm{Stab}(\mathcal{T}_X) = \mathrm{Amp}(X)$ *if and only if* $X = \mathbb{P}^2$
2 $\emptyset = \mathrm{Stab}(\mathcal{T}_X) \subsetneq \mathrm{sStab}(\mathcal{T}_X)$ *if and only if* $X \cong \mathbb{P}^1 \times \mathbb{P}^1$.
3 $\emptyset \subsetneq \mathrm{Stab}(\mathcal{T}_X) \subsetneq \mathrm{Amp}(X)$ *if and only if X is an iterated blow-up of $\mathbb{P}^2$, but not $\mathbb{P}^2$ itself,*
4 $\mathrm{Stab}(\mathcal{T}_X) = \emptyset$ *if and only if X is not an iterated blow-up of $\mathbb{P}^2$.*

In [2, Theorem 6.2], Biswas et al. show that when X is the Hirzebruch surface $\mathbb{F}_a$, $a \geq 2$ implies that $\mathrm{Stab}(\mathcal{T}_X) = \emptyset$ and for $a = 1$ they describe $\mathrm{Stab}(\mathcal{T}_X)$ in [2, Corollary 6.3].

Projectivisations of direct sums of line bundles on projective spaces yield examples of toric Fano varieties under some conditions, but are also interesting in their own right. By [18, Theorem 1] every smooth toric variety of Picard rank 2 is of the form $X = \mathbb{P}_{\mathbb{P}^s}(\mathcal{O} \oplus \bigoplus_{i=1}^r \mathcal{O}(a_i))$, and X is a blow-up of $\mathbb{P}^s$ if and only if $(a_1, \dots, a_r) = (0, \dots, 0, 1)$. Note that the polytopes corresponding to ample line bundles on these varieties are special cases of *Cayley polytopes*, see for example [4].

In general, the projectivization $X := \mathbb{P}_Y(\mathcal{E}) = \mathrm{Proj}_Y(\bigoplus_m S^m\mathcal{E})$ of a vector bundle $\mathcal{E}$ on a variety Y admits a relatively ample line bundle $\mathcal{O}_X(1)$ induced by the relative Proj construction and we have $\mathrm{Pic}(X) = \pi^* \mathrm{Pic}(Y) \oplus \mathbb{Z}\mathcal{O}_X(1)$. Here, $\pi : X \to Y$ is the structure morphism of the relative Proj-construction. In the following, we always have $Y = \mathbb{P}^s$ and every element of the Picard group of X can be uniquely written in the form $\mathcal{O}_X(\lambda) \otimes \pi^*\mathcal{O}(\mu)$ with $\lambda, \mu \in \mathbb{Z}$.

Theorem 1.4 *Consider the smooth projective variety*

$$X = \mathbb{P}(\mathcal{O}_{\mathbb{P}^s} \oplus \bigoplus_{i=1}^{r} \mathcal{O}_{\mathbb{P}^s}(a_i))$$

for $s, r \geq 1$ with $0 \leq a_1 \leq \dots \leq a_r$. For $a_r \geq 1$, we have $\mathrm{Stab}(\mathcal{T}_X) \neq \emptyset$ *if and only if* $\mathrm{sStab}(\mathcal{T}_X) \neq \emptyset$ *if and only if* $(a_1, \dots, a_r) = (0, \dots, 0, 1)$. *In this case,*

$\mathcal{T}_X$ is (semi-)stable with respect to a polarisation $\mathcal{L} = \mathcal{O}_X(\lambda) \otimes \pi^*\mathcal{O}(\mu)$ if and only if $p(\mu/\lambda) \overset{(\leq)}{<} 0$, where $p(x)$ is the following polynomial of degree s:

$$p(x) := -\left(\sum_{q=0}^{s-1} \binom{r+s-1}{q} x^q\right) + \frac{s(r+1)}{r}\binom{r+s-1}{s} x^s.$$

We note that $p(\mu/\lambda) < 0$ if and only if μ/λ is in the interval $(0, \gamma)$, where γ is the only positive root of $p(x)$. For $r = 1$ we have $\gamma = \frac{1}{(2s+1)^{1/s}-1}$, and for $s = 1$ we get $\gamma = \frac{1}{r+1}$.

One has $\emptyset = \mathrm{Stab}(\mathcal{T}_X) \subsetneq \mathrm{sStab}(\mathcal{T}_X)$ if and only if $(a_1, \ldots, a_r) = (0, \ldots, 0)$, i.e. if $X = \mathbb{P}^s \times \mathbb{P}^r$. In this case $\mathcal{T}_X$ is semistable only with respect to pluri-anticanonical polarisations.

This result has been independently proved by [11]. It is extending a result by Biswas et. al. [2, Theorem 8.1], who show that in the Fano case (when $0 < a \leq s$), and when $s \geq 2$, the tangent bundle on $X = \mathbb{P}_{\mathbb{P}^s}(\mathcal{O} \oplus \mathcal{O}(a))$ is not stable with respect to the anticanonical polarisation $\mathcal{O}(-K_X) = \mathcal{O}(2) \otimes \pi^*\mathcal{O}(s+1-a)$.

The tangent bundle to a smooth Fano surface is stable with respect to the anticanonical polarisation by [12]. Moreoever, in [31] all smooth Fano threefolds with stable (resp. semistable) tangent bundle are classified. Moreover, for smooth toric Fano varieties of dimension 4 and Picard rank 2, the (semi-)stability of the tangent bundle with respect to the anticanonical polarisation is treated in [2, Section 9], and for smooth toric Fano varieties of dimension 4 and Picard rank 3 in [11].

The above results motivate the following question:

Question 1.5 Are there only finitely many isomorphism classes of smooth projective toric varieties X of given dimension n and Picard number ρ with $\mathrm{Stab}(\mathcal{T}_X) \neq \emptyset$?

Corollary 1.6 *Question 1.5 has an affirmative answer for $n \leq 2$ or $\rho \leq 2$.*

Proof The cases $n = 1$ or $\rho = 1$ are trivial. For $n = 2$ this follows from Theorem 1.3(3). For $\rho = 2$ this follows from Theorem 1.4 (note that $\dim(X) = r + s$). □

1.1 Connections to the existence problem of Kähler-Einstein metrics

When X is a smooth Fano variety over the complex numbers, the existence of a Kähler-Einstein metric on the underlying complex manifold X implies that its tangent bundle is polystable, (in particular, semistable) with respect to the anticanonical polarisation [25], [22, Sec 5.8]. However, the converse does not hold for the blow-up of $\mathbb{P}^2$ in two points [27]. The recent proof of the Yau–Tian–Donaldson conjecture [5, 6, 7, 8] shows that a Fano manifold has a Kähler-Einstein metric if and only if it is K-polystable. For a general toric Fano variety K-stability is equivalent to the fact that for the polytope corresponding to the anticanonical polarisation the barycenter coincides with the origin [24], in the smooth case this was known before due to combining [32] and [26].

Thus we obtain the following combinatorial statement:

Corollary 1.7 *Let P be a smooth reflexive polytope with barycenter in the origin. Then P satisfies the non-strict inequality* (1) *for every proper linear subspace $F \subset N_{\mathbb{Q}}$.*

This statement has been known to combinatorialists in a more general setting that implies the statement for reflexive polytopes with barycenter in the origin (without the smoothness assumption). Conditions of this type are known in convex geometry under the name *subspace concentration conditions*. They play a distinguished role in several problems from convex geometry, see e.g. [17, 16, 15]. The fact that this condition holds for a reflexive polytope whenever the barycenter coincides with the origin is far from being obvious. Moreover, our argument via Kähler-Einstein metrics is valid only in the smooth case (since we have to rely on [25], [22, Sec 5.8]), but the fact turns out to be true for every reflexive polytope. This follows from an even more general result in [15, Thm 1.1], which applies to every polytope with barycentre at the origin. Their proof relies entirely on methods from convex geometry.

Acknowledgements

We would like to thank Carolina Araujo, Arend Bayer, Stefan Kebekus, Adrian Langer, and Will Reynolds for helpful discussions. We would also like to thank University of Edinburgh, Institute of Mathematics of the Polish Academy of Sciences Warsaw, and Mathematical Sciences Research Institute for their hospitality. BN is funded by the Deutsche Forschungsgemeinschaft (DFG, German Research Foundation) - 314838170, GRK 2297 MathCoRe. BN was also partially supported by the Vetenskapsrådet grant NT:2014-3991 (as an affiliated researcher with Stockholm University).

Thanks to Bill

The first author would like to thank Bill Fulton for his invaluable advice and for helping create such a fertile mathematical atmosphere at Michigan.

2 Stability conditions for equivariant sheaves

We fix our setting as follows. We consider a polarized toric variety $(X, \mathcal{O}(D))$ corresponding to a lattice polytope P. Let Σ be the normal fan of P and P^ρ the facet of P corresponding to a ray $\rho \in \Sigma$. Denote by $\Sigma(1)$ the set of rays in Σ.

Recall that a coherent sheaf $\mathcal{E}$ is called *reflexive* if $\mathcal{E} \cong \mathcal{E}^{\vee\vee}$, where $\mathcal{E}^\vee = \mathcal{H}om(\mathcal{E}, \mathcal{O}_X)$. In [19] equivariant vector bundles on smooth toric varieties were classified in terms of collections of filtrations of k-vector spaces indexed by the rays of Σ. This classficiation extends to equivariant reflexive sheaves on normal toric varieties, see for example, [21, 30].

More precisely, we fix a k-vector space E and for every ray $\rho \in \Sigma^{(1)}$ we consider a decreasing filtration by subspaces

$$E \supset \cdots \supset E^\rho(i-1) \supset E^\rho(i) \supset E^\rho(i+1) \supset \cdots \supset 0,$$

such that $E^\rho(i)$ differs from E and 0 only for finitely many values of $i \in \mathbb{Z}$. Given such a collection of filtrations for every cone $\sigma \subset \Sigma$ we may consider

$$E_u := \left(\bigcap_{\rho \in \Sigma^{(1)}} E^\rho(-\langle v_\rho, u \rangle) \right) \otimes \chi^u \subset E \otimes k[M].$$

Then $\bigoplus_{u \in M} E_u$ is equipped with the structure of an M-graded $k[U_\sigma]$-module via the natural multiplication with $\chi^u \in k[U_\sigma]$. Then setting $H^0(U_\sigma, \mathcal{E}) = \bigoplus_{u \in M} E_u$ for every $\sigma \in \Sigma$ defines an equivariant reflexive sheaf on X.

The collections of filtrations form an abelian category in a natural way. A morphism between a collection of filtrations $F^\rho(i)$ of a vector space F and another collection $E^\rho(i)$ of filtrations of a vector space E is a linear map $L \colon F \to E$ that is compatible with the filtrations, i.e. $L(F^\rho(i)) \subset E^\rho(i)$ for all $\rho \in \Sigma^{(1)}$ and all $i \in \mathbb{Z}$.

Theorem 2.1 *There is an equivalence of categories between equivariant reflexive sheaves on a toric variety $X = X_\Sigma$ and collections of filtrations of k-vector spaces indexed by the rays of Σ. Here, the rank of the reflexive sheaf equals the dimension of the filtered k-vector space.*

For a collection of filtrations $E^\rho(i)$, we set $e^{[\rho]}(i) = \dim E^\rho(i) - \dim E^\rho(i+1)$. Similarly, for other filtrations we will always use the lower letter version to denote the differences of dimensions between the steps of the filtration. Then we have the following formula.

Lemma 2.2 *Assume that X is smooth. With the notation above we have*

$$\mu(\mathcal{E}) = \frac{1}{\dim E} \sum_{i,\rho} i \cdot e^{[\rho]}(i) \cdot \mathrm{vol}(P^\rho).$$

Proof By [23, Corollary 3.18], we have $c_1(\mathcal{E}) = \sum_\rho \sum_{i \in \mathbb{Z}} i e^{[\rho]}(i) D_\rho$. Now, for a ray $\rho \in \Sigma^{(1)}$ the intersection number $D^{n-1} \cdot D_\rho$ is given by the volume of the corresponding facet P^ρ of P, see e.g. [9]. □

With the notation above we get the following characterisation of stability.

Proposition 2.3 *Let X be a smooth toric variety. A toric vector bundle $\mathcal{E}$ on X corresponding to filtrations $E^\rho(i)$ is (semi-)stable if and only if the following inequality holds for every linear subspace $F \subset E$ and $F^\rho(i) = E^\rho(i) \cap F$.*

$$\frac{1}{\dim F} \sum_{i,\rho} i \cdot f^{[\rho]}(i) \cdot \mathrm{vol}(P^\rho) \overset{(\leq)}{<} \frac{1}{\dim E} \sum_{i,\rho} i \cdot e^{[\rho]}(i) \cdot \mathrm{vol}(P^\rho) \qquad (2)$$

Proof By [23, Proposition 4.13] it is sufficient to consider equivariant reflexive subsheaves. It remains to show that it is sufficient to consider those subsheaves, which correspond to filtrations of the form $E^\rho(i) \cap F$. For every subsheaf $\mathcal{F}' \subset \mathcal{E}$ corresponding to filtrations $(F')^\rho(i) \subset E^\rho(i)$ of some subspace $F \subset E$ we may consider the subsheaf $\mathcal{F}$ corresponding to the filtrations $F^\rho(i) := E^\rho(i) \cap F$. Then $\dim F^\rho(i) \geq \dim(F')^\rho(i)$ for all i, ρ. Now, Lemma 2.5 implies that $\mu(\mathcal{F}) \geq \mu(F')$. □

Remark 2.4 A subsheaf $\mathcal{F}$ of a torsion-free sheaf $\mathcal{E}$ is called *saturated* if $\mathcal{E}/\mathcal{F}$ is torsion-free. The saturation of a subsheaf $\mathcal{F} \subset \mathcal{E}$ is the smallest saturated subsheaf of $\mathcal{E}$ containing $\mathcal{F}$. It is not hard to derive from the description of $H^0(U_\sigma, \mathcal{E})$ given above, that $\mathcal{F} \subset \mathcal{E}$ given by $F^\rho(i) \subset E^\rho(i)$ is saturated, if and only if $F^\rho(i) = E^\rho(i) \cap F$. Hence, Lemma 2.5 below can be seen as a combinatorial version of the well-known fact that replacing a subsheaf by its saturation increases the slope.

Lemma 2.5 *Given integer functions $f, g : \mathbb{Z} \to \mathbb{Z}$ with $f \geq g$ such that $\{i \in \mathbb{Z} \mid f(i) \neq g(i)\}$ is finite. Then also*

$$\sum_i i \cdot (f(i) - f(i+1)) \geq \sum_i i \cdot (g(i) - g(i+1))$$

holds.

Proof Note that the assumption implies that $A(f,g) := \sum_i (f(i) - g(i)) \geq 0$ is finite. We fix f and proceed by induction on $A(f,g)$. If $A(f,g) = 0$, $f = g$ and the statement is trivially true. Fix f and assume that the statement holds for all $g \leq f$ with $A(f,g) \leq A$. Let g' be such that $A(f,g') = A+1$. Since $A > 0$, there exists a k such that $f(k) > g'(k)$. Define

$$g(i) = \begin{cases} g'(i) & \text{if } i \neq k \\ g'(i)+1 & \text{if } i = k. \end{cases}$$

Then $A(f,g) = A$. We calculate $\sum_i i \cdot (g(i) - g(i+1)) = \sum_i i \cdot (g'(i) - g'(i+1)) + 1$. By induction hypothesis, we have $\sum_i i \cdot (f(i) - f(i+1)) \geq \sum_i i \cdot (g(i) - g(i+1)) > \sum_i i \cdot (g'(i) - g'(i+1))$. □

By [19] the filtrations of the tangent bundle on $\mathcal{T}_X$ on a smooth toric variety X have the following form.

$$E^\rho(j) = \begin{cases} N \otimes k & j < 1 \\ \operatorname{span}_k(v_\rho) & j = 1 \\ 0 & j > 1 \end{cases} \tag{3}$$

Proof of Proposition 1.2 Looking at the filtrations $E^\rho(i)$ for $\mathcal{T}_X$ from (3) we see that

$$e^{[\rho]}(i) = E^\rho(i) - E^\rho(i+1) = \begin{cases} n-1 & j = 0 \\ 1 & j = 1 \\ 0 & \text{else.} \end{cases}$$

Similarly, for a proper subspace $F \subset N_{\mathbb{R}}$ and $F^\rho(i) = E^\rho(i) \cap F$, we have

$$f^{[\rho]}(i) = \begin{cases} \dim(F) - 2 & j = 0 \\ 1 & j = 1 \\ 0 & \text{else,} \end{cases} \quad \text{or} \quad f^{[\rho]}(i) = \begin{cases} \dim(F) - 1 & j = 0 \\ 0 & \text{else,} \end{cases}$$

depending on whether v_ρ is contained in the subspace F or not. Now, Proposition 2.3 immediately implies that $\mathcal{T}_X$ is (semi-)stable if and only if (1) holds for every proper subspace $F \subset N_{\mathbb{R}}$. To see that it suffices to test (1) for subspaces of the form $F = \operatorname{span}_k R$ with $R \subset \Sigma(1)$, assume that $\mathcal{F}$, given by some $F \subset N \otimes k$, destabilises $\mathcal{T}_X$. Then we may choose $\mathcal{F}'$ corresponding to $F' := \operatorname{span}\{v_\rho \subset \Sigma(1) \mid v_\rho \subset F\}$. With this choice we have $\sum_{v_\rho \in F} \operatorname{vol}(P^\rho) = \sum_{v_\rho \in F'} \operatorname{vol}(P^\rho)$ and $\operatorname{rk} \mathcal{F}' = \dim F' \leq \dim F = \operatorname{rk} \mathcal{F}$. □

Example 2.6 For $\mathbb{P}^n$ a polarisation is given by $\mathcal{O}(d)$. The corresponding polytope is a d-fold dilation of the standard simplex $\Delta \subset \mathbb{R}^n$. Every facet of $d\Delta$ has lattice volume d^{n-1}. For every proper subset $R \subsetneq \Sigma(1)$ and

$F = \operatorname{span} R$ we have $\dim F = \#R$. Now (1) becomes $d^{n-1} < d^{n-1} \cdot (n+1)/n$. Hence, we recover the well-known fact that $\mathbb{P}^n$ has a stable tangent bundle.

Lemma 2.7 *Assume that X is $\mathbb{Q}$-factorial and* $\operatorname{Stab}(\mathcal{T}_X)$ *is non-empty. If there is a birational morphism $f : X' \to X$, then* $\operatorname{Stab}(\mathcal{T}_{X'})$ *is non-empty, as well.*

Proof Consider a polarisation $\mathcal{O}(D)$ of X, such that $\mathcal{T}_X$ is stable. Then $\mathcal{T}_{X'}$ is stable with respect to the nef and big bundle $\mathcal{O}(f^*D)$, since any destabilising subsheaf $\mathcal{F}' \subset \mathcal{T}_{X'}$ with respect to $\mathcal{O}(f^*D)$ would induce a subsheaf $(f_*\mathcal{F}') \subset \mathcal{T}_X$ which, by projection formula, would be destabilising with respect to $\mathcal{O}(D)$. Now, the openness property from [14, Thm 3.3] ensures the existence of a stabilising ample class, which is given as a small pertubation of $[\mathcal{O}(f^*D)]$. □

We also have the following equivalent for the strictly unstable case.

Lemma 2.8 *Assume that X is $\mathbb{Q}$-factorial and* $\operatorname{Amp}(X) \setminus \operatorname{sStab}(\mathcal{T}_X)$ *is non-empty. If there is a birational morphism $f : X' \to X$, then* $\operatorname{Amp}(X') \setminus \operatorname{sStab}(\mathcal{T}_{X'})$ *is also non-empty.*

Proof Assume that a subsheaf $\mathcal{F} \subset \mathcal{T}_X$ destabilises $\mathcal{T}_X$ strictly with respect to an ample polarisation $\mathcal{O}(D)$. Then we note that $f^*\mathcal{F}$ and $\mathcal{T}_{X'}$ are both subsheaves of $f^*\mathcal{T}_X$. Now, we claim that $\mathcal{F}' := f^*\mathcal{F} \cap \mathcal{T}_{X'}$ destabilises $\mathcal{T}_{X'}$ with respect to $\mathcal{O}(D') = f^*\mathcal{O}(D)$. Indeed, by the projection formula we obtain

$$c_1(\mathcal{T}_{X'}).(D')^{n-1} = c_1(\mathcal{T}_X).(D)^{n-1}$$
$$c_1(\mathcal{F}').(D')^{n-1} = c_1(\mathcal{F}).(D)^{n-1}.$$

The line bundle $\mathcal{O}(D')$ is only nef, but the condition that a subsheaf destabilises strictly is an open condition on the divisor class. Hence, we can find an ample divisor class with the same property as a small perturbation of $\mathcal{O}(D')$. □

3 Smooth toric surfaces

Every toric surface can be obtained via equivariant blow-ups from $\mathbb{P}^2$ or from a Hirzebruch surface $F_a = \mathbb{P}_{\mathbb{P}^1}(\mathcal{O}_{\mathbb{P}^1} \oplus \mathcal{O}_{\mathbb{P}^1}(a))$, see e.g. [28]. For $\mathbb{P}^2$ it is well-known that the tangent bundle is stable (see also Example 2.6). The following corollary, which can be also found e.g. in [2, Sec. 6], clarifies the situation for the Hirzebruch surfaces.

Corollary 3.1 *For a Hirzebruch surface $F_a = \mathbb{P}_{\mathbb{P}^1}(\mathcal{O}_{\mathbb{P}^1} \oplus \mathcal{O}_{\mathbb{P}^1}(a))$ the tangent bundle is semistable with respect to $\mathcal{O}_{F_a}(\lambda) \otimes \pi^*\mathcal{O}_{\mathbb{P}^1}(\mu)$ in the following cases*

1 $a = 0$ *and* $\lambda = \mu$,
2 $a = 1$ *and* $2\mu \leq \lambda$.

The tangent bundle is stable if and only if $a = 1$ *and* $2\mu < \lambda$.

Proof The claim follows directly from Theorem 1.4 for the case $r = s = 1$. □

Lemma 3.2 *A smooth toric surface* $X = X_\Sigma$ *is not a blowup of* $\mathbb{P}^2$ *or* $\mathbb{P}^1 \times \mathbb{P}^1$ *if and only if there are integers* a, c, e *fulfilling* $a \geq c > e + 1 \geq 1$ *such that after an appropriate choice of basis for* N*:*

1 Σ *contains the rays spanned by*

$$(0,1),(1,0), \quad (0,-1),(1,-e), \quad (-1,c),(-1,a)$$

2 all other rays are contained in the cones $\langle(-1,c),(-1,a)\rangle$ *and* $\langle(1,0),(1,-e)\rangle$.

Remark 3.3 Note, that in Lemma 3.2 we explicitly allow the cases $(1,-e) = (1,0)$ $(-1,c) = (-1,a)$.

Proof For example by [28, Thm. 1.28] we may find a ray $\langle v \rangle \in \Sigma(1)$ such that $-v$ spans another ray in $\Sigma(1)$. We then may number the ray generators

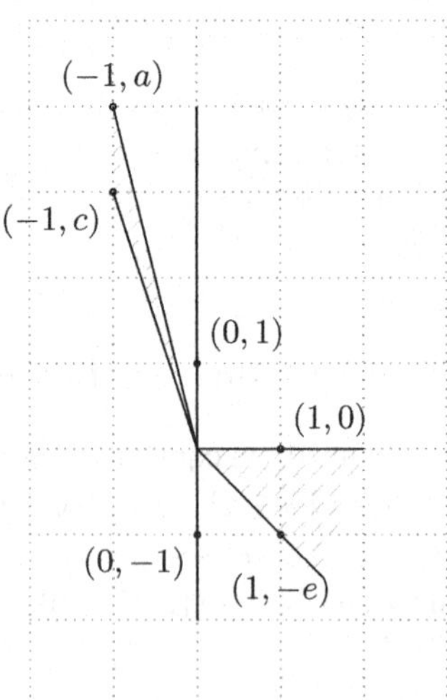

Figure 1 Schematic picture of a fan of a toric surface which does not blow down to $\mathbb{P}^2$ or $\mathbb{P}^1 \times \mathbb{P}^1$. All additional rays have to be contained in the shaded regions.

$v_0, \dots, v_r$ of the rays in $\Sigma(1)$ in consecutive order, such that $v_1 = v$ and $v_\ell = -v$ for some $\ell \in \{3, \dots, r\}$. Then by our smoothness condition v_1, v_2 form a basis of N. Hence, we may assume that $v_1 = (0,1)$ and $v_2 = (1,0)$. Again by smoothness $v_0 = (-1,a)$ for some $a \in \mathbb{Z}$. Since, X is assumed not to be a blowup of $\mathbb{P}^2$ or $\mathbb{P}^1 \times \mathbb{P}^1$ we have

$$\langle(-1,0)\rangle, \langle(-1,1)\rangle, \langle(-1,-1)\rangle \notin \Sigma^{(1)}. \tag{4}$$

Hence, $a \neq 0, 1, -1$. After possibly switching the role of v_1 and v_ℓ we may assume that $a \geq 2$. Now, assume $v_{\ell+1} = (b,c)$. The regularity of the cone $\langle(b,c),(0,-1)\rangle$ implies $b = -1$. Moreover, by [28, Prop. 1.19] the smoothness of X also implies that all the rays generated by vectors of the form $(-1,y)$ with $a \geq y \geq c$ have to be present in Σ. Now (4) implies that $a \geq c \geq 2$.

Similarly consider the ray $\langle v_{\ell-1}\rangle$ with $v_{\ell-1} = (d,-e)$. By regularity of the cone $\langle v_{\ell-1}, v_\ell\rangle$ we must have $d = 1$. Now, as before smoothness of X implies by [28, Prop. 1.19] that all the rays with generators $(1,-y)$ with $0 \leq y \leq e$ have to be contained in $\Sigma(1)$. On the other hand we must have $(1,-y) + (0,-1) \neq (1,-c)$ for all such y, since we assumed, that X is not a blowup of $\mathbb{P}^2$. Hence, $e \leq c - 2$ must hold. □

Proof of Theorem 1.3 The case of $\mathbb{P}^2$ is discussed in Example 2.6, and the case of Hirzebruch surfaces including $\mathbb{P}^1 \times \mathbb{P}^1$ in Corollary 3.1. For iterated blowups of $\mathbb{P}^2$ we see by Example 2.6 and Lemma 2.7 that $\operatorname{Stab}(\mathcal{T}_X) \neq \emptyset$. To show the strict inclusion $\operatorname{Stab}(\mathcal{T}_X) \subsetneq \operatorname{Amp}(X)$ as claimed in item (3) we refer to Corollary 3.1 together with Lemma 2.8. It remains to show that in the other cases there exists a subbundle of $\mathcal{T}_X$ which destabilises $\mathcal{T}_X$. For this we may assume that Σ has the form described in Lemma 3.2. We also fix the notation of the proof of that lemma, i.e. we may choose a basis of N order the primitive generators of rays in $\Sigma(1)$ clockwise, in such a way that

$$\begin{aligned} &v_1 = (0,1), v_2 = (1,0), v_{\ell-1} = (1,-e), \\ &v_\ell = (0,-1), v_{\ell+1} = (-1,c), v_0 = (-1,a). \end{aligned}$$

In the following we show that $F = \operatorname{span} v_1$ gives rise to a destabilising subbundle of $\mathcal{T}_X$.

Let us denote the torus invariant prime divisors corresponding to v_i by D_i and the maximal cones $\langle v_i, v_{i+1}\rangle$ by σ_i. Assume $D = \sum_i a_i D_i$ is an ample divisor. Then the corresponding polytope

$$P = \{u \in M_{\mathbb{R}} \mid \forall_{i=0,\dots,r} \colon \langle u, v_i\rangle \geq -a_i\}$$

has normal fan equal to Σ. Its facets are given by $P^{v_i} = P \cap \{u \mid \langle u, v_i \rangle = -a_i\}$ for $i = 0, \ldots, r$ and its vertices by

$$u_i = \{u \mid \langle u, v_i \rangle = -a_i\} \cap \{u \mid \langle u, v_{i+1} \rangle = -a_{i+1}\}.$$

To prove that $F = \operatorname{span} v_1$ gives rise to a destabilising subbundle by Proposition 1.2 we have to show that

$$\operatorname{vol} P^{v_1} + \operatorname{vol} P^{v_\ell} > \frac{1}{2} \sum_i \operatorname{vol} P^{v_i},$$

or equivalently that

$$\operatorname{vol} P^{v_1} + \operatorname{vol} P^{v_\ell} > \sum_{i \neq 1, \ell} \operatorname{vol} P^{v_i}.$$

Consider the trapezoid

$$Q = \{u \in M_{\mathbb{R}} \mid \forall_{i \in \{1, \ell-1, \ell, \ell+1\}} : \langle u, v_i \rangle \geq -a_i\}.$$

Then P is contained in Q. The lattice points $u_\ell, u_{\ell+1}$ are also vertices of Q and we have $Q^{v_\ell} = P^{v_\ell}$. We set $h = a_1 + a_\ell$. This is the lattice distance between the two parallel facets Q^{v_1} and Q^{v_ℓ} of the trapezoid. The two other facets of Q lie on the lines $\{\langle (1, -e), \cdot \rangle = -a_{\ell-1}\}$ and $\{\langle (-1, c), \cdot \rangle = -a_{\ell+1}\}$, respectively.

Then we have $\operatorname{vol} Q^{v_\ell} = \operatorname{vol} Q^{v_1} + h \cdot (c - e)$ for the lengths of these facets. Hence,

$$\operatorname{vol} P^{v_\ell} = \operatorname{vol} Q^{v_\ell} = \operatorname{vol} Q^{v_1} + h \cdot (c - e) \geq \operatorname{vol} Q^{v_1} + h \cdot 2 > 2h.$$

On the other hand the sum of the lattice length of the the remaining edges of P is at most $2h$. □

Remark 3.4 The argument that for $X \neq \mathbb{P}^2$, $\operatorname{Stab}(X)$ cannot be the full ample cone also follows from arguments in Pang's thesis [29]. Such a surface X admits a morphism to $\mathbb{P}^1$, and by [29, Theorem 0.09], the relative tangent

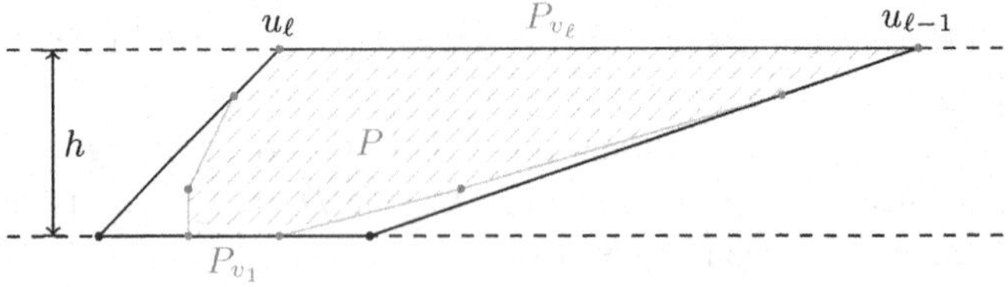

Figure 2 Schematic picture of the polytopes $P \subset Q$.

bundle of this morphism must be destabilising with respect to a suitable movable curve class. Note that Pang defines stability with respect to movable curve classes, but in the context of surfaces the definitions are equivalent.

4 Smooth toric varieties of Picard rank 2

We now turn to discussing toric varieties of Picard rank 2, i.e., of the form

$$X = \mathbb{P}_{\mathbb{P}^s}\left(\mathcal{O} \oplus \bigoplus_{i=1}^{r} \mathcal{O}(a_i)\right),$$

for $s, r \geq 1$. The lattice polytopes corresponding to line bundles on these varieties are special cases of Cayley polytopes, so we first review their construction and compute their volumes.

4.1 Cayley sums and volumes of Cayley sums of equal-dimensional dilated standard simplices

Let $P_0, \dots, P_r$ be $r+1$ polytopes in $\mathbb{R}^s$ and let $e_1, \dots, e_r$ denote the standard basis of $\mathbb{R}^r$. Then the convex hull of $(\{0\} \times P_0)$ and $(\{e_1\} \times P_1), \dots, (\{e_r\} \times P_r)$ in $\mathbb{R}^r \times \mathbb{R}^s$ is denoted by $P_0 * \cdots * P_r$ and is called the *Cayley sum* of these polytopes. In our situation, we consider Cayley polytopes where each P_i is a dilation of the s-dimensional standard simplex Δ_s. The normalised volumes of these Cayley sums are computed as follows.

Proposition 4.1 *Let $c_0, \dots, c_r \in \mathbb{Z}_{\geq 0}$, and $\nu > 0$. Then the normalized volume of $(\nu + c_0)\Delta_s * \cdots * (\nu + c_r)\Delta_s$ equals*

$$\sum_{k=0}^{s} \binom{s+r}{k} \left(\sum_{d_0+\cdots+d_r = s-k} c_0^{d_0} \cdots c_r^{d_r} \right) \nu^k.$$

The proof of Proposition 4.1 relies on the following general result:

Lemma 4.2 *For $(k_0, \dots, k_r) \in \mathbb{Q}^{r+1} \setminus \{(0, \dots, 0)\}$, we have that the normalized volume of $k_0\Delta_s * \cdots * k_r\Delta_s$ equals*

$$\sum_{m_0+\cdots+m_r=s} k_0^{m_0} \cdots k_r^{m_r},$$

where the sum is over $m_0, \dots, m_r \in \mathbb{N}$. For $r = 1$ this expression gets simplified to $(k_1^{s+1} - k_0^{s+1})/(k_1 - k_0)$ if $k_1 \neq k_0$, and to $(s+1)k_0^s$ if $k_1 = k_0$.

Proof By [10, 6.6], the normalized volume equals

$$\sum_{m_0+\cdots+m_r=s} \mathrm{MV}((k_0\Delta_s)^{(m_0)}, \ldots, (k_r\Delta_s)^{(m_r)}),$$

where MV denotes the (normalized) mixed volume of $r+1$ many s-dimensional polytopes and the exponents indicate the multiplicity of the polytope. Now, the statement follows from multilinearity of the mixed volume and $\mathrm{MV}(\Delta_s, \ldots, \Delta_s) = 1$. □

The following useful lemma is straightforward to prove by induction using a well-known identity of binomial coefficients, e.g., [13, (5.26), table 169].

Lemma 4.3 *Let $d_0, \ldots, d_r, k \in \mathbb{Z}_{\geq 0}$, and $r \geq 1$.*

$$\sum_{k_0+\cdots+k_r=k} \binom{d_0+k_0}{d_0} \cdots \binom{d_r+k_r}{d_r} = \binom{d_0+\cdots+d_r+r+k}{d_0+\cdots+d_r+r}$$
$$= \binom{d_0+\cdots+d_r+r+k}{k}$$

Proof of Proposition 4.1 Applying Lemma 4.2, we see that the normalized volume of $(\nu+c_0)\Delta_s * \cdots * (\nu+c_r)\Delta_s$ equals

$$\sum_{m_0+\cdots+m_r=s} (\nu+c_0)^{m_0} \cdots (\nu+c_r)^{m_r}$$
$$= \sum_{m_0+\cdots+m_r=s} \left(\sum_{k_0=0}^{m_0} \binom{m_0}{k_0} \nu^{k_0} c_0^{m_0-k_0}\right) \cdots \left(\sum_{k_r=0}^{m_r} \binom{m_r}{k_r} \nu^{k_r} c_r^{m_r-k_r}\right)$$
$$= \sum_{m_0+\cdots+m_r=s} \sum_{k=0}^{s} \sum_{k_0+\cdots+k_r=k} \binom{m_0}{k_0} \cdots \binom{m_0}{k_r} \cdot c_0^{m_0-k_0} \cdots c_r^{m_r-k_r} \nu^k$$
$$= \sum_{k=0}^{s} \left(\sum_{d_0+\cdots+d_r=s-k} \left(\sum_{k_0+\cdots+k_r=k} \binom{d_0+k_0}{d_0} \cdots \binom{d_r+k_r}{d_r}\right) \cdot c_0^{d_0} \cdots c_r^{d_r}\right) \nu^k.$$

By Lemma 4.3 this simplifies to

$$\sum_{k=0}^{s} \left(\sum_{d_0+\cdots+d_r=s-k} \binom{s+r}{k} \cdot c_0^{d_0} \cdots c_r^{d_r}\right) \nu^k.$$ □

4.2 Proof of Theorem 1.4

We start by reviewing the description of X in terms of the fan (see, for example, [4]). The primitive generators of the rays in the fan are vectors in $N_{\mathbb{R}} = \mathbb{R}^r \times \mathbb{R}^s$ of the following two types:

(i) $v_0 := \left(-\sum_{i=1}^r e_i, 0\right)$, and $v_i := (e_i, 0)$ for $i = 1, \ldots, r$;
(ii) $w_0 := \left(\sum_{i=1}^r a_i e_i, \; -\sum_{j=1}^s e_j\right)$, and $w_j := (0, e_j)$ for $j = 1, \ldots, s$.

Every ample line bundle on X has the form $\mathcal{L} = \mathcal{O}_X(\lambda) \otimes \pi^*\mathcal{O}(\mu)$ with $\lambda, \mu > 0$ and a lattice polytope corresponding to $\mathcal{L}$ is given by the Cayley sum

$$\lambda(\nu\Delta_s * (a_1 + \nu)\Delta_s * \cdots * (a_r + \nu)\Delta_s)) \subset M_{\mathbb{R}} = \mathbb{R}^r \times \mathbb{R}^s,$$

where $\nu = \mu/\lambda > 0$.

Since passing to multiples of polarisations has no effect on stability, we may equivalently consider the (rational) polarisation $\mathcal{O}_X(1) \otimes \pi^*\mathcal{O}(\nu)$. Hence, by Proposition 1.2 we are led to investigate whether for the lattice polytope

$$P := \nu\Delta_s * (a_1 + \nu)\Delta_s * \cdots * (a_r + \nu)\Delta_s$$

the subspace concentration condition (1) (strictly) holds. In this case, we say that P is *stable* (respectively, *semistable*).

We observe that

(i) for $i = 0, \ldots, r$ the facet of P corresponding to v_i is isomorphic to

$$\nu\Delta_s * (a_1 + \nu)\Delta_s * \cdots \overset{i}{\vee} \cdots * (a_r + \nu)\Delta_s,$$

i.e., to the polytope obtained from the Cayley sum representation of P by omitting the i-th summand. We denote its normalized volume by V_i.

(ii) for $j = 0, \ldots, s$ the facet of P corresponding to w_j is isomorphic to

$$\nu\Delta_{s-1} * (a_1 + \nu)\Delta_{s-1} * \cdots * (a_r + \nu)\Delta_{s-1}.$$

We denote its normalized volume by W.

We abbreviate

$$W_s := 0$$

$$W_k := \sum_{d_1 + \cdots + d_r = s-1-k} a_1^{d_1} \cdots a_r^{d_r}, \quad \text{for } k = 1, \ldots, s-1$$

$$D_k^0 := \sum_{d_1 + \cdots + d_r = s-k} a_1^{d_1} \cdots a_r^{d_r}$$

$$D_k^i := \sum_{d_1 + \cdots \overset{i}{\vee} \cdots + d_r = s-k} a_1^{d_1} \cdots \overset{i}{\vee} \cdots a_r^{d_r}, \quad \text{for } i = 1, \ldots, r, \text{ and } k = 0, \ldots, s.$$

Note that $D_s^i = 1$ and for $r = 1$, we use the convention that the interior expression equals 0^{s-k}, i.e., we have $D_s^i = 1$ even when $r = 1$.

The following Corollary follows from Proposition 4.1 and our notation.

Corollary 4.4 *The volumes of the facets of P are given by*

$$W = \sum_{k=0}^{s-1} \binom{s+r-1}{k} W_k \nu^k,$$

$$V_0 = \sum_{k=0}^{s} \binom{s+r-1}{k} D_k^0 \nu^k, \text{ and}$$

$$V_i = \sum_{k=0}^{s} \binom{s+r-1}{k} D_k^i \nu^k$$

for $i \in \{1, \ldots, r\}$. *Moreover, by convention we have* $V_1 = \nu^s$ *when* $r = 1$.

Proof of Theorem 1.4 If $a_1 = \cdots = a_r = 0$, then $X \cong \mathbb{P}^s \times \mathbb{P}^r$. Here, the tangent bundle splits and becomes semistable if and only if the summands have equal slope. It is straightfoward to check that this happens exactly for powers of the anti-canonical polarisation.

In the following we will *throughout* assume that $0 \le a_1 \le \cdots \le a_r \ge 1$. We first show that P is stable if and only if $a_1 = \cdots = a_{r-1} = 0, a_r = 1$, and $p(\mu/\lambda) < 0$ and treat the semi-stable situation in the end. Assume P is stable. We begin by showing that (1) for $F = \text{span}(\{v_0\})$ implies $a_r = 1$ and $a_{r-1} = 0$.

If $F = \text{span}(\{v_0\})$, then (1) multiplied through by $(r+s)$ becomes

$$(r+s-1)V_0 - V_1 - \cdots - V_r - (s+1)W < 0. \tag{5}$$

By Corollary 4.4, this implies

$$\sum_{k=0}^{s} \binom{s+r-1}{k} \alpha_k \nu^k < 0, \tag{6}$$

with

$$\alpha_k = (r+s-1)D_k^0 - D_k^1 - \cdots - D_k^r - (s+1)W_k.$$

Let $k \in \{0, \ldots, s\}$. Note that $D_k^i \le D_k^0$ for $i = 1, \ldots, r-1$ and

$$a_r W_k + D_k^r = \left(\sum_{d_1+\cdots+d_r=s-1-k} a_1^{d_1} \cdots a_{r-1}^{d_{r-1}} a_r^{d_r+1} \right) + \left(\sum_{d_1+\cdots+d_{r-1}=s-k} a_1^{d_1} \cdots a_{r-1}^{d_{r-1}} \right) \le D_k^0.$$

Thus we have

$$\begin{aligned}\alpha_k &\geq s D_k^0 - D_k^r - (s+1) W_k \\ &\geq s D_k^0 - D_k^r - \frac{s+1}{a_r}(D_k^0 - D_k^r) = \frac{(a_r-1)s-1}{a_r} D_k^0 + \frac{s+1-a_r}{a_r} D_k^r.\end{aligned}$$

Now assume $a_r \geq 2$. Then we get $(a_r - 1)s - 1 \geq 0$, and as $D_k^0 \geq D_k^r$, this yields

$$\alpha_k \geq \frac{(a_r-1)s-1}{a_r} D_k^r + \frac{s+1-a_r}{a_r} D_k^r = \frac{a_r(s-1)}{a_r} D_k^r \geq 0. \tag{7}$$

However, as $\nu > 0$, this implies

$$\sum_{k=0}^{s} \binom{s+r-1}{k} \alpha_k \, \nu^k \geq 0,$$

a contradiction to (6). Thus we have $a_r = 1$.

We now show that (6) also implies $a_{r-1} = 0$. Let us define the index $z \in \{0, \ldots, r-1\}$ such that

$$0 = a_1 = \cdots = a_z < a_{z+1} = \cdots = a_r = 1.$$

We claim that $z = r - 1$. Assume that $z < r - 1$. We show that in this case we also have $\alpha_k \geq 0$ for $k = 1, \ldots, s$, which is a contradiction to (5).

Our assumptions $a_r = 1$ and $z < r - 1$ imply

$$D_k^i = \begin{cases} \sum_{d_{z+1}+\cdots+d_r = s-k} 1 = \binom{r-z+s-k-1}{s-k} & \text{if } i = 0, \ldots, z \\ \sum_{d_{z+1}+\cdots \overset{i}{\vee} \cdots + d_r = s-k} 1 \cdots \overset{i}{\vee} \cdots 1 = \binom{r-z+s-k-2}{s-k} & \text{if } i = z+1, \ldots, r \end{cases}$$

and

$$W_k = \sum_{d_{z+1}+\cdots+d_r = s-1-k} 1 = \binom{r-z+s-k-2}{s-1-k}.$$

This implies that for $k = 0, \dots, s-1$

$$\begin{aligned}
\alpha_k &= (r+s-1-z)\binom{r-z+s-k-1}{s-k} - (r-z)\binom{r-z+s-k-2}{s-k} \\
&\quad - (s+1)\binom{r-z+s-k-2}{s-1-k} \\
&= ((r+s-1-z)-(r-z))\binom{r-z+s-k-2}{s-k} \\
&\quad + ((r+s-1-z)-(s+1))\binom{r-z+s-k-2}{s-1-k} \\
&= (s-1)\binom{r-2-z+s-k}{s-k} + (r-2-z)\binom{r-2-z+s-k}{s-1-k} \geq 0,
\end{aligned}$$

and for $k = s$ we have

$$\alpha_s = (r+s-1-z) - (r-z) = (s-1) \geq 0, \tag{8}$$

a contradiction to (6).

We have shown that if P is stable then $a_r = 1$ and $a_{r-1} = 0$. We now claim that in this case the left hand side of (1) is maximised for $F = \operatorname{span}(v_0, \dots, v_r)$, independent of ν. Note that in this case the volumes of the facets are $W = \sum_{k=0}^{s-1} \binom{s+r-1}{k} \nu^k$, $V_r = \binom{s+r-1}{s} \nu^s$, and for $i = 0, \dots, r-1$ we have $V_i = \sum_{k=0}^{s} \binom{s+r-1}{k} \nu^k$. In particular, we have

$$\frac{1}{r}\sum_{i=0}^{r} V_i = \left(\sum_{k=0}^{s-1}\binom{s+r-1}{k}\nu^k\right) + \frac{r+1}{r}\binom{s+r-1}{s}\nu^s > V_0 > W. \tag{9}$$

Now let F be a proper subspace of $\mathbb{R}^r \times \mathbb{R}^s$. By Proposition 1.2, we need to check (1) for subspaces generated by primitive vectors generating the rays of the fan of X. Then there are $I \subset \{0, \dots, r\}$ and $J \subset \{0, \dots, s\}$ such that $F = \operatorname{span}(\{v_i \mid i \in I\} \cup \{w_j \mid j \in J\})$.

Case 1: $|I| \geq r$. Then F contains $\{v_0, \dots, v_r\}$. Note that $\{w_0, \dots, w_s\}$ is linearly independent. As F is proper and contains $(\sum_{i=1}^r a_i e_i, 0)$, we get $|J| < s$ and $\dim(F) = r + |J|$. Therefore, the left hand side of (1) equals $\frac{1}{r+|J|}(V_0 + \cdots + V_r + |J|W) = \frac{1}{r+|J|}(r((V_0 + \cdots + V_r)/r) + |J|W) \leq \max((V_0 + \cdots + V_r)/r, W) = (V_0 + \cdots + V_r)/r$, and the claim follows.

Case 2: $|I| < r$. Here, $\{v_i : i \in I\}$ is linearly independent. If even $\{v_i : i \in I\} \cup \{w_j : j \in J\}$ is linearly independent, then the left hand side of (1) equals

$$\frac{1}{|I|+|J|}\left(\sum_{i\in I} V_i + |J|W\right) \le \max(V_0, W) = V_0 < (V_0+\cdots+V_r)/r.$$

Otherwise, we must have $\{0,\ldots,s\} \subseteq J$ and $(\sum_{i=1}^{r} a_i e_i, 0) = (e_r, 0) \in \operatorname{span}(\{v_i : i \in I\})$, which implies $r \in I$. We compute for $I \subseteq \{0,\ldots,r\}$ with $r \in I$ and $l := |I| < r$ that $\frac{1}{l+s}(\sum_{i\in I} V_i + (s+1)W)$ equals

$$\frac{1}{l+s}\left((l-1)\left(\sum_{k=0}^{s}\binom{s+r-1}{k}\nu^k\right)+\binom{s+r-1}{s}\nu^s+(s+1)\sum_{k=0}^{s-1}\binom{s+r-1}{k}\nu^k\right)$$

$$=\left(\sum_{k=0}^{s-1}\binom{s+r-1}{k}\nu^k\right)+\frac{l}{l+s}\binom{s+r-1}{s}\nu^s < V_0 < (V_0+\cdots+V_r)/r.$$

The claim implies that P is stable if and only if

$$\frac{s}{r(r+s)}\left(\sum_{i=0}^{r} V_i\right) - \frac{s+1}{r+s} W < 0,$$

equivalently,

$$\frac{s}{r}\left(r\left(\sum_{k=0}^{s}\binom{s+r-1}{k}\nu^k\right)+\binom{s+r-1}{s}\nu^s\right)-(s+1)\left(\sum_{k=0}^{s-1}\binom{s+r-1}{k}\nu^k\right)<0,$$

equivalently,

$$p(\nu) = -\left(\sum_{k=0}^{s-1}\binom{s+r-1}{k}\nu^k\right)+\frac{s(r+1)}{r}\binom{s+r-1}{s}\nu^s < 0.$$

We observe that by Descartes' rule of signs there is only one positive root of this polynomial. Note that for $r=1$ we get $p(x) = -\left(\sum_{q=0}^{s-1}\binom{s}{q}x^q\right)+2sx^s = -(x+1)^s+(2s+1)x^s$; and for $s=1$ we have $p(x) = -1+(r+1)x$.

Now assume P is semistable. We first claim that then (1) for $F = \operatorname{span}(v_0)$ implies $a_r = 1$ or $s = 1$, where in the latter case we have

$$(a_1,\ldots,a_r) \in \{(0,\ldots,0,1),(0,\ldots,0,2),(0,\ldots,0,1,1)\}.$$

Indeed, if P is semistable, inequality (6) becomes

$$\sum_{k=0}^{s}\binom{s+r-1}{k}\alpha_k\,\nu^k \le 0. \tag{10}$$

Assuming $a_r \geq 2$ and $s \geq 2$, we observe from (7) that $\alpha_k \geq 0$ for $k = 0, \ldots, s-1$, and $\alpha_s > 0$ as $D_s^r = 1$. Hence,

$$\sum_{k=0}^{s} \binom{s+r-1}{k} \alpha_k \nu^k \geq \binom{s+r-1}{s} \alpha_s \nu^s > 0,$$

a contradiction to (10). Hence, let $s = 1$. In this case, inequality (10) becomes $\alpha_0 + r\alpha_1\nu \leq 0$ with

$$\alpha_0 = r(a_1 + \cdots + a_r) - (r-1)(a_1 + \cdots + a_r) - 2 = a_1 + \cdots + a_r - 2,$$

and $\alpha_1 = 0$. Hence, $a_1 + \cdots + a_r \leq 2$ which implies the claim.

If $s \geq 2$, then $a_r = 1$ by the claim. The same arguments as in the stable case apply to show that this implies $a_{r-1} = 0$ (here, the contradiction is that (6) is nonpositive, while (8) is positive). Thus $(a_1, \ldots, a_r) = (0, \ldots, 0, 1)$, and the same arguments as in the stable case show that P is semistable if and only if $p(\nu) \leq 0$.

Now assume $s = 1$. We have already treated the case when $a_r = 1$ and $a_{r-1} = 0$. The remaining cases are $(a_1, \ldots, a_r) \in \{(0, \ldots, 0, 2), (0, \ldots, 0, 1, 1)\}$. One computes in both cases for the right hand side in (1) the value $r\nu + 2$, while the value in of the LHS of (1) for $F = \operatorname{span}\{v_0, \ldots, v_r\}$ is equal to $(r+1)\nu + 2$. Hence, P is never semistable in these cases. □

Proof of Theorem 1.1 This follows immediately from Theorem 1.3, Theorem 1.4, and [18, Theorem 1]. □

5 Two Examples of higher Picard rank

The following example shows that there exist toric Fano varieties that do not admit a morphism to projective space and whose tangent bundle is stable with respect to the anticanonical polarisation.

Example 5.1 Let X be the 3-dimensional smooth toric Fano variety that is the blow up at a line of $\mathcal{O}_{\mathbb{P}^1 \times \mathbb{P}^1} \oplus \mathcal{O}_{\mathbb{P}^1 \times \mathbb{P}^1}(1,1)$. Then X does not admit a morphism to $\mathbb{P}^3$ by [33]. However, the tangent bundle to X is stable with respect to the anticanonical polarisation by [31]. One can also check this by applying Proposition 1.2 to the polytope corresponding to the anticanonical polarisation, whose vertices are

$$\{(0,-1,-1),(-1,-1,-1),(-1,0,-1),(0,0,-1),(-1,-1,0),\\ (1,-1,0),(-1,2,1),(-1,0,1),(2,0,1),(2,2,1)\}.$$

See also [1, Example 4-11] and [3, ID#10].

The next example shows that the stability region for the tangent bundle inside the nef cone is usually neither convex nor polyhedral. At first glance this may look like a surprising fact for toric varieties. However, when replacing the self-intersection $(D)^{n-1}$ in the definition of the stability notion by an arbitrary class α of a movable curve, then for a fixed subbundle $\mathcal{F}$ the condition $\mu_\alpha(\mathcal{F}) < \mu_\alpha(\mathcal{T}_X)$ imposes a linear condition on α. Since, there are again only finitely many subbundles to consider, these conditions cut out a rational polyhedral subcone of the cone of movable curves (c.f. [29]). Then our stability region $\mathrm{Stab}(\mathcal{T}_X) \subset \mathrm{Nef}(X)$ is just the preimage of this polyhedral cone under the non-linear map

$$\mathrm{Nef}(X) \to \mathrm{Mov}(X), \quad [D] \to [(D)^{n-1}].$$

For a systematic treatment of stability with respect to curve classes in the toric setting see [29].

Example 5.2 We consider the iterated blowup $\phi\colon X \to \mathbb{P}^3$ in a point and the strict transform of a line through this point. Let us denote the pullback of a hyperplane by H and the exceptional divisors of the first and second blowup by E_1 and E_2 respectively.

The classes of $\mathcal{O}_X(H)$, $\mathcal{O}_X(E_1)$ and $\mathcal{O}_X(E_2)$ form a basis of the Picard group of X. The nef cone is spanned by $\mathcal{O}_X(H)$, $\mathcal{O}_X(H - E_1)$ and $\mathcal{O}_X(H - E_1 - E_2)$. Thus, a line bundle of the form

$$\mathcal{O}_X(\lambda H + \mu_1(H - E_1) - \mu_2(E_2))$$

is ample iff $\lambda, \mu_1, \mu_2 > 0$ and $\mu_1 > \mu_2$. We may rescale such a line bundle and obtain a $\mathbb{Q}$-line bundle of the form

$$\mathcal{O}_X(H + \nu_1(H - E_1) - \nu_2(E_2)).$$

To describe the corresponding polytope we consider the standard simplex $\Delta = \mathrm{conv}\{0, e_1, e_2, e_3\}$ and the following halfspaces in $\mathbb{R}^3$.

$$H_1^+(\nu) = \{u \in \mathbb{R}^3 \mid u_1 + u_2 \geq \nu\}, \quad H_2^+(\nu) = \{u \in \mathbb{R}^3 \mid u_3 \leq \nu\}.$$

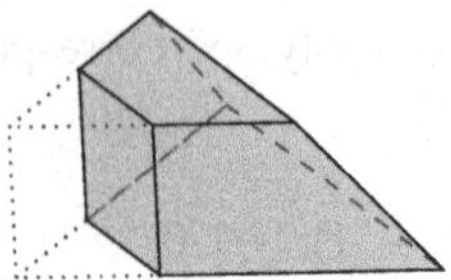

Figure 3 The polytope P_D corresponding to the polarisation.

Then the rational polytope corresponding to our ample line bundle from above is given by

$$(1+\nu_1)\Delta \cap H_1^+(\nu_1) \cap H_2^+(\nu_2).$$

We have two parallel facets $P^{\rho_0}, P^{-\rho_0} \prec P$, which are perpendicular to $\rho_0 = \mathbb{R}_{\geq 0} \cdot (0,0,1)$. These facets have volume $\nu_1^2 - \nu_2^2$ and $(1+\nu_1)^2 - \nu_2^2$, respectively. The remaining facets consist of a rectangle P^{ρ_1} and a trapezoid P^{ρ_2} opposite to each other and two more trapezoids P^{ρ_3}, P^{ρ_4}. Here, ρ_i denotes the corresponding rays in the normal fan of P. Elementary calculations show

$$\begin{aligned} \operatorname{vol}(P^{\rho_1}) &= 2\nu_2 \\ \operatorname{vol}(P^{\rho_2}) &= (1+\nu_1)^2 - \nu_1^2 \\ \operatorname{vol}(P^{\rho_3}) = \operatorname{vol}(P^{\rho_4}) &= (1+\nu_1)^2 - \nu_1^2 - 2\nu_2. \end{aligned}$$

We consider the subspaces $F_1 = \{\rho_0, -\rho_0\}$ and $F_2 = \operatorname{span}\{\rho_0, -\rho_0, \rho_1, \rho_2\}$. Note that $\dim F_1 = 1$ and $\dim F_2 = 2$. Now applying the stability codition (1) for F_1 gives

$$\operatorname{vol} P^{\rho_0} + \operatorname{vol} P^{-\rho_0} < \frac{\operatorname{vol} \partial P}{3},$$

where $\operatorname{vol} \partial P$ is the sum over all facet volumes. Using the facet volumes stated above, the inequality can be seen to be equivalent to

$$\frac{1}{3}(\nu_1+1)^2 - \frac{5}{3}\nu_1^2 + \frac{4}{3}\nu_2^2 - \frac{2}{3}\nu_2 > 0. \tag{11}$$

Similarly, evaluating (1) for F_2 leads to the inequality.

$$\operatorname{vol} P^{\rho_0} + \operatorname{vol} P^{-\rho_0} + \operatorname{vol} P^{\rho_1} + \operatorname{vol} P^{\rho_2} < \frac{\operatorname{vol} \partial P}{3},$$

which is equivalent to

$$\frac{1}{3}(\nu_1+1)^2 - \frac{2}{3}\nu_1^2 + \frac{1}{3}\nu_2^2 - \frac{5}{3}\nu_2 > 0. \tag{12}$$

A sketch of a cross-section of Nef(X) with the regions cut out by (11) and(12), respectively, is shown in Figure 4. By Proposition 1.2, there are four more

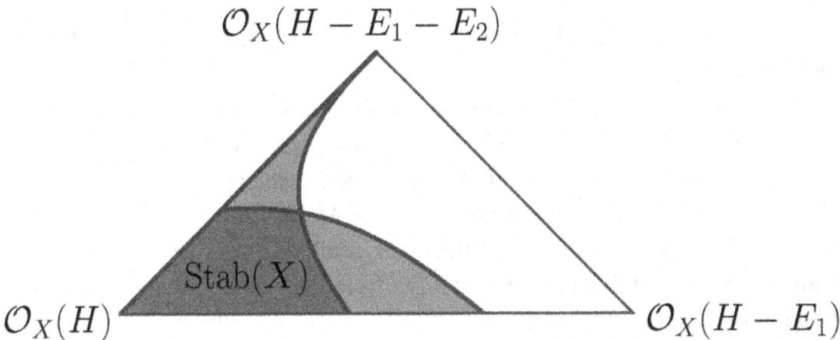

Figure 4 The stability region in Nef(X).

one-dimensional subspaces and three more two-dimensional subspaces which could provide additional obstructions for the stability of the tangent bundle. However, the corresponding subsheaves turn out to be not destabilising for any ample polarisation. Hence, Stab$(\mathcal{T}_X) \subset$ Nef(X) is cut out by the two inequalities (11) and (12).

References

[1] Pieter Belmans. Entry in Fanography, A tool to visually study the geography of Fano 3-folds. https://fanography.pythonanywhere.com/4-5.

[2] Indranil Biswas, Arijit Dey, Ozhan Genc, and Mainak Poddar. On stability of tangent bundle of toric varieties. *ArXiv e-prints*, August 2018.

[3] Gavin Brown and Alexander Kasprzyk. Graded Ring Database. A database of graded rings in algebraic geometry. http://www.grdb.co.uk.

[4] Eduardo Cattani, David Cox, and Alicia Dickenstein. Residues in toric varieties. *Compositio Math.*, 108(1):35–76, 1997.

[5] Xiuxiong Chen, Simon Donaldson, and Song Sun. Kähler-Einstein metrics and stability. *Int. Math. Res. Not.*, 2014(8):2119–2125, 2014.

[6] Xiuxiong Chen, Simon Donaldson, and Song Sun. Kähler-Einstein metrics on Fano manifolds. I: Approximation of metrics with cone singularities. *J. Am. Math. Soc.*, 28(1):183–197, 2015.

[7] Xiuxiong Chen, Simon Donaldson, and Song Sun. Kähler-Einstein metrics on Fano manifolds. II: Limits with cone angle less than 2π. *J. Am. Math. Soc.*, 28(1):199–234, 2015.

[8] Xiuxiong Chen, Simon Donaldson, and Song Sun. Kähler-Einstein metrics on Fano manifolds. III: Limits as cone angle approaches 2π and completion of the main proof. *J. Am. Math. Soc.*, 28(1):235–278, 2015.

[9] Vladimir I. Danilov. Geometry of toric varieties. *Russ. Math. Surv.*, 33(2):97–154, 1978.

[10] Vladimir I. Danilov and A. G. Khovanskiĭ. Newton polyhedra and an algorithm for calculating Hodge-Deligne numbers. *Izv. Akad. Nauk SSSR Ser. Mat.*, 50(5):925–945, 1986.

[11] Jyoti Dasgupta, Arijit Dey, and Bivas Khan. Stability of equivariant vector bundles over toric varieties. arXiv:1910.13964 [math.AG], 2019.

[12] Rachid Fahlaoui. Stabilité du fibré tangent des surfaces de Del Pezzo. (Stability of the tangent bundle of Del Pezzo surfaces). *Math. Ann.*, 283(1):171–176, 1989.

[13] Ronald L. Graham, Donald E. Knuth, and Oren Patashnik. *Concrete mathematics*. Addison-Wesley Publishing Company, Reading, MA, second edition, 1994. A foundation for computer science.

[14] Daniel Greb, Stefan Kebekus, and Thomas Peternell. Movable curves and semistable sheaves. *Int. Math. Res. Not.*, 2016(2):536–570, 2016.

[15] Martin Henk and Eva Linke. Cone-volume measures of polytopes. *Adv. Math.*, 253:50–62, 2014.

[16] Károly jun. Böröczky, Erwin Lutwak, Deane Yang, and Gaoyong Zhang. The logarithmic Minkowski problem. *J. Am. Math. Soc.*, 26(3):831–852, 2013.

[17] Károly jun. Böröczky, Erwin Lutwak, Deane Yang, and Gaoyong Zhang. Affine images of isotropic measures. *J. Differ. Geom.*, 99(3):407–442, 2015.

[18] Peter Kleinschmidt. A classification of toric varieties with few generators. *Aequationes Math.*, 35(2-3):254–266, 1988.

[19] Alexander A. Klyachko. Equivariant vector bundles on toral varieties. *Math. USSR-Izv.*, 35(2):337–375, 1990.

[20] Alexander A. Klyachko. Stable bundles, representation theory and Hermitian operators. *Selecta Math. (N.S.)*, 4(3):419–445, 1998.

[21] Allen Knutson and Eric Sharpe. Sheaves on toric varieties for physics. *Adv. Theor. Math. Phys.*, 2(4):873–961, 1998.

[22] Shoshichi Kobayashi. *Differential geometry of complex vector bundles*. Princeton, NJ: Princeton University Press; Tokyo: Iwanami Shoten Publishers, 1987.

[23] Martijn Kool. Fixed point loci of moduli spaces of sheaves on toric varieties. *Adv. Math.*, 227(4):1700–1755, 2011.

[24] Chi Li, Xiaowei Wang, and Chenyang Xu. Algebraicity of the metric tangent cones and equivariant k-stability. *arXiv:1805.03393*, 2018.

[25] Martin Lübke. Stability of Einstein-Hermitian vector bundles. *Manuscr. Math.*, 42:245–257, 1983.

[26] Toshiki Mabuchi. Einstein-Kähler forms, Futaki invariants and convex geometry on toric Fano varieties. *Osaka J. Math.*, 24(4):705–737, 1987.

[27] Yozô Matsushima. Sur la structure du groupe d'homéomorphismes analytiques d'une certaine variété kaehlérienne. *Nagoya Math. J.*, 11:145–150, 1957.

[28] Tadao Oda. *Convex bodies and algebraic geometry*, volume 15 of *Ergebnisse der Mathematik und ihrer Grenzgebiete (3) [Results in Mathematics and Related Areas (3)]*. Springer-Verlag, Berlin, 1988. An introduction to the theory of toric varieties, Translated from the Japanese.

[29] Thiam-Sun Pang. *The Harder-Narasimhan Filtrations and Rational Contractions*. PhD thesis, Universität Freiburg, Fakultät für Mathematik und Physik, 2015.

[30] Markus Perling. Graded rings and equivariant sheaves on toric varieties. *Math. Nachr.*, 263/264:181–197, 2004.

[31] Andreas Steffens. On the stability of the tangent bundle of Fano manifolds. *Math. Ann.*, 304(4):635–643, 1996.
[32] Xu-Jia Wang and Xiaohua Zhu. Kähler–Ricci solitons on toric manifolds with positive first Chern class. *Adv. Math.*, 188(1):87–103, 2004.
[33] Keiichi Watanabe and Masayuki Watanabe. The classification of Fano 3-folds with torus embeddings. *Tokyo J. Math.*, 5(1):37–48, 1982.

15

Tropical Cohomology with Integral Coefficients for Analytic Spaces

Philipp Jell[a]

The author was supported by the DFG Collaborative Research Center 1085 "Higher Invariants".

Abstract. We study tropical Dolbeault cohomology for Berkovich analytic spaces, as defined by Chambert-Loir and Ducros. We provide a construction that lets us pull back classes in tropical cohomology to classes in tropical Dolbeault cohomology as well as check whether those classes are non-trivial. We further define tropical cohomology with integral coefficients on the Berkovich space and provide some computations. Our main tool is extended tropicalization of toric varieties as introduced by Kajiwara and Payne.

To Bill Fulton on the occasion of his 80th birthday.

1 Introduction

Real valued-differential forms and currents on Berkovich analytic spaces were introduced by Chambert-Loir and Ducros in their fundamental preprint [3]. They provide a notion of bigraded differential forms and currents on these spaces that has striking similarities with the complex of smooth differential forms on complex analytic spaces. The definition works by formally pulling back Lagerberg's superforms on $\mathbb{R}^n$ along tropicalization maps. These tropicalization maps are induced by mapping open subsets of the analytic space to analytic tori and then composing with the tropicalization maps of the tori.

Payne, and independently Kajiwara [18, 26], generalized this tropicalization procedure from tori to general toric varieties and Payne showed

[a] University of Regensburg

that the Berkovich analytic space is the inverse limit over all these tropicalizations.

Shortly after the preprint by Chambert-Loir and Ducros, Gubler showed that one may, instead of considering arbitrary analytic maps to tori, restrict one's attention to algebraic closed embeddings if the analytic space is the Berkovich analytification of an algebraic variety [9].

Let K be a field that is complete with respect to a non-archimedean absolute value and let X be a variety over K. We write $\Gamma = \log|K^*|$ for the value group of K and X^{an} for the Berkovich analytification of X. Both the approach by Gubler and the one by Chambert-Loir and Ducros provide bigraded complexes of sheaves of differential forms $(\mathcal{A}^{\bullet,\bullet}, d', d'')$ on X^{an}. We denote by $\mathrm{H}^{*,*}$ (resp. $\mathrm{H}_c^{*,*}$) the cohomology of the complex of global sections (resp. global sections with compact support) with respect to d''.

In this paper, we generalize Gubler's approach, showing that one can define forms on Berkovich analytic spaces by using certain classes of embeddings of open subsets into toric varieties. Given a fine enough family of tropicalizations $\mathcal{S}$ (see Section 4 for the definition of this notion) we obtain a bigraded complex of sheaves $(\mathcal{A}_{\mathcal{S}}^{\bullet,\bullet}, d', d'')$ on X^{an}. We show that for many useful $\mathcal{S}$, our complex $\mathcal{A}_{\mathcal{S}}^{\bullet,\bullet}$ is isomorphic to $\mathcal{A}^{\bullet,\bullet}$.

For the rest of the introduction, we make the very mild assumption the X is normal and admits at least one closed embedding into a toric variety.

The general philosophy of this paper and also the definition of forms by Chambert-Loir and Ducros and Gubler is that we can transport constructions done for tropical varieties to Berkovich spaces by locally pulling back along tropicalizations. While Chambert-Loir, Ducros and Gubler used only tropicalization maps of tori, we will also allow tropicalization maps of general toric varieties. We will show advantages of this equivalent approach throughout the paper.

The definitions by both Chambert-Loir and Ducros as well as Gubler work with local embeddings. We show that we can also work with global embeddings. Our constructions provides us with the following: Let $\varphi\colon X \to Y_\Sigma$ be a closed embedding into a toric variety. Then we obtain pullback morphisms

$$\mathrm{trop}^*\colon\ \mathrm{H}^{p,q}(\mathrm{Trop}(\varphi(X))) \to \mathrm{H}^{p,q}(\mathrm{X}^{\mathrm{an}}) \text{ and} \tag{1}$$

$$\mathrm{trop}^*\colon\ \mathrm{H}_c^{p,q}(\mathrm{Trop}(\varphi(X))) \to \mathrm{H}_c^{p,q}(\mathrm{X}^{\mathrm{an}}) \tag{2}$$

in cohomology.

Note that (1) and (2) were not in general available in the approaches by Chambert-Loir and Ducros resp. Gubler. This construction allows us to explicitly construct classes in tropical Dolbeault cohomology.

For cohomology with compact support, we even obtain all classes this way:

Theorem 1.1 (Theorem 5.9) *We have*

$$\mathrm{H}_c^{p,q}(\mathrm{X}^{\mathrm{an}}) = \varinjlim_{\varphi\colon X \to Y_\Sigma} \mathrm{H}_c^{p,q}(\mathrm{Trop}(\varphi(X)).$$

where the limit runs over all closed embeddings of X into toric varieties.

We also show that the analogous result for $\mathrm{H}^{p,q}$ is not true (Remark 5.10).

Further, in certain cases, we can check whether one of these classes is non-trivial on the tropical side.

Theorem 1.2 (Theorem 10.3) *Assume that* $\mathrm{Trop}(\varphi(X))$ *is smooth. Then* (1) *and* (2) *are both injective.*

We exhibit three examples in Section 10, namely Mumford curve, curves of good reduction and toric varieties.

Another construction that we transport over from the tropical to the analytic world is cohomology with coefficients other than the real numbers. For a subring R of $\mathbb{R}$, we define a cohomology theory

$$\mathrm{H}_{\mathrm{trop}}^{*,*}(\mathrm{X}^{\mathrm{an}}, R) \text{ and } \mathrm{H}_{\mathrm{trop},c}^{*,*}(\mathrm{X}^{\mathrm{an}}, R)$$

with values in R-modules. Liu introduced in [21] a canonical rational subspace of $\mathrm{H}^{p,q}(\mathrm{X}^{\mathrm{an}})$. We show that this space agrees with $\mathrm{H}^{p,q}(\mathrm{X}^{\mathrm{an}}, \mathbb{Q})$ as defined in this paper (Proposition 9.3).

We obtain the analogue of Theorem 1.1 where on the right hand side we have tropical cohomology with coefficients in R (Proposition 8.4), and we provide an explicit isomorphism

$$\mathbb{R}\colon \mathrm{H}^{p,q}(\mathrm{X}^{\mathrm{an}}) \to \mathrm{H}_{\mathrm{trop}}^{p,q}(\mathrm{X}^{\mathrm{an}}, \mathbb{R}).$$

which is a version of de Rham's theorem in this context (Theorem 8.12). Liu introduced in [22] a *monodromy operator*

$$M\colon \mathrm{H}_{\mathrm{trop},c}^{p,q}(\mathrm{X}^{\mathrm{an}}) \to \mathrm{H}_{\mathrm{trop},c}^{p-1,q+1}(\mathrm{X}^{\mathrm{an}})$$

that respects rational classes if $\log|K^*| \subset \mathbb{Q}$ [22, Theorem 5.5 (1)]. Mikhalkin and Zharkov introduce in [25] a wave operator

$$W\colon \mathrm{H}_c^{p,q}(\mathrm{Trop}_\varphi(X), \mathbb{R}) \to \mathrm{H}_c^{p-1,q+1}(\mathrm{Trop}_\varphi(\mathrm{X}^{\mathrm{an}}), \mathbb{R}).$$

Note that both these operators are also available without compact support.

We show that W can be used to give an operator on $\mathrm{H}_c^{*,*}(\mathrm{X}^{\mathrm{an}})$ and that this operator agrees with M up to sign in Corollary 9.1.

We also obtain the following result regarding the interaction between the wave operator and the coefficients of the cohomology groups:

Theorem 1.3 *Let R be a subring of $\mathbb{R}$ and $R[\Gamma]$ the smallest subring of $\mathbb{R}$ that contains both R and Γ. Then the wave operator W restricts to a map*

$$W\colon \mathrm{H}_c^{p,q}(\mathrm{X}^{\mathrm{an}}, R) \to \mathrm{H}_c^{p-1,q+1}(\mathrm{X}^{\mathrm{an}}, R[\Gamma]).$$

As W and M agree up to sign and Liu's subspace of rational classes agrees with $\mathrm{H}^{*,*}(\mathrm{X}^{\mathrm{an}}, \mathbb{Q})$, this generalizes Liu's result for $\Gamma \subset R = \mathbb{Q}$.

We now sketch the organization of the paper. In Section 2 we recall background on toric varieties and their tropicalizations. Section 3 contains all constructions on tropical varieties that are needed for the paper. Most of these should be known to experts, however we still chose to list them for completeness. The main new result is the identification of the wave and monodromy operator, which is based on Lemma 3.13. In Section 4 we consider what we call *families of tropicalizations*, which is what we will use to define forms on Berkovich spaces. We give the definitions and some examples of families that we will consider. In Section 5 we define for a fine enough family of tropicalizations $\mathcal{S}$ a bigraded complex $\mathcal{A}_{\mathcal{S}}^{\bullet,\bullet}$ of sheaves of differential forms on Berkovich spaces. We also provide some conditions under which those complexes are isomorphic for different $\mathcal{S}$ and introduce tropical Dolbeault cohomology for X^{an}. In Section 6, we prove that for so-called admissible families $\mathcal{S}$, the complexes $\mathcal{A}_{\mathcal{S}}^{\bullet,\bullet}$ are isomorphic to $\mathcal{A}^{\bullet,\bullet}$. In Section 7 we discuss integration of top-dimensional differential forms with compact support. In Section 8 we introduce tropical cohomology with coefficients for X^{an} and compare it with $\mathrm{H}^{*,*}$. In Section 9, we discuss the relation between the wave and monodromy operators and consequences thereof. Section 10 provides partial computations of $\mathrm{H}^{p,q}(\mathrm{X}^{\mathrm{an}})$ and $\mathrm{H}^{p,q}(\mathrm{X}^{\mathrm{an}}, R)$ for curves and toric varieties, using our new approaches. In Section 11 we list open questions that one might ask as a consequence of our results.

Acknowledgments

The author would like to thank Walter Gubler, Johannes Rau and Kristin Shaw for helpful comments and suggestions. The idea for the proof of Theorem 3.13 came from joint work with Johannes Rau and Kristin Shaw on the paper [15]. The author would like to express his gratitude for being allowed to use those ideas.

Parts of this work already appeared in a more a hoc and less conceptual way in the author's PhD thesis [11].

The author would also like to thank the referee for their detailed report and specific remarks which greatly improved the paper.

Notations and conventions

Throughout K is a field that is complete with respect to a (possibly trivial) non-archimedean absolute value. We denote its value group by $\Gamma := \log |K^*|$. If the absolute value is non-trivial, we normalize it in such a way that $\mathbb{Z} \subset \Gamma$. A variety X is an geometrically integral separated K-scheme of finite type. For any variety X over K, we will throughout the paper denote by X^{an} the analytification in the sense of Berkovich [1].

2 Toric varieties and tropicalization

2.1 Toric varieties

Let N be a free abelian group of finite rank, M its dual and denote by $N_{\mathbb{R}}$ resp. $M_{\mathbb{R}}$ the respective scalar extensions to $\mathbb{R}$.

Definition 2.1 A *rational cone* $\sigma \in N_{\mathbb{R}}$ is a polyhedron defined by equations of the form $\varphi(.) \geq 0$ with $\varphi \in M$, that does not contain a positive dimensional linear subspace. A *rational fan* Σ in $N_{\mathbb{R}}$ is a polyhedral complex all of whose polyhedra are rational cones. For $\sigma \in \Sigma$ we define the monoid

$$S_\sigma := \{\varphi \in M \mid \varphi(v) \geq 0 \text{ for all } v \in \sigma\}.$$

We denote by $U_\sigma := \mathrm{Spec}(K[S_\sigma])$. For $\tau \prec \sigma$ we obtain an open immersion $U_\tau \to U_\sigma$. We define the toric variety Y_Σ to be the gluing of the $(U_\sigma)_{\sigma\in\Sigma}$ along these open immersions. For an introduction to toric varieties, see for example [7].

Remark 2.2 The toric variety Y_Σ comes with an open immersion $T \to Y_\Sigma$, where $T = \mathrm{Spec}(K[M])$ and a T-action that extends the group action of T on itself by translation. In fact any normal variety with such an immersion and action arises by the above described procedure ([4, Corollary 3.1.8]). This was shown by Sumihiro.

Choosing a basis of N gives an identification $N \simeq \mathbb{Z}^r \simeq M$ and $T \simeq \mathbb{G}_m^r$.

Definition 2.3 A map $\psi\colon Y_\Sigma \to Y_{\Sigma'}$ is called a *morphism of toric varieties* if it is equivariant with respect to the torus actions and restricts to a morphism of algebraic groups on dense tori. It is called an *affine map of toric varieties* if it is a morphism of toric varieties composed with a multiplicative torus translation.

Remark 2.4 A morphism of toric varieties $\psi: Y_\Sigma \to Y_{\Sigma'}$ is induced by a morphism of corresponding fans, meaning a linear map $N \to N'$ that maps cones in Σ to cones in Σ'.

2.2 Tropical toric varieties

Let Σ be a rational fan in $N_\mathbb{R}$. We write $\mathbb{T} := \mathbb{R} \cup \{-\infty\}$. For $\sigma \in \Sigma$ we define $N(\sigma) := N_\mathbb{R}/\langle\sigma\rangle_\mathbb{R}$. We write

$$N_\Sigma = \coprod_{\sigma\in\Sigma} N(\sigma).$$

We call the $N(\sigma)$ the *strata* of N_Σ. Note that N_Σ has a canonical action by N and $N_\mathbb{R}$ and the strata are the strata of the action of $N_\mathbb{R}$. We endow N_Σ with a topology in the following way:

For $\sigma \in \Sigma$ write $N_\sigma = \coprod_{\tau\prec\sigma} N(\tau)$. This is naturally identified with $\mathrm{Hom}_{\mathbb{M}\rtimes\ltimes\rtimes\sqsupset\sim}(S_\sigma, \mathbb{T})$. We equip $\mathbb{T}^{S_\sigma}$ with the product topology and give N_σ the subspace topology. For $\tau \prec \sigma$, the space $\mathrm{Hom}(S_\tau, \mathbb{T})$ is naturally identified with the open subspace of $\mathrm{Hom}_{\mathbb{M}\rtimes\ltimes\rtimes\sqsupset\sim}(S_\sigma, \mathbb{T})$ of maps that map $\tau^\perp \cap M$ to $\mathbb{R}$. We define the topology of N_Σ to be the one obtained by gluing along these identifications.

Definition 2.5 We call the space N_Σ a *tropical toric variety.*

We would like to remark that the tropical toric variety N_Σ can also be constructed by glueing its affine pieces along monomial maps [23, Section 3.2]

Note that N_Σ contains $N_\mathbb{R}$ as a dense open subset. For a subgroup Γ of $\mathbb{R}$ and each stratum $N(\sigma)$ we call the set $N(\sigma)_\Gamma := (N \otimes \Gamma)/\langle\sigma\rangle_\Gamma$ the *set of* Γ*-points*.

Let Σ and Σ' be fans in $N_\mathbb{R}$ and $N'_\mathbb{R}$ respectively. Let $L: N \to N'$ be a linear map such that $L_\mathbb{R}$ maps every cone in Σ into a cone in Σ'. Such a map canonically induces a map $N_\Sigma \to N_{\Sigma'}$ that is continuous and linear on each stratum.

Definition 2.6 A map $N_\Sigma \to N_{\Sigma'}$ that arises this way is called *morphism of tropical toric varieties.*

An *affine map* of tropical toric varieties is a map that is the composition of morphism of toric varieties with an $N_\mathbb{R}$-translation.

2.3 Tropicalization

Let Σ be a rational fan in $N_{\mathbb{R}}$. Denote by Y_Σ the associated toric variety and by N_Σ the associated tropical toric variety.

Definition 2.7 Payne defined in [26] a tropicalization map

$$\mathrm{trop}_\Sigma \colon Y_\Sigma^{\mathrm{an}} \to N_\Sigma$$

to the topological space N_Σ as follows: For $\tau \prec \sigma$, the space $\mathrm{Hom}(S_\tau, \mathbb{T})$ is naturally identified with the open subspace of $\mathrm{Hom}_{\mathbb{M}\rtimes\ltimes\rtimes\sqsupset\sim}(S_\sigma, \mathbb{T})$ of maps which map $\tau^\perp \cap M$ to $\mathbb{R}$. The map $\mathrm{trop}\colon U_\sigma^{\mathrm{an}} \to \mathrm{Trop}(U_\sigma)$ is then defined by mapping $|\,.\,|_x \in U_\sigma^{\mathrm{an}}$ to the homomorphism $u \mapsto \log |u|_x \in \mathrm{Trop}(U_\sigma) = \mathrm{Hom}(S_\sigma, \mathbb{T})$. We will often write $\mathrm{Trop}(Y_\Sigma) := N_\Sigma$.

For Z a closed subvariety of Y_Σ we define $\mathrm{Trop}(Z)$ to be the image of Z^{an} under $\mathrm{trop}\colon Y_\Sigma^{\mathrm{an}} \to \mathrm{Trop}(Y_\Sigma)$.

Definition 2.8 The construction $\mathrm{Trop}(Y_\Sigma)$ is functorial with respect to affine maps of toric varieties. In particular, for a morphism (resp. affine map) of toric varieties $\psi\colon Y_\Sigma \to Y_{\Sigma'}$, we obtain a morphism (resp. affine map) of tropical toric varieties $\mathrm{Trop}(\psi)\colon \mathrm{Trop}(Y_\Sigma) \to \mathrm{Trop}(Y_{\Sigma'})$.

If ψ is a closed immersion, then $\mathrm{Trop}(\psi)$ is a homeomorphism onto its image.

Example 2.9 Affine space $\mathbb{A}^r = \mathrm{Spec}\, K[T_1, \ldots, T_r]$ is the toric variety that arises from the cone $\{x \in \mathbb{R}^r \mid x_i \geq 0 \text{ for all } i \in [r]\}$. By definition the tropicalization in then $\mathbb{T}^r$ and the map $\mathrm{trop}\colon \mathbb{A}^{r,\mathrm{an}} \to \mathbb{T}^r;\ |\,.\,| \mapsto (\log |T_i|)_{i\in[r]}$.

Let σ be a cone in $N_{\mathbb{R}}$. We pick a finite generating set $b_1, \ldots, b_r$ of the monoid S_σ. Let U_σ be the affine toric variety associated to a cone σ. Then we have a surjective map $K[T_1, \ldots, T_r] \to K[S_\sigma]$, which induces a toric closed embedding $\varphi_B\colon U_\sigma \to \mathbb{A}^r$. By functoriality of tropicalization we also get a morphism of tropical toric varieties $\mathrm{Trop}(U_\sigma) \to \mathbb{T}^r$ that is a homeomorphism onto its image.

2.4 Tropical subvarieties of tropical toric varieties

In this section we fix a subgroup $\Gamma \subset \mathbb{R}$.

Definition 2.10 An integral Γ-affine polyhedron in $N_{\mathbb{R}}$ is a set defined by finitely many inequalities of the form $\varphi(\,.\,) \geq r$ for $\varphi \in M, r \in \Gamma$. An integral Γ-affine polyhedron on N_Σ is the topological closure of an integral Γ-affine polyhedron in $N(\sigma_\tau)$ for $\sigma_\tau \in \Sigma$.

Let τ be an integral Γ-affine polyhedron in N_Σ. For $\sigma \prec \sigma'$ we have that $N(\sigma') \cap \tau$ is a polyhedron in $N(\sigma')$ (that might be empty) and we consider this as a face of τ. Further we denote by $\mathbb{L}(\tau) = \{\lambda(u_1 - u_2) \mid u_1, u_2 \in \tau, \lambda \in \mathbb{R}\} \subset N(\sigma_\tau)$ the *linear space of* τ. If τ is integral Γ-affine, $\mathbb{L}(\tau)$ contains a canonical lattice that we denote by $\mathbb{Z}(e)$.

Definition 2.11 A tropical subvariety of a tropical toric variety is given the support of a integral Γ-affine polyhedral complex with weights attached to its top dimensional faces, satisfying the balancing condition.

Definition 2.12 Let Z be a closed subvariety of a toric variety Y_Σ. Then $\mathrm{Trop}(Z) := \mathrm{trop}(Z^{\mathrm{an}}) \subset N_\mathbb{R}$ is a tropical subvariety of $\mathrm{Trop}(Y_\Sigma)$. For a variety X and a closed embedding $\varphi\colon X \to Y_\Sigma$ we write $\mathrm{Trop}_\varphi(X) := \mathrm{Trop}(\varphi(X))$ and $\mathrm{trop}_\varphi := \mathrm{trop} \circ \varphi^{\mathrm{an}}\colon \mathrm{X}^{\mathrm{an}} \to \mathrm{Trop}_\varphi(X)$.

We will never explicitly use the weights nor the balancing condition, so the reader may be happy with the fact that there are weights and that they satisfy the balancing condition. If they are not happy with this, let us refer them to the excellent introduction [8]. The reader who already knows the balancing condition for tropical varieties in $N_\mathbb{R}$ will be glad to hear that the balancing condition for X is exactly the classical balancing condition for $X \cap N_\mathbb{R}$, there are no additional properties required at infinity.

3 Constructions in Cohomology of tropical varieties

In this section, R is a ring such that $\mathbb{Z} \subset R \subset \mathbb{R}$ and Γ is a subgroup of $\mathbb{R}$ that contains $\mathbb{Z}$. Further N is a free abelian group of finite rank, M is its dual, and Σ is a rational fan in $N_\mathbb{R}$. Additionally X is an integral Γ-affine tropical subvariety of N_Σ.

Definition 3.1 Let $U \subset N_\mathbb{R}$ an open subset. A *superform* of bidegree (p,q) is an element of

$$\mathcal{A}^{p,q}(U) = C^\infty(U) \otimes \Lambda^p M \otimes \Lambda^q M.$$

Remark 3.2 There are differential operators d' and d'' and a wedge product which are induced by the usual differential operator and wedge product on differential forms.

Definition 3.3 For an open subset U of N_Σ we write $U_\sigma := U \cap N_\sigma$. A superform of bidegree (p,q) on U is given by a collection $\alpha = (\alpha_\sigma)_{\sigma\in\Sigma}$ such that $\alpha_\sigma \in \mathcal{A}^{p,q}(U_\sigma)$ and for each σ and each $x \in U_\sigma$ there exists an open neighborhood U_x of x in U such that for each $\tau \prec \sigma$ we have $\pi^*_{\sigma,\tau}\alpha_\sigma = \alpha_\tau$. We call this the *condition of compatibility at the boundary.*

For a polyhedron σ in N_Σ we can define the restriction of a superform α to σ. Let Ω be an open subset of $|\,\mathcal{C}\,|$ for a polyhedral complex $\mathcal{C}$ in N_Σ. The space of superforms of bidegree (p,q) on Ω is defined as the set of pairs (U,α) where U is an open subset of N_Σ such that $U \cap |\,\mathcal{C}\,| = \Omega$ and $\alpha \in \mathcal{A}^{p,q}(U)$. Two such pairs are identified if their restrictions to $\sigma \cap \Omega$ agree for every $\sigma \in \mathcal{C}$.

Definition 3.4 For a tropical subvariety X of a tropical toric variety we obtain a double complex of sheaves $(\mathcal{A}^{\bullet,\bullet}, d', d'')$ on X. We define $\mathrm{H}^{p,q}(X)$ (resp. $\mathrm{H}^{p,q}_c(X)$) as the cohomology of the complex of global sections (resp. global sections with compact support) with respect to d''.

Definition 3.5 There exists an integration map $\int_X \colon \mathcal{A}^{n,n}_c(X) \to \mathbb{R}$ that satisfies Stokes' theorem and thus descends to cohomology.

Remark 3.6 Superforms on tropical subvarieties of tropical toric varieties are functorial with respect to affine maps of tropical toric varieties [16].

Definition 3.7 Let X be a tropical variety and $x \in X$ and denote by σ the cone of Σ such that $x \in N(\sigma)$. Then the tropical mutitangent space at x is defined to be

$$\mathbb{F}^R_p(\tau) = \left(\sum_{\sigma \in X \cap N(\sigma),\, x \in \sigma} \Lambda^p \mathbb{L}(\sigma) \right) \cap \Lambda^p R^n \subset \Lambda^p N(\sigma).$$

If ν is a face of τ, then there are transition maps $\iota_{\tau,\nu} \colon \mathbb{F}^R_p(\tau) \to \mathbb{F}^R_p(\nu)$ that are just inclusions if τ and ν live in the same stratum and compositions with projections to strata otherwise.

We denote by Δ_q the standard q-simplex.

Definition 3.8 A smooth stratified q-simplex is a map $\delta \colon \Delta_q \to X$ such that:

- If σ is a face of Δ_q then there exists a polyhedron τ in X such that $\mathring{\sigma}$ is mapped into $\mathring{\tau}_i$.
- Let $\Delta_q = [0, \ldots, q]$. If $\delta(i)$ is contained in the closure of a stratum of N_Σ, then so is $\delta(j)$ for $j \le i$.
- for each stratum X_i of X the map $\delta \colon \delta^{-1}(X_i) \to X_i$ is C^∞.

We denote the free abelian group of smooth stratified q-simplices δ satisfying $\delta(\mathring{\Delta}_q) \subset \mathring{\tau}$ by $C_q(\tau)$.

There is a boundary operator $\partial_{p,q}\colon C_{p,q}(X,R) \to C_{p,q-1}(X,R)$ that is given by the usual boundary operator on the simplex side and by the maps $\iota_{\tau,\nu}$ on the coefficient side, when necessary. Dually we have $\partial^{p,q}\colon C_{p,q}(X,R) \to C_{p,q+1}(X,R)$.

Definition 3.9 The groups of smooth tropical (p,q)-cell and cocells are respectively

$$C_{p,q}(X,R) := \bigoplus_{\tau \subset X} \mathbb{F}_p^R(\tau) \otimes C_q(\tau),$$
$$C^{p,q}(X,R) := \mathrm{Hom}_R(C_{p,q}(X,R),R).$$

We denote by $\underline{C}^{p,q}(R)$ the sheafification of $C^{p,q}(X,R)$ as defined in [16, Definition 3.13] and by $\mathcal{F}_R^p := \ker(\partial\colon \underline{C}^{p,0}(R) \to \underline{C}^{p,1}(R))$. Note that we have $\mathcal{F}_R^p = \mathcal{F}_{\mathbb{Z}}^p \otimes R$.

Definition 3.10 We denote by $\mathrm{H}_{\mathrm{trop}}^{p,q}(X) := \mathrm{H}^q(C^{p,\bullet}(X,R),\partial) = \mathrm{H}^q(\underline{C}^{p,\bullet}(X,R),\partial)$ and call this the *tropical cohomology of X with coefficients in R*. Similarly we define *tropical cohomology of X with coefficients in R with compact support.*

It was shown in [16] that the morphisms of complexes

$$\mathcal{F}_{\mathbb{R}}^p \to \mathcal{A}^{p,\bullet} \text{ and } \mathcal{F}_{\mathbb{R}}^p \to \underline{C}^{p,\bullet},$$

that are given by inclusion in degree 0 are in fact quasi isomorphisms. We now want to construct a de Rham morphism, meaning a quasi isomorphism $\mathbb{R}\colon \mathcal{A}^{p,\bullet} \to \underline{C}^{p,\bullet}$ that is compatible with the respective inclusions of $\mathcal{F}_{\mathbb{R}}^p$.

Remark 3.11 Let $v \otimes \delta$ be a smooth tropical (p,q)-cell on an open subset Ω of X. Then we define for a (p,q) form $\alpha \in \mathcal{A}^{p,q}(\Omega)$

$$\int_{v\otimes\delta} \alpha = \int_{\Delta_q} \delta^{-1}\langle \alpha; v\rangle.$$

We have to argue that this integral is well defined, since δ might map parts of Δ_q to infinity: Let X_0 be the stratum of X to which the barycenter of Δ_q is mapped. Let $\Delta_{q,0} := \delta^{-1}(X_0)$. Then we have $\mathrm{supp}(\delta^*\alpha) \subset \Delta_{q,0}$ by the condition of compatibility at the boundary for α, hence the integral is finite.

This defines a morphism

$$\mathbb{R}\colon \mathcal{A}^{p,q}(\Omega) \to C^{p,q}(\Omega)$$
$$\alpha \mapsto \left(v \otimes \delta \mapsto \int_{v\otimes\delta} \alpha\right)$$

and one directly verifies using the classical Stokes' theorem that this indeed induces a morphism of complexes

$$\mathbb{R}\colon \mathcal{A}^{p,\bullet} \to \underline{C}^{p,\bullet}.$$

that respects the respective inclusions of $\mathcal{F}^p_{\mathbb{R}}$. Since both $\underline{C}^{p,\bullet}$ and $\mathcal{A}^{p,\bullet}$ form acyclic resolutions of $\mathcal{F}^p_{\mathbb{R}}$, the map $\mathbb{R}$ is a quasi-isomorphism.

Definition 3.12 The *monodromy operator* is the unique $\mathcal{A}^{0,0}$-linear map such that

$$M\colon \mathcal{A}^{p,q} \to \mathcal{A}^{p-1,q+1};$$
$$d'x_I \wedge d''x_J \mapsto \sum_{k=1}^{p}(-1)^{p-k} d'x_{I\setminus i_k} \wedge d''x_{i_k} \wedge d''x_J.$$

The wave operator $W\colon C^{p,q}(\mathbb{R}) \to C^{p-1,q+1}(\mathbb{R})$ is the sheafified version of the map dual to

$$C_{p-1,q+1}(X,\mathbb{R}) \to C_{p,q}(X,\mathbb{R});$$
$$v \otimes \delta \mapsto (v \wedge (\iota(\delta(1)) - \delta(0))) \otimes \delta|_{[1,\ldots,q+1]}.$$

Lemma 3.13 *The diagram*

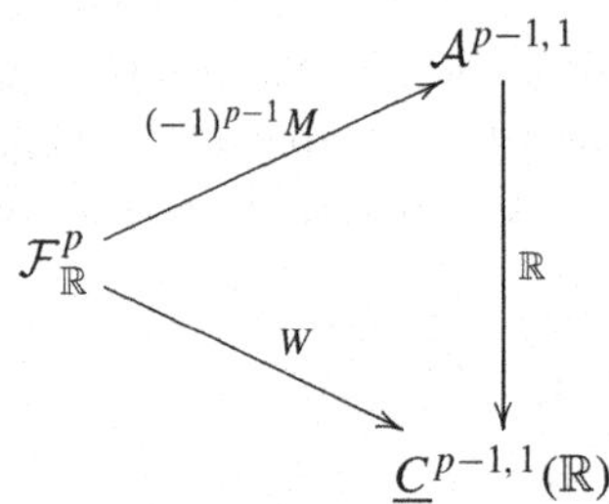

commutes.

Proof By the definition of the de Rham map we have to show that

$$\int_{[0,1]} \delta^*\langle M(\alpha), v\rangle = (-1)^{p-1}\langle \alpha, W(e \otimes v)\rangle$$

where $\delta\colon [0,1] \to X$ is a smooth stratified 1-simplex and $v \in \mathbb{F}^{p-1}(\tau)$, where $\delta((0,1)) \subset \mathring{\tau}$ and $\alpha \in \mathcal{F}^p$. After picking bases and using multilinearity for both v and α we may assume that $\alpha = d'x_1 \wedge \cdots \wedge d'x_p$ and $v = x_1 \wedge \cdots \wedge x_{p-1}$. Then we have

$$\int_{[0,1]} \delta^*\langle M(\alpha), v\rangle$$
$$= \sum_{i=1}^{p}(-1)^{p-i}\int_{[0,1]} \delta^*\langle d'x_1 \wedge \cdots \widehat{d'x_i} \wedge \cdots \wedge d'x_p, x_1 \wedge \cdots \wedge x_{p-1}\rangle \wedge d''x_i$$
$$= \int_{[0,1]} \delta^*\langle d'x_1 \wedge \cdots \wedge d''x_{p-1}, x_1 \wedge \cdots \wedge x_{p-1}\rangle \wedge d''x_p$$
$$= \int_{[0,1]} \delta^* d''x_p = dx_p(\delta(1)) - dx_p(\delta(0))$$

and

$$\langle d'x_1 \wedge \cdots \wedge d'x_p, \delta(1) - \delta(0) \wedge x_1 \wedge \cdots \wedge x_{p-1}\rangle$$
$$= (-1)^{p-1} dx_p(\delta(1)) - dx_p(\delta(0)),$$

where dx_p denotes the p-th coordinate functions with respect to $x_1, \ldots, x_n$. This calculation holds true as long as the p-th coordinate functions is bounded on $\delta([0,1])$. If this is not the case however, then $d''x_p$ vanishes in a neighborhood of $\delta(0)$ (resp. $\delta(1)$) by the compatibility condition and we may replace $[0,1]$ by $[\varepsilon, 1]$ resp. $[\varepsilon, 1-\varepsilon]$ resp. $[0, 1-\varepsilon]$. □

Theorem 3.14 *The wave and the monodromy operator agree on cohomology up to sign by virtue of the isomorphism $\mathbb{R}$, meaning that the diagram*

$$\begin{array}{ccc} \mathrm{H}^{p,q}(X) & \xrightarrow{(-1)^{p-1}M} & \mathrm{H}^{p-1,q+1}(X) \\ \downarrow{\scriptstyle \mathbb{R}} & & \downarrow{\scriptstyle \mathbb{R}} \\ \mathrm{H}^{p,q}_{\mathrm{trop}}(X, \mathbb{R}) & \xrightarrow{W} & \mathrm{H}^{p-1,q+1}_{\mathrm{trop}}(X, \mathbb{R}) \end{array}$$

commutes. The same is true for cohomology with compact support.

Proof The wave and monodromy operators give morphisms of complexes is a morphism of complexes

$$W\colon \underline{C}^{p,\bullet}(\mathbb{R}) \to \underline{C}^{p-1,\bullet}(\mathbb{R})[1] \text{ and } M\colon\ \mathcal{A}^{p,\bullet} \to \mathcal{A}^{p-1,\bullet}[1],$$

hence it is sufficient to show that

$$\begin{array}{ccc} \mathcal{A}^{p,\bullet} & \xrightarrow{(-1)^{p-1}M} & \mathcal{A}^{p-1,\bullet}[1] \\ \downarrow{\scriptstyle \mathbb{R}} & & \downarrow{\scriptstyle \mathbb{R}} \\ \underline{C}^{p,\bullet}(\mathbb{R}) & \xrightarrow{W} & \underline{C}^{p-1,\bullet}(\mathbb{R})[1] \end{array}$$

commutes in the derived category. Replacing both $\mathcal{A}^{p,\bullet}$ and $\underline{C}^{p,\bullet}$ with the quasi-isomorphic $\mathcal{F}^p_{\mathbb{R}}$, we have to show that

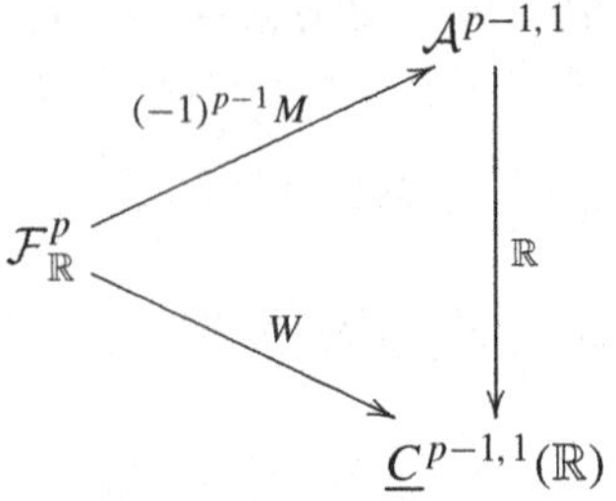

commutes. This follows directly from Theorem 3.13. □

Proposition 3.15 *The wave operator descends to an operator on cohomology*

$$W\colon \mathrm{H}^{p,q}_{\mathrm{trop}}(X,R) \to \mathrm{H}^{p-1,q+1}_{\mathrm{trop}}(X,R[\Gamma]) \text{ and}$$
$$W\colon \mathrm{H}^{p,q}_{\mathrm{trop},c}(X,R) \to \mathrm{H}^{p-1,q+1}_{\mathrm{trop},c}(X,R[\Gamma]),$$

where $R[\Gamma]$ is the smallest subring of $\mathbb{R}$ that contains both R and Γ.

Proof We pick a triangulation of X with smooth stratified simplices such that all vertices are Γ-points. We can now compute $\mathrm{H}^{p,q}_c(X,R)$ and the wave homomorphism using this triangulation by [25, Section 2.2]. For a (p,q)-chain δ with respect to this triangulation and with coefficients in R, we now have that $W(\delta)$ has coefficients in $R[\Gamma]$. Hence W restricts to a map $\mathrm{H}^{p,q}(X,R) \to \mathrm{H}^{p-1,q+1}(X,R[\Gamma])$ resp. $\mathrm{H}^{p,q}_c(X,R) \to \mathrm{H}^{p-1,q+1}_c(X,R[\Gamma])$. □

Definition 3.16 We denote by

$$\cap[X]_R\colon C^{n,n}_c(X,R) \to R$$

the *evaluation against the fundamental class*, as defined in [15, Definition 4.8] and also the induced map on cohomology $\mathrm{H}^{n,n}_{\mathrm{trop},c}(X,R) \to R$.

Proposition 3.17 *The following diagram commutes*

$$\begin{array}{ccc} \mathcal{A}^{n,n}_c(X) & \xrightarrow{\int_X} & \mathbb{R} \\ \downarrow{\scriptstyle \mathbb{R}} & \nearrow{\scriptstyle \cap[X]} & \\ C^{n,n}_c(X,\mathbb{R}). & & \end{array}$$

Proof This is a straightforward calculation using the definitions. □

4 Families of tropicalizations

In this section, K is a complete non-archimedean field and X is a K-variety.

4.1 Definitions

The philosophy throughout the paper will be that we can approximate non-archimedean analytic spaces through embedding them into toric varieties and tropicalizing. We will define families that approximate the analytic space well enough (fine enough families) as well as notions that tell us that two families basically contain the same amount of information (final and cofinal families).

Definition 4.1 A *family of tropicalizations* $\mathcal{S}$ of X consists of the following data:

1. A class $\mathcal{S}_{\text{map}}$ containing closed embeddings $\varphi\colon U \to Y_\Sigma$ for open subsets U of X and toric varieties Y_Σ.
2. For an element $\varphi\colon U \to Y_\Sigma$ of $\mathcal{S}_{\text{map}}$ a subclass $\mathcal{S}_\varphi$ of $\mathcal{S}_{\text{map}}$ that contains maps $\varphi'\colon U' \to Y_{\Sigma'}$ for open subsets $U' \subset U$ and such that there exists an affine map of toric varieties $\psi_{\varphi,\varphi'}$, such that

$$\begin{array}{ccc} U' & \xrightarrow{\varphi'} & Y_{\Sigma'} \\ \downarrow{\scriptstyle \iota} & & \downarrow{\scriptstyle \psi_{\varphi,\varphi'}} \\ U & \xrightarrow{\varphi} & Y_\Sigma \end{array}$$

commutes. Such a φ' is called *refinement* of φ. The map $\psi_{\varphi,\varphi'}$ induces an affine map of toric varieties $\text{Trop}(\psi_{\varphi,\varphi'})$. The restriction of $\text{Trop}(\psi_{\varphi,\varphi'})$ to $\text{Trop}_{\varphi'}(U')$ depends only on φ and φ', so we denote this map by $\text{Trop}(\varphi,\varphi')\colon \text{Trop}_{\varphi'}(U') \to \text{Trop}_\varphi(U)$.

We further require that if φ' is a refinement of φ and φ'' is a refinement of φ', then φ'' be a refinement of φ.

A *subfamily* of tropicalizations of $\mathcal{S}$ is a family of tropicalizations $\mathcal{S}'$ such that all embeddings and refinements in $\mathcal{S}$ are also in $\mathcal{S}'$. Further $\mathcal{S}'$ is said to be a *full subfamily* if whenever $\varphi, \varphi' \in \mathcal{S}'_{\text{map}}$ and φ' is a refinement of φ in $\mathcal{S}$, it is also a refinement of φ in $\mathcal{S}'$.

Foster, Gross and Payne study in [6] so-called "Systems of toric embeddings". They only consider the case where all of X is embedded into the toric variety, a case we later call *global families* of tropicalizations.

Definition 4.2 Let $\mathcal{S}$ be a family of tropicalizations on X. An *$\mathcal{S}$-tropical chart* is given by a pair (V,φ) where $\varphi\colon U \to Y_\Sigma \in \mathcal{S}_{\mathrm{map}}$ and $V = \mathrm{trop}_\varphi^{-1}(\Omega)$ is an open subset of X^{an} which is the preimage of an open subset Ω of $\mathrm{Trop}_\varphi(U)$.

Another $\mathcal{S}$ tropical chart (V',φ') is called an *$\mathcal{S}$-tropical subchart* of (V,φ) if φ' is a refinement of φ and $V' \subset V$.

Note that we have $\Omega = \mathrm{trop}_\varphi(V)$.

Example 4.3 The family of all tropicalizations is a family of tropicalizations in the sense of Definition 4.1. We denote it by $\mathcal{T}$.

In the following definition, we define the terms *final* and *cofinal* for two families of tropicalizations $\mathcal{S}$ and $\mathcal{S}'$. While the definitions are a bit on the technical side, the idea is that if $\mathcal{S}'$ is either final or cofinal for $\mathcal{S}$, then $\mathcal{S}$-tropical charts provide the same information as $\mathcal{S}'$-tropical charts.

Definition 4.4 Let $\mathcal{S}$ and $\mathcal{S}'$ be families of tropicalizations. We say $\mathcal{S}'$ is *cofinal* for $\mathcal{S}$ if for every embedding $\varphi\colon U \to Y_\Sigma$ in $\mathcal{S}_{\mathrm{map}}$ and every $x \in U^{\mathrm{an}}$ there exists $\varphi'\colon U' \to Y_{\Sigma'}$ in $\mathcal{S}'_{\mathrm{map}}$ with $x \in U'^{\mathrm{an}}$ such that $\varphi'|_{U'\cap U}$ restricts to a closed embedding of $U \cap U'$ into an open torus invariant subvariety of $Y_{\Sigma'}$, and that embedding is a refinement of φ in $\mathcal{T}$.

$\mathcal{S}'$ is said to be *final* for $\mathcal{S}$ for every $\varphi\colon U \to Y_\Sigma$ in $\mathcal{S}_{\mathrm{map}}$ and $x \in U^{\mathrm{an}}$, there exists a refinement $\varphi'\colon U' \to Y_{\Sigma'}$ with $x \in U'^{\mathrm{an}}$, a closed embedding $m\colon Y_{\Sigma'} \to Y_{\Sigma''}$, that is an affine map of toric varieties, such that $m \circ \varphi'$ is in $\mathcal{S}'_{\mathrm{map}}$ and φ' is a refinement of $m \circ \varphi'$ via m in $\mathcal{S}$.

Definition 4.5 A family of tropicalizations is called *fine enough* if the sets V such that there exist $\mathcal{S}$-tropical charts (V,φ) form a basis of the topology of X^{an} and for each pair of $\mathcal{S}$-tropical charts (V_1,φ_1) and (V_2,φ_2) there exists $\mathcal{S}$-tropical charts $(V_i,\varphi_i)_{i\in I}$, which are $\mathcal{S}$-tropical subcharts of both (V_1,φ_1) and (V_2,φ_2), such that $V_1 \cap V_2 = \bigcup V_i$.

Lemma 4.6 *Let $\mathcal{S}'$ be a full subfamily of $\mathcal{S}$ and assume that $\mathcal{S}$ is fine enough. If $\mathcal{S}'$ is final or cofinal in $\mathcal{S}$, then $\mathcal{S}'$ is also fine enough.*

Proof Let $x \in \mathrm{X}^{\mathrm{an}}$. Since $\mathcal{S}$ is fine enough and $\mathcal{S}'$ is a full subfamily, it is sufficient to prove that given an $\mathcal{S}$-tropical chart (V,φ) with $x \in V$, there exists an $\mathcal{S}$-tropical chart (V',φ') with $x \in V'$ that is a $\mathcal{S}$-tropical subchart of (V,φ).

If $\mathcal{S}'$ is cofinal in S, then we pick φ' as in Definition 4.4. Then $(V \cap U'^{\mathrm{an}},\varphi')$ is a tropical chart.

If $\mathcal{S}'$ is final, then $(V \cap U'^{\mathrm{an}}, m \circ \varphi')$ is a tropical chart. □

4.2 Examples

In the following we will give examples of families of tropicalizations for a variety X. We will always specify the class S_{map} and for $\varphi \in S_{\text{map}}$ simply define $\mathcal{S}_\varphi$ to be those φ' where an affine map of toric varieties $\psi_{\varphi,\varphi'}$ as required in Definition 4.1 (iii) exists. The exception to this rule are Example 4.15, where we require the map $\psi_{\varphi,\varphi'}$ to be a coordinate projection in order for φ' to be a refinement of φ and Example 4.9.

Example 4.7 The family $\mathbb{A}$ is the family whose embeddings are closed embeddings of affine open subsets of X into affine space. This family is fine enough by the definition of the topology of X^{an}.

Suppose we are given an embedding $\varphi\colon U \to Y_\Sigma$ of an open subset of X into a toric variety with $x \in U^{\text{an}}$. Let Y_σ be an open affine toric subvariety of Y_Σ such that $Y_{\sigma'}^{\text{an}}$ contains $\varphi^{\text{an}}(x)$. Let $\varphi' := \varphi|_{\varphi^{-1}(Y_\sigma)}$. Now we pick a toric embedding of $m\colon Y_\sigma \to \mathbb{A}^n$ as in Example 2.9. This shows that $\mathbb{A}$ is final in $\mathcal{T}$.

Example 4.8 The family $\mathbb{G}$ is the family whose embeddings are closed embeddings of very affine open subsets of X into $\mathbb{G}_m^n$.

This family is also fine enough if the base field is non-trivially valued [9, Proposition 4.16], but not when K is trivially valued [11, Example 3.3.1].

Example 4.9 Assume that K is algebraically closed. Let X be a variety and U a very affine open subset. Then $M = \mathcal{O}^*(U)/K^*$ is a free abelian group of finite rank and the canonical map $K[M] \to \mathcal{O}(U)$ induces a closed embedding $\varphi_U : U \hookrightarrow T$ for a torus T with character lattice M. The embedding φ_U is called the *canonical moment map of* U. We denote by $\mathbb{G}_{\text{can}}$ the family of tropicalizations where $\mathbb{G}_{\text{can, map}} = \{\varphi_U \mid U \subset X \text{ very affine}\}$ and refinements being the maps induced by inclusions.

It is easy to see that this family is cofinal in $\mathbb{G}$.

Definition 4.10 A family of tropicalizations $\mathcal{S}$ for a variety X is called *global* if all $\varphi \in \mathcal{S}_{\text{map}}$ are defined on all of X.

Global families of tropicalizations will play a special role, as they will allow us to construct classes in tropical Dolbeault cohomology.

Definition 4.11 We say that X satisfies condition (†) if X is normal and every two points in X have a common affine neighborhood.

By Włodarczyk's Embedding Theorem, Condition (†) is equivalent to X being normal and admitting a closed embedding into a toric variety (cf. [30]). Observe also that it is satisfied by any quasi-projective normal variety. It is however weaker then being normal and quasi-projective, as there exist proper toric varieties which are not projective.

Example 4.12 Let $\mathcal{S}$ be a global family of tropicalizations such that

$$X^{an} = \varprojlim_{\varphi \in \mathcal{S}_{map}} \operatorname{Trop}_{\varphi}(X) \tag{3}$$

and such that if $\varphi_1 \colon X \to Y_{\Sigma_1}$ and $\varphi_2 \colon X \to Y_{\Sigma_2}$ in $\mathcal{S}_{map}$ then also $\varphi_1 \times \varphi_2 \colon X \to Y_{\Sigma_1} \times Y_{\Sigma_2} \in \mathcal{S}_{map}$. Then $\mathcal{S}$ is fine enough. In fact it follows directly from (3) that $\mathcal{S}$-tropical charts form a basis of the topology and from the product property that we can always locally find common subcharts.

Families with the properties from Example 4.12 were extensively studied in [6].

When X satisfies condition (†), this helps us say even more:

Example 4.13 Assume that X satisfies condition (†). Then we may consider the family of tropicalizations $\mathcal{T}_{\text{global}}$, where $\mathcal{T}_{\text{global, map}}$ is the class of all embeddings of X into toric varieties.

Let $\varphi \colon U \to \mathbb{A}^n$ be a closed embedding of an open subset U of X into an affine space given by regular functions $f_1, \dots, f_n$ on U. Then by [6, Theorem 4.2] there exists an embedding $\overline{\varphi} \colon X \to Y_\Sigma$ such that U is the preimage of an open affine invariant subvariety U_σ and each f_i is the pullback of a character that is regular on U_σ. This exactly shows that $\overline{\varphi}|_U$ is a refinement of φ in $\mathcal{T}$, which shows that $\mathcal{T}_{\text{global}}$ is cofinal in $\mathbb{A}$.

Example 4.14 Let X be an affine variety and $\mathbb{A}_{\text{global}}$ the family of tropicalizations whose class of embeddings are embeddings of all of X into affine spaces.

We show that this family is cofinal in $\mathbb{A}$: Let U be an open subset of X and $\varphi \colon U \to \mathbb{A}^n$ an closed embedding given by $f_1/g_1, \dots, f_n/g_n$, where f_i, g_i are regular functions on X. In particular, $U = D(g_1, \dots, g_n)$. We pick regular functions $h_1, \dots, h_k$ on X such that the f_i, g_i, h_i generate $\mathcal{O}(X)$. We consider the embedding $\varphi' \colon X \to \mathbb{A}^{2n} \times \mathbb{A}^k$ given by $f_1, \dots, f_n, g_1, \dots, g_n, h_1, \dots, h_k$. Then $\varphi'|_U$ gives a closed embedding of U into $\mathbb{A}^n \times \mathbb{G}_m^n \times \mathbb{A}^k$, which is clearly a refinement of φ in $\mathcal{T}$.

Example 4.15 Assume that K is algebraically closed. Wanner and the author considered in [17] the class of *linear tropical charts* for $X = \mathbb{A}^1$. In the language of the present paper, they use the following family of tropicalizations, which we denote by $\mathbb{A}_{\text{linear}}$. $\mathbb{A}_{\text{linear, map}}$ is the set of linear embeddings i.e. those embeddings $\varphi \colon \mathbb{A}^1 \to \mathbb{A}^r$, where the corresponding algebra homomorphism $K[T_1, \dots, T_r] \to K[X]$ is given by mapping T_i to $(X - a_i)$ for $a_1, \dots, a_r \in K$.

A refinement of φ is then a map $\varphi'\colon \mathbb{A}^1 \to \mathbb{A}^s$ given by $(X-b_1,\dots,X-b_s)$ where $s>r$ and $\{a_i\}\subset\{b_j\}$. This family of tropicalizations is fine enough and global, for details see [17, Section 3.2]. By factoring the defining polynomials of any map $\mathbb{A}^1\to\mathbb{A}^n$ into linear factors, it follows that $\mathbb{A}_{\text{linear}}$ is cofinal in $\mathbb{A}_{\text{global}}$.

Example 4.16 Let K be algebraically closed and X be a smooth projective Mumford curve. Let $\mathcal{T}_{\text{Smooth}}$ be the class of embeddings of X into toric varieties such that $\text{Trop}_\varphi(X)$ is a smooth tropical curve. Then $\mathcal{T}_{\text{Smooth}}$ is cofinal in $\mathcal{T}_{\text{global}}$ by [13, Theorem A].

5 Differential forms on Berkovich spaces

In this section, K is a complete non-archimedean field and X is a variety over K. Further $\mathcal{S}$ is a fine enough family of tropicalizations.

5.1 Sheaves of differential forms

In this section, we define a sheaf of differential forms $\mathcal{A}_{\mathcal{S}}^{p,q}$ with respect to $\mathcal{S}$. We will also show that for final and cofinal families, these sheaves are isomorphic. We use the sheaves $\mathcal{A}^{p,q}$ of differential forms on tropical varieties which are recalled in Section 3.

Definition 5.1 Let $\mathcal{S}$ be a fine enough family of tropicalizations. For V an open subset of X^{an}. An element $\alpha\in\mathcal{A}_{\mathcal{S}}^{p,q}(V)$ is given by a family of triples $(V_i,\varphi_i,\alpha_i)_{i\in I}$, where

1. The V_i cover V, i.e. $V=\bigcup_{i\in I}V_i$.
2. For each $i\in I$ the pair (V_i,φ_i) is an $\mathcal{S}$-tropical chart.
3. For each $i\in I$ we have $\alpha_i\in\mathcal{A}^{p,q}(\text{trop}_\varphi(V))$.
4. For all $i,j\in I$ there exist $\mathcal{S}$-tropical subcharts $(V_{ijl},\varphi_{ijl})_{l\in L}$ that cover $V_i\cap V_j$ such that

$$\text{Trop}(\varphi_i,\varphi_{ijl})^*\alpha_i=\text{Trop}(\varphi_j,\varphi_{ijl})^*\alpha_j\in\mathcal{A}^{p,q}(\text{trop}_{\varphi_{ijl}}(V_{ijl})).$$

Another such family $(V_j,\varphi_j,\alpha_j)_{j\in J}$ defines the same form α if their union $(V_i,\varphi_i,\alpha_i)_{i\in I\cup J}$ still satisfies iv).

For an open subset W of V we can cover W by $\mathcal{S}$-tropical subcharts (V_{ij},φ_{ij}) of the (V_i,φ_i). Then we define $\alpha|_W\in\mathcal{A}_{\mathcal{S}}^{p,q}(W)$ to be defined by $(V_{ij},\varphi_{ij},\,\text{Trop}(\varphi_i,\varphi_{ij})^*(\alpha_i))$.

The differentials d' and d'' are well defined on $\mathcal{A}_{\mathcal{S}}^{p,q}$ and thus we obtain a complex $(\mathcal{A}_{\mathcal{S}}^{\bullet,\bullet}, d', d'')$ of differential forms on X^{an}.

Lemma 5.2 *Let $\mathcal{S}$ be a fine enough family of tropicalization. Let $\alpha \in \mathcal{A}_{\mathcal{S}}^{p,q}(V)$ be given by a single $\mathcal{S}$-tropical chart (V, φ, α'). Then $\alpha = 0$ if and only if $\alpha' = 0$.*

Proof This works the same as the proof in [3, Lemma 3.2.2]. □

Lemma 5.3 *Let $\mathcal{S}$ be a fine enough family of tropicalization. Let $\mathcal{S}'$ be a fine enough subfamily. Then there exists a unique morphism of sheaves*

$$\Psi_{\mathcal{S}',\mathcal{S}} \colon \mathcal{A}_{\mathcal{S}'} \to \mathcal{A}_{\mathcal{S}},$$

such that the image of a form given by a triple (V, φ, β) is given by the same triple. This morphism is injective. Furthermore if $\mathcal{S}'$ is final or cofinal, then this morphism is an isomorphism.

Recall that if $\mathcal{S}'$ is a full subfamily that is either final or cofinal in the fine enough family $\mathcal{S}$, then $\mathcal{S}'$ is itself fine enough by Lemma 4.6. Thus Lemma 5.3 implies that $\Psi_{\mathcal{S}',\mathcal{S}}$ is an isomorphism.

Proof Injectivity follows from Lemma 5.2.

Assume that $\mathcal{S}'$ is cofinal in $\mathcal{S}$ and that we are given a form α given by (V, φ, β). Fixing $x \in V$ and picking φ' as in Definition 4.4, we define a form α' to be given by $(V \cap U'^{\mathrm{an}}, \varphi', \mathrm{Trop}(\psi_{\varphi',\varphi})^* \beta)$. Since φ' was a refinement of φ in $\mathcal{T}$, we have

$$\Psi_{\mathcal{S},\mathcal{T}}(\alpha) = \Psi_{\mathcal{S}',\mathcal{T}}(\alpha') = \Psi_{\mathcal{S},\mathcal{T}}(\Psi_{\mathcal{S}',\mathcal{S}}(\alpha'))$$

which proves $\Psi(\mathcal{S}', \mathcal{S})(\alpha') = \alpha$ locally at x by injectivity of $\Psi_{\mathcal{S},\mathcal{T}}$.

Assume that $\mathcal{S}'$ is final in $\mathcal{S}$ and we are given a form α given by a tuple (V, φ, β). We pick m as in Definition 4.4 and define α' to be given by $(\varphi \circ m_2, V, \beta)$. Note here that m induces an isomorphism of the tropicalizations of V, thus "pushing β forward along φ" is possible. Then $\Psi_{\mathcal{S}',\mathcal{S}}(\alpha') = \alpha$ since φ is a refinement of $\varphi \circ m$ via m. □

Definition 5.4 Let $\mathcal{S}$ be a fine enough family of tropicalizations. We say that $\mathcal{S}$ is *admissible* if $\Psi_{\mathcal{S},\mathcal{T}}$ is an isomorphism.

Corollary 5.5 *We have an isomorphism*

$$\mathcal{A}_{\mathbb{A}} \cong \mathcal{A}_{\mathcal{T}}.$$

If X is affine, these are also isomorphic to $\mathcal{A}_{\mathbb{A}_{\mathrm{global}}}$ and if X satisfies condition (†), then these are also isomorphic to $\mathcal{A}_{\mathcal{T}_{\mathrm{global}}}$.

Theorem 5.6 *Let $\mathcal{S}$ be a fine enough global family of tropicalizations. Let $\alpha \in \mathcal{A}^{p,q}(V)$ be given by a* finite *family $(V_i, \varphi_i, \alpha_i)$, where $\varphi_i \colon X \to Y_{\Sigma_i}$ in $\mathcal{S}_{\text{map}}$. Then α can be defined by one $\mathcal{S}$-tropical chart.*

Proof Since $\mathcal{S}$ is fine enough and global, there exists a common refinement $\varphi\colon X \to Y_\Sigma$ for all the φ_i. Then (V_i, φ) is an $\mathcal{S}$-tropical subchart of (V_i, φ_i) for all i. Denote by $\alpha_i' := \text{Trop}(\varphi_i, \varphi)^*\alpha_i$ and $\Omega_i := \text{trop}_\varphi(V_i)$. Then $\alpha|_{V_i}$ is given by both $(V_i \cap V_j, \varphi, \alpha_i'|_{\Omega_i \cap \Omega_j})$ and $(V_i' \cap V_j', \varphi, \alpha_j'|_{\Omega_i \cap \Omega_j})$. By Lemma 5.2, the forms α_i' and α_j' agree on $\Omega_i \cap \Omega_j$, thus glue to give a form $\alpha' \in \mathcal{A}^{p,q}(\text{trop}_\varphi(V))$. The form $\alpha \in \mathcal{A}^{p,q}(V)$ is then defined by (V, φ, α'). □

Let $\mathcal{S}$ be a global admissible family of tropicalizations. Let $\varphi\colon X \to Y_\Sigma$ be a closed embedding in $\mathcal{S}_{\text{map}}$. We define a map $\text{trop}^*\colon \mathcal{A}^{p,q}(\text{Trop}_\varphi(X)) \to \mathcal{A}^{p,q}_{\mathcal{S}}(\text{X}^{\text{an}})$ by setting for $\beta \in \mathcal{A}^{p,q}(\text{Trop}_\varphi(X))$ the image $\text{trop}^*\beta$ to be the form given by the triple $(\varphi, \text{X}^{\text{an}}, \beta)$. One immediately checks that this is well defined. We define this similarly for forms with compact support.

Theorem 5.7 *Let $\mathcal{S}$ be a fine enough global family of tropicalizations. Let V be an open subset of X^{an} such that there exists an $\mathcal{S}$-tropical chart (V, φ). Then pullbacks along the tropicalization maps induce an isomorphism*

$$\varinjlim \mathcal{A}^{p,q}_c(\text{trop}_\varphi(V)) \to \mathcal{A}^{p,q}_{\mathcal{S},c}(V)$$

where the limit runs over all $\mathcal{S}$-tropical charts (V, φ).

Proof For any $\mathcal{S}$-tropical chart (V, φ), the pullback along the proper map trop_φ induces a well defined morphism $\mathcal{A}^{p,q}_c(\text{trop}_\varphi(V)) \to \mathcal{A}^{p,q}_{\mathcal{S},c}(V)$. By definition this map is compatible with pullback between charts. Thus the universal property of the direct limit leads to a morphism $\Psi\colon \varinjlim \mathcal{A}^{p,q}_{\mathcal{S},c}(\text{trop}_\varphi(V)) \to \mathcal{A}^{p,q}_c(V)$, where the limit runs over all $\mathcal{S}$-tropical charts of V.

This map is injective by construction. For surjectivity, let $\alpha \in \mathcal{A}^{p,q}_c(V)$ be given by $(V_i, \varphi_i, \alpha_i)_{i \in I}$. Let I' be a finite subset of I such that the V_i with $i \in I'$ cover the support of α. Then $(V_i, \varphi_i, \alpha_i)_{i \in I'}$ defines $\alpha|_{V'} \in \mathcal{A}^{p,q}_{\mathcal{S},c}(V')$, where $V' = \bigcup_{i \in I'} V_i$. Since $\mathcal{S}$ is global and fine enough, the conditions of Theorem 5.6 are satisfied and $\alpha|_{V'}$ can be defined by a triple (V', φ', α'). By passing to a common refinement with (V, φ) we may assume that (V', φ') is a subchart of (V, φ).

It follows from Lemma 5.2 that $\text{supp}(\alpha') = \varphi'_{\text{trop}}(\text{supp}(\alpha))$ (cf. [3, Corollaire 3.2.3]). Thus $\text{supp}(\alpha')$ is compact. We extend α' by zero to $\tilde{\alpha}' \in \mathcal{A}^{p,q}_{\text{Trop}_{\varphi'}(X),c}(\text{trop}_{\varphi'}(V))$. Then α is defined by $(V, \varphi', \tilde{\alpha}')$, which is in the image of Ψ. □

5.2 Tropical Dolbeault cohomology

In this section we assume that $\mathcal{S}$ is a admissible family of tropicalizations and often just write $\mathcal{A}^{p,q}$ for $\mathcal{A}^{p,q}_{\mathcal{S}}$.

Definition 5.8 We define *tropical Dolbeault cohomology* to be the cohomology of the complex $(\mathcal{A}^{p,\bullet}(\mathrm{X}^{\mathrm{an}}), d'')$, i.e.

$$\mathrm{H}^{p,q}(\mathrm{X}^{\mathrm{an}}) := \frac{\ker(d''\colon\ \mathcal{A}^{p,q}(\mathrm{X}^{\mathrm{an}}) \to \mathcal{A}^{p,q+1}(\mathrm{X}^{\mathrm{an}}))}{\operatorname{im}(d''\colon\ \mathcal{A}^{p,q-1}(\mathrm{X}^{\mathrm{an}}) \to \mathcal{A}^{p,q}(\mathrm{X}^{\mathrm{an}}))}.$$

Similarly we define cohomology with compact support $\mathrm{H}^{p,q}_c(\mathrm{X}^{\mathrm{an}})$ as the cohomology of forms with compact support.

Theorem 5.9 *Let $\mathcal{S}$ be an admissible global family of tropicalizations. Then pullbacks along tropicalization maps induce an isomorphism*

$$\varinjlim_{\varphi\in\mathcal{S}} \mathrm{H}^{p,q}_c(\mathrm{Trop}_\varphi(X)) \to \mathrm{H}^{p,q}_c(\mathrm{X}^{\mathrm{an}}).$$

Proof The map is induced by trop*. The theorem follows from the fact that taking cohomology commutes with forming direct limits and Theorem 5.7. □

Remark 5.10 The corresponding statement for $\mathrm{H}^{p,q}(\mathrm{X}^{\mathrm{an}})$ fails. We sketch the argument. If X is an affine variety, then we may pick $\mathcal{S} = \mathbb{A}_{\mathrm{global}}$, the class of closed embeddings into affine space.

One can show that $\mathrm{H}^{n,n}(Y) = 0$ for all tropical subvarieties of $\mathbb{T}^n = \mathrm{Trop}(\mathbb{A}^n)$. Thus we have $\varinjlim_{\varphi\in\mathcal{S}} \mathrm{H}^{p,q}(\mathrm{Trop}_\varphi(X)) = 0$.

Let K be algebraically closed, E be an elliptic curve of good reduction and let e be a rational point of E. Let V be an open neighborhood of e that is isomorphic to an open annulus and let $X = E \setminus e$.

Using the Mayer-Vietoris sequence for the cover $(\mathrm{X}^{\mathrm{an}}, V)$ for E^{an} gives the following exact sequence

$$\mathrm{H}^{1,0}(V \setminus e) \to \mathrm{H}^{1,1}(E^{\mathrm{an}}) \to \mathrm{H}^{1,1}(\mathrm{X}^{\mathrm{an}}) \oplus \mathrm{H}^{1,1}(V).$$

It was shown in [17, Theorem 5.7] that both $\mathrm{H}^{1,0}(V \setminus e)$ and $\mathrm{H}^{1,1}(V)$ are finite dimensional. Further it was shown in [14, Theorem B] that when the residue field of K is $\mathbb{C}$, then $\mathrm{H}^{1,1}(E^{\mathrm{an}})$ is infinite dimensional. This implies that $\mathrm{H}^{1,1}(\mathrm{X}^{\mathrm{an}})$ is infinite dimensional. In particular, it is not equal to $\varinjlim_{\varphi\in\mathcal{S}} \mathrm{H}^{p,q}(\mathrm{Trop}_\varphi(X)) = 0$.

6 Comparison theorems

6.1 Comparing with Gubler's definition

Since Gubler's definition only works when K is non-trivially valued and algebraically closed we assume for this subsection that this is the case.

Theorem 6.1 *The family of tropicalizations $\mathbb{G}$ from Example 4.8 is admissible.*

This theorem is not a formal consequence of the definition, since the family $\mathbb{G}$ does not see any boundary considerations. Since for algebraically closed K, Gubler defined in [9] the sheaf $\mathcal{A}$ like we here defined $\mathcal{A}_{\mathbb{G}_{\text{can}}}$ this theorem is crucial for us as it proves that what we show in later sections actually applies to $\mathcal{A}$.

Proof We have to show that the map $\Psi := \Psi_{\mathbb{G},\mathcal{T}}\colon \mathcal{A}_{\mathbb{G}} \to \mathcal{A}_{\mathcal{T}}$ is surjective. We argue locally around a point $x \in \mathrm{X}^{\text{an}}$. Let $\alpha \in \mathcal{A}_{\mathcal{T}}$ be locally given by (V, φ, β), where $\varphi\colon U \to Y_\Sigma$ is a closed embedding of an open subset U of X into a toric variety. We fix coordinates on the torus stratum T of Y_Σ that x is mapped to under φ^{an}, identifying T with $\mathbb{G}_m^n$ for some n and we denote by $Z := \varphi^{-1}(T)$. Since very affine open subset form a basis of the topology of X, there exists a closed embedding $\varphi'\colon U' \to \mathbb{G}_m^r$ for $x \in U'^{\text{an}} \subset U^{\text{an}}$ such that if $\pi : \mathbb{G}_m^r \to \mathbb{G}_m^n$ is the projection to the first n factors, then $\varphi|_Z = \pi \circ \varphi' \circ \iota_Z$.

After shrinking V we may assume that $\text{trop}_\varphi(V)$ is a neighborhood of $\text{trop}_\varphi(x)$, on which $\beta = \pi^*(\beta_0)$ holds, where π is the projection to the stratum $\text{Trop}(T)$ of $\text{Trop}(Y)$. We define the form $\alpha' \in \mathcal{A}_{\mathbb{G}}$ to be given by $(V', \varphi', \text{Trop}(\psi)^*\beta_0)$. Since $\Psi(\alpha')$ is by definition given by the same triple, we have to show that (V, φ, β) and $(V', \varphi', \text{Trop}(\psi)^*\beta_0)$ define the same form in a neighborhood of x. We do that by pulling back to a common subcharts, namely $\varphi \times \varphi'\colon U' \to Y_\Sigma \times \mathbb{G}_m^r$. When pulling back on the tropical side we find that both $\text{Trop}(\pi_1)^*\beta$ as well as $\text{Trop}(\pi_2)^*\beta_0$ are simply the pullback of β_0, hence these forms agree indeed. □

Corollary 6.2 *The family of tropicalizations $\mathbb{G}_{\text{can}}$ is admissible.*

Proof This follows from Lemma 5.3, the fact that $\mathbb{G}_{\text{can}}$ is cofinal in $\mathbb{G}$ and Theorem 6.1. □

6.2 Comparing with the analytic definition by Chambert-Loir and Ducros

We denote by $\mathcal{A}_{\text{an}}^{p,q}$ the analytically defined sheaf of differential forms by Chambert-Loir and Ducros [3].

Definition 6.3 We define a map

$$\Psi_{\text{an}}\colon \mathcal{A}_{\mathbb{A}}^{p,q} \to \mathcal{A}_{\text{an}}^{p,q}\,.$$

Let (V,φ,α) an $\mathbb{A}$-tropical chart, where φ is given by $f_1,\ldots,f_r$. Let $f_{i_1},\ldots,f_{i_k}$ be those f_i that do not vanish at X. Then those induce a map $\varphi'\colon V \to \mathbb{G}_m^k$ and $\operatorname{trop}_{\varphi'}(V) = \operatorname{trop}_{\varphi}(V) \cap \{z_{i_j} = -\infty \text{ for all } j = 1,\ldots,k\}$, which is a stratum of $\operatorname{trop}_{\varphi}(V)$ and denote by α_I the restriction of α to that stratum.. Then we define $\Psi(\alpha)$ to be given by $(V,\varphi_{\mathbb{G}_m},\alpha_I)$.

Theorem 6.4 *The map Ψ_{an} is an isomorphism.*

Proof We only have to prove the statement on stalks, so we fix $x \in \mathrm{X}^{\text{an}}$. By construction any form is around x given by one $\mathbb{A}$-tropical chart. Then $\Psi_{\text{an}}(\alpha)$ is also given around x by one chart. If now $\Psi_{\text{an}}(\alpha) = 0$, then α_I equals zero by [3, Lemma 3.2.2] and thus α equals zero. This shows injectivity.

To show surjectivity, we take a form α that is given locally around x by a map $V \to \mathbb{G}_m^{r,\text{an}}$ defined by $f_1,\ldots,f_r$ and a form α_1 on $\operatorname{Trop}_{f_1,\ldots,f_r}(V)$. The f_r can be expressed as Laurent series and around x we may cut them of in sufficiently high degree to obtain Laurent polynomials $f_1',\ldots,f_r'$ such that $|f_i| = |f_i'|$ for all i on a neighborhood V_1 of x. Now choosing functions $g_1,\ldots,g_s \in \mathcal{O}_X(U)$ for an open subset U of X such that $x \in U^{\text{an}}$, the f_i' and g_i define a closed embedding $\varphi\colon U \to \mathbb{A}^{r+s}$ with the property that there exists a neighborhood V_2 of x in V_1 and a tropical chart (V_2,φ). We may assume that $g_1,\ldots,g_t$ are non vanishing at x and $g_{t+1},\ldots,g_s$ are. Now $\alpha|_{V_2}$ is defined by a form α_2 on $\operatorname{Trop}_{f_1',\ldots,f_r',g_1,\ldots,g_t}(V_2)$. By construction we have $\operatorname{Trop}_{f_1',\ldots,f_r',g_1,\ldots,g_t}(V_2)$ is the stratum of $\operatorname{Trop}_{\varphi}(V_2)$, where the coordinates $t+1$-th to s-th coordinate are $-\infty$. Denote by $\pi\colon \operatorname{Trop}_{\varphi}(V_2) \to \operatorname{Trop}_{f_1',\ldots,f_r',g_1,\ldots,g_t}(V_2)$ the projection Let $\alpha_3 = \pi^*\alpha_2$. Then we define $\beta \in \mathcal{A}_{\mathbb{A}}^{p,q}(V_2)$ by (V_2,φ,α_3). Now $\Psi_{\text{an}}(\beta)$ is given by $V_2 \to \mathbb{G}_m^{r+t}$ given by $f_1',\ldots,f_r',g_1,\ldots,g_t$ and α_2. Since $\operatorname{trop}_{f_1,\ldots,f_r,g_1,\ldots,g_t} = \operatorname{trop}_{f_1',\ldots,f_r',g_1,\ldots,g_t}\colon V_2 \to \mathbb{R}^{r+t}$, the result follows from [3, Lemma 3.1.10]. □

Lemma 6.5 *Let K be non-trivially valued and $\alpha \in \mathcal{A}_{\mathbb{G}}^{p,q}(V)$ given by (V,φ,α'). Then $\Psi_{\text{an}} \circ \Psi_{\mathbb{A},\mathcal{T}}^{-1} \circ \Psi_{\mathbb{G},\mathcal{T}}(\alpha)$ is given by $(V,\varphi^{\text{an}},\alpha')$.*

Proof Let $\varphi\colon U \to \mathbb{G}_m^r$ be given by invertible functions $f_1,\ldots,f_r$. Then the corresponding embedding into $\mathbb{A}^{2r}$, which we denote by $\varphi_\pm$ is given by $f_1,f_1^{-1},\ldots,f_r,f_r^{-1}$. Denote by $\pi\colon \mathbb{R}^{2r} \to \mathbb{R}^r$ the projection to the odd coordinates. Then by construction $\Psi^{\text{an}} \circ \Psi_{\mathbb{G}_m}(\alpha)$ is given by $(V,\varphi_\pm^{\text{an}},\pi^*(\alpha'))$. Since $\varphi_\pm^{\text{an}}$ is a refinement of φ^{an} and the map induced on tropicalizations is precisely π, we find that $(V, \varphi_{\pm}^{\mathrm{an}}, \pi^*(\alpha)) = (V, \varphi^{\mathrm{an}}, \alpha') \in \mathcal{A}_{\mathrm{an}}^{p,q}(V)$, which proves the claim. □

7 Integration

In this section we denote by n the dimension of X and we let $\mathcal{S}$ be a fine enough family of tropicalizations.

Definition 7.1 Let $\alpha \in \mathcal{A}_{\mathcal{S},c}^{n,n}(\mathrm{X}^{\mathrm{an}})$ be an (n,n)-form with compact support. A *tropical chart of integration for* α is a $\mathcal{S}$-tropical chart $(U^{\mathrm{an}}, \varphi)$ where U is an open subset of X and $\varphi\colon U \to Y_\Sigma$ is a closed embedding of U into a toric variety, such that $\alpha|_{U^{\mathrm{an}}}$ is given by $(U^{\mathrm{an}}, \varphi, \beta)$ for $\beta_U \in \mathcal{A}_c^{n,n}(\mathrm{Trop}_\varphi(U))$.

Lemma 7.2 *There always exists an $\mathbb{A}$-tropical chart of integration. If K is non-trivially valued and algebraically closed, there always exist $\mathbb{G}$-tropical charts of integration. If $\mathcal{S}$ is global, then there always exist $\mathcal{S}$-tropical charts of integration.*

Proof The existence of $\mathbb{A}$-tropical charts of integration is proved in [11, Lemma 3.2.57] and the existence of $\mathbb{G}$-tropical charts of integration is proved in [9, Proposition 5.13]. Note that [11] assumes that K is algebraically closed, but the proof goes through here.

In the case of global $\mathcal{S}$, let α be given by $(V_i, \varphi_i, \alpha_i)_{i \in I}$. Then we may pick finitely many i such that $\mathrm{supp}(\alpha)$ is covered by the V_i and apply Theorem 5.6 to obtain an $\mathcal{S}$-tropical chart of integration. □

Theorem 7.3 *Let $\alpha \in \mathcal{A}_{\mathcal{S},c}^{n,n}(\mathrm{X}^{\mathrm{an}})$ and $(U^{\mathrm{an}}, \varphi, \alpha_U)$ an $\mathcal{S}$-tropical chart of integration for α. Then the value*

$$\int_{\mathrm{X}^{\mathrm{an}}}^{\mathcal{S}} \alpha := \int_{\mathrm{Trop}_\varphi(U)} \alpha_U$$

depends only on α, in the sense that it is independent of the triple (U, φ, α_U) representing α. Further it is also independent of $\mathcal{S}$, in the sense that

$$\int_{\mathrm{X}^{\mathrm{an}}}^{\mathcal{S}} \alpha = \int_{\mathrm{X}^{\mathrm{an}}}^{\mathcal{T}} \Psi_{\mathcal{S},\mathcal{T}}(\alpha).$$

Proof Assume first that K is non-trivially valued and algebraically closed. Then Gubler showed in [9, Lemma 5.15] that as a consequence of the Sturmfels–Tevelev formula [2, 28] the first part of the statement holds for $\mathcal{S} = \mathbb{G}$.

Let $\varphi \colon U \to Y_\Sigma$. We may assume that $\varphi(U)$ meets the dense torus T. Denote by $\mathring{U} = \varphi^{-1}(T)$ and $\mathring{\varphi} := \varphi|_{\mathring{U}}$. We have $\operatorname{supp}(\alpha) \subset \mathring{U}$ and $\operatorname{supp}(\alpha_U) \subset \operatorname{Trop}_{\varphi|_{\mathring{U}}}(\mathring{U})$ by [11, Proposition 3.2.56], which implies that $(\mathring{U}^{\mathrm{an}}, \varphi, \alpha_U|_{\operatorname{Trop}_\varphi(\mathring{U})})$ is a $\mathbb{G}$-tropical chart of integration for α. Thus we get

$$\int_{X^{\mathrm{an}}}^{\mathcal{S}} \alpha = \int_{\operatorname{Trop}_\varphi(U)} \alpha_U = \int_{\operatorname{Trop}_\varphi(\mathring{U})} \alpha_U|_{\operatorname{Trop}_\varphi(\mathring{U})} = \int_{X^{\mathrm{an}}}^{\mathbb{G}} \Psi_{\mathcal{S},\mathcal{T}} \circ \Psi_{\mathbb{G},\mathcal{T}}^{-1} \alpha. \quad (4)$$

The right-hand side of equation (4) does not depend on U, φ or α_U by Gubler's result, hence neither does the left hand side. So we proved the first part of the theorem when K is non-trivially valued. Then second part follows also from (4) applied once to $\mathcal{S}$ and once to $\mathcal{T}$.

We reduce to the non-trivially valued case by picking a non-archimedean, complete, algebraically closed non-trivially valued extension L of K. Let $p \colon X_L \to X$ be the canonical map. Since tropicalization is invariant under base field extension (cf. [26, Section 6 Appendix]), we can define $\alpha \in \mathcal{A}_c^{n,n}({X^{\mathrm{an}}}_L)$ to be given by $(U_L^{\mathrm{an}}, \varphi_L, \alpha_U)$. Now we have

$$\int_{\operatorname{Trop}_\varphi(U)} \alpha_U = \int_{\operatorname{Trop}_{\varphi_L}(U_L)} \alpha_U$$

and the right hand side depends only on α_L, which depends only on α. The last part follows because the maps Ψ are also compatible with base change. □

Definition 7.4 We define

$$\int_{X^{\mathrm{an}}} \alpha = \int_{X^{\mathrm{an}}}^{\mathcal{T}} \alpha_U.$$

Lemma 7.5 $\int_{X^{\mathrm{an}}}$ *does not change when extending the base field.*

Proof This follows from the last part of the proof of Theorem 7.3. □

Chambert-Loir and Ducros also define an integration for (n,n)-forms with compact support, which we denote by $\int^{\mathrm{CLD}}$.

Lemma 7.6 $\int^{\mathrm{CLD}}$ *does not change when extending the base field.*

Proof The base changes of any atlas of integration in the sense of [3] is still an atlas of integration for the base changed form. Then since tropicalizations and the multiplicity d_D from their definition does not change, we obtain the result. □

Theorem 7.7 *Let* $\alpha \in \mathcal{A}_c^{n,n}(X^{\mathrm{an}})$. *Then*

$$\int_{X^{\mathrm{an}}}^{\mathrm{CLD}} \Psi_{\mathrm{an}}(\alpha) = \int_{X^{\mathrm{an}}} \alpha.$$

Proof Let L be a non-trivially valued non-archimedean extension of K. After replacing K by L, X by X_L and α be α_L, we may, since neither integral changes by Lemmas 7.5 and 7.6, assume that K is non-trivially valued.

Then, by Corollary 6.2, $\mathbb{G}_{\text{can}}$ is an admissible family of tropicalizations and thus by Theorem 7.3, we may take $\alpha \in \mathcal{A}^{n,n}_{\mathbb{G}_{\text{can}},c}$, say given by $(V_i, \varphi_i, \alpha_i)_{i\in I}$. Then by Lemma 6.5 we have to show

$$\int_{\mathrm{X}^{\text{an}}}^{\text{CLD}} (V_i, \varphi_i^{\text{an}}, \alpha_i)_{i\in I} = \int_{\mathrm{X}^{\text{an}}} (V_i, \varphi_i, \alpha_i)_{i\in I}.$$

This was precisely shown in [9, Section 7]. □

8 Tropical cohomology

In this section we assume that $\mathcal{S}$ is a fine enough family of tropicalizations for X that is cofinal in $\mathcal{T}$. We let R be a subring of $\mathbb{R}$. We will use the sheaf of tropical cochains $\underline{C}^{p,q}(R)$ and the constructions from Section 3.

Definition 8.1 Let V be a open subset of X^{an}. An element of $\underline{C}^{p,q}_{\mathcal{S}}(V, R)$ is given by a family $(V_i, \varphi_i, \eta_i)_{i\in I}$ such that:

1. The V_i cover V, i.e. $V = \bigcup_{i\in I} V_i$.
2. For each $i \in I$ the pair (V_i, φ_i) is an $\mathcal{S}$-tropical chart.
3. For each $i \in I$ we have $\eta_i \in \underline{C}^{p,q}(\text{trop}_{\varphi_i}(V_i), R)$.
4. For all $i, j \in I$ there exist $\mathcal{S}$-tropical subcharts $(V_{ijl}, \varphi_{ijl})_{l\in L}$ that cover $V_i \cap V_j$ such that

$$\text{Trop}(\varphi_i, \varphi_{ijl})^* \eta_i = \text{Trop}(\varphi_j, \varphi_{ijl})^* \eta_j \in \underline{C}^{p,q}(\text{trop}_{\varphi_{ijl}}(V_{ijl}), R).$$

Another such family $(V_j, \varphi_j, \eta_j)_{j\in J}$ defines the same form if and only if their union $(V_i, \varphi_i, \eta_i)_{i\in I\cup J}$ still satisfies (iv).

For each p we obtain a complex of sheaves $(\underline{C}^{p,\bullet}_{\mathcal{S}}(R), \partial)$ on X^{an}. If $\mathcal{S} = \mathcal{T}$ we will drop the subscript and write $\underline{C}^{p,q}(R) := \underline{C}^{p,q}_{\mathcal{T}}(R)$.

Remark 8.2 It follows the same way as in Section 5 for $\mathcal{A}^{p,q}$ that $\underline{C}^{p,q}_{\mathcal{S}}$ is isomorphic to $\underline{C}^{p,q}_{\mathcal{S}'}$ when $\mathcal{S}'$ is final or cofinal for $\mathcal{S}$.

The author does not know whether one gets isomorphic sheaves when one applies Definition 8.1 with for example $\mathcal{S} = \mathbb{G}$. The missing piece here is that for differential forms on tropical toric varieties, the condition of compatibility requires every forms to locally be a pullback of a forms from a tropical torus (i.e. $\mathbb{R}^n$). The same is not true for tropical cocycles, hence the arguments form Theorem 6.1 do not work.

Lemma 8.3 *Let $\eta \in \underline{C}^{p,q}(V,R)$ be given by (V,φ,η'). Then $\eta = 0$ if and only if $\eta' = 0$.*

Proof The proof for forms [11, Lemma 3.2.12] works word for word. □

Proposition 8.4 *Comparing with tropical cohomology, we obtain*

$$\underline{C}_c^{p,q}(\mathrm{X}^{\mathrm{an}}, R) = \varinjlim_{\varphi \in \mathcal{S}} \underline{C}_{\mathrm{trop},c}^{p,q}(\mathrm{Trop}_\varphi(X), R)$$

and

$$\mathrm{H}_{\mathrm{trop},c}^{p,q}(\mathrm{X}^{\mathrm{an}}, R) = \varinjlim_{\varphi \in \mathcal{S}} \mathrm{H}_{\mathrm{trop},c}^{p,q}(\mathrm{Trop}_\varphi(X), R).$$

Proof This follows from Lemma 8.3 in the same way as for forms in Theorems 5.7 and 5.9 follow from Lemma 5.2. □

Definition 8.5 The maps $\mathbb{R}$ defined in Section 3 define maps

$$\mathbb{R}\colon \mathcal{A}^{p,q} \to \underline{C}^{p,q}(\mathbb{R})$$

the induces a morphism of complexes of sheaves.

Definition 8.6 The cohomology

$$\mathrm{H}_{\mathrm{trop}}^{p,q}(\mathrm{X}^{\mathrm{an}}, R) := \mathrm{H}^q(\underline{C}^{p,\bullet}(\mathrm{X}^{\mathrm{an}}, R), \partial)$$

is called *tropical cohomology with coefficients in R* of X^{an}. Similarly

$$\mathrm{H}_{\mathrm{trop},c}^{p,q}(\mathrm{X}^{\mathrm{an}}, R) := \mathrm{H}^q(\underline{C}_c^{p,\bullet}(\mathrm{X}^{\mathrm{an}}, R), \partial)$$

is called *tropical cohomology with coefficients in R with compact support* of X^{an}.

Definition 8.7 We denote by $\mathcal{F}_R^p := \ker(\partial\colon \underline{C}^{p,0}(R) \to \underline{C}^{p,1}(R))$.

Lemma 8.8 *The complex*

$$0 \to \mathcal{F}_R^p \to \underline{C}^{p,0}(R) \to \underline{C}^{p,1}(R) \to \cdots \to \underline{C}^{p,n}(R) \to 0$$

is exact.

Proof Exactness on the tropical side is true by [16, Proposition 3.11 & and Lemma 3.14] (with real coefficients, but the proof goes through here). It is then automatically true on the analytic side using the definitions (cf. the proof for forms [12, Theorem 4.5]). □

Remark 8.9 The sheaves $\mathcal{A}_{\mathcal{S}}^{p,q}$ admit partitions of unity, which can be shown the same way as it was shown by Gubler for $\mathcal{S} = \mathbb{G}$ in [9, Proposition 5.10]. This proof however uses the $\mathbb{R}$-structure of those sheaves.

The sheaves $\underline{C}^{p,q}(R)$ on a tropical variety (as defined in Section 3) are flasque sheaves [16, Lemma 3.14], hence in particular acyclic.

However, it is not clear whether this property also holds for $\underline{C}^{p,q}(R)$ on the analytic space X^{an} in general. We will prove some partial results in the next Lemma.

Recall condition (†) from Definition 4.11.

Lemma 8.10 *Assume that $R = \mathbb{R}$ or that X satisfies condition (†). Then the sheaves $\underline{C}^{p,q}(R)$ are acyclic with respect to the functor of global sections as well as global sections with compact support.*

Proof Using the map $\mathbb{R}\colon \mathcal{A}^{0,0} \to \underline{C}^{0,0}(\mathbb{R})$, we see that $\underline{C}^{0,0}(\mathbb{R})$ admits partitions of unity. Hence $\underline{C}^{0,0}(\mathbb{R})$ is a fine sheaf and since $\underline{C}^{p,q}(\mathbb{R})$ is a $\underline{C}^{0,0}(\mathbb{R})$-module (via the cap product) we see that $\underline{C}^{p,q}(\mathbb{R})$ is also a fine sheaf.

In general, if X satisfied condition (†) then $\mathcal{T}_{\text{global}}$ is a global family of tropicalizations that is cofinal in $\mathbb{A}$, which is final in $\mathcal{T}$. hence $\underline{C}^{p,q}_{\mathcal{T}_{\text{global}}}(R) \cong \underline{C}^{p,q}(R)$. Any section of $\underline{C}^{p,q}_{\mathcal{T}_{\text{global}}}(R)$ that is defined by finitely many charts (V_i, φ_i, η_i) can be defined by a single chart. This can be shown the same way as for forms in Theorem 5.6.

Since any section over a compact subset of X^{an} is defined by finitely many charts, each such section can be defined by one chart (X^{an}, φ, η). Now since the sheaf $\underline{C}^{p,q}(R)$ on $\text{Trop}_\varphi(X)$ is flasque, this section can be extended to a global section. This shows that the sheaf $\underline{C}^{p,q}(R)$ on X^{an} is c-soft in the sense of [19, Definition 2.5.5], which implies that it is acyclic for the functor of global sections with compact support [19, Proposition 2.58 & Corollary 2.5.9].

Since $\underline{C}^{p,q}(R)$ is c-soft, to show that it is acyclic for the functor of global sections, we have to show that X^{an} admits a countable cover by compact sets [19, Proposition 2.5.10]. Since $\mathbb{A}^{n,\text{an}}$ is covered by countably many discs, this holds if X is affine. Since general X is covered by finitely many affine varieties, the claim follows. □

Corollary 8.11 *If $R = \mathbb{R}$ or if X satisfies condition (†) we have*

$$H^{p,q}_{\text{trop}}(X^{an}, R) = H^q(X^{an}, \mathcal{F}^p_R) \text{ and } H^{p,q}_{\text{trop},c}(X^{an}, R) = H^q_c(X^{an}, \mathcal{F}^p_R).$$

Further, we have

$$H^{p,q}_{\text{trop}}(X^{an}, R) = H^{p,q}_{\text{trop}}(X^{an}, \mathbb{Z}) \otimes R. \text{ and}$$
$$H^{p,q}_{\text{trop},c}(X^{an}, R) = H^{p,q}_{\text{trop},c}(X^{an}, \mathbb{Z}) \otimes R.$$

Proof This follows from Lemmas 8.8 and 8.10, the fact that since R is torsion free, thus a flat $\mathbb{Z}$-module and $\mathcal{F}^p_R = \mathcal{F}^p_{\mathbb{Z}} \otimes \mathbb{Z}$. □

Theorem 8.12 (Tropical analytic de Rham theorem) *There exist isomorphisms*

$$\mathrm{H}^{p,q}(\mathrm{X}^{\mathrm{an}}) \cong \mathrm{H}^{p,q}_{\mathrm{trop}}(\mathrm{X}^{\mathrm{an}}, \mathbb{R}) \text{ and } \mathrm{H}^{p,q}_{c}(\mathrm{X}^{\mathrm{an}}) \cong \mathrm{H}^{p,q}_{\mathrm{trop},c}(\mathrm{X}^{\mathrm{an}}, \mathbb{R})$$

that are induced by the de Rham morphism on the tropical level, as defined in Remark 3.11.

Proof We have a map

$$\mathbb{R}\colon \mathcal{A}^{p,q} \to \underline{C}^{p,q}(\mathbb{R})$$

that is locally given by using the de Rham map on the tropical side constructed in Remark 3.11. This makes the following diagram commutative:

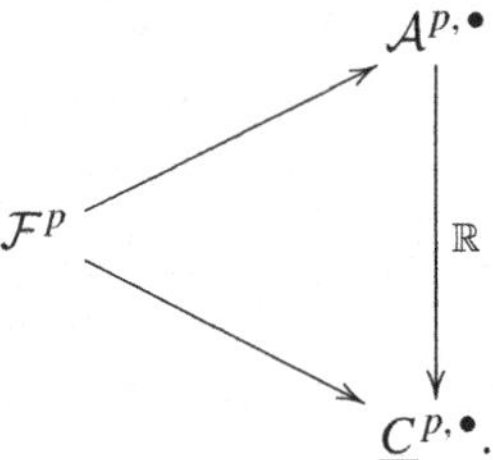

This is now a commutative diagram of acyclic resolutions of $\mathcal{F}^p$, which proves the theorem. □

9 Wave and monodromy operators

Since both the monodromy operator M on superforms and the wave operator defined in Section 3 on tropical cochains commute with pullbacks along affine maps on tropical toric varieties, we obtain maps

$$M\colon \mathrm{H}^{p,q}(\mathrm{X}^{\mathrm{an}}) \to \mathrm{H}^{p-1,q+1}(\mathrm{X}^{\mathrm{an}}) \text{ and}$$
$$W\colon \mathrm{H}^{p,q}_{\mathrm{trop}}(\mathrm{X}^{\mathrm{an}}, \mathbb{R}) \to \mathrm{H}^{p-1,q+1}_{\mathrm{trop}}(\mathrm{X}^{\mathrm{an}}, \mathbb{R}).$$

Theorem 9.1 *The wave and the monodromy operator agree on cohomology up to sign by virtue of the isomorphism* $\mathbb{R}$*, meaning that the diagram*

$$\begin{array}{ccc} \mathrm{H}^{p,q}(\mathrm{X}^{\mathrm{an}}) & \xrightarrow{(-1)^{p-1}M} & \mathrm{H}^{p-1,q+1}(\mathrm{X}^{\mathrm{an}}) \\ \downarrow \mathbb{R} & & \downarrow \mathbb{R} \\ \mathrm{H}^{p,q}_{\mathrm{trop}}(\mathrm{X}^{\mathrm{an}}, \mathbb{R}) & \xrightarrow{W} & \mathrm{H}^{p-1,q+1}_{\mathrm{trop}}(\mathrm{X}^{\mathrm{an}}, \mathbb{R}) \end{array}$$

commutes. The same is true for cohomology with compact support.

Proof The proof of Theorem 3.14 works word for word. □

In [22], Liu defined a $\mathbb{Q}$-subsheaf of $\mathcal{F}^p_{\mathbb{R}}$ and defined *rational classes* in tropical Dolbeault cohomology.

Definition 9.2 (Liu) Denote by $\mathcal{J}^p$ the $\mathbb{Q}$-subsheaf of $\mathcal{F}^p$ generated by sections of the form (V, φ, α), where $\varphi\colon U \to T$ and $\alpha \in \Lambda^p M$. The classes in $\mathrm{H}^{p,q}(\mathrm{X}^{\mathrm{an}}, \mathcal{J}^p) \subset \mathrm{H}^{p,q}(\mathrm{X}^{\mathrm{an}}, \mathbb{R})$ are called *rational classes*.

Proposition 9.3 *Assume that X satisfies condition (†). We have isomorphisms* $\mathcal{F}^p_{\mathbb{Q}} = \mathcal{J}^p$ *and* $\mathrm{H}^{p,q}(\mathrm{X}^{\mathrm{an}}, \mathbb{Q}) = \mathrm{H}^{p,q}(\mathrm{X}^{\mathrm{an}}, \mathcal{J}^p)$.

Proof The explicit computation [16, Proposition 3.11] of $\mathcal{F}^p_{\mathbb{R}}$ works also for rational coefficients. Then this follows directly from the definitions and Corollary 8.11. □

The following statement in particular shows that we have a strict inclusion $\mathrm{H}^{p,q}(\mathrm{X}^{\mathrm{an}}, \mathbb{Z}) \subsetneq \mathrm{H}^{p,q}(\mathrm{X}^{\mathrm{an}}, \mathbb{Q})$.

Theorem 9.4 *Assume that X satisfies condition (†). Then is a non-trivial R-linear map*

$$\cap[\mathrm{X}^{\mathrm{an}}]_R\colon \mathrm{H}^{n,n}_{\mathrm{trop},c}(\mathrm{X}^{\mathrm{an}}, R) \to R.$$

If $R = \mathbb{R}$, *then this agrees with the map induced by integration via* $\mathbb{R}$.

Proof The maps $[\mathrm{Trop}_\varphi(X)]_R$ as defined in Definition 3.16 are compatible with pullback along refinements, so by Proposition 8.4 we get a well defined map on $\mathrm{H}^{n,n}_c(\mathrm{X}^{\mathrm{an}}, R)$.

The last part of the statement follows from Proposition 3.17. □

Liu showed that if the value group of K is equal to $\mathbb{Q}$, then his monodromy map M maps rational classes to rational classes [22, Theorem 5.5 (1)]. We generalize to the following statement:

Theorem 9.5 *Assume that X satisfies condition (†). The wave operator W (and by virtue of Corollary 9.1 also the monodromy map M) restricts to a map*

$$W\colon \mathrm{H}^{p,q}_{\mathrm{trop},c}(\mathrm{X}^{\mathrm{an}}, R) \to \mathrm{H}^{p-1,q+1}_{\mathrm{trop},c}(\mathrm{X}^{\mathrm{an}}, R[\Gamma]).$$

Proof By Proposition 8.4, it is sufficient to prove this theorem for $\mathrm{Trop}_\varphi(X)$. Since $\mathrm{Trop}_\varphi(X)$ is an integral Γ-affine tropical variety, this follows from Proposition 3.15. □

Mikhalkin and Zharkov conjectured that for a smooth tropical variety X, the iterated wave operator

$$W^{p-q}\colon \mathrm{H}^{p,q}_{\mathrm{trop}}(X,\mathbb{R}) \to \mathrm{H}^{q,p}_{\mathrm{trop}}(X,\mathbb{R})$$

is an isomorphism for all $p \geq q$ [25, Conjecture 5.3].

Liu conjectured that if K is such that the residue field $\tilde{K}$ is the algebraic closure of a finite field and X is smooth and proper, then the iterated monodromy operator

$$M^{p-q}\colon \mathrm{H}^{p,q}(X) \to \mathrm{H}^{q,p}(X)$$

is an isomorphism for all $p \geq q$ [22, Conjecture 5.2].

As a consequence of Theorem 9.1 we can tie together both of these conjectures.

Proposition 9.6 *Let X be a proper variety. Assume there exists a global admissible family of tropicalizations $\mathcal{S}$ for X such that for all $\varphi \in \mathcal{S}_{\mathrm{map}}$ the tropical variety $\mathrm{Trop}_\varphi(X)$ satisfies Mikhalkin's and Zharkov's conjecture. Then X satisfies Liu's conjecture.*

Proof This follows directly from Theorem 9.1 and Theorem 5.9. □

10 Non-trivial classes

In this section we (partially) compute tropical cohomology with coefficients in three examples: Curves of good reduction, toric varieties and Mumford curves.

For the first theorem assume that the value group Γ of K is a subring of $\mathbb{R}$. We denote by $\log|\mathcal{O}_X^\times|$ the sheaf of real valued functions on X^{an} that are locally of the form $\log|f|$ for an invertible function f on X.

Theorem 10.1 *Let K be algebraically closed and X be a smooth projective curve of good reduction. Then there exists an injective morphism*

$$\mathrm{Pic}^0(\tilde{X}) \to \mathrm{H}^{1,1}(X,\mathbb{Z}),$$

where $\mathrm{Pic}^0(\tilde{X})$ denotes the group of degree 0 line bundles on the reduction $\tilde{X}$ of X.

Proof We have the following exact sequence

$$0 \to \Gamma \to \log|\mathcal{O}_X^\times| \to \mathcal{F}^1_{\mathbb{Z}} \to 0, \tag{5}$$

which is a non-archimedean version of a well-known exponential sequence from tropical geometry [24, Definition 4.1]. This induces the following exact sequence in cohomology groups

$$0 \to \mathrm{H}^{1,0}_{\mathrm{trop}}(\mathrm{X}^{\mathrm{an}}, \mathbb{Z}) \to \mathrm{H}^{0,1}_{\mathrm{trop}}(\mathrm{X}^{\mathrm{an}}, \Gamma) \to \tag{6}$$

$$\mathrm{H}^1(\mathrm{X}^{\mathrm{an}}, \log|\mathcal{O}_X^\times|) \to \mathrm{H}^{1,1}_{\mathrm{trop}}(\mathrm{X}^{\mathrm{an}}, \mathbb{Z}) \to 0. \tag{7}$$

In particular, since X is a curve of good reduction, X^{an} is contractible and $\mathrm{H}^{0,1}_{\mathrm{trop}}(\mathrm{X}^{\mathrm{an}}, \Gamma) = \mathrm{H}^1(\mathrm{X}^{\mathrm{an}}, \Gamma) = 0$. Hence we have that

$$\mathrm{H}^1(\mathrm{X}^{\mathrm{an}}, \log|\mathcal{O}_X^\times|) \to \mathrm{H}^{1,1}(\mathrm{X}^{\mathrm{an}}, \mathbb{Z}).$$

is an isomorphism. Therefore it is sufficient to prove that there exists an injective morphism $\mathrm{Pic}^0(\tilde{X}) \to \mathrm{H}^1(\mathrm{X}^{\mathrm{an}}, \log|\mathcal{O}_X^\times|)$. We have an exact sequence

$$0 \to \log|\mathcal{O}_X^\times| \to \mathcal{H}_{\mathbb{Z}} \to \iota_* \mathrm{Pic}^0(\tilde{X}) \to 0 \tag{8}$$

where $\mathcal{H}_{\mathbb{Z}}$ is the sheaf of real valued functions on X^{an} that locally factor as the retraction to a skeleton composed with a piecewise linear function with integer slopes and values in Γ on the edges of said skeleton. This is sequence is the integral version of [29, Lemma 2.3.22]. Further, Thuillier showed that every harmonic function on a compact Berkovich analytic space is constant [29, Proposition 2.3.13]. Thus we obtain the following long exact sequence:

$$0 \to \Gamma \to \Gamma \to \mathrm{Pic}^0(\tilde{X}) \to \mathrm{H}^1(\mathrm{X}^{\mathrm{an}}, \log|\mathcal{O}_X^\times|). \tag{9}$$

This shows the existence of an injective morphism

$$\mathrm{Pic}^0(\tilde{X}) \to \mathrm{H}^1(\mathrm{X}^{\mathrm{an}}, \log|\mathcal{O}_X^\times|).$$ □

Remark 10.2 Since the Picard group of a smooth projective curve of positive genus over an algebraically closed field contains torsion, Theorem 10.1 implies that $\mathrm{H}^{1,1}(\mathrm{X}^{\mathrm{an}}, \mathbb{Z})$ can contain torsion. In other word the map $\mathrm{H}^{p,q}(\mathrm{X}^{\mathrm{an}}, \mathbb{Z}) \to \mathrm{H}^{p,q}(\mathrm{X}^{\mathrm{an}}, \mathbb{R}) = \mathrm{H}^{p,q}(\mathrm{X}^{\mathrm{an}}, \mathbb{Z}) \otimes \mathbb{R}$ need not be injective.

It is very possible that one can drop the assumption for K to be algebraically closed in Theorem 10.1.

Theorem 10.3 *Let $\varphi\colon X \to Y$ be a closed embedding of X into a toric variety Y. Assume that $\mathrm{Trop}_\varphi(X)$ is a smooth tropical variety. Then $\mathrm{trop}^*\colon \mathrm{H}^{p,q}(\mathrm{Trop}_\varphi(X)) \to \mathrm{H}^{p,q}(\mathrm{X}^{\mathrm{an}})$ and $\mathrm{trop}^*\colon \mathrm{H}^{p,q}_c(\mathrm{Trop}_\varphi(X)) \to \mathrm{H}^{p,q}_c(\mathrm{X}^{\mathrm{an}})$ are injective.*

Proof By [16, Theorem 4.33], since $\mathrm{Trop}_\varphi(X)$ is smooth there is a perfect pairing

$$\mathrm{H}^{p,q}(\mathrm{Trop}_\varphi(X)) \times \mathrm{H}^{n-p,n-q}_c(\mathrm{Trop}_\varphi(X)) \to \mathbb{R}$$

induced by the wedge product and integration of superforms. Thus given a d''-closed α in $\mathcal{A}^{p,q}(\mathrm{Trop}_\varphi(X))$ whose class $[\alpha] \in \mathrm{H}^{p,q}(\mathrm{Trop}_\varphi(X))$ is non-trivial, there exists $[\beta] \in \mathrm{H}_c^{n-p,n-q}(\mathrm{Trop}_\varphi(X))$ such that $\int_{\mathrm{Trop}_\varphi(X)} \alpha \wedge \beta \neq 0$. Thus we have

$$\int_{\mathrm{X}^{\mathrm{an}}} \mathrm{trop}_\varphi^* \alpha \wedge \mathrm{trop}_\varphi^* \beta \neq 0.$$

Since integration and the wedge product are well defined on cohomology this means that $[\mathrm{trop}_\varphi^* \alpha \wedge \mathrm{trop}_\varphi^* \beta]$ and consequently $[\mathrm{trop}_\varphi^* \alpha]$ is not trivial. The argument for $\alpha \in \mathrm{H}_c^{p,q}(\mathrm{Trop}_\varphi(X))$ works the same except $[\beta] \in \mathrm{H}^{n-p,n-q}(\mathrm{Trop}_\varphi(X))$. □

Example 10.4 Let Y_Σ be a smooth toric variety. Then Y_Σ is locally isomorphic to $\mathbb{A}^n$ and hence $\mathrm{Trop}(Y)$ is locally isomorphic to $\mathrm{Trop}(\mathbb{A}^n)$ and hence is a smooth tropical variety. Thus

$$\mathrm{trop}^*\colon \mathrm{H}^{p,q}(\mathrm{Trop}(Y)) \to \mathrm{H}^{p,q}(Y^{\mathrm{an}})$$

is injective by Theorem 10.3. Let $Y_\Sigma(\mathbb{C})$ be the complex toric variety associated with Σ. Then $\mathrm{H}^{p,q}_{\mathrm{Hodge}}(Y_\Sigma(\mathbb{C})) \cong \mathrm{H}^{p,q}(\mathrm{Trop}(Y_\Sigma), \mathbb{C})$ [10, Corollary 2]. In particular we have $\dim_{\mathbb{R}} \mathrm{H}^{p,q}(Y_\Sigma) \geq \dim_{\mathbb{C}} \mathrm{H}^{p,q}_{\mathrm{Hodge}}(Y_\Sigma(\mathbb{C}))$. One may figure out the latter in terms of Σ using [7, Section 5.2] or with the help of a computer and in terms of the polytope of Y using the package cellularSheaves for polymake [20]. Note that $\mathrm{H}^{p,q}_{\mathrm{Hodge}}(Y(\mathbb{C})) = 0$ if $p \neq q$ by [5, Corollary 12.7].

Example 10.5 Let K be algebraically closed, X be a smooth projective curve of genus g and $\varphi\colon X \to Y$ be a closed embedding of X into a toric variety such that $\mathrm{Trop}_\varphi(X)$ is a smooth tropical variety (this exists if and only if X is a Mumford curve by [13]). Then

$$\mathrm{H}^{p,q}_{\mathrm{trop}}(\mathrm{Trop}_\varphi(X), R) \to \mathrm{H}^{p,q}_{\mathrm{trop}}(\mathrm{X}^{\mathrm{an}}, R)$$

is an isomorphism. In particular we have $\mathrm{H}^{0,0}_{\mathrm{trop}}(\mathrm{X}^{\mathrm{an}}, R) \cong \mathrm{H}^{1,1}_{\mathrm{trop}}(\mathrm{X}^{\mathrm{an}}, R) \cong R$ and $\mathrm{H}^{1,0}_{\mathrm{trop}}(\mathrm{X}^{\mathrm{an}}, R) \cong \mathrm{H}^{0,1}_{\mathrm{trop}}(\mathrm{X}^{\mathrm{an}}, R) \cong R^g$.

Proof Assume that $\mathrm{Trop}_\varphi(X)$ is smooth. Then trop_φ is a homeomorphism from a skeleton of X^{an} onto $\mathrm{Trop}_\varphi(X)$ [12, Theorem 5.7]. Using comparison with singular cohomology we obtain $\mathrm{H}^{0,0}_{\mathrm{trop}}(\mathrm{Trop}_\varphi(X), R) = R$ and $\mathrm{H}^{0,1}_{\mathrm{trop}}(\mathrm{Trop}_\varphi(X), R) = R^g$. Using duality with coefficients in R as proven in [15, Theorem 5.3] and comparison with singular homology, we also obtain $\mathrm{H}^{1,1}_{\mathrm{trop}}(\mathrm{X}^{\mathrm{an}}, R) = R$ and $\mathrm{H}^{1,0}_{\mathrm{trop}}(\mathrm{Trop}_\varphi(X), R) = R^g$. One immediately verifies all transition maps induced by refinements in the family $\mathcal{T}_{\mathrm{Smooth}}$

defined in Example 4.16 are isomorphisms and hence the claim follows from Theorem 5.9. □

11 Open questions

In this section, we let X be a variety over K.

When X is smooth, Liu constructed cycles class maps, meaning maps $\mathrm{cyc}_k\colon \mathrm{CH}(X)^k \to \mathrm{H}^{k,k}(\mathrm{X}^{\mathrm{an}})$ that are compatible with the product structure on both sides and have the expected integration property [21].

Question 11.1 What is the image of cyc_k?

In light of the tropical Hodge conjecture and Corollary 9.1, one might conjecture that the image of $\mathrm{CH}(X)_{\mathbb{Q}}$ is $\mathrm{H}^{k,k}(\mathrm{X}^{\mathrm{an}}, \mathbb{Q}) \cap \ker(M)$. One might start with the case $k = \dim X - 1$. Here one knows the answer tropically [15], but the non-archimedean analogue is not a direct consequence.

The following question was asked by a referee and the author thinks it is worthwhile to include it here along with a partial answer.

Question 11.2 Is there an analogue of Theorem 10.1 when X has semistable reduction?

Let X be a smooth projective curve, let $\mathcal{X}_s$ be the special fiber of a strictly semistable model of X and let $C_1, \ldots, C_n$ be the irreducible components of $\mathcal{X}_s$. Then we can construct a map $\mathrm{Pic}^0(\mathcal{X}_s) \to \mathrm{H}^{1,1}(\mathrm{X}^{\mathrm{an}}, \mathbb{Z})$ by the composition

$$\mathrm{Pic}^0(\mathcal{X}_s) \to \bigoplus_{i=1}^{n} \mathrm{Pic}^0(C_i) \to \mathrm{H}^{1,1}(\mathrm{X}^{\mathrm{an}}, \log|\mathcal{O}_X^\times|) \to \mathrm{H}^{1,1}(\mathrm{X}^{\mathrm{an}}, \mathbb{Z}).$$

Here the first map is pullback along normalization, and the second and third maps are induced by (9) and (6), which remain valid when replacing $\mathrm{Pic}^0(\tilde{X})$ with $\bigoplus_{i=1}^n \mathrm{Pic}^0(C_i)$. It is well known that the first map need not be injective, hence the composition need not be injective. Whether the map

$$\bigoplus_{i=1}^{n} \mathrm{Pic}^0(C_i) \to \mathrm{H}^{1,1}(\mathrm{X}^{\mathrm{an}}, \log|\mathcal{O}_X^\times|) \to \mathrm{H}^{1,1}(\mathrm{X}^{\mathrm{an}}, \mathbb{Z})$$

is injective is unclear to the author. The map $\mathrm{H}^{1,1}(\mathrm{X}^{\mathrm{an}}, \log|\mathcal{O}_X^\times|) \to \mathrm{H}^{1,1}(\mathrm{X}^{\mathrm{an}}, \mathbb{Z})$ will not be injective when X^{an} is not contractible, since the map $\mathrm{H}^{1,0}(\mathrm{X}^{\mathrm{an}}, \mathbb{Z}) \to \mathrm{H}^{0,1}(\mathrm{X}^{\mathrm{an}}, \Gamma)$ from (6) will not be surjective. But that of course does not imply that the composition can not be injective.

Question 11.3 Does there exists a toric variety Y and a closed embedding $\varphi\colon X \to Y$ such that

$$\mathrm{trop}^*\colon \mathrm{H}^{p,q}(\mathrm{Trop}_\varphi(X)) \to \mathrm{H}^{p,q}(\mathrm{X}^{\mathrm{an}}) \text{ and}$$
$$\mathrm{trop}^*\colon \mathrm{H}_c^{p,q}(\mathrm{Trop}_\varphi(X)) \to \mathrm{H}_c^{p,q}(\mathrm{X}^{\mathrm{an}})$$

are isomorphisms?

The statement for $\mathrm{H}_c^{p,q}(\mathrm{X}^{\mathrm{an}})$ is implied by the finite dimensionality of $\mathrm{H}_c^{p,q}(\mathrm{X}^{\mathrm{an}})$ via Theorem 5.9. It is in fact equivalent to the finite dimensionality of $\mathrm{H}^{p,q}(\mathrm{X}^{\mathrm{an}})$ if one knew that $\mathrm{H}_c^{p,q}(\mathrm{Trop}_\varphi(X))$ is always finite dimensional, though the author is not aware of such a result (without regularity assumptions on $\mathrm{Trop}_\varphi(X)$).

Other questions related to this concern smoothness of the tropical variety.

Question 11.4 Let $\varphi\colon X \to Y$ be a closed embedding of X into a toric variety Y such that $\mathrm{Trop}_\varphi(X)$ is smooth. Are then

$$\mathrm{trop}^*\colon \mathrm{H}^{p,q}(\mathrm{Trop}_\varphi(X)) \to \mathrm{H}^{p,q}(\mathrm{X}^{\mathrm{an}}) \text{ and}$$
$$\mathrm{trop}^*\colon \mathrm{H}_c^{p,q}(\mathrm{Trop}_\varphi(X)) \to \mathrm{H}_c^{p,q}(\mathrm{X}^{\mathrm{an}})$$

isomorphisms?

This is certainly a natural question and "optimistically expected" to be true by Shaw [27, p.3]. We now know it holds for curves, as we showed in Example 10.5, but even the case $X = Y$ is open in dimension ≥ 2.

Question 11.5 Let $\varphi\colon X \to Y$ be a closed embedding of X into a toric variety Y such that $\mathrm{Trop}_\varphi(X)$ is smooth. Does the diagram

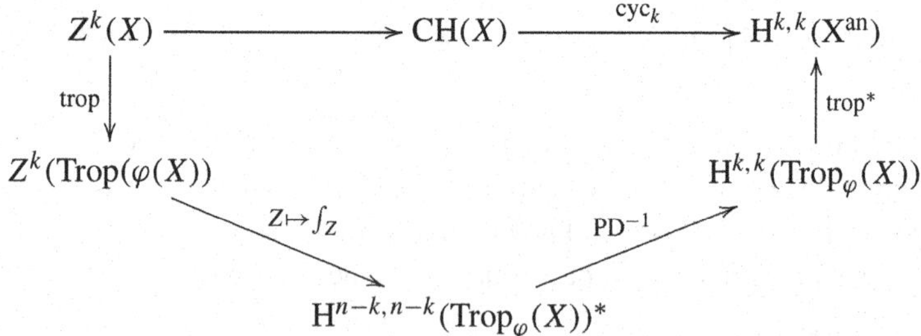

commute? Here PD denote the Poincaré duality isomorphism on tropical varieties [16] and cyc_k denotes Liu's cycles class map [21].

Let us finish with the remark that the author does not know of any variety X with $\dim(X) \geq 2$ and any $0 < p \leq \dim(X)$ and $0 < q \leq \dim(X)$ with $(p,q) \neq (1,1)$ where we know $\dim_{\mathbb{R}} \mathrm{H}^{p,q}(\mathrm{X}^{\mathrm{an}})$. (No, not even $\mathrm{H}^{2,2}(\mathbb{P}^{2,\mathrm{an}})$ or $\mathrm{H}^{2,2}(\mathbb{A}^{2,\mathrm{an}})$.)

References

[1] Vladimir G. Berkovich. *Spectral theory and analytic geometry over non-Archimedean fields*, volume 33 of *Mathematical Surveys and Monographs*. American Mathematical Society, Providence, RI, 1990.

[2] Matthew Baker, Sam Payne, and Joseph Rabinoff. Nonarchimedean geometry, tropicalization, and metrics on curves. *Algebr. Geom.*, 3(1):63–105, 2016.

[3] Antoine Chambert-Loir and Antoine Ducros. Formes différentielles réelles et courants sur les espaces de Berkovich. 2012. http://arxiv.org/abs/1204.6277.

[4] David A. Cox, John B. Little, and Henry K. Schenck. *Toric varieties*, volume 124 of *Graduate Studies in Mathematics*. American Mathematical Society, Providence, RI, 2011.

[5] V. I. Danilov. The geometry of toric varieties. *Uspekhi Mat. Nauk*, 33(2(200)):85–134, 247, 1978.

[6] Tyler Foster, Philipp Gross, and Sam Payne. Limits of tropicalizations. *Israel J. Math.*, 201(2):835–846, 2014.

[7] William Fulton. *Introduction to toric varieties*, volume 131 of *Annals of Mathematics Studies*. Princeton University Press, Princeton, NJ, 1993. The William H. Roever Lectures in Geometry.

[8] Walter Gubler. A guide to tropicalizations. In *Algebraic and combinatorial aspects of tropical geometry*, volume 589 of *Contemp. Math.*, pages 125–189. Amer. Math. Soc., Providence, RI, 2013.

[9] Walter Gubler. Forms and currents on the analytification of an algebraic variety (after Chambert-Loir and Ducros). In Matthew Baker and Sam Payne, editors, *Nonarchimedean and Tropical Geometry*, Simons Symposia, pages 1–30, Switzerland, 2016. Springer.

[10] Ilia Itenberg, Ludmil Katzarkov, Grigory Mikhalkin, and Ilia Zharkov. Tropical homology. *Math. Ann.*, 374(1-2):963–1006, 2019.

[11] Philipp Jell. Differential forms on Berkovich analytic spaces and their cohomology. 2016. PhD Thesis, availible at http://epub.uni-regensburg.de/34788/1/ThesisJell.pdf.

[12] Philipp Jell. A Poincaré lemma for real-valued differential forms on Berkovich spaces. *Math. Z.*, 282(3-4):1149–1167, 2016.

[13] Philipp Jell. Constructing smooth and fully faithful tropicalizations for Mumford curves. 2018. https://arxiv.org/abs/1805.11594.

[14] Philipp Jell. Tropical Hodge numbers of non-archimedean curves. *Israel J. Math.*, 229(1):287–305, 2019.

[15] Philipp Jell, Johannes Rau, and Kristin Shaw. Lefschetz (1,1)-theorem in tropical geometry. *Épijournal Geom. Algébrique*, 2:Art. 11, 2018.

[16] Philipp Jell, Kristin Shaw, and Jascha Smacka. Superforms, tropical cohomology, and Poincaré duality. *Adv. Geom.*, 19(1):101–130, 2019.

[17] Philipp Jell and Veronika Wanner. Poincaré duality for the tropical Dolbeault cohomology of non-archimedean Mumford curves. *J. Number Theory*, 187:344–371, 2018.

[18] Takeshi Kajiwara. Tropical toric geometry. In *Toric topology*, volume 460 of *Contemp. Math.*, pages 197–207. Amer. Math. Soc., Providence, RI, 2008.

[19] Masaki Kashiwara and Pierre Schapira. *Sheaves on manifolds*, volume 292 of *Grundlehren der Mathematischen Wissenschaften [Fundamental Principles of Mathematical Sciences]*. Springer-Verlag, Berlin, 1994. With a chapter in French by Christian Houzel, Corrected reprint of the 1990 original.

[20] Lars Kastner, Kristin Shaw, and Anna-Lena Winz. Cellular sheaf cohomology of *polymake*. In *Combinatorial algebraic geometry*, volume 80 of *Fields Inst. Commun.*, pages 369–385. Fields Inst. Res. Math. Sci., Toronto, ON, 2017.

[21] Yifeng Liu. Tropical cycle classes for non-archimedean spaces and weight decomposition of de Rham cohomology sheaves. 2017. https://users.math.yale.edu/~yl2269/deRham.pdf, to appear in *Ann. Sci. Éc. Norm. Supér.*

[22] Yifeng Liu. Monodromy map for tropical Dolbeault cohomology. *Algebr. Geom.*, 6(4):384–409, 2019.

[23] Grigory Mikhalkin and Johannes Rau. Tropical geometry. Draft of a book available at: https://www.dropbox.com/s/9lpv86oz5f4za75/main.pdf.

[24] Grigory Mikhalkin and Ilia Zharkov. Tropical curves, their Jacobians and theta functions. In *Curves and abelian varieties*, volume 465 of *Contemp. Math.*, pages 203–230. Amer. Math. Soc., Providence, RI, 2008.

[25] Grigory Mikhalkin and Ilia Zharkov. Tropical eigenwave and intermediate Jacobians. In *Homological mirror symmetry and tropical geometry*, volume 15 of *Lect. Notes Unione Mat. Ital.*, pages 309–349. Springer, Cham, 2014.

[26] Sam Payne. Analytification is the limit of all tropicalizations. *Math. Res. Lett.*, 16(3):543–556, 2009.

[27] Kristin Shaw. Superforms and tropical cohomology. 2017. https://web.ma.utexas.edu/users/sampayne/pdf/Shaw-Simons2017.pdf

[28] Bernd Sturmfels and Jenia Tevelev. Elimination theory for tropical varieties. *Math. Res. Lett.*, 15(3):543–562, 2008.

[29] Amaury Thuillier. Théorie du potentiel sur les courbes en géométrie analytique non archimédienne. Applications à la théorie d'Arakelov. 2005. https://tel.archives-ouvertes.fr/file/index/docid/48750/filename/tel-00010990.pdf.

[30] Jarosław Włodarczyk. Embeddings in toric varieties and prevarieties. *J. Algebraic Geom.*, 2(4):705–726, 1993.

16

Schubert Polynomials, Pipe Dreams, Equivariant Classes, and a Co-transition Formula

Allen Knutson[a]

Abstract. We give a new proof that three families of polynomials coincide: the double Schubert polynomials of Lascoux and Schützenberger defined by divided difference operators, the pipe dream polynomials of Bergeron and Billey, and the equivariant cohomology classes of matrix Schubert varieties. All three families are shown to satisfy a "co-transition formula" which we explain to be some extent projectively dual to Lascoux' transition formula. We comment on the K-theoretic extensions.

1 Overview

Let $S_\infty := \cup_{n=1}^{\infty} S_n$ be the permutations π of $\mathbb{N}_+$ that are eventually the identity, i.e. $\pi(i) = i$ for $i \gg 0$. We define three families of polynomials in $\mathbb{Z}[x_1, x_2, \ldots, y_1, y_2, \ldots]$, named A(lgebra), C(ombinatorics), and G(eometry), and each indexed by S_∞:

1. *Double Schubert polynomials* A_π. These were defined by Lascoux and Schützenberger (15), using a recurrence relation based on divided difference operators. We recapitulate the definition in §2, with a mildly novel approach.
2. *Pipe dream polynomials* C_π. These were introduced (in the (x_i) variables only, and not called this) by N. Bergeron and Billey (2); we recall them in §3.
3. *Matrix Schubert classes* G_π. These were introduced by Fulton (5; 7) (and again, not called this) to give universal formulæ for the classes of degeneracy loci of generic maps between flagged vector bundles. This concept was reinterpreted cohomologically in (8; 11), as giving the

[a] Cornell University, New York

equivariant cohomology classes associated to matrix Schubert varieties; we recall this interpretation in §4.

In this paper we give an expeditious proof of the following known results (2; 11):

Theorem 1.1 *For all* $\pi \in S_\infty$, $A_\pi = C_\pi = G_\pi$.

This will follow from a base case w_0^n they share, where $w_0^n(i) := \begin{cases} i & \text{if } i > n \\ n+1-i & \text{if } i \leq n, \end{cases}$

Lemma 1.2 (The base case) *For each of* $P \in \{A, C, G\}$, *we have*

$$P_{w_0^n} = \prod_{i,j \in [n],\ i+j \leq n} (x_i - y_j),$$

along with a recurrence they each enjoy:

Lemma 1.3 (The co-transition formula) *For each of* $P \in \{A, C, G\}$, *and* $\pi \in S_n \setminus \{w_0^n\}$, *there exist i such that* $i + \pi(i) < n$. *Pick i minimum such. Then*

$$(x_i - y_{\pi(i)})\, P_\pi = \sum \{P_\sigma \ :\ \sigma \in S_n,\ \sigma \gtrdot \pi,\ \sigma(i) \neq \pi(i)\}$$

where $\gtrdot$ *indicates a cover in the Bruhat order.*

The derivations of the co-transition formula in the three families are to some extent parallel. For $P = A$ we define the "support" of a polynomial and remove one point from the support of A_π. In $P = C$ we (implicitly) study a subword complex (11) whose facets correspond to pipe dreams for π, and delete a cone vertex from the complex. In $P = G$ we study a hyperplane section of the matrix Schubert variety $\overline{X_\pi}$, which removes one T-fixed point from $\overline{X_\pi}/T$.

In the remainder we recall the polynomials and prove the lemmata for each of them. The word "transition" will appear in §2, but the "co-" will only be explained in §6. To further illuminate the connection between the pipe dream formula and the co-transition formula, we include in §5 an inductive formula that generalizes both and can be derived from either.

Acknowledgments

It is a great pleasure to get to thank Bill for so much mathematics, encouragement, and guidance (especially in the practice and the importance of writing well; while my success has been limited I have at least always striven to emulate his example). I thank Ezra Miller for his many key insights in (11), even as I now obviate some of them here, and Bernd Sturmfels for his early input to that project. I thank the referee for catching some embarassing errors. Finally, this is my chance once more to thank Nantel Bergeron, Sara Billey, Sergei Fomin, and Anatol Kirillov for graciously accepting the terminology "pipe dream". (See (1)!)

2 The double Schubert polynomials (A_π)

Define the **nil Hecke algebra** $\mathbb{Z}[\partial]$ as having a $\mathbb{Z}$-basis $\{\partial_\pi \; : \; \pi \in S_\infty\}$ and the following very simple product structure:

$$\partial_\pi \partial_\rho := \begin{cases} \partial_{\pi \circ \rho} & \text{if } \ell(\pi\rho) = \ell(\pi) + \ell(\rho) \\ 0 & \text{otherwise, i.e. } \ell(\pi\rho) < \ell(\pi) + \ell(\rho). \end{cases}$$

Here $\ell(\pi) := \#\left\{(i,j) \in (\mathbb{N}_+)^2 \; : \; i < j, \; \pi(i) > \pi(j)\right\}$ denotes the number of inversions of π. So this algebra $\mathbb{Z}[\partial]$ is graded by $\deg \partial_\pi = \ell(\pi)$, and plainly is generated by the degree 1 elements $\{\partial_i := \partial_{r_i}\}$, modulo the **nil Hecke relations**

$$\partial_i^2 = 0 \qquad [\partial_i, \partial_j] = 0, \quad |i - j| > 2 \qquad \partial_i \partial_{i+1} \partial_i = \partial_{i+1} \partial_i \partial_{i+1}.$$

This algebra has a module $\mathbb{Z}[\underline{x}, \underline{y}] := \mathbb{Z}[x_1, x_2, \ldots, y_1, y_2, \ldots]$ where the action is by **divided difference operators** in the x variables:

$$\partial_i p := \frac{p - r_i(p)}{x_i - x_{i+1}}.$$

Here $r_i \circlearrowright \mathbb{Z}[\underline{x}, \underline{y}]$ is the ring automorphism exchanging $x_i \leftrightarrow x_{i+1}$ and leaving all other variables alone. Since the numerator of $\partial_i p$ negates under switching x_i and x_{i+1}, the long division algorithm for polynomials shows that numerator to be a multiple of $x_i - x_{i+1}$, so $\partial_i p$ is again a polynomial. To confirm that the above defines an action, one has to check the nil Hecke relations, which is straightforward.

The action is linear in the (y_i) variables, and the module comes with a $\mathbb{Z}[\underline{y}]$-linear augmentation $|_e \, : \, \mathbb{Z}[\underline{x}, \underline{y}] \to \mathbb{Z}[\underline{y}]$ setting each $x_i \mapsto y_i$. With this, we can define a pairing

$$\begin{aligned} \mathbb{Z}[\partial] \times \mathbb{Z}[\underline{x}, \underline{y}] &\to \mathbb{Z}[\underline{y}], \\ (\partial_w, p) &\mapsto (\partial_w(p)) \, |_e. \end{aligned}$$

Since the ∂_w act $\mathbb{Z}[\underline{y}]$-linearly, it is safe to extend the scalars in the nil Hecke algebra from $\mathbb{Z}$ to $\mathbb{Z}[\underline{y}]$, and regard $\mathbb{Z}[\underline{y}]$ as our base ring for the two spaces being paired, as well as the target of their pairing.

Proposition 2.1 *This $\mathbb{Z}[\underline{y}]$-valued pairing of $\mathbb{Z}[\underline{y}][\partial]$ and $\mathbb{Z}[\underline{x},\underline{y}]$ is perfect, so the basis $\{\partial_\pi : \pi \in S_\infty\}$ has a dual $\mathbb{Z}[\underline{y}]$-basis $\{A_\pi : \pi \in S_\infty\}$ of $\mathbb{Z}[\underline{x},\underline{y}]$, called the* **double Schubert polynomials**. *These are homogeneous with $\deg(A_\pi) = \ell(\pi)$.*

In this basis, the $\mathbb{Z}[\partial]$-module structure becomes

$$\partial_\pi A_\rho = \begin{cases} A_{\rho\circ\pi^{-1}} & \text{if } \ell(\rho\circ\pi^{-1}) = \ell(\rho) - \ell(\pi), \\ 0 & \text{otherwise, i.e. if } \ell(\rho\circ\pi^{-1}) > \ell(\rho) - \ell(\pi). \end{cases}$$

There are enough fine references for Schubert polynomials (e.g. (6)) that we don't further recapitulate the basics here. Dual bases are always unique, and perfection of the pairing is equivalent to existence of the dual basis. The usual proof of the existence starts with the base case $A_{w_0^n}$ as an *axiom,* defining the other double Schubert polynomials using the module action stated in the proposition.

It remains to prove the co-transition formula (for $P = A$), which in the "single" situation (setting all $y_i \equiv 0$) is plainly a Monk's rule calculation. Since the "double" Monk rule is not a standard topic, and the references we found to it (e.g. notes by D. Anderson from a course by Fulton) use theory beyond the algebra definition above, we include a proof of the co-transition formula appropriate to $P = A$.

One tool for studying double Schubert polynomials is a $\mathbb{Z}[\underline{y}]$-algebra homomorphism $\mathbb{Z}[\underline{x},\underline{y}] \to \mathbb{Z}[\underline{y}]$, $x_i \mapsto y_{\rho(i)} \forall i$, called **restriction to the point** ρ. We'll write this as $f \mapsto f|_\rho$, generalizing the case $\rho = e$ (the identity) we used above to define the pairing. Here is how it interacts with divided difference operators:

$$(\partial_i f)|_\rho = \left.\frac{f - r_i f}{x_i - x_{i+1}}\right|_\rho = \frac{f|_\rho - (r_i f)|_\rho}{x_i|_\rho - x_{i+1}|_\rho} = \frac{f|_\rho - f|_{\rho r_i}}{y_{\rho(i)} - y_{\rho(i+1)}}. \qquad (*)$$

Define the **support** $supp(f)$ of a polynomial $f \in \mathbb{Z}[\underline{x},\underline{y}]$ by $supp(f) := \{\sigma \in S_\infty : f|_\sigma \neq 0\}$. It has a couple of obvious properties: $supp(pq) = supp(p) \cap supp(q)$, and $supp(p+q) \subseteq supp(p) \cup supp(q)$.

Proposition 2.2 *1. $supp(\partial_i f) \subseteq supp(f) \cup supp(f) \cdot r_i$*

2. If $supp(f) \subseteq \{\tau : \tau \geq \sigma\}$, then $supp(\partial_i f) \subseteq \{\tau : \tau \geq \min(\sigma, \sigma r_i)\}$.

3. $supp(A_{w_0^n}) \cap S_n = \{w_0^n\}$.

4. $A_\pi|_\rho = 0$ unless $\rho \geq \pi$ in Bruhat order. (The converse holds, but we won't show it.)

5. Let $\pi \in S_n$. Then $A_\pi|_\pi \neq 0$. (A small converse to (4).)

6. If $f|_\rho = 0$ for all $\rho \in S_\infty$, then $f = 0$.

7. *There is an algorithm to expand a polynomial p as a* $\mathbb{Z}[\underline{y}]$*-combination of double Schubert polynomials: look for a Bruhat-minimal element* σ *of the support, subtract off* $\frac{p|_\sigma}{A_\sigma|_\sigma}A_\sigma$ *from p (recording the coefficient* $\frac{p|_\sigma}{A_\sigma|_\sigma}$*), and recurse until p becomes* 0.

Proof 1. Use $(\partial_i f)|_\rho = f|_\rho/(y_{\rho(i)} - y_{\rho(i+1)}) - f|_{\rho r_i}/(y_{\rho(i)} - y_{\rho(i+1)})$ from equation $(*)$.
2. This follows from (1) and the subword characterization of Bruhat order.
3. This follows trivially from $A_{w_0^n} = \prod_{i,j\in[n],\ i+j\leq n}(x_i - y_j)$.
4. Fix n such that $\pi, \rho \in S_n$, so $A_\pi = \partial_{\pi^{-1}w_0^n} A_{w_0^n}$. Let Q be a reduced word for $\pi^{-1}w_0^n$. Apply (2); by the reducedness of Q the $min(\sigma,\sigma r_i)$ is always σr_i. By induction on $\#Q$ we learn $supp(\partial_{\pi^{-1}w_0^n} A_{w_0^n}) \subseteq \{\tau : \tau \geq \pi\}$, which is the result we seek.
5. We use downward induction in weak Bruhat order from the easy base case w_0^n. If $\pi r_i > \pi$, then $A_\pi|_\pi = (\partial_i A_{\pi r_i})|_\pi \propto (A_{\pi r_i}|_\pi - A_{\pi r_i}|_{\pi r_i}) = -A_{\pi r_i}|_{\pi r_i} \neq 0$ using equation $(*)$ for the $\propto$, part (4) to kill the first term, and induction.
6. Expand $f = \sum_{\pi\in S_\infty} c_\pi A_\pi$ in the $\mathbb{Z}[\underline{y}]$-basis $\{A_\pi\}$ and, if $f \neq 0$, let A_ρ be a summand appearing (i.e. $c_\rho \neq 0$) with ρ minimal in Bruhat order. Then $f|_\rho = \sum_\pi c_\pi A_\pi|_\rho = c_\rho A_\rho|_\rho$ by (4), and this is $\neq 0$ by (5).
7. In the finite $\mathbb{Z}[\underline{y}]$-expansion $p = \sum_\rho d_\rho A_\rho$, if σ is chosen minimal such that $d_\sigma \neq 0$, then $p|_\sigma = \sum_\rho d_\rho A_\rho|_\sigma = d_\sigma A_\sigma|_\sigma \neq 0$, so σ lies in p's support. Meanwhile, by (4) σ must also be Bruhat-minimal in p's support. When we perform the subtraction in the algorithm, the coefficient is d_σ, and the number of terms in p decreases. □

When we later learn $A = C = G$, then properties (4), (5) of the (A_π) will also hold for (C_π), (G_π), and we leave the reader to seek direct proofs of them.

Proposition 2.3 (Equivariant Monk's rule) *Let* $\pi \in S_\infty$, $i > 0$. *Then*

$$(x_i - y_{\pi(i)})A_\pi = \sum_{\rho \gtrdot \pi} A_\rho \begin{cases} +1 & \text{if } \rho(i) > \pi(i), \\ -1 & \text{if } \rho(i) < \pi(i), \\ 0 & \text{if } \rho(i) = \pi(i). \end{cases}$$

Proof Using the algorithm from proposition 2.2(7), and also proposition 2.2(4), we know that the expansion $f = \sum_\rho c_\rho A_\rho$ can only involve those $\rho \geq$ elements of f's support. The support of $(x_i - y_{\pi(i)})A_\pi$ lies in $\{\rho \in S_\infty : \rho \geq \pi\} \setminus \{\pi\} = \{\rho \in S_\infty : \rho > \pi\}$. The only elements of that set with length $\leq \deg\left((x_i - y_{\pi(i)})A_\pi\right)$ are $\{\rho \in S_\infty : \rho \gtrdot \pi\}$. Hence the left-hand side, expanded in double Schubert polynomials, must have constant coefficients, not higher-degree polynomials in $\mathbb{Z}[\underline{y}]$. (This is the sense in which the "right" extension of Monk's nonequivariant rule concerns multiplication by $x_i - y_{\pi(i)}$ not just x_i. There is of course another, equally "right", extension, computing $A_{r_i} A_\pi$.)

If a polynomial f is in the common kernel of the ∂_j operators, it must be symmetric in *all* the $\underline{x}$ variables...which means f must involve no $\underline{x}$ variables at all, i.e. $f \in \mathbb{Z}[\underline{y}]$. If we also insist that $f|_e = 0$ then we may infer $f = 0$.

Both sides of our desired equation are homogeneous polynomials of the same degree, $\ell(\pi) + 1$, and with $f|_e = 0$. By the argument above it suffices to show that ∂_j LHS $= \partial_j$ RHS for all j. There are five cases: $(j = i$ or $j = i - 1) \times (\pi(i) > \pi(i+1)$ or $\pi(i) < \pi(i+1))$, and $j \neq i, i-1$, each of which one can check using the (itself easily checked) twisted Leibniz rule $\partial_i(fg) = (\partial_i f)g + (r_i f)(\partial_i g)$ along with induction on $\ell(\pi)$. We explicitly check the most unpleasant of the five cases: $j = i, \pi(i) > \pi(i+1)$.

$$\begin{aligned} \partial_i (x_i - y_{\pi(i)}) A_\pi &= A_\pi + (x_{i+1} - y_{\pi(i)}) \partial_i A_\pi \\ &= A_\pi + (x_{i+1} - y_{(\pi r_i)(i+1)}) A_{\pi r_i} \qquad \text{now use induction} \\ &= A_\pi + \sum_{\sigma \gtrdot \pi r_i} A_\sigma \begin{cases} +1 & \text{if } \sigma(i+1) > (\pi r_i)(i+1) = \pi(i) \\ -1 & \text{if } \sigma(i+1) < (\pi r_i)(i+1) = \pi(i) \\ 0 & \text{if } \sigma(i+1) = (\pi r_i)(i+1) = \pi(i) \end{cases} \end{aligned}$$

$$\sum_{\rho \gtrdot \pi} \partial_i A_\rho \begin{cases} +1 & \text{if } \rho(i) > \pi(i) \\ -1 & \text{if } \rho(i) < \pi(i) \\ 0 & \text{if } \rho(i) = \pi(i) \end{cases} = \sum_{\rho \gtrdot \pi,\ \rho(i) > \rho(i+1)} A_{\rho r_i} \begin{cases} +1 & \text{if } \rho(i) = (\rho r_i)(i+1) > \pi(i) \\ -1 & \text{if } \rho(i) = (\rho r_i)(i+1) < \pi(i) \\ 0 & \text{if } \rho(i) = (\rho r_i)(i+1) = \pi(i). \end{cases}$$

Each term σ in the first corresponds to the $\rho = \sigma r_i$ term in the second. □

Proof of the co-transition formula for $P = A$. We need to check that each ρ term in the equivariant Monk rule has $\rho(i) \in (\pi(i), n]$, so as to only get positive terms and only from $\rho \in S_n$.

Since π has only descents before i (by choice of i), we know $\rho = \pi \circ (i \leftrightarrow b)$ with $i < b$, i.e. $\rho(i) = \pi(b) > \pi(i)$.

By choice of i, we have $\pi = n\, n\text{-}1 \ldots n\text{-}i\text{+}2\, \pi(i) \ldots \pi(n)$ with $\pi(i) < n - i + 1$. Hence $\exists j \in (i, n]$ with $\pi(j) = n - i + 1 \in (\pi(i), n+1)$. The covering relations in S_∞ don't allow us to switch positions $i, n+k$ if some position $j \in (i, n+k)$ has $\pi(j) \in (\pi(i), \pi(n+k) = n+k)$. □

Lascoux' *transition formula* (14) for double Schubert polynomials is also based on Monk's rule, but doesn't include implicit division like the co-transition formula does. (It is worth noting that *each* of the summands on the right-hand side of the co-transition formula is divisible by $x_i - y_{\pi(i)}$, not merely their total.) We discuss the connection in §6.

3 The pipe dream polynomials (C_π)

Index the squares in the Southeastern quadrant of the plane using matrix coordinates $\{(a,b) \ : \ a,b \in \{1,2,3,\dots\}\}$. A **pipe dream** is a filling of that quadrant with two kinds of tiles, mostly ╯╭ but finitely many ┼, such that no two pipes cross twice[1]. We label the pipes $1,2,3,\dots$ across the top side, and speak of "the 1-pipe of D", "the 2-pipe of D", and so on. For example, the left two diagrams below are pipe dreams, the right one not:

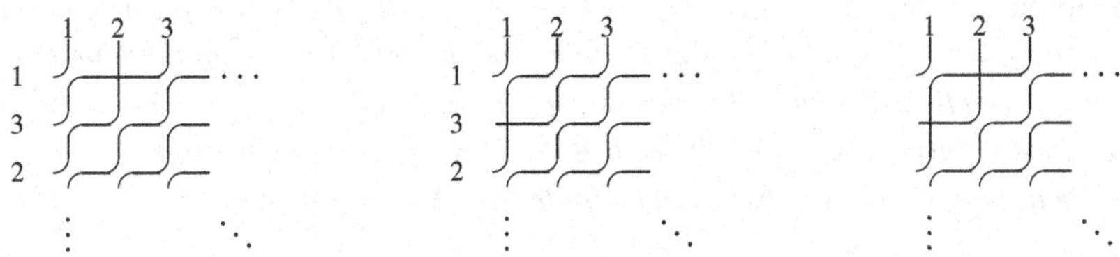

Because of the no-double-crossing rule, if we regard a pipe dream D for π as a wiring diagram for π, it's easy to see that the number of ┼ is exactly $\ell(\pi)$.

To a pipe dream D we can associate a permutation π by reading off the pipe labels down the left side, and say that D is a "pipe dream for π". With this we can define the **pipe dream polynomials**:

$$C_\pi := \sum_{\text{pipe dreams } D \text{ for } \pi} \ \prod_{\text{crosses } + \text{ in } D} (x_{\text{row}(+)} - y_{\text{col}(+)})$$

Example: C_{1423}

$$= (x_2 - y_1)(x_2 - y_2) \ + \ (x_2 - y_1)(x_1 - y_3) \ + \ (x_1 - y_2)(x_1 - y_3)$$

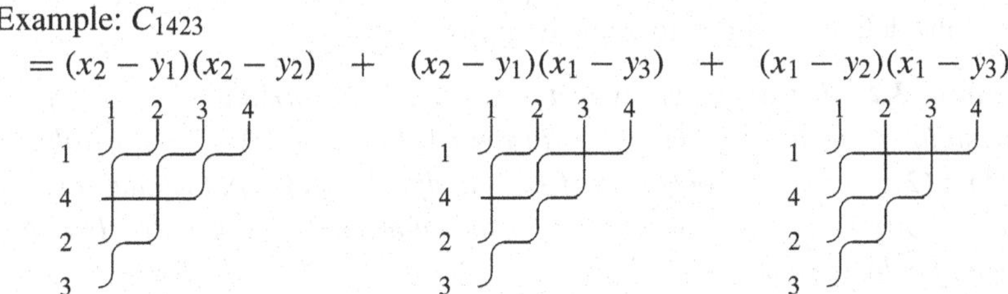

where we skip drawing any of the pipes outside the triangle $\{(a,b) \ : \ a+b \leq n\}$, as will be justified by lemma 3.1 below.

The main idea of the proof of the co-transition formula for the $\{C_w\}$ polynomials is easy to explain. Let $\mathcal{D}_1$ be the set of pipe dreams for w, and

$$\mathcal{D}_2 := \coprod \{\text{the pipe dreams for } w' \ : \ w' \text{ occurs in the co-transition formula}\}.$$

[1] In subtler contexts than considered here, one *does* allow pipes to cross twice, and instead refers to the pipe dreams without double crossings as "reduced pipe dreams". See §7.

Our goal (which will take some doing) is to show that the maps

$\mathcal{D}_1 \to \mathcal{D}_2$ $\qquad\qquad$ $\mathcal{D}_2 \to \mathcal{D}_1$

D╯╭ $\mapsto$ D┼ $\qquad\qquad$ D┼ $\mapsto$ D╯╭

that place, or remove, a ┼ at position $(i,\pi(i))$ have the claimed targets $\mathcal{D}_2$, $\mathcal{D}_1$. The maps are then obviously inverse, and the co-transition formula will follow easily.

Lemma 3.1 *Let $w \in S_{\mathbb{N}}$, and $i \in \{1,2,\ldots\}$ such that $\forall j \in \{1,2,\ldots\}$, $i < j \iff w(i) < w(j)$. Let D be a pipe dream for w. Then the pipe that enters from the North in column i only goes through* ╯╭ *tiles, no* ┼, *coming out at row $w(i)$. Consequently, if $w \in S_n$ and D is a pipe dream for w, then there are no* ┼ *tiles outside the triangle $\{(a,b) \,:\, a+b \leq n\}$.*

Proof If $i < j$ and $w(i) < w(j)$, then the i-pipe starts and ends Northwest of the j-pipe. By the Jordan curve theorem these two pipes cross an even number of times, and since D is a pipe dream, that even number is 0. The opposite argument (Southeast) works if $i > j$ and $w(i) > w(j)$. Doing this for all $j \neq i$, we find that the i-pipe crosses no other pipe, i.e. it goes only through ╯╭ tiles, ruling out ┼ tiles in the adjacent diagonals $\{(a,b) \,:\, a+b = i-1, i\}$. Finally, if $w \in S_n$ then each $i > n$ satisfies the condition. □

Proof of the base case for $P = C$. The number of squares in the triangle $\{(a,b) \,:\, a+b \leq n\}$ is $\binom{n}{2}$, which is also $\ell(w_0^n)$. As such, every one of them must have a ┼ in a pipe dream for w_0^n, making the pipe dream for w_0^n unique. Then the definition of C_π gives the base case. □

Lemma 3.2 *Let π, i, ρ be as in the co-transition formula. If D is a pipe dream for π, then the leftmost* ╯╭ *in rows $1,2,\ldots,i$ of D occurs in column $\pi(1), \pi(2), \ldots, \pi(i)$ respectively. If D' is a pipe dream for ρ, then the same is true in rows $1,2,\ldots,i-1$ but in row i the leftmost* ╯╭ *occurs strictly to the right of column $\pi(i)$.*

Proof Assume that the claim is established for each row above the jth. Start on the North side of D in column $\pi(j)$, and follow that pipe down. By our inductive knowledge of rows $1\ldots j-1$, and the fact that $\pi(1) > \pi(2) > \cdots > \pi(i)$ by choice of i, this pipe will go straight down through $j-1$ crosses to the jth row. Since it then needs to exit on the jth row, it must turn West in matrix position $(j, \pi(j))$, and go due West through only ┼ in columns $1, \ldots, \pi(j)-1$ of that row.

Exactly the same analysis holds for ρ, except that $\rho(i) > \pi(i)$. □

The following technical lemma is key.

Lemma 3.3 *Let D be a filling of the Southeastern quadrant with finitely many* ┼*, the rest* ╯╭*, except* with an empty square *at* (a,b). *Let* N, S, E, W *denote the four pipes coming out of* (a,b) *in those respective compass directions and call the remaining pipes the "old pipes". Let* D╯╭, D┼ *denote D with the respective tile inserted at* (a,b)*; these have "new pipes"* WN, ES *in* D╯╭ *and* NS, EW *in* D┼. *Assume that:*

1. *Every square West of E except the hole* (a,b) *(so, in rows* $1,\dots,a$*) has a* ┼*, so in particular, the pipes N and W are straight.*

Then if D╯╭*, is a pipe dream, so is* D┼. *If in addition we assume*

2. *No old pipe has North end between N and E's North ends while also having West end between W and S's West ends*

then D┼ *being a pipe dream implies* D╯╭ *is a pipe dream.*

We give an example to refer to while following the case analysis in the proof.

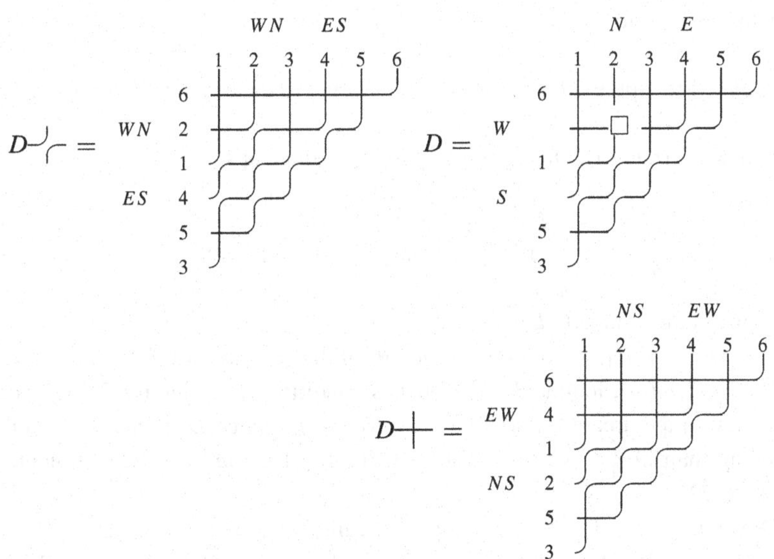

Proof Say D╯╭ is a pipe dream, i.e. its new pipes WN and ES don't cross any other pipe twice; in particular no old pipe crosses any of N, S, E, W twice. We need to make sure that in D┼ the two new pipes NS and EW don't cross any old pipe twice. Equivalently, no old pipe should cross both W and N, or both E and S. Exactly the same analysis will hold for the opposite direction: if

$D{+}$ at (a,b) is a pipe dream, we need show that no old pipe crosses both E and S, or both N and W.

If a pipe (in either $D{\lrcorner\ulcorner}$ or $D{+}$) crosses N going West, then by condition (1) it goes straight West from there and cannot cross W or S. Similarly, if a pipe crosses W going North, then by condition (1) it goes straight North from there and cannot cross N or E.

That rules out double-crossing NS, EW, and WN, so is already enough to establish our first conclusion ($D{\lrcorner\ulcorner}$ a pipe dream $\implies$ $D{+}$ a pipe dream). What remains for the second conclusion is to show that, if $D{+}$ is a pipe dream, then no old pipe should cross both E and S.

Let i, j denote the respective columns of the tops of N, E. If $h < i$, then the h-pipe stays West of E. If $h > j$, and the h-pipe crosses E, then it does so horizontally, at which point it continues due West and stays above S. Finally, if $i < h < j$, then by condition (2) the h-pipe has West end either above W's West end or below S's West end. In the first case, the h-pipe stays above W hence above S. In the latter case, the h-pipe begins and ends Southeast of the NS pipe in $D{+}$, so doesn't cross it at all, hence doesn't cross S. □

Proof of the co-transition formula for $P = C$. Let $\mathcal{D}_1$ be the set of pipe dreams for w, and

$$\mathcal{D}_2 := \bigcup \{\text{the pipe dreams for } w' \ : \ w' \text{ occurs in the co-transition formula}\}.$$

Let $(a,b) = (i, w(i))$. Our goal is to show that the maps

$$\begin{array}{ccc} \mathcal{D}_1 & \to & \mathcal{D}_2 \\ D{\lrcorner\ulcorner} & \mapsto & D{+} \end{array} \qquad \begin{array}{ccc} \mathcal{D}_2 & \to & \mathcal{D}_1 \\ D{+} & \mapsto & D{\lrcorner\ulcorner} \end{array} \qquad \text{as in lemma 3.2}$$

have the claimed targets $\mathcal{D}_2$, $\mathcal{D}_1$.

Let $D{\lrcorner\ulcorner} \in \mathcal{D}_1$. By our choice of i from the co-transition formula, and of $(a,b) = (i, w(i))$, lemma 3.2 establishes condition (1) of lemma 3.3. Hence $D{+}$ is a pipe dream for some $w' = w(i \leftrightarrow j)$. Since $D{+}$ has one more crossing than $D{\lrcorner\ulcorner}$, we infer $\ell(w') = \ell(w) + 1$, so $w' > w$. Consequently $D{+} \in \mathcal{D}_2$.

Now start from $D{+} \in \mathcal{D}_2$, a pipe dream for some w'; we want to show that $D{\lrcorner\ulcorner} \in \mathcal{D}_1$. Again our choices of i and (a,b) establish condition (1) of lemma 3.3. Define j so that the EW pipe of $D{+}$ is the j-pipe, i.e. E exits the North side in column j. Since $w' > w$, we verify condition (2) of lemma 3.3. Hence $D{\lrcorner\ulcorner} \in \mathcal{D}_1$.

Each $+$ inserted at $(i, \pi(i))$ contributes a factor of $x_i - y_{\pi(i)}$ in the formula for C-polynomials, so while the bijection above corresponds pipe dreams for

C_π to those for $\{C_\rho\}$, the induced equality of polynomials is between $\sum_\rho C_\rho$ and $(x_i - y_{\pi(i)})\, C_\pi$, giving the co-transition formula. □

4 The matrix Schubert classes (G_π)

Define a **matrix Schubert variety** $\overline{X_\pi} \subseteq M_n(\mathbb{C})$, for $\pi \in S_n$ or more generally[2] a *partial* permutation matrix, by

$$\overline{X_\pi} := \overline{B_- \pi B_+} \qquad \text{closure taken in } M_n(\mathbb{C})$$

where B_-, B_+ are respectively the groups of lower and upper triangular matrices intersecting in the diagonal matrices T. The equations defining $\overline{X_\pi}$ were determined in (5, §3).

Define the **matrix Schubert class**

$$\begin{aligned} G_\pi := [\overline{X_\pi}] &\in H^*_{B_- \times B_+}(M_n(\mathbb{C})) && \text{using the smoothness of } M_n(\mathbb{C}) \\ &\cong H^*_{B_- \times B_+}(pt) && \text{using the contractibility of } M_n(\mathbb{C}) \\ &\cong H^*_{T \times T}(pt) && \text{since } B_-, B_+ \text{ retract to } T \\ &\cong \mathbb{Z}[x_1, \ldots, x_n, y_1, \ldots, y_n] && \text{using the usual isomorphism } T \cong (\mathbb{C}^\times)^n \end{aligned}$$

in equivariant cohomology.

Though we won't use it, we recall the connection to degeneracy loci (8). If we follow the Borel definition of $(B_- \times B_+)$-equivariant cohomology, based on the "mixing space" construction $Z(N) := N \times^{B_- \times B_+} E(B_- \times B_+)$, the maps $\overline{X_\pi} \hookrightarrow M_n(\mathbb{C}) \to pt$ give a triangle

$$\begin{array}{ccccc} Z(\overline{X_\pi}) & \hookrightarrow & Z(M_n(\mathbb{C})) & & \\ & \searrow & \downarrow & & \\ & & Z(pt) & = & B(B_- \times B_+) \end{array}$$

where $B(B_- \times B_+)$ is the classifying space for principal $(B_- \times B_+)$-bundles.

With this, $[Z(\overline{X_\pi})]$ defines a class in $H^*(Z(M_n(\mathbb{C})))$. Since $\downarrow$ is a vector bundle hence a homotopy equivalence, we can also take $[Z(\overline{X_\pi})]$ as a class in $H^*(B(B_- \times B_+)) =: H^*_{B_- \times B_+}$.

Now consider a space N bearing a flagged vector bundle $V_1 = V_1^{(n)} \hookleftarrow V_1^{(n-1)} \hookleftarrow \cdots \hookleftarrow V_1^{(1)}$ and a co-flagged vector bundle $V_2 = (V_2)_{(n)} \twoheadrightarrow (V_2)_{(n-1)} \twoheadrightarrow \cdots (V_2)_{(1)}$ (of course, in finite dimensions flagged and co-flagged are the same concept), plus a generic map $\sigma : V_1 \to V_2$. Since such pairs of bundles are classified by maps into $B(B_- \times B_+)$, we can enlarge the diagram to

[2] Indeed, once one allows partial permutation matrices there's no need for the matrices to be square, but square will suffice for our application.

$$\begin{array}{ccccc} Z(\overline{X}_\pi) & \hookrightarrow & Z(M_n(\mathbb{C})) & & \\ & \searrow & \downarrow & \nwarrow \sigma & \\ & & B(B_- \times B_+) & \leftarrow & N \end{array}$$

and the genericity of σ becomes its transversality to $Z(\overline{X}_\pi)$. Consequently, and using the equations from (5, §3) defining $\overline{X}_\pi$,

$$\sigma^*([\overline{X}_\pi]) = \left[\left[\left\{ x \in N \ : \ \forall i,j, \ \begin{array}{r} \text{rank}\left(V_1^{(i)} \hookrightarrow V_1 \xrightarrow{\sigma} V_2 \twoheadrightarrow (V_2)_{(j)} \text{ over the point } x\right) \\ \leq \text{rank(NW } i \times j \text{ submatrix of } \pi) \end{array} \right\}\right]\right]$$

i.e. $G_\pi = \left[\overline{X}_\pi\right] \in \mathbb{Z}[x_1, \ldots, x_n, y_1, \ldots, y_n]$ is providing a universal formula for the class of the π degeneracy locus of the generic map σ. The principal insight is the dual role of the space $B(B_- \times B_+)$, as the classifying space for pairs of bundles and also as the base space of equivariant cohomology.

Lemma 4.1 *The definition above is independent of n, so long as $\pi \in S_n$.*

Proof The equations defining $\overline{X}_\pi$, determined in (5, §3), depend only on the matrix entries northwest of the Fulton essential set of π, which is independent of n. Hence enlarging n to m amounts to crossing both $M_n(\mathbb{C})$, and $\overline{X}_\pi$, by the same irrelevant vector space consisting of matrix entries $\{(i,j) \ : \ i > n \text{ or } j > n\}$. □

Proof of the base case for $P = G$ The Rothe diagram of w_0^n is the triangle $\{(a,b) \ : \ a + b \leq n\}$, so by (5, §3) the equations defining $\overline{X}_{w_0^n}$ are that each entry m_{ab} in that triangle must vanish. This $\overline{X}_{w_0^n}$ thus being a complete intersection, its class is the product of the $(T \times T)$-weights $x_a - y_b$ of its defining equations $m_{ab} = 0$, giving the base case formula. □

The following geometric interpretation of the Rothe diagram seems under-appreciated:

Lemma 4.2 *The tangent space $T_\pi \overline{X}_\pi$ is $(T \times T)$-invariant (even though the point π itself is not!), spanned by the matrix entries* not *in the Rothe diagram of π. In particular* $\deg G_\pi = \#$*(the Rothe diagram), which is in turn* $\min\{\ell(\rho) :$ *ρ a permutation matrix with the partial permutation matrix π in its NW corner*$\}$.

Proof The tangent space to a group orbit is the image of the Lie algebra, $\mathfrak{b}_- \pi + \pi\, \mathfrak{b}_+$. The diagonal matrices (from either side) scale the nonzero entries of π, and the $\mathfrak{n}_-$, $\mathfrak{n}_+$ copy those entries to the South and East, recovering the usual death-ray definition of the complement of the Rothe diagram.

For the "in turn" claim, observe that if $\text{rank}(\pi) = n - k$ then there is a unique way to extend π to $\rho \in S_{n+k}$ without adding any boxes to the Rothe diagram, and $\ell(\rho)$ is the size of that diagram (of ρ or of π). □

That lemma also gives a nice proof of proposition 2.2(5) for G_π, though we won't need an independent one.

To compute other tangent spaces of $\overline{X}_\pi$, soon, we prepare a technical lemma.

Lemma 4.3 *Let $\rho \geq \pi \in S_n$. For $i, j \leq n$ denote the NW $i \times j$ rectangle of M by $M_{[i][j]}$. Let a', b' be such that* $\operatorname{rank} \pi_{[a'][b']} = \operatorname{rank} \rho_{[a'][b']}$. *Let (a,b) be in that rectangle, such that the a row and b column of $\rho_{[a'][b']}$ vanish. Then the (a,b) entry vanishes on every element of $T_\rho \overline{X}_\pi$.*

Proof Let R be the nonzero rows and C the nonzero columns of $\rho_{[a'][b']}$ (so $\#R = \#C = \operatorname{rank} \rho_{[a'][b']} = \operatorname{rank} \pi_{[a'][b']}$, and $a \notin R, b \notin C$ by the assumption on (a', b')). Consider the determinant that uses rows $R \cup \{a\}$ and columns $S \cup \{b\}$; it is one of Fulton's required equations for $\overline{X}_\pi$. We apply it to the infinitesimal perturbation $\rho + \varepsilon Z$. By construction this is $\varepsilon Z_{ab} + O(\varepsilon^2)$, so for Z to be in $T_\rho \overline{X}_\pi$ we must have $Z_{ab} = 0$. □

This allows for a second proof of lemma 4.2, when $\rho = \pi$; we can take $(a', b') = (a, b)$ for each (a, b) in the Rothe diagram. These equations are already enough to cut down $\dim T_\pi \overline{X}_\pi$ to the right dimension, and the tangent space can't get any lower-dimensional than that, so we have successfully determined it from these determinants. Having two proofs shows that the equations from (5, §3) define a *generically* reduced scheme supported on $\overline{X}_\pi$, unlike Fulton's stronger result that they actually define $\overline{X}_\pi$.

Lemma 4.4 *Let $\rho = \pi \circ (i \leftrightarrow j) > \pi$, with $i < j \leq n$. Then $T_\rho \overline{X}_\pi \cap \{M : m_{i,\pi(i)} = 0\} = T_\rho \overline{X}_\rho$.*

Proof The diagrams of π and ρ's agree except on the boundary of the "flipping rectangle" with NW corner $(i, \pi(i))$ and SE corner $(j, \pi(j))$. Let (a, b) in ρ's diagram; we need to find an (a', b') to apply lemma 4.3 to.

For (a, b) outside the flipping rectangle, hence also in π's diagram, we can use $(a', b') = (a, b)$ as explained directly after lemma 4.3. In other cases we will need to move Southeast from (a, b) to (a', b'), without hitting the entries $(a, \pi(a))$ or $(\pi^{-1}(b), b)$ making lemma 4.3 inapplicable.

For (a, b) in the interior of the flipping rectangle, we have $i < a < j$. Since (a, b) is in π's diagram, $a < \pi^{-1}(b)$. We know that $(\pi^{-1}(b), b)$ isn't in the flipping rectangle since $\rho > \pi$, so $\pi^{-1}(b) < i$ or $\pi^{-1}(b) > j$. That first case is impossible since we'd have $a < \pi^{-1}(b) < i < a$, so we know $\pi^{-1}(b) > j$. This means we can safely go below (a, b) to $(a', b') := (a, j)$, with the benefit that $\operatorname{rank} \pi_{[a][j]} = \operatorname{rank} \rho_{[a][j]}$ and we can apply lemma 4.3.

It remains to handle the boundary of the flipping rectangle. The South edge $(j,*)$ and East edge $(*,\pi(i))$ are not in ρ's diagram, so not at issue. Across the top edge $a = i$ and $\pi(i) < b < \pi(j)$, if (i,b) is in ρ's diagram then $\rho^{-1}(b) > i$, and similarly to the above, we learn $\rho^{-1}(b) > j$. So once again we can safely go below (a,b) to $(a',b') := (a,j)$, with the benefit that $\operatorname{rank} \pi_{[a][j]} = \operatorname{rank} \rho_{[a][j]}$ and we can apply lemma 4.3.

Finally, $(i,\pi(i))$ is in ρ's diagram, but is killed by the intersection with $\{M : m_{i,\pi(i)} = 0\} = T_\rho \overline{X}_\rho$ rather than by a determinantal condition.

This defines a vector space of dimension $\dim \overline{X}_\pi - 1 = \dim \overline{X}_\rho$, and $\dim T_\rho(\overline{X}_\pi \cap \{M : m_{i,\pi(i)} = 0\})$ has at least that dimension, so we have found it. $\square$

It will actually be more convenient to prove a slightly more general formula than the co-transition formula as stated in §1. Define the **dominant part of** π's Rothe diagram to be the boxes connected to the NW corner (this may be empty, when $\pi(1) = 1$). These are exactly the matrix entries (a,b) such that $m_{ab} \equiv 0$ on $\overline{X}_\pi$. (A permutation is "dominant" in the usual sense if the dominant part is the entire Rothe diagram, hence the terminology.) The $(i,\pi(i))$ of the co-transition formula was picked to be

- just outside of the dominant part of π's diagram
- while still in the NW triangle,

and to be the Northernmost such (i least such). However, the co-transition formula holds *for any* $(i,\pi(i))$ *satisfying the two bulleted conditions.* This generalization would have made the proof in §3 more complicated, but of course once we know $C = G$ then we know that the (C_π) also satisfy this more general formula. Notice that this formula is stable under incrementing n while not changing the Rothe diagram (e.g. replacing $\pi \in S_n$ by $\pi \oplus I_1 \in S_{n+1}$, or a more complicated possibility if π is a partial permutation).

Lemma 4.5 *Let* $\pi \in S_n \setminus \{w_0^n\}$ *stabilize to* $\rho \in S_{n+1}$, *so* $\rho(n+1) = n+1$ *and* π, ρ *have the same Rothe diagram. Pick* $(i,\pi(i))$ *just outside the dominant part of this diagram, such that* $i + \pi(i) \leq n$. *Then this more general co-transition formula is "stable" in the sense being the same for* π *as for* ρ; *there aren't extra terms in* S_{n+1} *for the formula for* $(x_i - y_{\rho(i)}) P_\rho$.

(Of course this independence *follows* from the co-transition formula and the linear independence of the G polynomials, neither of which we've proven yet.)

Proof Let $\sigma > \rho$, so $\sigma = \rho \circ (a \leftrightarrow b)$ with $a < b$, $\rho(a) < \rho(b)$, and $c \in (a,b) \implies \rho(c) \notin (\rho(a),\rho(b))$ ("no position c is in the way when swapping positions a, b"). For σ to appear in the co-transition formula, we also have $\sigma(i) \neq \pi(i)$, hence $i \in \{a,b\}$. Finally, for $(i, \pi(i))$ to be just outside the dominant part, we need $\pi(c) < \pi(i) \implies c > i$.

The case we need to rule out is $a = i, b > n$. Since $\pi \in S_n$, we can't have $b \geq n+2$ ($n+1$ would be in the way). What remains is to rule out $b = n+1$. For each $c \in (i, n+1)$ to not be in the way, we would need $\pi(c) > n+1$ (impossible since $\pi \in S_n$) or $\pi(c) < \pi(i)$. So $\pi(c) < \pi(i) \implies c > i \implies \pi(c) < \pi(i)$, setting up a correspondence between the $\pi(i) - 1$ numbers $< \pi(i)$ and the $n - i$ numbers $> i$. But then $i + \pi(i) = n+1$, contradicting our choice of $(i, \pi(i))$. □

Proof of this more general co-transition formula, for $P = G$ For π, i as in this more general co-transition formula, we have

$$\overline{X}_\pi \cap \{M : m_{i\pi(i)} = 0\} = \overline{X}_\pi \cap \{M : m_{ab} = 0 \ \forall (a,b) \text{ weakly NW of } (i, \pi(i))\}$$

since all those (a,b) entries other than $(i, \pi(i))$ itself are already zero.

The first description shows that the intersection is a hyperplane section (and nontrivial: $m_{i,\pi(i)} \not\equiv 0$ on $\overline{X}_\pi$) of the irreducible $\overline{X}_\pi$, so each component of the intersection is codimension 1 in $\overline{X}_\pi$. Moreover, $\left[\{M : m_{i\pi(i)} = 0\} \cap \overline{X}_\pi\right] = \left[\{M : m_{i\pi(i)} = 0\}\right] [\overline{X}_\pi] = (x_i - y_{\pi(i)}) G_\pi$.

The benefit of the second description is that the two varieties being intersected are plainly $(B_- \times B_+)$-invariant. Hence that intersection is a union of $(B_- \times B_+)$-invariant subvarieties, each of which is necessarily a matrix Schubert variety $\overline{X}_\rho$ by the Bruhat decomposition of $M_n(\mathbb{C})$.

So far we know set-theoretically that the intersection is some union of $\overline{X}_\rho \subseteq \{M : m_{i\pi(i)} = 0\}$ (for, as yet, *partial* permutation matrices ρ) with $\dim \overline{X}_\rho = \dim \overline{X}_\pi + 1$.

Hence $\rho(i) \neq \pi(i)$, with $\rho \gtrdot \pi$. What remains is to show that every such $\rho \in S_n$ occurs, with multiplicity 1, and that genuinely partial permutations ρ (i.e. not in S_n) don't occur. Then we'll know that $\left[\{M : m_{i\pi(i)} = 0\} \cap \overline{X}_\pi\right] = \sum \left\{1 \cdot [\overline{X}_\rho] \ : \ \rho \in S_n, \ \rho \gtrdot \pi, \ \rho(i) \neq \pi(i)\right\}$.

Certainly the permutation matrix ρ is in $\{M : m_{i\pi(i)} = 0\}$ and $\overline{X}_\pi$. If a partial permutation ρ of corank k were to give a component, then upon stabilizing π to $\pi^+ := \pi \oplus I_k$, the permutation matrix ρ^+ (chosen to have the same diagram as ρ) would give a component. But then $\rho^+ \gtrdot \pi^+$, and by the same argument as in lemma 4.5 $\rho^+ \in S_n$, i.e. $k = 0$.

Finally, we need to show the multiplicity of the component $\overline{X}_\rho$ is 1, i.e. the tangent space to $\{M : m_{i\pi(i)} = 0\} \cap \overline{X}_\pi$ at the point ρ is just $T_\rho \overline{X}_\rho$. This was lemma 4.4. □

5 An inductive pipe dream formula

The formula defining C_π as a sum over pipe dreams, and the co-transition formula writing $(x_i - y_{\pi(i)})C_\pi$ as a sum of other C_ρs, have a common generalization. We include it here, though it's not actually required for the main theorems.

Take $\pi \in S_n$, and let λ be an English partition fitting in the strict Northwest triangle (i.e. $\lambda_i \leq n - i$ for $i \in [1,n]$). Define a **partial pipe dream** for the pair (π, λ) to be

- a tiling of λ with the two tiles as usual, and
- a chord diagram in the complement $\square/\lambda$ of λ in the square $\square$, whose n chords have endpoints at the centers of the North and West edges of $\square/\lambda$, considered up to isotopy and braid moves,

such that

- each chord has positive slope, hence connects a West end to a North end Northeast of it, and
- the combination of the pipes in λ and chords in $\square/\lambda$ connect $1 \ldots n$ on the North side to $\pi(1) \ldots \pi(n)$ down the West side. In this combination, no two pipe-chord combinations should cross twice.

Some examples are given in Figure 1. As the pictures suggest, one can consider the λ region as the "crystalline" part of the diagram, and the $\square/\lambda$ complement as the "molten" region. One uses the co-transition formula to freeze more, increasing λ.

Associate a second permutation ρ to a partial pipe dream D for π, λ by

$$\rho(D) := \left(\prod_{i \lrcorner \ulcorner j \in D} (\pi(\text{row}) \leftrightarrow j) \right) \circ \pi \qquad \begin{array}{c}\text{product over elbows in } D\\ \text{in any order NW to SE}\end{array}$$

Theorem 5.1 *Fix $\pi \in S_n$ and λ a partition with $\lambda_i \leq n - i, i \in [1,n]$. Then*

$$C_\pi = \sum_D \frac{1}{\prod_{\lrcorner\ulcorner \in D}(x_{row} - y_{col})} C_{\rho(D)}$$

where D varies over the partial pipe dreams for (π, λ). The dominant part of the Rothe diagram of each $\rho(D)$ contains λ, so each summand is polynomial.

Proof outline. Call two pipe dreams for π **λ-equivalent** if they agree (in tiles and labels on pipes) inside λ. There is an obvious map from equivalence classes to partial pipe dreams, which we baldly assert to be bijective. If we take the

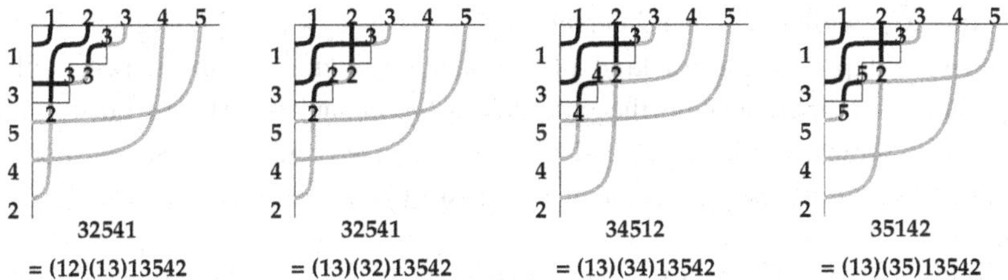

Figure 1 The four partial pipe dreams D for ($\pi = 13542$, $\lambda = 2 + 1$), with each pipe's $\rho(D)$ written below. Note that the tiles alone do not characterize the partial pipe dream; one must number the pipes. (When λ is the full staircase every pipe connects to the North and West boundaries, determining its number.)

pipe dreams in an equivalence class D and replace all the ╯╭s in λ with ┼s, we further assert that we get exactly the pipe dreams for $\rho(D)$. The result follows. □

If we take λ to be the dominant part of π, then there is only one partial pipe dream D for (π, λ), where λ is solid ┼s, and theorem 5.1 says $C_\pi = C_\pi$ (since $\rho(D) = \pi$). If we take λ to have *one* more square at $(i, \pi(i))$, then the only freedom in D is the choice ╯╭$_j$ of pipe label j on that square, and theorem 5.1 becomes the generalized co-transition formula from §4. Finally, if we take λ to be the full staircase $(n-1) + (n-2) + \ldots + 1$, then every D has $\rho(D) = w_0$, and theorem 5.1 recovers the definition of C_π as a sum over pipe dreams.

6 Transition vs. co-transition

In (11) the Fulton determinants defining $\overline{X_\pi}$ were shown to be a Gröbner basis for antidiagonal term orders $<$, and the components of $init_< \overline{X_\pi}$ to be in obvious correspondence with π's pipe dreams. There are four natural sources of antidiagonal term orders:

1. lexicographic, where the matrix entries are ordered from NE to SW (more precisely, by some linear extension of that partial order),
2. lexicographic, where the matrix entries are ordered from SW to NE,
3. reverse lexicographic, where the matrix entries are ordered from NW to SE, and
4. reverse lexicographic, where the matrix entries are ordered from SE to NW.

Slicing $\overline{X}_\pi$ with the hyperplane $m_{i,\pi(i)} = 0$ is a way of doing the first nontrivial step of the third kind of Gröbner degeneration, and hence, will a priori be compatible with the pipe dream combinatorics. (It is from there that the co-transition formula, and §3, were reverse-engineered. Stated more bluntly: after this insight, producing the rest of the paper was essentially an exercise.)

Define the **co-dominant part outside** π's Rothe diagram as the set of matrix entries (a, b) such that no Fulton determinant defining $\overline{X}_\pi$ involves m_{ab}. This is always connected to the SE corner of the square. Its complement is the boxes NW of at least one diagram box, or equivalently NW of at least one essential box. The (i, j) in Lascoux' transition formula was picked to be just outside the co-dominant part outside π's Rothe diagram. See (12) for this view of the transition formula.

In unpublished work, Alex Yong and I gave a Gröbner-degeneration-based proof of Lascoux' transition formula, based on one step of a *lex* order from SE to NW (so, not one of the orders above compatible with pipe dreams). For this reason, one might expect it to be very difficult to connect the pipe dream formula to the transition formula, requiring "little bumping algorithms" and the like (see (3)), and essentially impossible if one wants to include the $\underline{y}$ variables. Indeed, it should be about as difficult as giving a bijective proof that two unimodular triangulations of a polytope should have the same number of simplices. (See (4) where polytopes arise from some matrix Schubert varieties, and this becomes more than an analogy.)

Recall the **conormal variety** CX of a closed subvariety $X \subseteq V$ of a vector space:

$$CX := \overline{\{(x, f) \in V \times V^* \ :\ x \in X \text{ a smooth point}, \vec{v} \perp T_x X\}} \quad \subseteq V \times V^*$$

Use the trace form to identify $M_n(\mathbb{C})^*$ with $M_n(\mathbb{C})$, and call two matrix Schubert varieties $\overline{X}_\pi$, $\overline{X}_\rho$ **projective dual** if $C\overline{X}_\pi \subseteq M_n(\mathbb{C}) \times M_n(\mathbb{C})$ becomes $C\overline{X}_\rho$ upon switching the two $M_n(\mathbb{C})$ factors and rotating both matrices by 180°. (This is essentially the statement that the projective varieties $\mathbb{P}(\overline{X}_\pi), \mathbb{P}(\overline{X}_\rho)$ are projective dual in the 19th-century sense; our reference is (16).) It is a fun exercise to determine ρ from π; note that at least one of the two must be partial, not a permutation.

If $\overline{X}_\pi$ and $\overline{X}_\rho$ are projectively dual, then the dominant part of π's diagram is the 180° rotation of the co-dominant part outside ρ's diagram – projective duality swaps zeroed-out coordinates with free coordinates.

Projective duality also exchanges lex term orders with revlex term orders. So finally, in this sense, the co-transition formula is related to the transition formula by projective duality. (The relation would be exact were to consider

Gröbner degenerations of the conormal varieties, rather than of the matrix Schubert varieties themselves; since we only see the components in one $M_n(\mathbb{C})$ or the other the relation is more of an analogy.)

The reader may wonder, since the lex-from-NE term order was useful (this is effectively the approach in (9)) and the revlex-from-NW term order was useful (in §4), why are the other two (at 180° from these) left out? The 180° symmetry is achieved if we refine the matrix Schubert variety stratification on $M_n(\mathbb{C})$ to the pullback of the *positroid stratification* on $Gr(n;\ \mathbb{C}^{2n})$ along the inclusion $graph\ :\ M_n(\mathbb{C}) \hookrightarrow Gr(n;\ \mathbb{C}^{2n})$ regarding $M_n(\mathbb{C})$ as the big cell.

In (13) was introduced an alternative family $\{C'_\pi\}$ of "bumpless pipe dream" polynomials, and a proof that they match the double Schubert polynomials. The bijection from §3 deriving the co-transition formula for the pipe dream polynomials $\{C_\pi\}$ has a tightly analogous bijective proof of the transition formula for the $\{C'_\pi\}$, in the recent preprint (17, §5).

7 Grothendieck polynomials, nonreduced pipe dreams, and equivariant K-classes

All three families of polynomials A, C, G have extensions to inhomogeneous Laurent polynomials A', C', G' in $\mathbb{Z}[\exp(\pm x_1), \exp(\pm x_2), \ldots, \exp(\pm y_1), \exp(\pm y_2), \ldots]$:

1. *Double Grothendieck polynomials* A'_π. These satisfy recurrence relations based on isobaric Demazure operators.
2. *Nonreduced pipe dream polynomials* C'_π. These allow pipes to cross twice. To read a permutation off of a (nonreduced) pipe dream, one follows the pipes, ignoring the second (and later) crossings of any two pipes.
3. *Equivariant K-classes of matrix Schubert varieties* G'_π. The subvariety $\overline{X}_\pi \subseteq M_n(\mathbb{C})$ defines a class in $(T \times T)$-equivariant K-theory of $M_n(\mathbb{C})$.

Betraying our predilection towards geometry, we call each the "K-theoretic version" of the unprimed family, with the original being the "cohomological".

Each K-theoretic family satisfies the same new base case

Lemma 7.1 (K-theoretic base case) *For each family P' we have* $P'_{w_0^n} = \prod_{i,j\in[n],\ i+j\leq n} (1 - \exp(y_i - x_j))$.

and the recurrence.

Lemma 7.2 (the K-theoretic co-transition formula) *Let $\pi, i, \{\rho\}$ be as in the cohomological co-transition formula. Let S vary over the nonempty subsets of the set of such ρ. Then*

$$(1 - \exp(y_{\pi(i)} - x_i))\, P'_\pi = \sum_S (-1)^{\#S-1}\, P'_{l.u.b.(S)}$$

where $l.u.b.(S)$ is the (unique) least upper bound of S in Bruhat order, automatically of length $\ell(\pi) + \#S$.

Intriguingly, this "boolean lattice inside Bruhat order" phenomenon shows up in the K-theoretic transition formula (14; 17) as well.

We won't prove these two lemmata for A', C', G', but comment on the changes necessary from the cohomological proofs. (Of course, it is already known that $A' = C' = G'$, see e.g. (11), so it suffices to prove these results for, say, just G'.) The bijection in C', placing a $+$ at $(i, \pi(i))$ where there was always a ╯╭, is the same. For the G' co-transition formula one needs to know that the intersection $\overline{X}_\pi \cap \{M : m_{ab} = 0\ \forall (a,b)$ weakly NW of $(i, \pi(i))\}$ is reduced, and that each intersection $\cap_S \overline{X}_\rho = \overline{X}_{l.u.b.(S)}$ is likewise reduced. The swiftest way to confirm this is to observe that there is a Frobenius splitting on the space of matrices (over each $\mathbb{F}_p$, rather than $\mathbb{C}$), with respect to which each $\overline{X}_\pi$ is compatibly split; as at the end of §6, one can infer this from the compatible splitting of the positroid varieties in the Grassmannian (10).

References

[1] Nantel Bergeron, Cesar Ceballos, Vincent Pilaud, *Hopf dreams*, preprint 2018. https://arxiv.org/abs/1807.03044

[2] Nantel Bergeron and Sara Billey, *RC-graphs and Schubert polynomials*, Experimental Math. **2** (1993), no. 4, 257–269.

[3] Sara Billey, Alexander Holroyd, Benjamin Young, *A bijective proof of Macdonald's reduced word formula*, Algebraic Combinatorics 2(2) . February 2017.

[4] Laura Escobar and Karola Mészáros, *Toric matrix Schubert varieties and their polytopes*, Proc. Amer. Math. Soc., 144(12):5081–5096, 2016.

[5] William Fulton, *Flags, Schubert polynomials, degeneracy loci, and determinantal formulas*, Duke Math. J.,**65** (1992), 381–420.

[6] William Fulton, *Young Tableaux, With Applications to Representation Theory and Geometry*, Cambridge University Press, 1996.

[7] William Fulton, *Universal Schubert polynomials*. Duke Math. J. 96 (1999), no. 3, 575–594.

[8] M. E. Kazarian, *Characteristic classes of singularity theory, The Arnold-Gelfand mathematical seminars*, Birkhäuser Boston, Boston, MA, 1997, pp. 325–340.

[9] Allen Knutson, *Schubert patches degenerate to subword complexes*, Transform. Groups 13 (2008), no. 3–4, 715–726.

[10] Allen Knutson, Thomas Lam, and David E Speyer, *Positroid varieties: juggling and geometry*, Compositio Mathematica, 149, (2013), no. 10, 1710–1752.

[11] Allen Knutson and Ezra Miller, *Gröbner geometry of Schubert polynomials*, Annals of Mathematics, Volume 161 (2005), Issue 3, pp 1245–1318.

[12] Allen Knutson and Alex Yong, *A formula for K-theory truncation Schubert calculus*, International Mathematics Research Notices, 70 (2004), 3741–3756.

[13] Thomas Lam, Seung Jin Lee, Mark Shimozono, *Back stable Schubert calculus*, preprint 2018.

[14] Alain Lascoux, *Transition on Grothendieck polynomials*, Physics and combinatorics, 2000 (Nagoya), 164–179, World Sci. Publishing, River Edge, NJ, 2001.

[15] Alain Lascoux, *Polynômes de Schubert: une approche historique*, Discrete Mathematics, 139 (1): 303–317.

[16] Jenia Tevelev, *Projective duality and homogeneous spaces*, Encyclopaedia of Mathematical Sciences, 133. Invariant Theory and Algebraic Transformation Groups, IV. Springer-Verlag, Berlin, 2005.

[17] Anna Weigandt, *Bumpless pipe dreams and alternating sign matrices*, preprint 2020.

17

Positivity Certificates via Integral Representations

Khazhgali Kozhasov[a], Mateusz Michałek[b] and Bernd Sturmfels[c]

Abstract. Complete monotonicity is a strong positivity property for real-valued functions on convex cones. It is certified by the kernel of the inverse Laplace transform. We study this for negative powers of hyperbolic polynomials. Here the certificate is the Riesz kernel in Gårding's integral representation. The Riesz kernel is a hypergeometric function in the coefficients of the given polynomial. For monomials in linear forms, it is a Gel'fand–Aomoto hypergeometric function, related to volumes of polytopes. We establish complete monotonicity for sufficiently negative powers of elementary symmetric functions. We also show that small negative powers of these polynomials are not completely monotone, proving one direction of a conjecture by Scott and Sokal.

1 Introduction

A real-valued function $f : \mathbb{R}^n_{>0} \to \mathbb{R}$ on the positive orthant is *completely monotone* if

$$(-1)^k \frac{\partial^k f}{\partial x_{i_1} \partial x_{i_2} \cdots \partial x_{i_k}}(x) \geq 0 \quad \text{for all } x \in \mathbb{R}^n_{>0}, \tag{1}$$

and for all index sequences $1 \leq i_1 \leq i_2 \leq \cdots \leq i_k \leq n$ of arbitrary length k. In words, the function f and all of its signed derivatives are nonnegative on the open cone $\mathbb{R}^n_{>0}$.

While this definition makes sense for C^∞ functions f, we here restrict ourselves to a setting that is the natural one in real algebraic geometry. Namely, we consider functions

[a] Technical University of Braunschweig
[b] University of Konstanz
[c] Max Planck Institute for Mathematics in the Sciences, Leipzig

$$f = p_1^{s_1} p_2^{s_2} \cdots p_m^{s_m}, \qquad (2)$$

where each p_i is a homogeneous polynomial in n variables that is positive on $\mathbb{R}^n_{>0}$. The alternating sign in (1) implies that positive powers of a polynomial cannot be completely monotone. The real numbers $s_1, s_2, \ldots, s_m$ will therefore be negative in most cases.

The signed k-th order derivative in (1) for $f = p^s$ has the form $P_{i_1 \cdots i_k} \cdot p^{s-k}$ where $P_{i_1 \cdots i_k}$ is a homogeneous polynomial. Saying that f is completely monotone means that infinitely many polynomials $P_{i_1 \cdots i_k}$ are nonnegative on $\mathbb{R}^n_{>0}$. How can this be certified?

To answer this question we apply the inverse Laplace transform to the function $f(x)$. Our goal is to find a function $q(y)$ that is nonnegative on $\mathbb{R}^n_{>0}$ and that satisfies

$$f(x) = \int_{\mathbb{R}^n_{\geq 0}} e^{-\langle y,x \rangle} q(y) dy \qquad \text{for all } x \in \mathbb{R}^n_{>0}. \qquad (3)$$

Here dy denotes Lebesgue measure and $\langle y,x \rangle = \sum_{i=1}^n y_i x_i$ is the usual dot product. This integral representation certifies that $f(x)$ is completely monotone, since it implies

$$(-1)^k \frac{\partial^k f}{\partial x_{i_1} \partial x_{i_2} \cdots \partial x_{i_k}}(x) = \int_{\mathbb{R}^n_{\geq 0}} e^{-\langle y,x \rangle} y_{i_1} y_{i_2} \cdots y_{i_k} q(y) dy \geq 0.$$

Our interest in the formula (3) derives from statistics and polynomial optimization. Certifying that a polynomial is nonnegative is an essential primitive in optimization. This can be accomplished by finding a representation as a sum of squares (SOS) or as a sum of nonnegative circuit polynomials (SONC). See e.g. [3] and [6] for introductions to these techniques. We here introduce an approach that may lead to new certificates.

Algebraic statistics [21] is concerned with probabilistic models that admit polynomial representations and are well suited for data analysis. The *exponential families* in [14] enjoy these desiderata. Using the formulation above, the distributions in an exponential family have supports in $\mathbb{R}^n_{>0}$ and probability density functions $y \mapsto e^{-\langle y,x \rangle} \frac{q(y)}{f(x)}$.

One natural class of instances f arises when the factors $p_1, \ldots, p_m$ in (2) are linear forms with nonnegative coefficients. An easy argument, presented in Section 3, furnishes the integral representation (3) and shows that f is completely monotone. However, even this simple case is important for applications, notably in the algorithms for efficient volume computations due to Lasserre and Zerron [12, 13]. While that work was focused on polyhedra, the theory presented here promises a natural generalization to spectrahedra, spectrahedral shadows, and feasible regions in hyperbolic programming.

The present article is organized as follows. In Section 2 we replace $\mathbb{R}^n_{>0}$ by an arbitrary convex cone C, and we recall the Bernstein–Hausdorff–Widder-Choquet Theorem. This theorem says that a certifcate (3) exists for every completely monotone function on C, provided we replace $q(y)dy$ by a Borel measure supported in the dual cone C^*. We also discuss operations that preserve complete monotonicity, such as restricting $f(x)$ to a linear subspace. Dually, this corresponds to pushing forward the representing measure.

In Section 3 we study the case of monomials in linear forms, seen two paragraphs above. Here the function $q(y)$ is piecewise polynomial, for integers $s_1, \ldots, s_m < 0$, it measures the volumes of fibers under projecting of a simplex onto a polytope. In general, we interpret $q(y)dy$ as pushforward of a Dirichlet distribution with concentration parameters $-s_i$, thus offering a link to Bayesian statistics and machine learning.

Section 4 concerns the $m = 1$ case in (2). In order for $f = p^s$ to be completely monotone, the polynomial p must be hyperbolic and the exponent s must be negative. The function $q(y)$ is known as the *Riesz kernel*. It can be computed from the complex integral representation in Theorem 4.8 which is due to Gårding [7]. Following [14, Section 3] and [19, Section 4], we study the case when p is the determinant of a symmetric matrix of linear forms. Here the Riesz kernel is related to the Wishart distribution.

Our primary objective is to develop tools for computing Riesz kernels, and thereby certifying positivity as in (1). Section 5 brings commutative algebra into our tool box. The relevant algebra arises from the convolution product on Borel measures. We are interested in finitely generated subalgebras and the polynomial ideals representing these. Elements in such ideals can be interpreted as formulas for Riesz kernels. In the setting of Section 3 we recover the Orlik–Terao algebra [15] of a hyperplane arrangement, but now realized as a convolution algebra of piecewise polynomial volume functions.

In Section 6 we give partial answers to questions raised in the literature, namely in [14, Conjecture 3.5] and in [19, Conjecture 1.11]. Scott and Sokal considered elementary symmetric polynomials p, and they conjectured necessary and sufficient conditions on negative s for complete monotonicity of p^s. We prove the necessity of their conditions. We also show that p^s is completely monotone for sufficiently negative exponents s, and we explain how to build the Riesz kernel in this setting.

In Section 7 we view the Riesz kernel q of $f = p^s$ as a function of the coefficients of the hyperbolic polynomial p. In the setting of Section 3, when $f = \ell_1^{s_1} \cdots \ell_m^{s_m}$ for linear forms ℓ_i, we examine the dependence of q on the coefficients of the ℓ_i. These dependences are hypergeometric in nature. In the former case, the Riesz kernel satisfies the $\mathcal{A}$-hypergeometric system [17, Section 3.1], where $\mathcal{A}$ is the support of p. In the latter case, it is a

Gel'fand–Aomoto hypergeometric function [17, Section 1.5]. This explains the computation in [14, Example 3.11], and it opens up the possibility to use D-modules as in [17, Section 5.3] or [22] for further developing the strand of research initiated here.

2 Complete Monotonicity

We now start afresh, by offering a more general definition of complete monotonicity. Let V denote a finite-dimensional real vector space and C an open convex cone in V. The closed dual cone to C is denoted $C^* = \{y \in V^* : \langle y,x\rangle \geq 0 \text{ for all } x \in C\}$. Points in C^* are linear functions that are nonnegative on C. We are interested in differentiable functions on the open cone C that satisfy the following very strong notion of positivity.

Definition 2.1 (Complete monotonicity) A function $f: C \to \mathbb{R}$ is *completely monotone* if f is C^∞-differentiable and, for all $k \in \mathbb{N}$ and all vectors $v_1, \ldots, v_k \in C$, we have

$$(-1)^k D_{v_1} \cdots D_{v_k} f(x) \geq 0 \quad \text{for all } x \in C. \tag{4}$$

Here D_v is the directional derivative along the vector v. In this definition it suffices to assume that $v_1, \ldots, v_k$ are extreme rays of C. This is relevant when C is polyhedral.

Example 2.2 For $V = \mathbb{R}^n$ and $C = \mathbb{R}^n_{>0}$, the positive orthant, (4) is equivalent to (1). Moreover, if $C \supset \mathbb{R}^n_{>0}$ holds in Definition 2.1 then f is completely monotone on $\mathbb{R}^n_{>0}$ too.

The following lemma will be used frequently in this paper.

Lemma 2.3 *Let W be a linear subspace of V such that $C \cap W$ is nonempty. If f is completely monotone on C, then the restriction $f|_W$ is completely monotone on $C \cap W$.*

Proof This follows from the fact that the derivative of $f|_W$ along a vector v in $C \cap W$ coincides with the restriction of $D_v f : C \to \mathbb{R}$ to the subspace W. Here D_v is the directional derivative used in (4). □

No polynomial of positive degree is completely monotone. Indeed, by Lemma 2.3, it suffices to consider the case $n = 1$ and $C = \mathbb{R}_{>0}$. Derivatives of univariate polynomials cannot alternate their sign when $x \gg 0$. But, this is easily possible for rational functions.

Example 2.4 Fix arbitrary negative real numbers $s_1, s_2, \ldots, s_n$. Then the function

$$f \;=\; x_1^{s_1} x_2^{s_2} \cdots x_n^{s_n}$$

is completely monotone on $C = \mathbb{R}^n_{>0}$. Lemma 2.3 says that we may replace the x_i by linear forms. We conclude that products of negative powers of linear forms are completely monotone on their polyhedral cones. This is the theme we study in Section 3.

The Bernstein–Hausdorff–Widder Theorem [23, Thm. 12a] characterizes completely monotone functions in one variable. They are obtained as Laplace transforms of Borel measures on the positive reals. This result was generalized by Choquet [4, Thm. 10]. He proved the following theorem for higher dimensions and arbitrary convex cones C.

Theorem 2.5 (Bernstein–Hausdorff–Widder–Choquet theorem) *Let f be a real-valued function on the open cone C. Then f is completely monotone if and only if it is the Laplace transform of a unique Borel measure μ supported on the dual cone C^*, that is,*

$$f(x) \;=\; \int_{C^*} e^{-\langle y,x\rangle}\, d\mu(y). \tag{5}$$

Remark 2.6 By the formula (5), any completely monotone function $f : C \to \mathbb{R}$ extends to a holomorphic function on the complex tube $C + i\cdot V = \{x + i\cdot v : x \in C, v \in V\} \subset V_{\mathbb{C}}$.

The if direction in Theorem 2.5 is seen by differentiating under the integral sign. Indeed, the left hand side of (4) equals the following function which is nonnegative:

$$(-1)^k D_{v_1} \ldots D_{v_k} f(x) \;=\; \int_{C^*} e^{-\langle y,x\rangle} \prod_{j=1}^{k} \langle y, v_i\rangle\, d\mu(y).$$

The integrand is a positive real number whenever $x, v_1, \ldots, v_k \in C$ and $y \in C^*$. The more difficult part is the only-if direction. Here one needs to construct the integral representation. This is precisely our topic of study in this paper, for certain functions f.

The Borel measure μ in the integral representation (5) is called the *Riesz measure* when $f = p^s$. As we shall see, its support can be a lower-dimensional subset of C^*. But in many cases the Riesz measure is absolutely continuous with respect to Lesbesgue measure on C^*, that is, there exists a measurable function $q : C^* \to \mathbb{R}_{\geq 0}$ such that

$$d\mu(y) = q(y)dy.$$

The nonnegative function $q(y)$ is called the *Riesz kernel* of $f = p^s$. It serves as the certificate for complete monotonicity of f, and our aim is to derive formulas for $q(y)$.

We next present a formula for the Riesz kernel in the situation of Example 2.4.

Proposition 2.7 *For any positive real numbers $\alpha_1, \ldots, \alpha_n$, we have*

$$x_1^{-\alpha_1} \cdots x_n^{-\alpha_n} = \int_{\mathbb{R}^n_{\geq 0}} e^{-\langle y, x\rangle} \frac{y_1^{\alpha_1 - 1} \cdots y_n^{\alpha_n - 1}}{\Gamma(\alpha_1) \cdots \Gamma(\alpha_n)} dy \qquad \text{for all } x \in \mathbb{R}^n_{>0}. \tag{6}$$

In particular, the Riesz kernel of the function $f = \prod_{i=1}^n x_i^{-\alpha_i}$ is equal to $q(y) = \prod_{i=1}^n \Gamma(\alpha_i)^{-1} y_i^{\alpha_i - 1}$.

Proof Recall the definition of the gamma function from its integral representation:

$$\Gamma(a) = \int_{\mathbb{R}_{\geq 0}} t^{a-1} e^{-t} dt.$$

Regarding $x > 0$ as a constant, we perform the change of variables $t = yx$. This gives

$$x^{-a} = \frac{1}{\Gamma(a)} \int_{\mathbb{R}_{\geq 0}} y^{a-1} e^{-yx} dy. \tag{7}$$

This is the $n = 1$ case of the desired formula (6). The general case $n > 1$ is obtained by multiplying n distinct copies of the univariate formula (7). □

The Riesz kernel offers a certificate for a function f to be completely monotone. This is very powerful because it proves that infinitely many functions are nonnegative on C.

Example 2.8 Let $V = \mathbb{R}^3$. Note that $p = x_1x_2 + x_1x_3 + x_2x_3$ is positive on $C = \mathbb{R}^3_{>0}$. We consider negative powers $f = p^{-\alpha}$ of this polynomial. The function f is not completely monotone for $\alpha < 1/2$. For instance, if $\alpha = 5/11$ then this is proved by

$$\frac{\partial^{12} f}{\partial x_1^4 \partial x_2^4 \partial x_3^4} = p^{-137/11} \cdot \text{a polynomial of degree 12 with 61 terms.}$$

The value of this function at the positive point $(1, 1, 1)$ is found to be

$$-\frac{16652440985600}{762638095543203}3^{6/11} = -0.0397564287 < 0.$$

Suppose now that $\alpha > 1/2$. We claim that no such violation exists, i.e. the function $f = p^{-\alpha}$ is completely monotone. This is certified by exhibiting the Riesz kernel:

$$q(y) = \frac{2^{1-\alpha}}{(2\pi)^{\frac{1}{2}}\Gamma(\alpha)\Gamma(\alpha - \frac{1}{2})}\left(y_1y_2 + y_1y_3 + y_2y_3 - \frac{1}{2}(y_1^2 + y_2^2 + y_3^2)\right)^{\alpha-\frac{3}{2}}.$$

Then the integral representation (5) holds with the Borel measure $d\mu(y) = q(y)dy$. A proof can be found in [19, Corollary 5.8]. We invite our readers to verify this formula.

We next discuss the implications of Lemma 2.3 for the measure $d\mu$ in (5). Suppose that $f(x) = \int_{C^*} e^{-\langle y,x\rangle}\, d\mu(y)$ holds for a cone C in a vector space V. We want to restrict f to a subspace W. The inclusion $W \subset V$ induces a *projection* $\pi : V^* \to W^*$, by restricting the linear form. The dual of the cone $C \cap W$ equals $\pi(C^*)$ in the space W^*. We are looking for a representation $f(x) = \int_{(C\cap W)^*} e^{-\langle y,x\rangle}\, d\mu'(y)$, $x \in C \cap W$. The measure μ' is obtained from the measure μ by the general *push-forward* construction.

Definition 2.9 Let μ be a measure on a set S and $\pi : S \to S'$ a measurable function. The push-forward measure $\pi_*\mu$ on S' is defined by setting $(\pi_*\mu)(U) := \mu(\pi^{-1}(U))$, where $U \subset S'$ is any measurable subset.

Example 2.10 Let us suppose we are given a (measurable) function $\pi : V_1 \to V_2$ of real vector spaces and a measure with density g on V_1. Then $(\pi_*\mu)(U) = \int_{\pi^{-1}(U)} g(x)dx$.

The following results are straightforward from the definitions.

Proposition 2.11 *Let $f = p^s$ be completely monotone on a convex cone $C \subset V$ and μ the Riesz measure of f. Then the Riesz measure of the restriction of f to the subspace W is the push-forward $\pi_*\mu$, where $\pi : V^* \to W^*$ is the projection.*

Corollary 2.12 *A function f is completely monotone on the cone $C \subset V$ if and only if, for every $g \in \mathrm{GL}(V)$, the function $f \circ g^{-1}$ is completely monotone on $g(C)$. The Riesz measures transform accordingly under the linear transformation dual to g.*

3 Monomials in Linear Forms

Let $\ell_1, \ell_2, \ldots, \ell_m$ be linear forms in n variables and consider the open polyhedral cone

$$C = \{v \in \mathbb{R}^n : \ell_i(v) > 0 \text{ for } i = 1, 2, \ldots, m\}. \tag{8}$$

The dual cone C^* is spanned by $\ell_1, \ell_2, \ldots, \ell_m$ in $(\mathbb{R}^n)^*$. A generator ℓ_i is an extreme ray of the cone C^* if and only if the face $\{v \in C : \ell_i(v) = 0\}$ is a facet of C.

Proposition 3.1 *For any positive real numbers $\alpha_1, \alpha_2, \ldots, \alpha_m$, the function*

$$f = \ell_1^{-\alpha_1} \ell_2^{-\alpha_2} \cdots \ell_m^{-\alpha_m}$$

is completely monotone on the cone C. The Riesz measure of f is the pushforward of the Dirichlet measure $\frac{y_1^{\alpha_1 - 1} \cdots y_m^{\alpha_m - 1}}{\Gamma(\alpha_1) \cdots \Gamma(\alpha_m)} dy$ on the orthant $\mathbb{R}^m_{>0}$ under the linear map $L : \mathbb{R}^m_{\geq 0} \to C^*$ that takes the standard basis $e_1, \ldots, e_m$ to the linear forms $\ell_1, \ldots, \ell_m$.

Proof The monomial $x_1^{-\alpha_1} \cdots x_m^{-\alpha_m}$ is completely monotone on $\mathbb{R}^m_{>0}$ by Example 2.4. Our function f is its pullback under the linear map $C \to \mathbb{R}^m_{>0}$ that is dual to the map $L : \mathbb{R}^m_{\geq 0} \to C^*$. The result follows from a slight extension of Proposition 2.11. In that proposition, the map L would have been surjective but this is not really necessary. □

In what follows we assume that the dual cone C^* is full-dimensional in $\mathbb{R}^n$. Then $\ell_1, \ldots, \ell_m$ span $(\mathbb{R}^n)^*$. The case $\dim(C^*) < n$ is discussed at the end of this section. There is a canonical polyhedral subdivision of C^*, called the *chamber complex*. It is the common refinement of all cones spanned by linearly independent subsets of $\{\ell_1, \ldots, \ell_m\}$. Thus, for a general vector ℓ in C^*, the chamber containing ℓ is the intersection of all halfspaces defined by linear inequalities $\det(\ell_{i_1}, \ldots, \ell_{i_{n-1}}, y) > 0$ that hold for $y = \ell$. The collection of all chambers and their faces forms a polyhedral fan with support C^*. That fan is the chamber complex of the map L, which we regard as an $n \times m$-matrix.

Example 3.2 Let $n = 3$, $m = 5$ and consider the matrix

$$L = \begin{pmatrix} 1 & 1 & 1 & 1 & 1 \\ 0 & 1 & 2 & 1 & 0 \\ 0 & 0 & 1 & 2 & 1 \end{pmatrix}.$$

The corresponding linear forms $\ell_1, \ldots, \ell_5$ are the entries of the row vector $(x_1, x_2, x_3)L$. For any positive real numbers $\alpha_1, \alpha_2, \alpha_3, \alpha_4, \alpha_5$, we are interested in the function

$$f = x_1^{-\alpha_1}(x_1+x_2)^{-\alpha_2}(x_1+2x_2+x_3)^{-\alpha_3}(x_1+x_2+2x_3)^{-\alpha_4}(x_1+x_3)^{-\alpha_5}.$$

This is completely monotone on the pentagonal cone C, consisting of all vectors (x_1, x_2, x_3) where the five linear forms are positive. Its dual cone C^* is also a pentagonal cone. The elements (y_1, y_2, y_3) of C^* are nonnegative linear combinations of the columns of L.

The chamber complex of L consists of 11 full-dimensional cones. Ten of them are triangular cones. The remaining cone, right in the middle of C^*, is pentagonal. It equals

$$\left\{ y \in \mathbb{R}^3 \,:\, y_1 \geq y_2,\ y_1 \geq y_3,\ 2y_3 \geq y_2,\ 2y_2 \geq y_3 \text{ and } y_2 + y_3 \geq y_1 \right\}. \quad (9)$$

Returning to the general case, for each y in the interior of C^*, the fiber $L^{-1}(y)$ of the map $L : \mathbb{R}^m_{\geq 0} \to C^*$ is a convex polytope of dimension $m - n$. The polytope $L^{-1}(y)$ is simple provided y lies in an open cone of the chamber complex. All fibers $L^{-1}(y)$ lie in affine spaces that are parallel to kernel$(L) \simeq \mathbb{R}^{m-n}$. In the sequel we denote by $|L|$ the square root of the *Gram determinant* of a matrix L, that is, $|L| = \sqrt{\det(LL^T)}$.

Theorem 3.3 *The Riesz kernel $q(y)$ of $f = \prod_{i=1}^m \ell_i^{-\alpha_i}$ is a continuous function on the polyhedral cone C^* that is homogeneous of degree $\sum_{i=1}^m \alpha_i - n$. Its value $q(y)$ at a point $y \in C^*$ equals the integral of $|L|^{-1} \prod_{i=1}^m y_i^{\alpha_i - 1}\Gamma(\alpha_i)^{-1}$ over the fiber $L^{-1}(y)$ with respect to $(m-n)$-dimensional Lebesgue measure. If $\alpha_1, \ldots, \alpha_m$ are integers then the Riesz kernel $q(y)$ is piecewise polynomial and differentiable of order $\sum_{i=1}^m \alpha_i - n - 1$. To be precise, $q(y)$ is a homogeneous polynomial on each cone in the chamber complex.*

Proof We first consider the case $\alpha_1 = \cdots = \alpha_m = 1$. Here Proposition 2.7 says that

$$x_1^{-1} \cdots x_m^{-1} = \int_{\mathbb{R}^m_{\geq 0}} e^{-\langle y, x \rangle}\, dy.$$

The Riesz measure of this function is Lebesgue measure on $\mathbb{R}^m_{\geq 0}$, and its Riesz kernel is the constant function 1. The pushforward of this measure under the linear map L is absolutely continuous with respect to Lebesgue measure on C^*. The Riesz kernel of this pushforward is the function $q(y)$ whose value at $y \in C^*$ is the volume of the fiber $L^{-1}(y)$ divided by $|L|$. This volume is a function in y that is homogeneous of degree $m - n$. Moreover, the volume function is a polynomial on each chamber and it is differentiable of order $m - n - 1$. This is known from the theory of splines; see e.g. the book by De Concini and Procesi [5]. This proves the claim since $m = \sum_{i=1}^m \alpha_i$.

Next we let $\alpha_1, \ldots, \alpha_m$ be arbitrary positive integers. We form a new matrix $L^{(\alpha)}$ from L by replicating α_i identical columns from the column ℓ_i of L. Thus the number of columns of $L^{(\alpha)}$ is $m' = \sum_{i=1}^{m} \alpha_i$. The matrix $L^{(\alpha)}$ represents a linear map from $\mathbb{R}^{m'}_{\geq 0}$ onto the same cone C^* as before, and also the chamber complex of $L^{(\alpha)}$ remains the same. However, its fibers are polytopes of dimension $m' - n$, which is generally larger than $m - n$. We now apply the argument in the previous paragraph to this map. This gives the asserted result, namely the Riesz kernel $q(y)$ is piecewise polynomial of degree $m' - n$, and is differentiable of order $m' - n - 1$ across walls of the chamber complex.

We finally consider the general case when the α_i are arbitrary positive real numbers. The formula for the Riesz kernel follows by Proposition 3.1. The homogeneity and continuity properties of the integral over $x_1^{\alpha_1 - 1} \cdots x_m^{\alpha_m - 1}$ hold here as well, because this function is well-defined and positive on each fiber $L^{-1}(y)$. □

Example 3.4 Let $n = 2$, $m = 4$ and $L = \begin{pmatrix} 3 & 2 & 1 & 0 \\ 0 & 1 & 2 & 3 \end{pmatrix}$. Here $C^* = \mathbb{R}^2_{\geq 0}$ and the chamber complex is the division into three cones defined by the lines $y_2 = 2y_1$ and $y_1 = 2y_2$. The corresponding monomial in linear forms has the integral representation

$$\begin{aligned} &(3x_1)^{-\alpha_1}(2x_1 + x_2)^{-\alpha_2}(x_1 + 2x_2)^{-\alpha_3}(3x_2)^{-\alpha_4} \\ &\quad = \int_{\mathbb{R}^2_{\geq 0}} e^{-y_1x_1 - y_2x_2} q(y_1, y_2) dy_1 dy_2. \end{aligned}$$

For $\alpha_1 = \alpha_2 = \alpha_3 = \alpha_4 = 1$, the Riesz kernel is the piecewise quadratic function

$$q(y_1, y_2) \;=\; \frac{1}{108} \cdot \begin{cases} 3y_1^2 & \text{if } y_2 \geq 2y_1, \\ 4y_1y_2 - y_1^2 - y_2^2 & \text{if } y_2 \leq 2y_1 \text{ and } y_1 \leq 2y_2, \\ 3y_2^2 & \text{if } y_1 \geq 2y_2. \end{cases}$$

The three binary quadrics measure the areas of the convex polygons $L^{-1}(y)$. These polygons are triangles, quadrilaterals and triangles, in the three cases. Note that $q(y)$ is differentiable. For small values of $k \geq 1$, it is instructive to work out the piecewise polynomials $q(y)$ of degree $2 + k$ for all $\binom{3+k}{3}$ cases when $\alpha_i \in \mathbb{N}_+$ with $\sum_{i=1}^{4} \alpha_i = 4 + k$.

We have shown that the reciprocal product of linear forms is the Laplace transform of the piecewise polynomial function $q(y)$ which measures the volumes of fibers of L:

$$\ell_1^{-1}(x) \cdots \ell_m^{-1}(x) \quad = \quad \int_{C^*} e^{\langle y,x\rangle} q(y) dy.$$

The volume function can be recovered from this formula by applying the inverse Laplace transform. This technique was applied by Lasserre and Zerron [12, 13] to develop a surprisingly efficient algorithm for computing the volumes of convex polytopes.

Example 3.5 Fix the matrix L in Example 3.2, and consider any vector y in the central chamber (9). The fiber $L^{-1}(y)$ is a pentagon. Its vertices are the rows of

$$\frac{1}{6}\begin{pmatrix} 0 & 3y_1+3y_2-3y_3 & 0 & -3y_1+3y_2+3y_3 & 6y_1-6y_2 \\ 0 & 6y_1-6y_3 & -3y_1+3y_2+3y_3 & 0 & 3y_1-3y_2+3y_3 \\ 6y_1-6y_3 & 0 & 3y_2 & 0 & -3y_2+6y_3 \\ 6y_1-2y_2-2y_3 & 0 & 4y_2-2y_3 & -2y_2+4y_3 & 0 \\ 6y_1-6y_2 & 6y_2-3y_3 & 0 & 3y_3 & 0 \end{pmatrix}.$$

The Riesz kernel is given by the area of this pentagon:

$$q(y) = \tfrac{1}{6}\text{area}(L^{-1}(y)) = \tfrac{1}{24}(6y_1y_2 + 6y_1y_3 + 2y_2y_3 - 3y_1^2 - 5y_2^2 - 5y_3^2).$$

Similar quadratic formulas hold for the Riesz kernel $q(y)$ on the other ten chambers.

Remark 3.6 Proposition 3.1 also applies in the case when $\ell_1, \ldots, \ell_m$ do not span $\mathbb{R}^n$. Here the cone C^* is lower-dimensional. The Riesz measure is supported on that cone. There is no Riesz kernel $q(y)$ in the sense above. But, we can still consider the volume function on the cone C^*. If $m = 1$, then $f = \ell^{-\alpha}$ is a negative power of a single linear form. This function f is completely monotone on the half-space $C = \{\ell > 0\}$. The cone $C^* = \{t\ell : t \geq 0\}$ is the ray spanned by ℓ. The Riesz measure is absolutely continuous with respect to Lebesgue measure dt on that ray, with density $t^{\alpha-1}/\Gamma(\alpha)$, cf. (7).

4 Hyperbolic Polynomials

We now return to the setting of Section 2 where we considered negative powers $f = p^{-\alpha}$ of a homogeneous polynomial. We saw in Section 3 that f is completely monotone if p is a product of linear forms. For which polynomials p and which α can we expect such a function to be completely monotone? We begin with a key example.

Fix an integer $m \geq 0$, set $n = \binom{m+1}{2}$, and let $x = (x_{ij})$ be a symmetric $m \times m$ matrix of unknowns. Its determinant $\det(x)$ is a homogeneous polynomial of degree m in the x_{ij}. Denote by C the open cone in $\mathbb{R}^n$ consisting of all positive definite matrices. Up to closure, this cone is self-dual with respect to the trace inner product $\langle x, y\rangle = \text{trace}(xy)$. Thus, C^* is the closed cone of positive semidefinite (psd) matrices in $\mathbb{R}^n$.

Theorem 4.1 (Scott-Sokal [19]) *Let $\alpha \in \mathbb{R}$. The function* $\det(x)^{-\alpha}$ *is completely monotone on the cone C of positive definite symmetric $m \times m$ matrices if and only if*

$$\alpha \in \{0, \tfrac{1}{2}, 1, \tfrac{3}{2}, \ldots, \tfrac{m-1}{2}\} \quad \text{or} \quad \alpha > \tfrac{m-1}{2}.$$

Proof The if-direction of this theorem is a classical result in statistics, as we explain below. We define a probability distribution N on the space of $m \times r$ matrices as follows: each column is chosen independently at random with respect to the m-variate normal distribution with mean zero and covariance matrix Σ. The *Wishart distribution* $W_m(\Sigma, r)$ is the push-forward of the distribution N to the space of symmetric $m \times m$ matrices, by the map $Z \to ZZ^T$. Since ZZ^T is always psd, the Wishart distribution is supported on the closure of C. There are two main cases one needs to distinguish.

If $r \geq m$, then the matrix ZZ^T has full rank m with probability one. Hence, the support of the Wishart distribution coincides with C. An explicit formula for the Riesz measure can be derived from the Wishart distribution. This is explained in detail in [14, Proposition 3.8]. Setting $\alpha = r/2$, the Riesz kernel for $\det(x)^{-\alpha}$ is equal to

$$q(y) = \left(\pi^{\frac{m(m-1)}{4}} \prod_{j=0}^{m-1} \Gamma\left(\alpha - \frac{j}{2}\right)\right)^{-1} \det(y)^{\alpha - \frac{m+1}{2}}. \tag{10}$$

This formula holds whenever $\alpha > (m-1)/2$. It proves that $\det(x)^{-\alpha}$ is completely monotone on C for this range of α.

If $r < m$ then the matrix ZZ^T has rank at most r. Hence the Wishart distribution is supported on the subset of $m \times m$ psd matrices of rank at most r. We have

$$1 = \int_{C^*} W_m(\Sigma, r) = \left((2\pi)^m \det \Sigma\right)^{-r/2} \int_{(\mathbb{R}^m)^r} \exp\left(-\frac{1}{2}\sum_{i=1}^{r} z_i^T \Sigma^{-1} z_i\right) dz_1 \cdots dz_r.$$

Let $Q := (2\Sigma)^{-1}$. After multiplying both sides in the formula above by $(\det Q)^{-r/2}$, we obtain

$$(\det Q)^{-r/2} = \int_{(\mathbb{R}^m)^r} \exp\left(-\sum_{i=1}^{r} z_i^T Q z_i\right) \pi^{-mr/2} dz_1 \cdots dz_r.$$

Here z_i are the columns of an $m \times r$ matrix Z. Hence we may rewrite this formula as

$$(\det Q)^{-r/2} \;=\; \int_{\mathbb{R}^{m\times r}} \exp\left(-\operatorname{tr}(QZZ^T)\right) \pi^{-mr/2}\, dZ.$$

This establishes complete monotonicity of $\det(x)^{-r/2}$, $x \in C$, for all $r = 0, 1, \ldots, m-1$. Indeed, we realized the Riesz measure on C^* as the push-forward of the (scaled) Lebesgue measure $dZ/\pi^{mr/2}$ on the space $\mathbb{R}^{m\times r}$ of $m\times r$ matrices under the map $Z \mapsto ZZ^T$. The only-if direction of Theorem 4.1 is due to Scott and Sokal [19, Theorem 1.3]. □

According to Lemma 2.3, the restriction of a completely monotone function to a linear subspace gives a completely monotone function. We thus obtain completely monotone functions by restricting the determinant function to linear spaces of symmetric matrices. This can be constructed as follows. Fix linearly independent symmetric $m \times m$ matrices $A_1, \ldots, A_n$ such that the following *spectrahedral cone* is non-empty:

$$C \;=\; \left\{ x \in \mathbb{R}^n \,:\, x_1A_1 + \cdots + x_nA_n \text{ is positive definite} \right\}.$$

Its dual C^* is the image of the cone of $m \times m$ psd matrices under the linear map L dual to the inclusion $x \in \mathbb{R}^n \mapsto x_1A_1 + \cdots + x_nA_n$. Such a cone is known as a *spectrahedral shadow*. The following polynomial vanishes on the boundary of C:

$$p(x) \;=\; \det(x_1A_1 + \cdots + x_nA_n). \tag{11}$$

Corollary 4.2 *Let $\alpha \in \{0, \frac{1}{2}, 1, \frac{3}{2}, \ldots, \frac{m-1}{2}\}$ or $\alpha > \frac{m-1}{2}$. Then the function $p^{-\alpha}$ is completely monotone on the spectrahedral cone C and its Riesz measure is the push-forward of the Riesz measure from the proof of Theorem 4.1 under the map L.*

Consider the case when $A_1, \ldots, A_n$ are diagonal matrices. Here the polynomial $p(x)$ is a product of linear forms, and we are in the situation of Section 3. For general symmetric matrices A_i, the fibers $L^{-1}(y)$ are spectrahedra and not polytopes. If $\alpha > \frac{m-1}{2}$ then the Riesz kernel exists, and its values are found by integrating (10) over the spectrahedra $L^{-1}(y)$. In particular, if $\alpha = \frac{m+1}{2}$, then the value of the Riesz kernel $q(y)$ equals, up to a constant, the volume of the spectrahedron $L^{-1}(y)$. The role of the chamber complex is now played by a nonlinear branch locus. It would be desirable to compute explicit formulas, in the spirit of Examples 3.4 and 3.5, for these spectrahedal volume functions.

Example 4.3 Let $m = n = 3$ and write C for the cone of positive definite matrices

$$\begin{pmatrix} x_1 + x_2 + x_3 & x_3 & x_2 \\ x_3 & x_1 + x_2 + x_3 & x_1 \\ x_2 & x_1 & x_1 + x_2 + x_3 \end{pmatrix}. \qquad (12)$$

This space of matrices is featured in [20, Example 1.1] where it serves the prominent role of illustrating the convex algebraic geometry of Gaussians. Let $p(x)$ be the determinant of (12). Cross sections of the cone C and its dual C^* are shown in the middle and right of [20, Figure 1]. The fibers $L^{-1}(y)$ are the 3-dimensional spectrahedra shown on the left in [20, Figure 1]. Their volumes are computed by the Riesz kernel $q(y)$ for $\alpha = 2$.

We have seen that determinants of symmetric matrices of linear forms are a natural class of polynomials $p(x)$ admitting negative powers that are completely monotone. Which other polynomials have this property? The section title reveals the answer.

Definition 4.4 A homogeneous polynomial $p \in \mathbb{R}[x_1, \ldots, x_n]$ is *hyperbolic* for a vector $e \in \mathbb{R}^n$ if $p(e) > 0$ and, for any $x \in \mathbb{R}^n$, the univariate polynomial $t \mapsto p(t \cdot e - x)$ has only real zeros. Let C be the connected component of the set $\mathbb{R}^n \setminus \{p = 0\}$ that contains e. If p is hyperbolic for e, then it is hyperbolic for all vectors in C. In that case, C is an open convex cone, called the *hyperbolicity cone* of p. Equivalently, a homogeneous polynomial $p \in \mathbb{R}[x_1, \ldots, x_n]$ is hyperbolic with hyperbolicity cone C if and only if $p(z) \neq 0$ for any vector z in the tube domain $C + i \cdot \mathbb{R}^n$ in the complex space $\mathbb{C}^n$.

Example 4.5 The canonical example of a hyperbolic polynomial is the determinant $p(x) = \det(x)$ of a symmetric $m \times m$ matrix x. Here we take e to be the identity matrix. Then C is the cone of positive definite symmetric matrices. Hyperbolicity holds because the roots of $p(t \cdot e - x)$ are the eigenvalues of x, and these are all real.

Example 4.6 The restriction of a hyperbolic polynomial p to a linear subspace V is hyperbolic, provided V intersects the hyperbolicity cone C of p. Therefore the polynomials in (11) are all hyperbolic. In particular, any product of linear forms ℓ_i is hyperbolic, if we take the hyperbolicity cone to be the polyhedral cone C in (8).

The following result states that hyperbolic polynomials are precisely the relevant class of polynomials for our study of positivity certificates via integral representations.

Theorem 4.7 *Let $p \in \mathbb{R}[x_1, \ldots, x_n]$ be a homogeneous polynomial that is positive on an open convex cone C in $\mathbb{R}^n$, and such that the power $f = p^{-\alpha}$ is completely monotone on C for some $\alpha > 0$. Then p is hyperbolic and its hyperbolicity cone contains C.*

Proof This follows from [19, Corollary 2.3] (see [14, Theorem 3.3]). □

A hyperbolic polynomial p with hyperbolicity cone $C \subset \mathbb{R}^n$ is called *complete* if C is pointed, that is, the dual cone C^* is n-dimensional. In the polyhedral setting of Proposition 3.1, this was precisely the condition for the Riesz kernel to exist. This fact generalizes. The following important result by Gårding [7, Theorem 3.1] should give a complex integral representation of Riesz kernels for arbitrary hyperbolic polynomials.

Theorem 4.8 (Gårding) *Let p be a complete hyperbolic polynomial with hyperbolicity cone $C \subset \mathbb{R}^n$. Fix a vector $e \in C$ and a complex number α with* $\mathrm{Re}(\alpha) > n$*, and define*

$$q_\alpha(y) = (2\pi)^{-n} \int_{\mathbb{R}^n} p(e + i \cdot x)^{-\alpha} e^{\langle y, e+i\cdot x\rangle}\, dx, \quad y \in \mathbb{R}^n, \tag{13}$$

where $p(e + i \cdot x)^{-\alpha} = e^{-\alpha(\log|p(e+i\cdot x)| + i \arg p(e+i\cdot x))}$. Then $q_\alpha(y)$ is independent of the choice of $e \in C$ and vanishes if $y \notin C^*$. If $\mathrm{Re}(\alpha) > n + k$ *then $q_\alpha(y)$ has continuous derivatives of order k as a function of y, and is analytic in α for fixed y. Moreover,*

$$p(x)^{-\alpha} = \int_{C^*} e^{-\langle y, x\rangle} q_\alpha(y)\, dy \qquad \text{for } x \in C. \tag{14}$$

The function $q_\alpha(y)$ looks like a Riesz kernel for $f = p^{-\alpha}$. However, it might lack one crucial property: we do not yet know whether $q_\alpha(y)$ is nonnegative for $y \in C^*$. If this holds then $f = p^{-\alpha}$ is completely monotone, (13) gives a formula for the Riesz kernel of f, and (14) is the integral representation of f that is promised in Theorem 2.5.

Remark 4.9 The condition $\mathrm{Re}(\alpha) > n$ is only sufficient for q_α to be a well-defined function. However, for any hyperbolic p and any $\alpha \in \mathbb{C}$, the formula (13) defines a distribution on C^*; see [2, Section 4]. If nonnegative, then this is the Riesz measure.

The following conjecture is important for statistical applications of the models in [14].

Conjecture 4.10 (Conjecture 3.5 in [14]) Let $p \in \mathbb{R}[x_1, \ldots, x_n]$ be a complete hyperbolic polynomial with hyperbolicity cone C. Then there exists a real $\alpha > n$ such that $q_\alpha(y)$ in (13) is nonnegative on the dual cone C^*. In particular, $q_\alpha(y)$ is a Riesz kernel.

Conjecture 4.10 holds true for all hyperbolic polynomials that admit a symmetric determinantal representation, by Corollary 4.2. This raises the question whether such a representation exists for every hyperbolic polynomial. The answer is negative.

Example 4.11 ($n = 4$) Consider the *specialized Vámos polynomial* in [10, Section 3]:

$$p(x) = x_1^2 x_2^2 + 4(x_1 + x_2 + x_3 + x_4)(x_1 x_2 x_3 + x_1 x_2 x_4 + x_1 x_3 x_4 + x_2 x_3 x_4).$$

It is known that no power of $p(x)$ admits a symmetric determinantal representation.

Example 4.12 For any $m = 0, 1, \ldots, n$ the mth elementary symmetric polynomial $E_{m,n}(x) = \sum_{1 \le i_1 < \cdots < i_m \le n} x_{i_1} \cdots x_{i_m}$ is hyperbolic with respect to $e = (1, \ldots, 1) \in \mathbb{R}^n$. The hyperbolicity cone of $E_{m,n}$ contains the orthant $\mathbb{R}^n_{>0}$. It is known that $E_{m,n}$ has no symmetric determinantal representation when $2 \le m \le n-2$, see [11, Example 5.10] for a proof. Exponential varieties associated with $E_{m,n}$ are studied in [14, Section 6].

In Section 6 we prove Conjecture 4.10 for all elementary symmetric polynomials. This is nontrivial because Corollary 4.2 does not apply to $E_{m,n}$ when $2 \le m \le n-2$. Scott and Sokal [19] gave a conjectural description of the set of parameters $\alpha > 0$ for which the function $E_{m,n}^{-\alpha}$ is completely monotone on $\mathbb{R}^n_{>0}$. For a warm-up see Example 2.8.

Conjecture 4.13 (Conjecture 1.11 in [19]) Let $2 \le m \le n$. Then $E_{m,n}^{-\alpha}$ is completely monotone on the positive orthant $\mathbb{R}^n_{>0}$ if and only if $\alpha = 0$ or $\alpha \ge (n - m)/2$.

Scott and Sokal settled Conjecture 4.13 for $m = 2$. However, they "have been unable to find a proof of either the necessity or the sufficiency" in general, as they stated in [19, page 334]. In Section 6 we prove the only if direction of Conjecture 4.13.

Remark 4.14 The positive orthant $\mathbb{R}^n_{>0}$ is strictly contained in the hyperbolicity cone of the elementary symmetric polynomial $E_{m,n}$ unless $m = n$. The next proposition ensures that we can replace $\mathbb{R}^n_{>0}$ by the hyperbolicity cone in the formulation of Conjecture 4.13.

Proposition 4.15 *Let p be a hyperbolic polynomial with hyperbolicity cone C and let $K \subset C$ be an open convex subcone. Then, for any fixed $\alpha > 0$,* *the function $p^{-\alpha}$ is completely monotone on C if and only if it is completely monotone on the subcone K.*

Proof The only if direction is obvious. For the if direction, we proceed as follows. For any α, by results in [2, Section 2], the function $p^{-\alpha} : C \to \mathbb{R}_{>0}$ is the Laplace transform of a distribution supported on C^*. For $\alpha > n$ this already follows from Theorem 4.8. If $p^{-\alpha}$ is completely monotone on an open subcone $K \subset C$, then, by Theorem 2.5, it is the Laplace transform of a unique Borel measure supported on K^*. But since $C^* \subset K^*$ and since the Laplace transform on the space of distributions supported on K^* is injective [18, Proposition 6], the above implies that this Borel measure is supported on C^*. We thus conclude that $p^{-\alpha}$ is completely monotone on C by Theorem 2.5. □

5 Convolution Algebras

In this section we examine the convolution of measures supported in a cone. This is a commutative product. It is mapped to multiplication of functions under the Laplace transform. We obtain isomorphisms of commutative algebras between convolution algebras of Riesz kernels and algebras of functions they represent. This allows to derive relations among Riesz kernels of completely monotone functions. In the setting of Section 3, we obtain a realization of the *Orlik–Terao algebra* [15] as a convolution algebra.

As before, we fix an open convex cone C in $\mathbb{R}^n$. We write $\mathcal{M}_+(C)$ for the set of locally compact Borel measures that are supported in the closed dual cone $C^* \subset (\mathbb{R}^n)^*$. These hypotheses on the measures $\mu \in \mathcal{M}_+(C)$ stipulate that $0 \leq \mu(K) < +\infty$ for any compact set $K \subset (\mathbb{R}^n)^*$, and $\mu(E) = 0$ for any Borel set $E \subset (\mathbb{R}^n)^* \backslash C^*$.

Remark 5.1 The set $\mathcal{M}_+(C)$ is a convex cone, i.e. if $\mu, \nu \in \mathcal{M}_+(C)$ and $a, b \in \mathbb{R}_{\geq 0}$, then $a \cdot \mu + b \cdot \nu \in \mathcal{M}_+(C)$.

Given two measures $\mu, \nu \in \mathcal{M}_+(C)$, one defines their *convolution* $\mu * \nu$ as follows:

$$(\mu * \nu)(E) \;=\; \int_{\mathbb{R}^n} \int_{\mathbb{R}^n} \chi_E(y+z)\, d\mu(y) d\nu(z), \tag{15}$$

where $E \subset (\mathbb{R}^n)^*$ is any Borel subset and χ_E denotes its characteristic function.

Lemma 5.2 *If $\mu, \nu \in \mathcal{M}_+(C)$ then $\mu * \nu \in \mathcal{M}_+(C)$. The convolution is commutative and associative, that is, $\mu * \nu = \nu * \mu$ and $(\mu * \nu) * \xi = \mu * (\nu * \xi)$ for $\mu, \nu, \xi \in \mathcal{M}_+(C)$.*

Proof The first statement holds because the preimage of a compact set under the addition map $C^* \times C^* \to C^*$, $(y,z) \mapsto y+z$ is compact. This ensures that $\mu * \nu$ is a locally compact Borel measure supported in C^*. The second statement, namely commutativity and associativity of the convolution, follows from Tonelli's Theorem. □

We can also check that the convolution product is distributive with respect to addition of measures. In light of Remark 5.1 and Lemma 5.2, this means that $(\mathcal{M}_+(C), +, *)$ is a semiring. We turn it into a ring by the following standard construction. Let $\mathcal{M}(C) = \mathcal{M}_+(C) - \mathcal{M}_+(C)$ denote the set of all $\mathbb{R}$-linear combinations of measures in $\mathcal{M}_+(C)$. Then we extend the convolution (15) by bilinearity to $\mathcal{M}(C)$.

We conclude that $(\mathcal{M}(C), +, *)$ is a commutative $\mathbb{R}$-algebra. If we are given a finite collection of m measures in $\mathcal{M}_+(C)$, then the subalgebra they generate is the quotient $\mathbb{R}[z_1, \dots, z_m]/I$ of a polynomial ring modulo an ideal I. In this representation, the tools of computer algebra, such as *Gröbner bases*, can be applied to the study of measures.

Example 5.3 Let $n = 1$ and $C = \mathbb{R}_{>0}$. Fix rational numbers $\alpha_1, \dots, \alpha_m > 0$ and let μ_i be the measure in $\mathcal{M}_+(\mathbb{R}_{>0})$ with density $y^{\alpha_i}/\Gamma(\alpha_i)$. Additive relations among the α_i translate into multiplicative relations among the μ_i. This follows from Example 5.8 below. The algebra generated by these measures is isomorphic to that generated by the monomials $x^{-\alpha_1}, \dots, x^{-\alpha_m}$. For instance, if $m = 3$ and $\alpha_i = i + 2$ for $i = 1, 2, 3$ then

$$\begin{aligned} (\mathbb{R}[\mu_1, \mu_2, \mu_3], +, *) &\simeq \mathbb{R}[x^{-3}, x^{-4}, x^{-5}] \\ &\simeq \mathbb{R}[z_1, z_2, z_3]/\langle z_2^2 - z_1 z_3,\, z_1^2 z_2 - z_3^2,\, z_1^3 - z_2 z_3 \rangle. \end{aligned}$$

The convolution product has the following nice interpretation in probability theory.

Remark 5.4 Let X_1 and X_2 be independent random variables with values in the cone C, and let μ_1 and μ_2 be their probability measures. Then the probability measure $\mu_{X_1+X_2}$ of their sum $X_1 + X_2$ is the convolution of the two measures, that is,

$$\mu_{X_1+X_2} = \mu_{X_1} * \mu_{X_2}.$$

The convolution of probability measures corresponds to adding random variables.

It is instructive to verify this statement when the random variables are discrete.

Example 5.5 ($n = 1$) Let X_1 and X_2 be independent *Poisson random variables* with parameters $\lambda_1, \lambda_2 > 0$ respectively. Thus μ_{X_1} and μ_{X_2} are atomic measures supported on the set of nonnegative integers, with $\mu_{X_i}(\{k\}) = \lambda_i^k e^{-\lambda_i}/k!$ for $k = 0, 1, \ldots$. Then

$$\begin{aligned} \mu_{X_1+X_2}(\{k\}) &= (\mu_{X_1} * \mu_{X_2})(\{k\}) = \sum_{k_1+k_2=k} \mu_{X_1}(\{k_1\}) \cdot \mu_{X_2}(\{k_2\}) \\ &= \sum_{k_1+k_2=k} \frac{\lambda_1^{k_1} e^{-\lambda_1}}{k_1!} \frac{\lambda_2^{k_2} e^{-\lambda_2}}{k_2!} = \frac{(\lambda_1+\lambda_2)^k e^{-(\lambda_1+\lambda_2)}}{k!}. \end{aligned}$$

Hence $X_1 + X_2$ is a Poisson random variable with parameter $\lambda_1 + \lambda_2$. This fact is well-known in probability. In this example, $C = \mathbb{R}_{>0}$ and $\mu_{X_1}, \mu_{X_2}, \mu_{X_1} * \mu_{X_2} \in \mathcal{M}_+(C)$.

Let $\mu \in \mathcal{M}_+(C)$. If the integral

$$\mathcal{L}\{\mu\}(x) := \int_{C^*} e^{-\langle y,x\rangle} d\mu(y)$$

converges for all $x \in C$, we say that the measure μ has a *Laplace transform* $\mathcal{L}\{\mu\} : C \to \mathbb{R}_{>0}$.

Remark 5.6 By the Bernstein–Hausdorff–Widder–Choquet Theorem 2.5, the Laplace transform is a completely monotone function on C, and if $\mathcal{L}\{\mu\} = \mathcal{L}\{\nu\}$ then $\mu = \nu$.

The Laplace transform takes convolutions of measures to products of functions.

Proposition 5.7 *If $\mu, \nu \in \mathcal{M}_+(C)$ have Laplace transforms, then so does $\mu * \nu$, and*

$$\mathcal{L}\{\mu * \nu\}(x) = \mathcal{L}\{\mu\}(x) \cdot \mathcal{L}\{\nu\}(x) \quad \textit{for all } x \in C.$$

In particular, the product of two completely monotone functions on C is completely monotone, and its Riesz measure is the convolution of the individual Riesz measures.

Proof The definition in (15) and Tonelli's Theorem imply that, for any $x \in C$,

$$\begin{aligned} \mathcal{L}\{\mu * \nu\}(x) &= \int_{C^*} e^{-\langle w,x\rangle} d(\mu * \nu)(w) = \int_{C^*} \left(\int_{C^*} e^{-\langle y,x\rangle} d\mu(y) \right) e^{-\langle z,x\rangle} d\nu(z) \\ &= \int_{C^*} e^{-\langle y,x\rangle} d\mu(y) \int_{C^*} e^{-\langle z,x\rangle} d\nu(z) = \mathcal{L}\{\mu\}(x) \cdot \mathcal{L}\{\nu\}(x). \end{aligned}$$

The assertion in the second sentence follows from Theorem 2.5 and Remark 5.6. □

Example 5.8 Fix $n = 1$ and let μ_i be the measure on $\mathbb{R}_{>0}$ with density $y^{\alpha_i - 1}/\Gamma(\alpha_i)$. By equation (7), its Laplace transform is the monomial $\mathcal{L}\{\mu_i\}(x) = x^{-\alpha_i}$. Thus the assignment $\mu_i \mapsto \mathcal{L}\{\mu_i\}$ gives the isomorphism of $\mathbb{R}$-algebras promised in Example 5.3.

Here we are tacitly using the following natural extension of the Laplace transform from the semiring $\mathcal{M}_+(C)$ to the full $\mathbb{R}$-algebra $\mathcal{M}(C)$. If $\mu = \mu_+ - \mu_- \in \mathcal{M}(C)$, where the measures $\mu_+, \mu_- \in \mathcal{M}_+(C)$ have Laplace transforms, then $\mathcal{L}\{\mu\} := \mathcal{L}\{\mu_+\} - \mathcal{L}\{\mu_-\}$.

Now, let $\mu_1, \ldots, \mu_m$ be measures in $\mathcal{M}_+(C)$ that have Laplace transforms. We write $\mathbb{R}[\mu_1, \ldots, \mu_m]$ for the $\mathbb{R}$-algebra they generate with respect to the convolution product $*$. This is a subalgebra of the commutative algebra $(\mathcal{M}(C), +, *)$. By Proposition 5.7, the Laplace transform is an algebra homomorphism from $\mathbb{R}[\mu_1, \ldots, \mu_m]$ into the algebra of $\mathcal{C}^\infty$-functions on C. Moreover, by Remark 5.6, this homomorphism has trivial kernel.

This construction allows us to transfer polynomial relations among completely monotone functions to polynomial relations among their Riesz kernels, and vice versa. In the remainder of this section, we demonstrate this for the scenario in Section 3.

Fix m linear forms $\ell_1, \ldots, \ell_m \in (\mathbb{R}^n)^*$ and let us denote by $C = \{\ell_1 > 0, \ldots, \ell_m > 0\} \subset \mathbb{R}^n$ the polyhedral cone they define. For all $i = 1, \ldots, m$, and all x with $\ell_i(x) > 0$, we have

$$\ell_i^{-1}(x) = \int_{\mathbb{R}_{\geq 0}\ell_i} e^{-\langle y, x\rangle} d\mu_i(y) = \int_{\mathbb{R}_{\geq 0}} e^{-t\ell_i(x)}\, dt. \tag{16}$$

Let μ_i be the Lebesgue measure dt on the ray $\mathbb{R}_{\geq 0}\ell_i = \{t\ell_i : t > 0\}$, viewed as a measure on C^*. By (16), the Laplace transform of this measure is the reciprocal linear form:

$$\mathcal{L}\{\mu_i\} = \ell_i^{-1}.$$

The subalgebra of $\mathcal{M}(C)$ generated by $\mu_1, \ldots, \mu_m$ is isomorphic, via Laplace transform, to the subalgebra of the algebra of rational functions on $\mathbb{R}^n$ generated by $\ell_1^{-1}, \ldots, \ell_m^{-1}$. This algebra was introduced in [15]. It is known as the *Orlik–Terao algebra* of $\ell_1, \ldots, \ell_m$.

Corollary 5.9 *The convolution algebra $\mathbb{R}[\mu_1, \ldots, \mu_m]$ is isomorphic to the Orlik–Terao algebra, and therefore to $\mathbb{R}[z_1, \ldots, z_m]/I$, where I is the Proudfoot-Speyer ideal in [16]. Its monomials $\mu_{i_1} * \mu_{i_2} * \cdots * \mu_{i_s}$, where*

$\ell_{i_1}, \ell_{i_2}, \ldots, \ell_{i_s}$ runs over multisubsets of linear forms that span $(\mathbb{R}^n)^*$, are the piecewise polynomial volume functions in Theorem 3.3.

Indeed, Proudfoot and Speyer [16] gave an excellent presentation of the Orlik–Terao algebra by showing that the circuit polynomials form a universal Gröbner basis of I.

Example 5.10 (n=3, m=5) Let $\ell_1, \ldots, \ell_5$ be the linear forms in Example 3.2. Then

$$\mathbb{R}[\mu_1, \ldots, \mu_5] \quad = \quad \mathbb{R}[z_1, \ldots, z_5]/I. \tag{17}$$

The Proudfoot-Speyer ideal I is generated by its universal Gröbner basis

$$\begin{gathered}\{z_1z_2z_3 - 2z_1z_2z_4 + 3z_1z_3z_4 - 2z_2z_3z_4,\ z_1z_2z_3 - z_1z_2z_5 + 2z_1z_3z_5 - 2z_2z_3z_5, \\ 2z_1z_2z_4 - z_1z_2z_5 + z_1z_4z_5 - 2z_2z_4z_5,\ 3z_1z_3z_4 - 2z_1z_3z_5 + z_1z_4z_5 - 2z_3z_4z_5, \\ z_2z_3z_4 - z_2z_3z_5 + z_2z_4z_5 - z_3z_4z_5\}.\end{gathered}$$

Note that these five cubics are the circuits in I. The monomial $z_1z_2z_3z_4z_5$ in the convolution algebra (17) represents the piecewise quadratic function $q(y)$ in Example 3.5.

6 Elementary Symmetric Polynomials

In this section we study complete monotonicity of inverse powers of the elementary symmetric polynomials $E_{m,n}$. In Theorem 6.4 we prove Conjecture 4.10 for this special class of hyperbolic polynomials. In Theorem 6.6 we prove the only if direction of Conjecture 4.13. These results resolve questions raised by Scott and Sokal in [19].

Our first goal is to show that sufficiently negative powers of $E_{m,n}$ are completely monontone. We begin with a lemma by Scott and Sokal which is derived from Theorem 2.5.

Lemma 6.1 (Lemma 3.3 in [19]) *Fix a positive $\alpha > 0$, an open convex cone $C \subset \mathbb{R}^n$, and C^∞ functions $A, B : C \to \mathbb{R}_{>0}$. The function $(x, y) \mapsto (A(x) + B(x)y)^{-\alpha}$ is completely monotone on $C \times \mathbb{R}_{>0}$ if and only if $B^{-\alpha}e^{-tA/B}$ is completely monotone on C for all $t \geq 0$.*

Remark 6.2 If $A + By$ is a homogeneous polynomial, then both conditions above are equivalent to complete monotonicity of the function $B^{-\alpha}e^{-tA/B}$: $C \to \mathbb{R}_{>0}$ for $t = 0$ and some $t > 0$.

We also need the following generalization of Lemma 3.9 in [19].

Lemma 6.3 *Fix two cones C and C' and $\alpha > 0$. Let A, B be C^∞ functions on C' such that $g = (A + By)^{-\alpha}$ is completely monotone on $C' \times \mathbb{R}_{>0}$ with Riesz kernel q'. Let $f(x,r)$ be a completely monotone function on $C \times \mathbb{R}_{>0}$ with Riesz kernel q. Then the function $B^{-\alpha} f(x, A/B)$ is completely monotone on $C \times C'$, with Riesz kernel*

$$(z, w) \mapsto \int_{\mathbb{R}_{\geq 0}} \frac{\Gamma(\alpha)}{s^{\alpha-1}} q(z,s) q'(w,s) ds. \tag{18}$$

Proof By Theorem 2.5, our functions admit the following integral representations:

$$f(x,r) = \int_{C^* \times \mathbb{R}_{\geq 0}} e^{-\langle z,x\rangle - s\cdot r}\, q(z,s) dz ds,$$

$$(A + By)^{-\alpha} = \int_{C'^* \times \mathbb{R}_{\geq 0}} e^{-\langle w,u\rangle - s\cdot y}\, q'(w,s) dw ds.$$

From the first equation we get

$$B^{-\alpha} f(x, A/B) = \int_{C^* \times \mathbb{R}_{\geq 0}} e^{-\langle z,x\rangle} (B^{-\alpha} e^{-s\cdot A/B})\, q(z,s) dz ds. \tag{19}$$

We have $(A + By)^{-\alpha} = \Gamma(\alpha)^{-1} \int_{\mathbb{R}_{\geq 0}} e^{-s\cdot y} s^{\alpha-1} B^{-\alpha} e^{-s\cdot A/B} ds$ for fixed A, B. This follows from (7) by setting $x = A + By$ and changing the variable of integration to s/B. By comparing the two integral representations of $(A + By)^{-\alpha}$, and by using the injectivity of the Laplace transform, we find

$$B^{-\alpha} e^{-s\cdot A/B} = \int_{C'^*} e^{-\langle w,u\rangle} \frac{\Gamma(\alpha)}{s^{\alpha-1}} q'(w,s) dw.$$

Substituting this expression into (19), we obtain

$$B^{-\alpha} f(x, A/B) = \int_{C^* \times \mathbb{R}_{\geq 0}} e^{-\langle z,x\rangle} \left(\int_{C'^*} e^{-\langle w,u\rangle} \frac{\Gamma(\alpha)}{s^{\alpha-1}} q'(w,s) dw \right) q(z,s) dz ds.$$

All functions we consider are nonnegative, so we can apply Tonelli's Theorem and get

$$B^{-\alpha} f(x, A/B) = \int_{C^* \times C'^*} e^{-\langle z,x\rangle - \langle w,u\rangle} \left(\int_{\mathbb{R}_{\geq 0}} \frac{\Gamma(\alpha)}{s^{\alpha-1}} q'(w,s) q(z,s) ds \right) dw dz.$$

The parenthesized expression is the desired Riesz kernel in (18). □

We are now ready to prove Conjecture 4.10 for elementary symmetric polynomials. Our proof is constructive, i.e., it yields an explicit formula for the associated Riesz kernel (13). However, the construction is quite complicated, as Example 6.5 shows.

Theorem 6.4 *For any elementary symmetric polynomial $E_{m,n}$, where $1 \leq m \leq n$, there exists a real number $\alpha' > 0$ such that $E_{m,n}^{-\alpha}$ is completely monotone for all $\alpha \geq \alpha'$.*

Proof If $m = 1$ or $m = n$, then $E_{m,n}^{-\alpha}$ is completely monotone on $\mathbb{R}^n_{>0}$ for any $\alpha \geq 0$ (see, e.g., Proposition 2.7). Also $E_{2,n}^{-\alpha}$ is completely monotone for $\alpha \geq (n-2)/2$, by [19, Corollary 1.10]. For $2 < m < n$ we proceed by induction on m. We have

$$E_{m,n} = E_{m,n-1} + E_{m-1,n-1} \cdot y \tag{20}$$

where $y = x_n$ and the other variables in (20) are $x_1, \ldots, x_{n-1}$. We apply Lemma 6.1.

We must prove that there exists $\alpha_{m,n} > 0$ such that, for all $\alpha \geq \alpha_{m,n}$ and all $t \geq 0$,

$$E_{m-1,n-1}^{-\alpha} e^{-tE_{m,n-1}/E_{m-1,n-1}} \text{ is completely monotone on } \mathbb{R}^{n-1}_{>0}.$$

One derives the following factorization, which holds for any fixed $t \geq 0$:

$$E_{m-1,n-1}^{-(n-1)\alpha} e^{-tmE_{m,n-1}/E_{m-1,n-1}} = \prod_{i=1}^{n-1} E_{m-1,n-1}^{-\alpha} e^{-tQ_i/E_{m-1,n-1}}, \tag{21}$$

where $Q_i = x_i E_{m-1,n-2}(x_1, \ldots, \hat{x_i}, \ldots, x_{n-1})$. The hat means that x_i is omitted. We claim that, for each i, the function $E_{m-1,n-1}^{-\alpha} e^{-tQ_i/E_{m-1,n-1}}$ is completely monotone on $\mathbb{R}^{n-1}_{>0}$ provided $\alpha \geq \max(1/2, \alpha_{m-1,n-1})$. Then, by Proposition 5.7, the product (21) is completely monotone on $\mathbb{R}^{n-1}_{>0}$ for $t \geq 0$, and we take $\alpha_{m,n} = (n-1)\max(1/2, \alpha_{m-1,n-1})$.

Now, by symmetry, it suffices to show that $E_{m-1,n-1}^{-\alpha} e^{-tQ_{n-1}/E_{m-1,n-1}}$ is completely monotone. This is equivalent, by Lemma 6.1, to complete monotonicity of $P^{-\alpha}$, where $P = Q_{n-1} + x_n E_{m-1,n-1}$. By (20), $E_{m-1,n-1} = E_{m-1,n-2} + x_{n-1} E_{m-2,n-2}$. This implies

$$\begin{aligned} P &= Q_{n-1} + x_n E_{m-1,n-1} = x_{n-1} x_n E_{m-2,n-2} + (x_{n-1} + x_n) E_{m-1,n-2} \\ &= E_{m-2,n-2} E_{2,3}(x_{n-1}, x_n, E_{m-1,n-2}/E_{m-2,n-2}). \end{aligned}$$

Let us fix any $\alpha \geq \max(1/2, \alpha_{m-1,n-1})$. We apply Lemma 6.3 to the functions $f = E_{2,3}^{-\alpha}$ and $g = E_{m-1,n-1}^{-\alpha} = (E_{m-1,n-2} + x_{n-1}E_{m-2,n-2})^{-\alpha}$. By [19, Corollary 1.10] and the induction hypothesis respectively, these functions are completely monotone. This implies, in particular, that $E_{m-2,n-2}^{-\alpha} E_{2,3}(x_{n-1}, x_n, E_{m-1,n-2}/E_{m-2,n-2})^{-\alpha} = P^{-\alpha}$ is completely monotone. □

The first new case of complete monotonicity concerns large negative powers of $E_{3,5}$.

Example 6.5 We here illustrate our proof of Theorem 6.4 by deriving the Riesz kernel for $E_{3,5}^{-\beta}$ from its steps. By [19, Corollary 1.10], the functions $E_{2,4}^{-\alpha}$ and $E_{2,3}^{-\alpha}$ are completely monotone for any $\alpha > 1$. By [19, Corollary 5.8], their Riesz kernels are

$$q_1(\alpha)(y_1, y_2, y_3, y_4) = \frac{3^{\frac{3}{2}-\alpha}}{2\pi\Gamma(\alpha)\Gamma(\alpha-1)}\left(E_{2,4}(y_1, y_2, y_3, y_4) - (y_1^2 + y_2^2 + y_3^2 + y_4^2)\right)^{\alpha-2},$$

$$q_2(\alpha)(y_1, y_2, y_3) = \frac{2^{1-\alpha}}{(2\pi)^{\frac{1}{2}}\Gamma(\alpha)\Gamma(\alpha-\frac{1}{2})}\left(E_{2,3}(y_1, y_2, y_3) - \frac{1}{2}(y_1^2 + y_2^2 + y_3^2)\right)^{\alpha-\frac{3}{2}}.$$

We apply Lemma 6.3 to

$$g = E_{2,4}^{-\alpha} = (x_1x_2 + x_1x_3 + x_2x_3 + (x_1 + x_2 + x_3)x_4)^{-\alpha}$$

with $y = x_4$, and to $f = E_{2,3}^{-\alpha} = (x_4x_5 + (x_4 + x_5)x_3)^{-\alpha}$ with $y = x_3$. We conclude that

$$\begin{aligned}(x_1 + x_2 + x_3)^{-\alpha}&(x_4x_5 + (x_4 + x_5)(x_1x_2 + x_1x_3 + x_2x_3)/(x_1 + x_2 + x_3))^{-\alpha}\\ &= (x_1x_2x_4 + x_1x_3x_4 + x_2x_3x_4 + x_5(x_1x_4 + x_2x_4 + x_3x_4 + x_1x_2 + x_1x_3 + x_2x_3))^{-\alpha}\\ &= (Q_4 + x_5E_{2,4})^{-\alpha}\end{aligned} \tag{22}$$

is completely monotone for $\alpha > 1$. It has the Riesz kernel

$$q(\alpha)(y_1, \ldots, y_5) = \int_{\mathbb{R}_{\geq 0}} \frac{\Gamma(\alpha)}{s^{\alpha-1}} q_1(\alpha)(y_1, y_2, y_3, s) q_2(\alpha)(y_4, y_5, s) ds.$$

By Lemma 6.1 applied to (22), we derive, for any $t \geq 0$, the complete monotonicity of

$$E_{2,4}^{-\alpha} \cdot e^{-tQ_4/E_{2,4}} = E_{2,4}^{-\alpha} \cdot e^{-t(x_1x_2x_4 + x_1x_3x_4 + x_2x_3x_4)/E_{2,4}}.$$

Proceeding as in the proof of Lemma 6.3, we express its Riesz kernel $R_4(\alpha, t)$ as follows

$$R_4(\alpha, t)(y_1, y_2, y_3, y_4) = \frac{\Gamma(\alpha)}{t^{\alpha-1}} q(\alpha)(y_1, y_2, y_3, y_4, t). \tag{23}$$

In the same way one obtains complete monotonicity, for any $t \geq 0$, of the functions

$$\begin{aligned}E_{2,4}^{-\alpha}e^{-tQ_1/E_{2,4}} &= E_{2,4}^{-\alpha}e^{-t(x_1x_2x_3 + x_1x_2x_4 + x_1x_3x_4)/E_{2,4}},\\ E_{2,4}^{-\alpha}e^{-tQ_2/E_{2,4}} &= E_{2,4}^{-\alpha}e^{-t(x_1x_2x_3 + x_1x_2x_4 + x_2x_3x_4)/E_{2,4}},\\ E_{2,4}^{-\alpha}e^{-tQ_3/E_{2,4}} &= E_{2,4}^{-\alpha}e^{-t(x_1x_2x_3 + x_1x_3x_4 + x_2x_3x_4)/E_{2,4}}.\end{aligned}$$

To obtain their Riesz kernels $R_1(\alpha,t), R_2(\alpha,t), R_3(\alpha,t)$, we exchange y_4 with y_1, y_2, y_3 in (23). Multiplying the four functions together, we obtain complete monotonicity of

$$E_{2,4}^{-4\alpha} e^{-3tE_{3,4}/E_{2,4}} = \prod_{i=1}^{4} E_{2,4}^{-\alpha} e^{-tQ_i/E_{2,4}}. \tag{24}$$

The Riesz kernel of (24) is written as $R(4\alpha, 3t)$. By Proposition 5.7, this is the convolution of $R_1(\alpha,t)$, $R_2(\alpha,t)$, $R_3(\alpha,t)$ and $R_4(\alpha,t)$. By applying Lemma 6.1 to (24) we derive complete monotonicity of $E_{3,5}^{-\beta}$ for $\beta = 4\alpha > 4$. Lemma 6.3 yields the formula

$$q_\beta(y_1,\ldots,y_4,y_5) = \frac{y_5^{\beta-1}}{\Gamma(\beta)} R(\beta, y_5)(y_1,\ldots,y_4)$$

for the Riesz kernel of $E_{3,5}^{-\beta}$. We note that q_β is symmetric in its five arguments.

We now come to our second main result in this section, namely the only if direction in Conjecture 4.13. This was posed by Scott and Sokal. Note that Conjecture 4.13 holds for $m = n$ since $E_{n,n}(x) = x_1 x_2 \cdots x_n$, with Riesz kernel for all negative powers given in Proposition 2.7. We now prove that the condition $\alpha \geq (n-m)/2$ from Conjecture 4.13 is necessary for $E_{m,n}^{-\alpha}$ to be completely monotone on $\mathbb{R}^n_{>0}$.

Theorem 6.6 *Let $2 \leq m < n$. If $E_{m,n}^{-\alpha}$ is completely monotone, then $\alpha = 0$ or $\alpha \geq \frac{n-m}{2}$.*

Proof The proof is by induction on m. The base case $m = 2$ was already established in [19, Corollary 1.10]. Assume that $E_{m,n}^{-\alpha}$ is completely monotone on $\mathbb{R}^n_{>0}$. Then

$$E_{m,n}^{-\alpha}(x_1,x_2,\ldots,x_n) = (x_1 E_{m-1,n-1}(x_2,\ldots,x_n) + E_{m,n-1}(x_2,\ldots,x_n))^{-\alpha}.$$

For large $x_1 > 0$, the sign of any derivative with respect to $x_2,\ldots,x_n$ of the functions $E_{m-1,n-1}^{-\alpha}(x_2,\ldots,x_n)$, $x_2,\ldots,x_n > 0$, and $E_{m,n}^{-\alpha}(x_1,\ldots,x_n)$, $x_1,\ldots,x_n > 0$, is the same (see [19, Lemma 3.1]). It follows that $E_{m-1,n-1}^{-\alpha}$ is completely monotone. By induction, we have $\alpha \geq \frac{n-1-(m-1)}{2} = \frac{n-m}{2}$. This completes the proof of Theorem 6.6. □

7 Hypergeometric Functions

In (2) we started with $f = p_1^{s_1} p_2^{s_2} \cdots p_m^{s_m}$, where p_i is a polynomial in $x = (x_1,\ldots,x_n)$. This expression can be viewed as a function in three different ways. First of all, it is a function in x, with domain C. Second, it is a function in $s = (s_1, \ldots, s_m)$, with domain a subset of $\mathbb{R}^m_{<0}$. And, finally, we can view f as function in the coefficients of the polynomials p_i. It is this third interpretation which occupies us in this final section.

Let us begin with the case $m = 1$ and consider $f = p^s$, for some hyperbolic polynomial

$$p = \sum_{a \in \mathcal{A}} z_a \cdot x_1^{a_1} x_2^{a_2} \cdots x_n^{a_n}. \tag{25}$$

Here $\mathcal{A}$ is a subset of $\mathbb{N}^n$ whose elements have a fixed coordinate sum $d =$ degree(f). We fix $s = -\alpha$ such that $p = f^s$ is completely monotone. We assume that p has Riesz kernel $q(z; y)$, which we consider as a function of the coefficient vector $z = (z_a : a \in \mathcal{A})$.

Let us denote by $\mathcal{D} = \mathbb{C}\langle z_a, \partial_a : a \in \mathcal{A}\rangle$ the *Weyl algebra* on the $|\mathcal{A}|$-dimensional affine space $\mathbb{C}^{\mathcal{A}}$ whose coordinates are the coefficients z_a in (25). We now briefly recall (e.g. from [17]) the definition of the *$\mathcal{A}$-hypergeometric system $H_{\mathcal{A}}(\beta)$* with parameters $\beta \in \mathbb{C}^n$.

The system $H_{\mathcal{A}}(\beta)$ is the left ideal in $\mathcal{D}$ generated by two sets of differential operators:

- the n *Euler operators* $\sum_{a \in \mathcal{A}} a_i z_a \partial_a - \beta_i$, where $i = 1, 2, \ldots, n$;
- the *toric operators* $\prod_{a \in \mathcal{A}} \partial_a^{u_a} - \prod_{a \in \mathcal{A}} \partial_a^{v_a}$, where u_a, v_a are nonnegative integers satisfying $\sum_{a \in \mathcal{A}} (u_a - v_a) a = 0$. Here it suffices to take a *Markov basis* [21] for $\mathcal{A}$.

It is known (cf. [17, Chapter 4]) that $H_{\mathcal{A}}(\beta)$ is regular holonomic and its holonomic rank equals vol$(\mathcal{A})$ for generic parameters β. A sufficiently differentiable function on an open subset of $\mathbb{R}^{\mathcal{A}}$ or $\mathbb{C}^{\mathcal{A}}$ is *$\mathcal{A}$-hypergeometric* if it is annihilated by the toric operators. It is called *$\mathcal{A}$-homogeneous of degree β* if it annihilated also by the n Euler operators.

Proposition 7.1 *The Riesz kernel $q(z; y)$ of $p^{-\alpha}$ is $\mathcal{A}$-hypergeometric in the coefficients z of the polynomial p as in* (25). *However, it is generally not $\mathcal{A}$-homogeneous.*

Proof We use Gårding's integral representation of $q(z; y)$ given in Theorem 4.8. The toric operators annihilate $q(z; y)$ since we can differentiate with respect to z under the integral sign. The fact that the Riesz kernel is generally not $\mathcal{A}$-homogeneous in z can be seen from the explicit formula for quadratic forms p given in [19, Proposition 5.6]. □

We now start afresh and develop an alternative approach for products of linear forms, as in Section 3. The history of hypergeometric functions dates back to 17th century, and there are numerous possible definitions. We describe the approach of Aomoto and Gel'fand [1, 8, 9], albeit in its simplified form via local coordinates on the Grassmannian.

Consider an $n \times m$ matrix $A = (a_{ij})$, with $n \leq m$, where the first $n \times n$ submatrix is the identity. Fix complex numbers $\alpha_1, \ldots, \alpha_m$ that sum to $m - n$. The *hypergeometric function* with parameters α_i is the function in the $n(m-n)$ unknowns a_{ij} defined by:

$$\Phi(\alpha; a_{ij}) := \int_{\mathbb{S}^{n-1}} \prod_{i=1}^{n} (x_i)_+^{\alpha_i - 1} \prod_{j=n+1}^{m} (a_{1j}x_1 + \cdots + a_{nj}x_n)_+^{\alpha_j - 1} \, dx. \quad (26)$$

We integrate over the unit sphere $\mathbb{S}^{n-1} \subset \mathbb{R}^n$ against the standard measure dx. Here, $(x)_+ = \max\{0, x\}$ for $x \in \mathbb{R}$. The integral in (26) is convergent if $\mathrm{Re}(\alpha_i) > 0$. For other values of α_i the hypergeometric function Φ is defined via analytic continuation, see [8] for details. One checks that the following partial differential operators annihilate Φ:

1. Column homogeneity gives the operators $\sum_{i=1}^{n} a_{ij}\partial_{ij} - \alpha_j + 1$ for $n < j \leq m$.
2. Row homogeneity gives the operators $\sum_{j=n+1}^{m} a_{ij}\partial_{ij} + \alpha_i$, $1 \leq i \leq n$.
3. We have the toric operators $\partial_{ij}\partial_{i'j'} - \partial_{ij'}\partial_{i'j}$, $1 \leq i, i' \leq n < j, j' \leq m$.

This means that the function Φ is $\mathcal{A}$-hypergeometric, in the sense defined above, if we take $\mathcal{A}$ to be the vertex set of the product of standard simplices $\Delta_n \times \Delta_{m-n}$.

Example 7.2 ($n = 2, m = 4$) Fix $\alpha_1 = 2 - \alpha_2 - \alpha_3 - \alpha_4$ and consider the matrix

$$A = \begin{pmatrix} 1 & 0 & a & b \\ 0 & 1 & c & d \end{pmatrix}.$$

The hypergeometric function Φ is obtained by integrating a product of four functions, each of which is zero on a half-plane. Hence the integrand is supported on a cone $C \subset \mathbb{R}^2$ defined by two out of four linear functions. Which functions these are, depends on the values of a, b, c, d. The integral over the circle $\mathbb{S}^1$ is an integral over a circular arc, specified by a, b, c, d. This can be written as an integral over a segment in $\mathbb{R}^2$. For instance, consider the range of parameters given by $0 < -\frac{c}{a} < -\frac{d}{b}$ and $0 < c, d$. The boundary lines of the cone C are $x_1 = 0$ and $ax_1 + cx_2 = 0$. This is shown in Figure 1.

Integrating along the segment between $(0, 1)$ and $(-\frac{c}{a}, 1)$, we obtain the formula

$$\Phi(\alpha_1, \alpha_3, \alpha_4; a, b, c, d) = \int_0^{-\frac{c}{a}} x_1^{\alpha_1 - 1}(ax_1 + c)^{\alpha_3 - 1}(bx_1 + d)^{\alpha_4 - 1} dx_1.$$

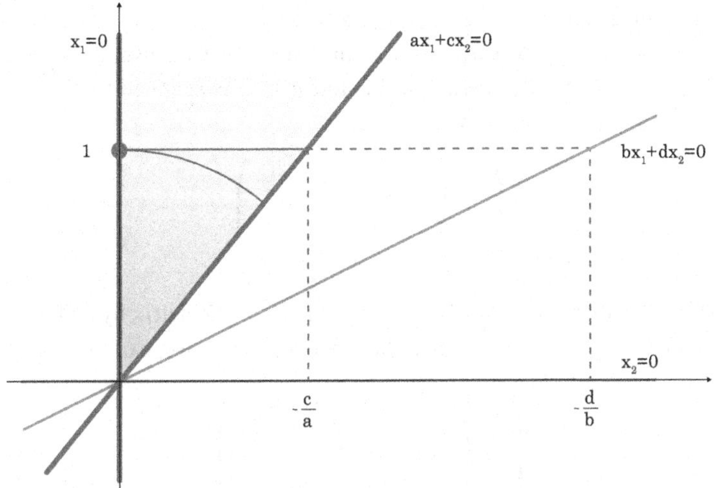

Figure 1 Four linear forms in two variables. They are positive in the shaded region. Integration over the unit circle reduces to integration over the displayed circular arc. We change the integration contour to the horizontal segment.

This integral can be expressed via the classical *Gauss hypergeometric function*

$$_2F_1(a,b,c;z) \;:=\; \frac{\Gamma(c)}{\Gamma(b)\Gamma(c-b)} \int_0^1 x^{b-1}(1-x)^{c-b-1}(1-zx)^{-a}dx.$$

We assume for simplicity that $c > b > 0$. Performing easy integral transformations, given that α_i and a,b,c,d satisfy the assumed inequalities, we obtain:

$$\Phi(\alpha_1,\alpha_3,\alpha_4;a,b,c,d) = (-a)^{-\alpha_1}c^{\alpha_1+\alpha_3-1}d^{\alpha_4-1}\tfrac{\Gamma(\alpha_1)\Gamma(\alpha_3)}{\Gamma(\alpha_1+\alpha_3)}\,{}_2F_1(1-\alpha_4,\alpha_1,\alpha_1+\alpha_3;\tfrac{bc}{ad}).$$

Below we present an example involving three linear forms in two variables. Here the Riesz kernel is expressed in terms of the classical Gauss hypergeometric function $_2F_1$.

Example 7.3 Let $p(x) = x_1x_2(x_1+vx_2)$ with hyperbolicity cone $C = \mathbb{R}^2_{>0}$, where $v > 0$. We consider the function $f = p^{-\alpha}$, where $\alpha > 0$. Then

$$q(y) = \begin{cases} \frac{(y_2/v)^{2\alpha-1}(vy_1-y_2)^{\alpha-1}}{\Gamma(\alpha)\Gamma(2\alpha)}\,{}_2F_1\left(1-\alpha,\alpha;2\alpha;\frac{y_2}{y_2-vy_1}\right), & \text{if } 0 \le y_2 \le vy_1, \\ \frac{y_1^{2\alpha-1}(y_2-vy_1)^{\alpha-1}}{\Gamma(\alpha)\Gamma(2\alpha)}\,{}_2F_1\left(1-\alpha,\alpha,2\alpha;\frac{-vy_1}{y_2-vy_1}\right), & \text{if } 0 \le y_1 \le y_2/v. \end{cases}$$

For the general case, let $\ell_1,\ldots,\ell_m$ be linear forms on $\mathbb{R}^n$ which span a full-dimensional pointed cone C^* in $(\mathbb{R}^n)^*$. After a linear change of coordinates, we can assume that $\ell_{m-n+1},\dots,\ell_m$ is a basis of $(\mathbb{R}^n)^*$, and that each other ℓ_i has coordinates $y_i = (y_{i1},\dots,y_{in})$ in that basis. Consider the projection $\mathbb{R}^m \to \mathbb{R}^n$, $e_i \mapsto \ell_i$. The kernel of this linear map is spanned by the rows of the following $(m-n)\times m$ matrix:

$$\begin{pmatrix} 1 & 0 & \cdots & 0 & -y_1 \\ 0 & 1 & \cdots & 0 & -y_2 \\ \vdots & & \ddots & & \vdots \\ 0 & 0 & \cdots & 1 & -y_{m-n} \end{pmatrix}.$$

We extend the above matrix to an $(m-n+1)\times(m+1)$ matrix by adding a first column $(1,0,\dots,0)$ and a first row that encodes the vector y of unknowns:

$$\begin{pmatrix} 1 & 0 & 0 & \cdots & 0 & y \\ 0 & 1 & 0 & \cdots & 0 & -y_1 \\ 0 & 0 & 1 & \cdots & 0 & -y_2 \\ \vdots & \vdots & & \ddots & & \vdots \\ 0 & 0 & 0 & \cdots & 1 & -y_{m-n} \end{pmatrix}.$$

The $(i+1)$-st column in the above matrix is associated to the linear form y_i, and hence we may associate to it the parameter α_i. We finally define α_0 by the equality $\sum_{i=0}^m \alpha_i = n$. The following formula gives an alternative perspective on Theorem 3.3.

Theorem 7.4 *Using the notation above, the Riesz kernel for the function* $f = \prod_{i=1}^m \ell_i^{-\alpha_i}$ *equals*

$$q(y) \;=\; \frac{\Phi(\alpha; y, -y_1, \dots, -y_{m-n})}{\prod_{i=1}^m \Gamma(\alpha_i)},$$

where the numerator is the Aomoto-Gel'fand hypergeometric function defined in (26).

Proof By the results of Gel'fand and Zelevinsky in [9], the function Φ satisfies

$$\Phi(\alpha; y, -y_1, \dots, -y_{m-n}) \;=\; \frac{1}{|L|}\int_{L^{-1}(y)} \prod_{i=1}^m x_i^{\alpha_i - 1}\, dx.$$

This derivation is non-trivial. Here, $L : \mathbb{R}^m_{\geq 0} \to C^*$ is the linear projection taking the standard basis $e_1,\dots,e_m$ to the linear forms $\ell_1,\dots,\ell_m$. Comparing this expression with the formula for the Riesz kernel given in Theorem 3.3, we obtain the claimed result. □

Theorem 7.4 serves a blueprint for other completely monotone functions (2). We are optimistic that future formulas for Riesz kernels will be inspired by Proposition 7.1.

References

[1] K. Aomoto and M. Kita: *Theory of Hypergeometric Functions*, Springer, New York, 1994.

[2] M. Atiyah, R. Bott, and L. Gårding: Lacunas for hyperbolic differential operators with constant coefficients. I, *Acta Math.* **131** (1973) 145–206.

[3] G. Blekherman, P. Parrilo and R. Thomas: *Semidefinite Optimization and Convex Algebraic Geometry*, MOS-SIAM Series on Optimization **13**, 2012.

[4] G. Choquet: Deux exemples classiques de représentation intégrale, *L'Enseignement Mathématique* **15** (1969) 63–75.

[5] C. De Concini and C. Procesi: *Topics in Hyperplane Arrangements, Polytopes and Box-Splines*, Universitext, Springer, New York, 2010.

[6] M. Dressler, S. Iliman and T. de Wolff: A positivstellensatz for sums of nonnegative circuit polynomials, *SIAM Journal on Applied Algebra and Geometry* **1** (2017) 536–555.

[7] L. Gårding: Linear hyperbolic partial differential equations with constant coefficients, *Acta Mathematica* **85** (1951) 1–62.

[8] I.M. Gel'fand: General theory of hypergeometric functions, *Dokl. Akad. Nauk SSSR* **288** (1986) 14–18.

[9] I.M. Gel'fand and A.V. Zelevinskii: Algebraic and combinatorial aspects of the general theory of hypergeometric functions, *Functional Analysis and Its Applications* **20** (1986) 183–197.

[10] M. Kummer: A note on the hyperbolicity cone of the specialized Vámos polynomial, *Acta Appl. Math.* **144** (2016) 11–15.

[11] M. Kummer, D. Plaumann and C. Vinzant: Hyperbolic polynomials, interlacers, and sums of squares, *Mathematical Programming*, Series B **153** (2015) 223–245.

[12] J. Lasserre and E. Zerron: A Laplace transform algorithm for the volume of a convex polytope, *Journal of the ACM* **48** (2001) 1126–1140.

[13] J. Lasserre and E. Zerron: Solving a class of multivariate integration problems via Laplace transforms, *Applicationes Mathematicae* **28** (2001) 391–405.

[14] M. Michałek, B. Sturmfels, C. Uhler, and P. Zwiernik: Exponential varieties, *Proceedings of the London Mathematical Society* **112** (2016) 27–56.

[15] P. Orlik and H. Terao: Commutative algebras for arrangements, *Nagoya Mathematical Journal*, **134** (1994) 65–73.

[16] N. Proudfoot and D. Speyer: A broken circuit ring, *Beiträge Algebra Geom.* **47** (2006) 161–166.

[17] M. Saito, B. Sturmfels and N. Takayama: *Gröbner Deformations of Hypergeometric Differential Equations*, Algorithms and Computations in Mathematics **6**, Springer, Berlin, 2000.

[18] L. Schwartz: *Theórie des Distributions*, Hermann, Strasbourg, 1966.

[19] A. Scott and A. Sokal: Complete monotonicity for inverse powers of some combinatorially defined polynomials, *Acta Mathematica* **212** (2014) 323–392.

[20] B. Sturmfels and C. Uhler: Multivariate Gaussians, semidefinite matrix completion, and convex algebraic geometry, *Annals of the Institute of Statistical Mathematics* **62** (2010) 603–638.

[21] S. Sullivant: *Algebraic Statistics*, Graduate Studies in Math. **194**, American Math. Society, 2018.

[22] U. Walther: Survey on the D-module f^s, *Commutative Algebra and Noncommutative Algebraic Geometry*, Math. Sci. Res. Inst. Publ. **67**, 391–430, Cambridge Univ. Press, New York, 2015.

[23] D. Widder: *The Laplace Transform*, Princeton University Press, 1946, or Franklin Classics, 2017.

18

On the Coproduct in Affine Schubert Calculus

Thomas Lam[a], Seung Jin Lee[b] and Mark Shimozono[c]

Dedicated to Bill Fulton on the occasion of his 80th birthday.
Thank you, Bill, for your inspirational and visionary work!

Abstract. The cohomology of the affine flag variety $\hat{\mathrm{Fl}}_G$ of a complex reductive group G is a comodule over the cohomology of the affine Grassmannian Gr_G. We give positive formulae for the coproduct of an affine Schubert class in terms of affine Stanley classes and finite Schubert classes, in (torus-equivariant) cohomology and K-theory. As an application, we deduce monomial positivity for the affine Schubert polynomials of the second author.

1 Introduction

Let G be a complex reductive group with maximal torus T and flag variety G/B, and denote by $\xi^v_{G/B}$ the Schubert classes of $H^*_T(G/B)$ (all cohomology rings are taken with integer coefficients), indexed by the finite Weyl group W. Let $\hat{\mathrm{Fl}}_G$ denote the affine flag variety of G and Gr_G denote the affine Grassmannian of G. There is a coaction map

$$\Delta : H^*_T(\hat{\mathrm{Fl}}_G) \to H^*_T(\mathrm{Gr}_G) \otimes_{H^*_T(\mathrm{pt})} H^*_T(\hat{\mathrm{Fl}}_G).$$

It is induced via pullback from the product map of topological spaces $\Omega K \times LK/T_{\mathbb{R}} \to LK/T_{\mathbb{R}}$, where $K \subset G$ is a maximal compact subgroup and $T_{\mathbb{R}} = K \cap T$ is the maximal compact torus. The cohomology ring $H^*_T(\hat{\mathrm{Fl}}_G)$ has Schubert classes ξ^w indexed by the affine Weyl group $\hat{W}$. The inclusion $\varphi : \Omega K \hookrightarrow LK/T_{\mathbb{R}}$ induces a "wrongway" pullback map

$$\varphi^* : H^*_T(\hat{\mathrm{Fl}}_G) \to H^*_T(\mathrm{Gr}_G).$$

[a] University of Michigan
[b] Seoul National University
[c] Virginia Tech

By definition, the equivariant affine Stanley class $F^w \in H_T^*(\mathrm{Gr}_G)$ is given by $F^w := \varphi^*(\xi^w)$. We refer the reader to [18] for further background.

Theorem 1.1 *Let $w \in \hat{W}$. Then we have*

$$\Delta(\xi^w) = \sum_{w \doteq uv} F^u \otimes \xi^v$$

and under the isomorphism $H_T^*(\hat{\mathrm{Fl}}_G) \cong H_T^*(\mathrm{Gr}_G) \otimes_{H_T^*(\mathrm{pt})} H_T^*(G/B)$,

$$\xi^w = \sum_{w \doteq uv} F^u \otimes \xi^v_{G/B}$$

where $u \in \hat{W}$ and $v \in W$ and we write $w \doteq uv$ if $w = uv$ and $\ell(w) = \ell(u) + \ell(v)$.

The class $\xi^v_{G/B}$ is considered an element of $H_T^*(\hat{\mathrm{Fl}}_G)$ via pullback under evaluation at the identity (see (20)). The same formulae hold in non-equivariant cohomology.

In the majority of this article (Sections 2–4) we will work in torus-equivariant K-theory $K_T^*(\hat{\mathrm{Fl}})$ of the affine flag variety. The coproduct formula for holds in (torus-equivariant) K-theory with Demazure product replacing length-additive products (see Theorem 4.7). Our proof relies heavily on the action of the affine nilHecke ring on $K_T^*(\hat{\mathrm{Fl}})$. Let us note that there are a number of different geometric approaches [8, 10, 19] for constructing Schubert classes in $K_T^*(\hat{\mathrm{Fl}})$; see [17, Section 3] for a comparison. However, our results holds at the level of Grothendieck groups and the precise geometric model (thick affine flag variety, thin affine flag variety, or based loop group) is not crucial.

In Section 5, the proofs for the cohomology case are indicated.

There is a long tradition of combinatorial formulae for Schubert classes in cohomology and K-theory using reduced factorizations or Hecke factorizations, dating at least back to [23], see also the references in Section 6. In particular, [2] gives a formula for Schubert polynomials using reduced factorizations and [14] gives a formula for affine Stanley symmetric functions using cyclically decreasing reduced factorizations. In Section 6 we combine these formulae with our Theorem 1.1 to prove (Theorem 6.1) that the affine Schubert polynomials [24] are monomial positive. We explain how the Billey–Haiman formula [3] for type C or D Schubert polynomials (see also [6]) is a consequence of our coproduct formula.

By taking an appropriate limit (see Section 6), the coproduct formula for backstable (double) Schubert polynomials [16] can be deduced from Theorem 1.1. Whereas the proofs in [16] are essentially combinatorial, the present work relies heavily on equivariant localization and the nilHecke algebra.

Acknowledgements. The authors thank an anonymous referee for comments and corrections. T.L. was supported by NSF DMS-1464693 and DMS-1953852, and by a von Neumann Fellowship from the Institute for Advanced Study. S. J. Lee was supported by the National Research Foundation of Korea(NRF) grant funded by the Korea government(MEST) (No. 2019R1C1C1003473). M.S. was supported by NSF DMS-1600653.

2 Affine nilHecke ring and the equivariant K-theory of the affine flag variety

The proofs of our results for a complex reductive group easily reduces to that of a semisimple simply-connected group. To stay close to our main references [8, 19], we work with the latter. Henceforth, we fix a complex semisimple simply-connected group G.

The results of this section are due to Kostant and Kumar [8]. Our notation follows that of [19].

2.1 Small-torus affine K-nilHecke ring

Let $T \subset G$ be the maximal torus with character group, or weight lattice P. We have $P = \bigoplus_{i\in I} \mathbb{Z}\omega_i$ where ω_i denotes a fundamental weight and I denotes the finite Dynkin diagram of G. Let $\hat{P} = \mathbb{Z}\delta \oplus \bigoplus_{i\in\hat{I}} \mathbb{Z}\Lambda_i$ be the affine weight lattice with fundamental weights Λ_i for i in the affine Dynkin node set $\hat{I} = I \cup \{0\}$, and let δ denote the null root. Let $\hat{P}^* = \mathrm{Hom}_{\mathbb{Z}}(\hat{P}, \mathbb{Z})$ be the affine coweight lattice. It has basis dual to $\{\delta\} \cup \{\Lambda_i \mid i \in \hat{I}\}$ given by $\{d\} \cup \{\alpha_i^\vee \mid i \in \hat{I}\}$ where d is the degree generator and the $\alpha_i^\vee$ are the simple coroots. The Cartan matrix $(a_{ij} \mid i, j \in \hat{I})$ is defined by $a_{ij} = \langle \alpha_i^\vee, \alpha_j \rangle$ using the evaluation pairing $\hat{P}^* \times \hat{P} \to \mathbb{Z}$. Let $c \in \hat{P}^*$ be the canonical central element [7, §6.2]. The level of $\Lambda \in \hat{P}$ is defined by $\mathrm{level}(\Lambda) = \langle c, \Lambda \rangle$. The natural projection $\mathrm{cl} : \hat{P} \to P$ has kernel $\mathbb{Z}\delta \oplus \mathbb{Z}\Lambda_0$ and satisfies $\mathrm{cl}(\Lambda_i) = \omega_i$ for $i \in I$. In particular $\mathrm{cl}(\alpha_0) = -\theta$ where θ is the highest root. This induces a map $\mathrm{cl} : \mathbb{Z}[\hat{P}] \to \mathbb{Z}[P]$ between the representation ring of the maximal torus of the affine Kac-Moody group and that of the torus T. Let $\mathrm{af} : P \to \hat{P}$ be the section of cl given by $\mathrm{af}(\omega_i) = \Lambda_i - \mathrm{level}(\Lambda_i)\Lambda_0$ for $i \in I$.

The finite Weyl group W acts naturally on P and on $R(T)$, where $R(T) \cong \mathbb{Z}[P] = \bigoplus_{\lambda\in P} \mathbb{Z}e^\lambda$ is the Grothendieck group of the category of finite-dimensional T-modules, and for $\lambda \in P$, e^λ is the class of the one-dimensional T-module with character λ. Let $Q(T) = \mathrm{Frac}(R(T))$. The affine Weyl group

$\hat{W}$ also acts on P, $R(T)$, and $Q(T)$ via the level-zero action, that is, via the homomorphism $\mathrm{cl}_{\hat{W}} : \hat{W} \cong Q^\vee \rtimes W \to W$ given by $t_\mu v \mapsto v$ for μ in the coroot lattice $Q^\vee$ and $v \in W$. In particular, $s_0 = t_{\theta^\vee} s_\theta$ satisfies $\mathrm{cl}(s_0) = s_\theta$ where $\theta^\vee$ is the coroot associated to θ.

We let $u * v \in \hat{W}$ denote the Demazure (or 0-Hecke) product of $u, v \in \hat{W}$. It is the associative product determined by

$$s_i * v = \begin{cases} s_i v & \text{if } s_i v > v, \\ v & \text{otherwise.} \end{cases} \qquad v * s_i = \begin{cases} v s_i & \text{if } v s_i > v, \\ v & \text{otherwise.} \end{cases}$$

Let $\hat{\mathbb{K}}_{Q(T)}$ be the smash product of the group algebra $\mathbb{Q}[\hat{W}]$ and $Q(T)$, defined by $\hat{\mathbb{K}}_{Q(T)} = Q(T) \otimes_{\mathbb{Q}} \mathbb{Q}[\hat{W}]$ with multiplication

$$(q \otimes w)(p \otimes v) = q(w \cdot p) \otimes wv$$

for $p, q \in Q(T)$ and $v, w \in W$. We write qw instead of $q \otimes w$. Define the elements $T_i \in \hat{\mathbb{K}}_{Q(T)}$ by

$$T_i = (1 - e^{\alpha_i})^{-1}(s_i - 1). \tag{1}$$

In particular $T_0 = (1 - e^{-\theta})^{-1}(t_{\theta^\vee} s_\theta - 1)$. The T_i satisfy

$$T_i^2 = -T_i \quad \text{and} \quad \underbrace{T_i T_j \cdots}_{m_{ij} \text{ factors}} = \underbrace{T_j T_i \cdots}_{m_{ij} \text{ factors}} \tag{2}$$

where m_{ij} is related to the Cartan matrix entries a_{ij} by

$a_{ij}a_{ji}$	0	1	2	3	≥ 4
m_{ij}	2	3	4	6	∞

We have the commutation relation in $\hat{\mathbb{K}}_{Q(T)}$

$$T_i\, q = (T_i \cdot q) + (s_i \cdot q) T_i \qquad \text{for } q \in Q(T). \tag{3}$$

Let $T_w = T_{i_1} T_{i_2} \cdots T_{i_N} \in \hat{\mathbb{K}}_{Q(T)}$ where $w = s_{i_1} s_{i_2} \cdots s_{i_N}$ is a reduced decomposition; it is well-defined by (2). It is easily verified that

$$T_i T_w = \begin{cases} T_{s_i w} & \text{if } s_i w > w \\ -T_w & \text{if } s_i w < w \end{cases} \qquad \text{and} \qquad T_w T_i = \begin{cases} T_{w s_i} & \text{if } w s_i > w \\ -T_w & \text{if } w s_i < w \end{cases}$$

where $<$ denotes the Bruhat order on $\hat{W}$. The algebra $\hat{\mathbb{K}}_{Q(T)}$ acts naturally on $Q(T)$. In particular, one has

$$T_i \cdot (qq') = (T_i \cdot q)q' + (s_i \cdot q) T_i \cdot q' \qquad \text{for } q, q \in Q(T). \tag{4}$$

The 0-Hecke ring $\hat{\mathbb{K}}_0$ is the subring of $\hat{\mathbb{K}}_{Q(T)}$ generated by the T_i over $\mathbb{Z}$. It can also be defined by generators $\{T_i \mid i \in \hat{I}\}$ and relations (2). We have $\hat{\mathbb{K}}_0 = \bigoplus_{w \in \hat{W}} \mathbb{Z} T_w$.

Lemma 2.1 *The ring $\hat{\mathbb{K}}_0$ acts on $R(T)$.*

Proof $\hat{\mathbb{K}}_0$ acts on $Q(T)$, and T_i preserves $R(T)$ by (4) and the following formulae for $\lambda \in P$:

$$T_i \cdot e^\lambda = \begin{cases} e^\lambda(e^{-\alpha_i} + e^{-2\alpha_i} \cdots + e^{-\langle \alpha_i^\vee, \lambda\rangle \alpha_i}) & \text{if } \langle \alpha_i^\vee, \lambda\rangle > 0 \\ 0 & \text{if } \langle \alpha_i^\vee, \lambda\rangle = 0 \\ -e^\lambda(1 + e^{\alpha_i} + \cdots + e^{(-\langle \alpha_i^\vee, \lambda\rangle - 1)\alpha_i}) & \text{if } \langle \alpha_i^\vee, \lambda\rangle < 0. \end{cases} \quad (5)$$

□

Define the *K-NilHecke ring* $\hat{\mathbb{K}}$ to be the subring of $\hat{\mathbb{K}}_{Q(T)}$ generated by $\hat{\mathbb{K}}_0$ and $R(T)$. We have $\hat{\mathbb{K}}_{Q(T)} \cong Q(T) \otimes_{R(T)} \hat{\mathbb{K}}$. By (3), we have

$$\hat{\mathbb{K}} = \bigoplus_{w \in \hat{W}} R(T) T_w. \quad (6)$$

2.2 $\hat{\mathbb{K}}$-$\hat{\mathbb{K}}$-bimodule structure on equivariant K-theory of affine flag variety

We have an isomorphism $K_T^*(\mathrm{pt}) \cong R(T)$. Let $\mathrm{Fun}(\hat{W}, R(T))$ be the $R(T)$-algebra of functions $\hat{W} \to R(T)$ under pointwise multiplication $(\phi\psi)(w) = \phi(w)\psi(w)$ for $\phi, \psi \in \mathrm{Fun}(\hat{W}, R(T))$ and $w \in \hat{W}$, and action $(s\psi)(w) = s\psi(w)$ for $s \in R(T)$, $\psi \in \mathrm{Fun}(\hat{W}, R(T))$ and $w \in \hat{W}$. There is an injective $R(T)$-algebra homomorphism $\mathrm{loc} : K_T^*(\hat{\mathrm{Fl}}) \to \mathrm{Fun}(\hat{W}, R(T))$ sending a class ψ to the function $w \mapsto \psi(w)$ where $\psi(w)$ denotes the localization of ψ at $w \in \hat{W}$. The image of the map loc is characterized by the small torus affine GKM condition of [19, Section 4.2].

There is a perfect left $Q(T)$-bilinear pairing

$$\langle \cdot, \cdot \rangle : \hat{\mathbb{K}}_{Q(T)} \times \mathrm{Fun}(\hat{W}, Q(T)) \to Q(T)$$

defined by evaluation:

$$\langle w, \psi \rangle = \psi(w) \quad (7)$$

for $w \in \hat{W}$ and $\psi \in \mathrm{Fun}(\hat{W}, Q(T))$. Abusing notation, we regard every $\psi \in \mathrm{Fun}(\hat{W}, Q(T))$ as an element of $\mathrm{Hom}_{Q(T)}(\hat{\mathbb{K}}_{Q(T)}, Q(T))$ by formal left $Q(T)$-linearity: for $a = \sum_{w \in \hat{W}} a_w w \in \hat{\mathbb{K}}_{Q(T)}$ with $a_w \in Q(T)$, let

$$\psi(a) = \langle a\,, \psi \rangle = \sum_w a_w \psi(w).$$

Thinking of $K_T^*(\hat{\mathrm{Fl}})$ as an $R(T)$-subalgebra of $\mathrm{Fun}(\hat{W}, Q(T))$, a function ψ lies in $K_T^*(\hat{\mathrm{Fl}})$ if and only if $\psi(\hat{\mathbb{K}}) \subseteq R(T)$. The pairing (7) restricts to a perfect left $R(T)$-bilinear pairing (see [19, (2.10)])

$$\hat{\mathbb{K}} \times K_T^*(\hat{\mathrm{Fl}}) \to R(T). \quad (8)$$

There is a left action $\psi \mapsto a \cdot \psi$ of $\hat{\mathbb{K}}$ on $K^*_T(\hat{\mathrm{Fl}})$ given by the formulae (see [18, Chapter 4, Proposition 3.16] for the very similar cohomology case)

$$(q \cdot \psi)(b) = q\, \psi(b) \tag{9}$$

$$(T_i \cdot \psi)(b) = T_i \cdot \psi(s_i b) + \psi(T_i b) \tag{10}$$

$$(w \cdot \psi)(b) = w\, \psi(w^{-1} b) \tag{11}$$

for $b \in \hat{\mathbb{K}}$, $\psi \in K^*_T(\hat{\mathrm{Fl}})$, $q \in R(T)$, $i \in \hat{I}$, and $w \in \hat{W}$. Here, T_i acts on $R(T)$ as in (4) and (5).

There is another left action $\psi \mapsto a \bullet \psi$ of $\hat{\mathbb{K}}$ on $K^*_T(\hat{\mathrm{Fl}})$ given by [19, §2.4]

$$(a \bullet \psi)(b) = \psi(ba) \tag{12}$$

for $a, b \in \hat{\mathbb{K}}$ and $\psi \in K^*_T(\hat{\mathrm{Fl}})$.

Remark 2.2 For those familiar with the double Schubert polynomial $\mathfrak{S}_w(x;a)$ (or also the double Grothendieck polynomial), the $\cdot$ action is on the equivariant variables a_i and the $\bullet$ action is on the x_i variables.

Let $p : \hat{\mathrm{Fl}} \to \mathrm{Gr}$ be the natural projection and $p^* : K^*_T(\mathrm{Gr}) \to K^*_T(\hat{\mathrm{Fl}})$ the pullback map, which is an injection. A class $\psi \in K^*_T(\hat{\mathrm{Fl}})$ lies in the image of p^* if and only if $\psi(wv) = \psi(w)$ for all $w \in \hat{W}$ and $v \in W$. We abuse notation by frequently identifying a class $\psi_{\mathrm{Gr}} \in K^*_T(\mathrm{Gr})$ with its image under p^*.

Let $[\mathcal{L}_\lambda] \in K^*_T(\hat{\mathrm{Fl}})$ denote the class of the T-equivariant line bundle on $\hat{\mathrm{Fl}}$ of weight λ. Using the level zero action of $\hat{W}$ on $R(T)$ we have [10, (2.5)]

$$\langle t_\mu v\,,\, [\mathcal{L}_\lambda]\rangle = v \cdot e^\lambda = e^{v\lambda} \qquad\qquad \mu \in Q^\vee,\ v \in W. \tag{13}$$

Lemma 2.3 *For any $\lambda \in P$ and $\psi \in K^*_T(\hat{\mathrm{Fl}})$,*

$$e^\lambda \bullet \psi = [\mathcal{L}_\lambda] \cup \psi. \tag{14}$$

Proof Localizing at $t_\mu v$ for $\mu \in Q^\vee$ and $v \in W$, we compute that $\langle t_\mu v\,,\, e^\lambda \bullet \psi\rangle$ is equal to

$$\begin{aligned}\langle t_\mu v e^\lambda\,,\, \psi\rangle &= \langle e^{v\lambda} t_\mu v\,,\, \psi\rangle = e^{v\lambda}\langle t_\mu v\,,\, \psi\rangle \\ &= \langle t_\mu v\,,\, [\mathcal{L}_\lambda]\rangle\langle t_\mu v\,,\, \psi\rangle = \langle t_\mu v\,,\, [\mathcal{L}_\lambda] \cup \psi\rangle.\end{aligned}$$

□

3 Endomorphisms of $K^*_T(\hat{\mathrm{Fl}})$

3.1 Wrong-way map and Peterson subalgebra

Recall that $K \subset G$ is the maximal compact subgroup and $T_{\mathbb{R}} := K \cap T$ is the maximal compact torus. We have $T_{\mathbb{R}}$-equivariant homotopy equivalences between G/B and $K/T_{\mathbb{R}}$, between Gr and the based loop group ΩK, and between $\hat{\mathrm{Fl}}$ and the space $LK/T_{\mathbb{R}}$ [25]. For an ind-variety X with T-action let $K_*^T(X)$ be the T-equivariant K-homology of X, the Grothendieck group of finitely supported T-equivariant coherent sheaves on X [11] [19]. There is a left $R(T)$-module isomorphism $\tau : K_*^T(\hat{\mathrm{Fl}}) \cong \hat{\mathbb{K}}$ given by $\tau(\psi_w) = T_w$, where ψ_w is the ideal sheaf Schubert class for the affine flag ind-variety (see Section 4.1). We give $K_*^T(\hat{\mathrm{Fl}})$ the structure of a noncommutative ring so that τ is a ring isomorphism. This ring structure can also be obtained geometrically from convolution; see [4] for the corresponding statements for $H_*(G/B)$. The K-group $K_*^T(\mathrm{Gr})$ has the structure of a commutative Hopf $R(T)$-algebra. The product is induced from the $T_{\mathbb{R}}$-equivariant product map of the topological group ΩK.

There is a $T_{\mathbb{R}}$-equivariant map $\varphi : \Omega K \to LK \to LK/T_{\mathbb{R}}$ given by inclusion followed by projection. The map φ induces an injective ring and left $R(T)$-module homomorphism $\varphi_* : K_*^T(\mathrm{Gr}) \to K_*^T(\hat{\mathrm{Fl}})$. It also induces an $R(T)$-algebra homomorphism $\varphi^* : K_T^*(\hat{\mathrm{Fl}}) \to K_T^*(\mathrm{Gr})$ which is called the *wrong-way map*, and characterized by (see Lemma 3.3)

$$\varphi^*(\psi)(v) = \psi(t_\mu) \qquad \text{for } v \in \hat{W}^0 \text{ and } t_\mu \in vW.$$

Let $\mathbb{L} = Z_{\hat{\mathbb{K}}}(R(T))$ be the centralizer of $R(T)$ in $\hat{\mathbb{K}}$, called the K-Peterson subalgebra. We have the following basic result [19, Lemma 5.2].

Lemma 3.1 *We have* $\mathbb{L} = \left(\bigoplus_{\mu \in Q^\vee} Q(T) t_\mu\right) \cap \hat{\mathbb{K}}$.

Theorem 3.2 ([19, Theorem 5.3]) *There is an isomorphism* $k : K_*^T(\mathrm{Gr}) \to \mathbb{L}$ *making the following commutative diagram of ring and left* $R(T)$*-module homomorphisms:*

$$\begin{array}{ccc} K_*^T(\mathrm{Gr}) & \xrightarrow{k} & \mathbb{L} \\ \downarrow{\scriptstyle \varphi_*} & & \downarrow \\ K_*^T(\hat{\mathrm{Fl}}) & \xrightarrow[\tau]{} & \hat{\mathbb{K}} \end{array}$$

3.2 Pullback from affine Grassmannian

Recall that $p : \hat{\mathrm{Fl}} \to \mathrm{Gr}$ denotes the natural projection. Define $\theta := p^* \circ \varphi^*$, so that $\theta : K_T^*(\hat{\mathrm{Fl}}) \to K_T^*(\hat{\mathrm{Fl}})$ is the pullback map in equivariant K-theory of the following composition

$$LK/T_{\mathbb{R}} \xrightarrow{p} \Omega K \xrightarrow{\varphi} LK/T_{\mathbb{R}} \tag{15}$$

where abusing notation, we are denoting also by p the natural quotient map $LK/T_{\mathbb{R}} \to LK/K \simeq \Omega K$.

Lemma 3.3 *For all $\mu \in Q^\vee$, $v \in W$, and $\psi \in K_T^*(\hat{\mathrm{Fl}})$ we have*

$$(\theta\psi)(t_\mu v) = \psi(t_\mu). \tag{16}$$

Proof The translation element t_μ defines the based loop given by the cocharacter $\mu \in \mathrm{Hom}_{\mathrm{alg.\,gp}}(\mathbb{C}^*, T)$ evaluated on the unit circle $S^1 \subset \mathbb{C}^*$. The unique $T_{\mathbb{R}}$-fixed point in $t_\mu vK \cap \Omega K$ is t_μ. Thus under the composition (15) $t_\mu v$ maps to t_μ. The Lemma follows by the definition of pullback. □

3.3 Coaction

The inclusion $\Omega K \hookrightarrow LK$ induces an action $\Omega K \times LK/T_{\mathbb{R}} \to LK/T_{\mathbb{R}}$ of ΩK on $LK/T_{\mathbb{R}}$. This action is $T_{\mathbb{R}}$-equivariant where $T_{\mathbb{R}}$ acts diagonally on the direct product, acting on ΩK by conjugation and on $LK/T_{\mathbb{R}}$ by left translation. Applying the covariant functor $K_*^{T_{\mathbb{R}}}$ we obtain the map $K_*^T(\mathrm{Gr}) \otimes_{R(T)} K_*^T(\hat{\mathrm{Fl}}) \cong K_*^T(\mathrm{Gr} \times \hat{\mathrm{Fl}}) \to K_*^T(\hat{\mathrm{Fl}})$. We have the commutative diagram

$$\begin{array}{ccc} K_*^T(\mathrm{Gr}) \otimes_{R(T)} K_*^T(\hat{\mathrm{Fl}}) & \longrightarrow & K_*^T(\hat{\mathrm{Fl}}) \\ \Big\downarrow {\scriptstyle k\otimes\tau} & & \Big\downarrow {\scriptstyle \tau} \\ \mathbb{L} \otimes_{R(T)} \hat{\mathbb{K}} & \xrightarrow[\text{mult}]{} & \hat{\mathbb{K}} \end{array}$$

Via the pairing (8) the dual map is the coproduct

$$\Delta : K_T^*(\hat{\mathrm{Fl}}) \longrightarrow K_T^*(\mathrm{Gr}) \otimes_{R(T)} K_T^*(\hat{\mathrm{Fl}}).$$

Note that $\Delta|_{K_T^*(\mathrm{Gr})}$ is the usual coproduct of $K_T^*(\mathrm{Gr})$, part of the $R(T)$-Hopf algebra structure of $K_T^*(\mathrm{Gr})$, and abusing notation we often denote $\Delta|_{K_T^*(\mathrm{Gr})}$ by Δ. Often, we will think of the image of Δ inside $K_T^*(\hat{\mathrm{Fl}}) \otimes_{R(T)} K_T^*(\hat{\mathrm{Fl}})$ via the inclusion $p^* : K_T^*(\mathrm{Gr}) \to K_T^*(\hat{\mathrm{Fl}})$.

Proposition 3.4 *For all $a \in \mathbb{L}$, $b \in \hat{\mathbb{K}}$, and $\psi \in K_T^*(\hat{\mathrm{Fl}})$ we have*

$$\langle ab\,,\,\psi\rangle = \sum_{(\psi)} \langle a\,,\,\psi_{(1)}\rangle\langle b\,,\,\psi_{(2)}\rangle \tag{17}$$

where

$$\Delta(\psi) = \sum_{(\psi)} \psi_{(1)} \otimes \psi_{(2)}. \tag{18}$$

Proof By definition and using (18) we have

$$\langle ab\,,\,\psi\rangle = \sum_{(\psi)} \langle a\,,\,\psi_{(1)}^{\mathrm{Gr}}\rangle_{\mathrm{Gr}} \langle b\,,\,\psi_{(2)}\rangle$$

where $\langle a\,,\,\psi\rangle_{\mathrm{Gr}}$ is the pairing between $\mathbb{L}$ and $K_T^*(\mathrm{Gr})$ induced by Theorem 3.2 and the duality between $K_*^T(\mathrm{Gr})$ and $K_T^*(\mathrm{Gr})$. But then since $\varphi^* \circ p^*$ is the identity, we have that $\langle a\,,\,\psi^{\mathrm{Gr}}\rangle_{\mathrm{Gr}} = \langle a\,,\,(\varphi^* \circ p^*)(\psi^{\mathrm{Gr}})\rangle_{\mathrm{Gr}} = \langle a\,,\,p^*(\psi^{\mathrm{Gr}})\rangle$. In the second equality, we have used the projection formula

$$\langle a\,,\,\varphi^*(b)\rangle_{\mathrm{Gr}} = \langle \varphi_*(a)\,,\,b\rangle_{\hat{\mathrm{Fl}}} \qquad \text{for } a \in K_*^T(\mathrm{Gr}),\, b \in K_T^*(\hat{\mathrm{Fl}}). \tag{19}$$

This gives the desired formula. □

Lemma 3.5 *Let $\psi \in K_T^*(\hat{\mathrm{Fl}})$. If $\Delta(\psi) = \sum_{(\psi)} \psi_{(1)} \otimes \psi_{(2)}$, then $\theta(\psi_{(1)}) = \psi_{(1)}$.*

Proof This follows from the fact that the elements in the first tensor factor are in fact in the image of $K_T^*(\mathrm{Gr})$ inside $K_T^*(\hat{\mathrm{Fl}})$. □

3.4 Loop evaluation at identity

Let $\mathrm{ev}_1 : LK/T_{\mathbb{R}} \to K/T_{\mathbb{R}}$ be induced by evaluation of a loop at the identity. Since this is a $T_{\mathbb{R}}$-equivariant map (via left translation) it induces an $R(T)$-algebra homomorphism

$$\mathrm{ev}_1^* : K_T^*(G/B) \to K_T^*(\hat{\mathrm{Fl}}). \tag{20}$$

Let $q : K/T_{\mathbb{R}} \to LK/T_{\mathbb{R}}$ be the natural inclusion; it is $T_{\mathbb{R}}$-equivariant for left translation. The algebraic analogue of q identifies G/B with the finite-dimensional Schubert variety $X_{w_0} \subset \hat{\mathrm{Fl}}$.

Define $\eta := \mathrm{ev}_1^* \circ q^*$ so that $\eta : K_T^*(\hat{\mathrm{Fl}}) \to K_T^*(\hat{\mathrm{Fl}})$ is the pullback map in equivariant K-theory of the following composition

$$LK/T_{\mathbb{R}} \xrightarrow{\mathrm{ev}_1} K/T_{\mathbb{R}} \xrightarrow{q} LK/T_{\mathbb{R}}. \tag{21}$$

Lemma 3.6 *For all $\mu \in Q^\vee$, $v \in W$, and $\psi \in K_T^*(\hat{\mathrm{Fl}})$ we have*

$$(\eta\psi)(t_\mu v) = \psi(v). \tag{22}$$

Proof Recalling the description of the based loop defined by t_μ from the proof of Lemma 3.3, evaluating the loop $t_\mu v$ at the identity yields the value v. Thus the $T_{\mathbb{R}}$-fixed point $t_\mu v$ is sent to v under the composition (21). □

Lemma 3.7 *For all $\lambda \in P$,*

$$\mathrm{ev}_1^*([\mathcal{L}_\lambda^{G/B}]) = [\mathcal{L}_\lambda]. \tag{23}$$

Proof For all $\mu \in Q^\vee$ and $u \in W$ we have

$$i_{t_\mu v}^*(\mathrm{ev}_1^*([\mathcal{L}_\lambda^{G/B}])) = i_v^*([\mathcal{L}_\lambda^{G/B}]) = v \cdot e^\lambda = (t_\mu v) \cdot e^\lambda = i_{t_\mu v}^*([\mathcal{L}_\lambda]).$$ □

3.5 Coproduct identity

The following identity is the main result of this section.

Proposition 3.8 *For $\psi \in K_T^*(\hat{\mathrm{Fl}})$ and $a \in \hat{\mathbb{K}}$, we have*

$$a \bullet \psi = \sum_{(\psi)} \psi_{(1)} \cup \eta(a \bullet \psi_{(2)})$$

where $\Delta(\psi) = \sum_{(\psi)} \psi_{(1)} \otimes \psi_{(2)}$. In particular, taking $a = 1$, we have the identity

$$\cup \circ (1 \otimes \eta) \circ \Delta = 1$$

in $\mathrm{End}_{R(T)}(K_T^*(\hat{\mathrm{Fl}}))$.

Proof For $\mu \in Q^\vee$ and $v \in W$, we compute

$$\begin{aligned}
\langle t_\mu v \,,\, a \bullet \psi\rangle &= \langle t_\mu v a \,,\, \psi\rangle \\
&= \sum_{(\psi)} \langle t_\mu \,,\, \psi_{(1)}\rangle \langle va \,,\, \psi_{(2)}\rangle \\
&= \sum_{(\psi)} \langle t_\mu \,,\, \psi_{(1)}\rangle \langle v \,,\, a \bullet \psi_{(2)}\rangle \\
&= \sum_{(\psi)} \langle t_\mu v \,,\, \psi_{(1)}\rangle \langle t_\mu v \,,\, \eta(a \bullet \psi_{(2)})\rangle \\
&= \langle t_\mu v \,,\, \sum_{(\psi)} \psi_{(1)} \cup \eta(a \bullet \psi_{(2)})\rangle.
\end{aligned}$$

In the first equality we have used Proposition 3.4. In the fourth equality we have used Lemmas 3.3, 3.5, and 3.6. □

3.6 Commutation relations

We record additional commutation relations involving the nilHecke algebra actions, and the endomorphisms θ and η.

Let $\kappa : K_T^*(\hat{\mathrm{Fl}}) \to K_T^*(\hat{\mathrm{Fl}})$ be the pullback map in equivariant K-theory induced by the composition

$$\hat{\mathrm{Fl}} \longrightarrow \mathrm{id} \longrightarrow \hat{\mathrm{Fl}} \tag{24}$$

where id denotes the basepoint of $\hat{\mathrm{Fl}}$. It is an $R(T)$-algebra homomorphism.

Lemma 3.9 *For all $\mu \in Q^\vee$, $v \in W$, and $\psi \in K_T^*(\mathrm{Gr})$ we have*

$$\kappa(\psi)(t_\mu v) = \psi(\mathrm{id}). \tag{25}$$

Lemma 3.10 *As $R(T)$-module endomorphisms of $K_T^*(\hat{\mathrm{Fl}})$, we have the relations*

$$\theta^2 = \theta, \qquad \eta^2 = \eta, \qquad \kappa^2 = \kappa;$$

$$\theta\eta = \eta\theta = \theta\kappa = \kappa\theta = \eta\kappa = \kappa\eta = \kappa.$$

Proof Straightforward from Lemmas 3.3, 3.6, and 3.9. □

For $w \in \hat{W}$, define the endomorphism

$$w\odot := (w\cdot) \circ (w\bullet) = (w\bullet) \circ (w\cdot)$$

of $K_T^*(\hat{\mathrm{Fl}})$.

Proposition 3.11 *The map θ interacts with the two actions $\cdot$ and $\bullet$ of $\hat{\mathbb{K}}$ on $K_T^*(\hat{\mathrm{Fl}})$ in the following way:*

1. $(q\cdot) \circ \theta = \theta \circ (q\cdot)$
2. $(t_\mu\cdot) \circ \theta = \theta \circ (t_\mu\cdot)$
3. $(w\cdot) \circ \theta = \theta \circ (w\odot)$
4. $(w\bullet) \circ \theta = \theta$

where $q \in R(T)$, $w \in W$, and $\mu \in Q^\vee$. By (1), (2), (3), we see that $\theta(K_T^*(\hat{\mathrm{Fl}})) = p^*(K_T^*(\mathrm{Gr}))$ is a $\hat{\mathbb{K}}$-submodule of $K_T^*(\hat{\mathrm{Fl}})$ under the $\cdot$ action.

Proposition 3.12 *The map η interacts with the two actions $\cdot$ and $\bullet$ of $\hat{\mathbb{K}}$ on $K_T^*(\hat{\mathrm{Fl}})$ in the following way:*

1. $(q\cdot) \circ \eta = \eta \circ (q\cdot)$
2. $(t_\mu\cdot) \circ \eta = \eta$
3. $(w\cdot) \circ \eta = \eta \circ (w\cdot)$
4. $(q\bullet) \circ \eta = \eta \circ (q\bullet)$

5. $(t_\mu \bullet) \circ \eta = \eta$
6. $(w \bullet) \circ \eta = \eta \circ (w \bullet)$

where $q \in R(T)$, $w \in W$, *and* $\mu \in Q^\vee$. *By (1)-(6) we see that* $\eta(K_T^*(\hat{\mathrm{Fl}})) = \mathrm{ev}_1^*(K_T^*(G/B))$ *is a* $\hat{\mathbb{K}}$*-submodule of* $K_T^*(\hat{\mathrm{Fl}})$ *under either the* $\cdot$ *or the* $\bullet$ *action.*

Proposition 3.13 *The map* κ *interacts with the two actions* $\cdot$ *and* $\bullet$ *of* $\hat{\mathbb{K}}$ *on* $K_T^*(\hat{\mathrm{Fl}})$ *in the following way:*

1. $(q \cdot) \circ \kappa = \kappa \circ (q \cdot)$
2. $(t_\mu \cdot) \circ \kappa = \kappa$
3. $(w \cdot) \circ \kappa = \kappa \circ (w \odot)$
4. $(t_\mu \bullet) \circ \kappa = \kappa$
5. $(w \bullet) \circ \kappa = \kappa$

where $q \in R(T)$, $w \in W$, *and* $\mu \in Q^\vee$.

3.7 Action of $\hat{\mathbb{K}}$ on tensor products

Define $\hat{\mathbb{K}}_{Q(T)} \otimes_{Q(T)} \hat{\mathbb{K}}_{Q(T)}$ to be the left $Q(T)$-bilinear tensor product such that

$$q(a \otimes b) = qa \otimes b = a \otimes qb \tag{26}$$

for all $a, b \in \hat{\mathbb{K}}_{Q(T)}$ and $q \in Q(T)$. Define $\Delta: \hat{\mathbb{K}}_{Q(T)} \to \hat{\mathbb{K}}_{Q(T)} \otimes_{Q(T)} \hat{\mathbb{K}}_{Q(T)}$ by

$$\Delta\left(\sum_{w \in \hat{W}} a_w w\right) = \sum_w a_w w \otimes w \tag{27}$$

for $a_w \in Q(T)$. Then for all $i \in \hat{I}$ we have

$$\Delta(T_i) = T_i \otimes 1 + 1 \otimes T_i + (1 - e^{\alpha_i}) T_i \otimes T_i. \tag{28}$$

This restricts to a left $R(T)$-bilinear tensor product $\Delta : \hat{\mathbb{K}} \to \hat{\mathbb{K}} \otimes_{R(T)} \hat{\mathbb{K}}$. If M and N are left $\hat{\mathbb{K}}$-modules then $M \otimes_{R(T)} N$ is a left $\hat{\mathbb{K}}$-module via

$$a(m \otimes n) = \sum_{(a)} a_{(1)}(m) \otimes a_{(2)}(n) \tag{29}$$

for all $a \in \hat{\mathbb{K}}$, $m \in M$ and $n \in N$.

Lemma 3.14 *For $\psi_1, \psi_2 \in K_T^*(\hat{\mathrm{Fl}})$ and $a \in \hat{\mathbb{K}}$, we have*

$$a \cdot (\psi_1 \cup \psi_2) = \sum_{(a)} (a_{(1)} \cdot \psi_1) \cup (a_{(2)} \cdot \psi_{(2)})$$
$$a \bullet (\psi_1 \cup \psi_2) = \sum_{(a)} (a_{(1)} \bullet \psi_1) \cup (a_{(2)} \bullet \psi_{(2)}).$$

Proof We have

$$w \cdot (\psi_1 \cup \psi_2)(x) = w(\psi_1(w^{-1}x)\psi_2(w^{-1}(x))) = ((w \cdot \psi_1) \cup (w \cdot \psi_2))(x)$$
$$w \bullet (\psi_1 \cup \psi_2)(x) = \psi_1(xw)\psi_2(xw) = ((w \bullet \psi_1) \cup (w \bullet \psi_2))(x),$$

consistent with $\Delta(w) = w \otimes w$. Next, we check that the formulae are compatible with $R(T)$-linearity. It is enough to work with the algebra generators e^λ of $R(T)$. We have $\Delta(e^\lambda w) = e^\lambda w \otimes w$ and

$$\begin{aligned}(e^\lambda w) \cdot (\psi_1 \cup \psi_2) &= e^\lambda \cdot (w \cdot (\psi_1 \cup \psi_2)) \\ &= e^\lambda \cdot ((w \cdot \psi_1) \cup (w \cdot \psi_2)) = ((e^\lambda w) \cdot \psi_1) \cup (w \cdot \psi_2).\end{aligned}$$

Using Lemma 2.3 we have

$$\begin{aligned}(e^\lambda w) \bullet (\psi_1 \cup \psi_2) &= e^\lambda \bullet (w \bullet (\psi_1 \cup \psi_2)) \\ &= [\mathcal{L}_\lambda] \cup (w \bullet \psi_1) \cup (w \bullet \psi_2) = ((e^\lambda w) \bullet \psi_1) \cup (w \bullet \psi_2).\end{aligned}$$

□

3.8 Finite nilHecke algebra

The finite nilHecke ring $\mathbb{K}$ is the subring of $\hat{\mathbb{K}}$ generated by $R(T)$ and T_i for $i \in I$. There are left actions $\cdot$ and $\bullet$ of $\mathbb{K}$ on $K_T^*(G/B)$ that are similarly to the actions of $\hat{\mathbb{K}}$ on $K_T^*(\hat{\mathrm{Fl}})$.

There is a $\mathbb{K}$-$\mathbb{K}$-bimodule and ring homomorphism $\mathrm{cl}_{\hat{\mathbb{K}}} : \hat{\mathbb{K}} \to \mathbb{K}$ defined (for convenience from $\hat{\mathbb{K}}_{Q(T)} \to \mathbb{K}_{Q(T)}$) by

$$\mathrm{cl}_{\hat{\mathbb{K}}}(t_\mu a) = a \qquad \text{for } \mu \in Q^\vee \text{ and } a \in \mathbb{K}. \tag{30}$$

In particular,

$$\begin{aligned}\mathrm{cl}_{\hat{\mathbb{K}}}(T_0) &= \mathrm{cl}_{\hat{\mathbb{K}}}((1-e^{-\theta})^{-1}(s_0 - 1)) \\ &= \mathrm{cl}_{\hat{\mathbb{K}}}((1-e^{-\theta})^{-1}(t_{\theta^\vee}s_\theta - 1)) = (1-e^{-\theta})^{-1}(s_\theta - 1) =: T_{-\theta}.\end{aligned}$$

Thus we have $\cdot$ and $\bullet$ actions of $\hat{\mathbb{K}}$ on $K_T^*(G/B)$ that factor through $\mathrm{cl}_{\hat{\mathbb{K}}} : \hat{\mathbb{K}} \to \mathbb{K}$.

3.9 Tensor product decomposition of $K_T^*(\hat{\mathrm{Fl}})$

The equivariant K-theory ring $K_T^*(\mathrm{Gr})$ is a left $\hat{\mathbb{K}}$-submodule of $K_T^*(\hat{\mathrm{Fl}})$ under the $\cdot$-action. Thinking of $\psi_{\mathrm{Gr}} \in K_T^*(\mathrm{Gr})$ as a function from cosets $\hat{W}/W$ to $R(T)$, we have $(w \cdot \psi_{\mathrm{Gr}})(xW) = w(\psi_{\mathrm{Gr}}(w^{-1}xW))$.

The left $\hat{\mathbb{K}}$-module structures via $\cdot$ on $K_T^*(\mathrm{Gr})$ and $K_T^*(G/B)$ give a left $\hat{\mathbb{K}}$-module structure on $K_T^*(\mathrm{Gr}) \otimes_{R(T)} K_T^*(G/B)$ via (29).

Theorem 3.15 *There is an $R(T)$-algebra isomorphism*

$$K_T^*(\mathrm{Gr}) \otimes_{R(T)} K_T^*(G/B) \cong K_T^*(\hat{\mathrm{Fl}}) \tag{31}$$

$$a \otimes b \mapsto p^*(a) \cup \mathrm{ev}_1^*(b) \tag{32}$$

with componentwise multiplication on the tensor product. This map is also an isomorphism of left $\hat{\mathbb{K}}$-modules under the $\cdot$ action.

The proof is delayed to after Theorem 4.7.

4 Affine Schubert classes

4.1 Schubert bases

The $R(T)$-algebras $K_T^*(\hat{\mathrm{Fl}})$, $K_T^*(\mathrm{Gr})$, and $K_T^*(G/B)$ have equivariant Schubert bases $\{\psi^x \mid x \in \hat{W}\}$, $\{\psi^u_{\mathrm{Gr}} \mid u \in \hat{W}^0\}$, and $\{\psi^w_{G/B} \mid w \in W\}$ respectively. The basis $\{\psi^x \mid x \in \hat{W}\} \subset K_T^*(\hat{\mathrm{Fl}})$ is uniquely characterized by

$$\psi^v(T_w) = \delta_{v,w}. \tag{33}$$

We have

$$p^*(\psi^z_{\mathrm{Gr}}) = \psi^z \qquad \text{for all } z \in \hat{W}^0, \tag{34}$$

$$q^*(\psi^x) = \begin{cases} \psi^x_{G/B} & \text{for } x \in W, \\ 0 & \text{for } x \in \hat{W} \setminus W. \end{cases} \tag{35}$$

In particular $\eta(\psi^x) = \mathrm{ev}_1^*(\psi^x_{G/B})$ for $x \in W$.

Similarly, let $\{\psi_x \mid x \in \hat{W}\}$, $\{\psi_u^{\mathrm{Gr}} \mid u \in \hat{W}^0\}$, and $\{\psi_w^{G/B} \mid w \in W\}$ denote homology Schubert bases of $K_*^T(\hat{\mathrm{Fl}})$, $K_*^T(\mathrm{Gr})$, and $K_*^T(G/B)$. We write $\langle\cdot,\cdot\rangle_{\hat{\mathrm{Fl}}}$, $\langle\cdot,\cdot\rangle_{\mathrm{Gr}}$, and $\langle\cdot,\cdot\rangle_{G/B}$ for the $R(T)$-bilinear pairings between T-equivariant K-homology and K-cohomology, so that for example $\langle\psi_x, \psi^y\rangle_{\hat{\mathrm{Fl}}} = \delta_{xy}$. For the precise geometric interpretations of ψ^x and ψ_x we refer the reader to [17, §3].

Remark 4.1 The map p^* is an isomorphism of $K_T^*(\mathrm{Gr})$ with its image $\bigoplus_{u\in\hat{W}^0} R(T)\psi^u$, whose elements are $W\bullet$-invariant by Proposition 3.11.

The localization values of Schubert classes are determined by the following triangular relation. For all $w \in \hat{W}$, in $\hat{\mathbb{K}}$ we have [8] [19, Proposition 2.4]

$$w = \sum_{v\leq w} \langle w\,,\, \psi^v\rangle T_v. \tag{36}$$

The Schubert basis $\{\psi^w \mid w \in \hat{W}\}$ interacts with the $\cdot$ and $\bullet$ actions of $\hat{\mathbb{K}}$ as follows. For $i \in \hat{I}$, define

$$y_i := 1 + T_i = \frac{1}{1-e^{-\alpha_i}}(1 - e^{-\alpha_i} s_i) \tag{37}$$

$$\tilde{y}_i := 1 - e^{\alpha_i} T_i = \frac{1}{1-e^{\alpha_i}}(1 - e^{\alpha_i} s_i). \tag{38}$$

Proposition 4.2 *For $\lambda \in P$ and T_i for $i \in \hat{I}$, on the Schubert basis element $\psi^w \in K_T^*(\hat{\mathrm{Fl}})$ for $w \in \hat{W}$, we have:*

$$\tilde{y}_i \cdot \psi^w = \begin{cases} \psi^{s_i w} & \text{if } s_i w < w \\ \psi^w & \text{otherwise.} \end{cases} \tag{39}$$

$$y_i \bullet \psi^w = \begin{cases} \psi^{w s_i} & \text{if } w s_i < w \\ \psi^w & \text{otherwise.} \end{cases} \tag{40}$$

$$e^\lambda \cdot \psi^w = e^\lambda \psi^w \tag{41}$$

$$e^\lambda \bullet \psi^w = [\mathcal{L}_\lambda] \cup \psi^w. \tag{42}$$

Proof (40) is [19, Lemma 2.2]. Equation (39) has a straightforward proof starting with $\langle T_v\,,\, \tilde{y}_i \cdot \psi^w\rangle$ and using (10) and the duality of the two bases $\{T_v\}$ with $\{\psi^w\}$.

Equation (41) follows from the definition and (42) follows from (14). □

4.2 Equivariant affine K-Stanley classes

Theorem 3.2 interacts with Schubert classes as follows.

Theorem 4.3 *[19, Theorem 5.4] For every $u \in \hat{W}^0$, $k_u := k(\psi_u^{\mathrm{Gr}})$ is the unique element of $\mathbb{L}$ of the form*

$$k_u = \sum_{z\in\hat{W}} k_u^z T_z \tag{43}$$

for some $k_u^z \in R(T)$, *where*

$$k_u^z = \delta_{z,u} \qquad \text{for } z \in \hat{W}^0. \tag{44}$$

Remark 4.4 It follows from Theorem 3.2 that

$$t_\mu = \sum_{u \in \hat{W}^0} \langle t_\mu \,,\, \psi^u \rangle k_u. \tag{45}$$

Taking $\mathrm{cl}_{\hat{\mathbb{K}}}$ of both sides and using (30), we have

$$T_{\mathrm{id}} = \sum_{u \in \hat{W}^0} \langle t_\mu \,,\, \psi^u \rangle \mathrm{cl}_{\hat{\mathbb{K}}}(k_u). \tag{46}$$

Now $\langle t_\mu \,,\, \psi^u \rangle$ is zero unless $u \le t_\mu$, which by the assumption $u \in \hat{W}^0$ is equivalent to $uW \le t_\mu W$. Since both the t_μ and the k_u are $Q(T)$-bases of $\mathbb{L}$ it follows that $\langle t_\mu \,,\, \psi^u \rangle \neq 0$ for $\mu \in Q^\vee$ and $u \in \hat{W}^0$ such that $t_\mu W = uW$. It follows by induction that

$$\mathrm{cl}_{\hat{\mathbb{K}}}(k_x) = \delta_{\mathrm{id},x} T_{\mathrm{id}} \qquad \text{for all } x \in \hat{W}^0. \tag{47}$$

For $w \in \hat{W}$ the equivariant affine K-Stanley class $G^w \in K_T^*(\mathrm{Gr})$ is defined by

$$G^w := \varphi^*(\psi^w). \tag{48}$$

We will also consider G^w an element of $K_T^*(\hat{\mathrm{Fl}})$ via p^*.

Lemma 4.5 *For* $w \in \hat{W}$, *we have*

$$G^w = \sum_{u \in \hat{W}^0} k_u^w \, \psi_{\mathrm{Gr}}^u \tag{49}$$

where the k_u^w *are defined in Theorem 4.3.*

Proof For $u \in \hat{W}^0$, by (19) and Theorems 3.2 and 4.3 we have

$$\begin{aligned} \langle \psi_u^{\mathrm{Gr}} \,,\, G^w \rangle_{\mathrm{Gr}} &= \langle \psi_u^{\mathrm{Gr}} \,,\, \varphi^*(\psi^w) \rangle_{\mathrm{Gr}} = \langle \varphi_*(\psi_u^{\mathrm{Gr}}) \,,\, \psi^w \rangle_{\hat{\mathrm{Fl}}} \\ &= \left\langle \sum_{z \in \hat{W}} k_u^z T_z \psi^w \right\rangle = k_u^w. \end{aligned}$$

□

Recall that $u * v$ denotes the Demazure product of u and v.

Proposition 4.6 *For* $w \in \hat{W}$, *we have*

$$\Delta(\psi^w) = \sum_{w = w_1 * w_2} (-1)^{\ell(w_1) + \ell(w_2) - \ell(w)} G^{w_1} \otimes \psi^{w_2}. \tag{50}$$

Proof For $u \in \hat{W}^0$ and $v \in \hat{W}$, we have

$$k_u T_v = \sum_{x \in \hat{W}} k_u^x T_x T_v = \sum_{w \in \hat{W}} \sum_{\substack{x \in \hat{W} \\ w = x * v}} (-1)^{\ell(x)+\ell(v)-\ell(w)} k_u^x T_w. \tag{51}$$

This gives a formula for the matrix of the multiplication map $\mathbb{L} \otimes_{R(T)} \hat{\mathbb{K}} \to \hat{\mathbb{K}}$ with respect to the bases $k_u \otimes T_v$ and T_w. The dual map $K_T^*(\hat{\mathrm{Fl}}) \xrightarrow{\Delta} K_T^*(\mathrm{Gr}) \otimes_{R(T)} K_T^*(\hat{\mathrm{Fl}})$ has the transposed matrix of Schubert matrix coefficients. That is, for all $w \in \hat{W}$, using Lemma 4.5 we have

$$\begin{aligned}
\Delta(\psi^w) &= \sum_{(u,v) \in \hat{W}^0 \times \hat{W}} \sum_{\substack{x \in \hat{W} \\ w = x * v}} (-1)^{\ell(x)+\ell(v)-\ell(w)} k_u^x \psi_{\mathrm{Gr}}^u \otimes \psi^v \\
&= \sum_{\substack{v, x \in \hat{W} \\ w = x * v}} (-1)^{\ell(x)+\ell(v)-\ell(w)} G^x \otimes \psi^v.
\end{aligned}$$

□

4.3 Coproduct formula for affine Schubert classes

The following formula decomposes ψ^w according to the tensor product isomorphism of Theorem 3.15.

Theorem 4.7 *For $w \in \hat{W}$, we have*

$$\psi^w = \sum_{\substack{(w_1, w_2) \in \hat{W} \times W \\ w_1 * w_2 = w}} (-1)^{\ell(w_1)+\ell(w_2)-\ell(w)} G^{w_1} \cup \mathrm{ev}_1^*(\psi_{G/B}^{w_2}) \tag{52}$$

Proof Apply Proposition 3.8 with $a = 1$ and $\psi = \psi^w$, and use Proposition 4.6. □

Proof of Theorem 3.15 As the maps p^* and ev_1^* are $R(T)$-algebra homomorphisms, so is (32). Note that for $u \in \hat{W}^0$, $G^u = \psi_{\mathrm{Gr}}^u$. To show that (32) is an isomorphism, it suffices to show that the image of the basis $\{\psi_{\mathrm{Gr}}^u \otimes \psi_{G/B}^v \mid (u,v) \in \hat{W}^0 \times W\}$ of $K_T^*(\mathrm{Gr}) \otimes_{R(T)} K_T^*(G/B)$, namely, $\{G^u \cup \mathrm{ev}_1^*(\psi_{G/B}^v) \mid (u,v) \in \hat{W}^0 \times W\}$, is an $R(T)$-basis of $K_T^*(\hat{\mathrm{Fl}})$. But the latter collection of elements is unitriangular with the Schubert basis of $K_T^*(\hat{\mathrm{Fl}})$, by Theorem 4.7. Thus (32) is a $R(T)$-algebra isomorphism.

Finally, (31) is a left $(\hat{\mathbb{K}}\,\cdot)$-module homomorphism, due to Lemma 3.14 and the fact that ev_1^* and p^* are left $(\hat{\mathbb{K}}\,\cdot)$-module homomorphisms. □

Corollary 4.8 *For $i \in \hat{I}$, we have*

$$\psi^{s_i} = \begin{cases} G^{s_0} & \text{if } i = 0 \\ G^{s_i} + \mathrm{ev}_1^*(\psi^{s_i}_{G/B}) - G^{s_i} \cup \mathrm{ev}_1^*(\psi^{s_i}_{G/B}) & \text{otherwise.} \end{cases} \tag{53}$$

Proposition 4.9 *For all $i \in \hat{I}$ we have*

$$1 - G^{s_i} = (1 - G^{s_0})^\ell \tag{54}$$

where $\ell = \mathrm{level}(\Lambda_i)$.

Proof Let $\{\psi^x_{\hat{T}} \mid x \in \hat{W}\}$ denote the equivariant Schubert basis of $K^*_{\hat{T}}(\hat{\mathrm{Fl}})$, where $\hat{T} \cong T \times \mathbb{C}^\times$ denotes the affine maximal torus. For all $i \in \hat{I}$, in $K^*_{\hat{T}}(\hat{\mathrm{Fl}})$ we have [10][1]

$$\psi^{s_i}_{\hat{T}}(w) = 1 - e^{\Lambda_i - w \cdot \Lambda_i} \qquad \text{for all } w \in \hat{W}.$$

For all $\mu \in Q^\vee$ and $v \in W$ we have

$$G^{s_i}(t_\mu v) = \psi^{s_i}(t_\mu) = \mathrm{cl}(\psi^{s_i}_{\hat{T}}(t_\mu)) = \mathrm{cl}(1 - e^{\Lambda_i - t_\mu \cdot \Lambda_i}).$$

Applying this equation twice, we have

$$\frac{1 - G^{s_i}(t_\mu v)}{(1 - G^{s_0}(t_\mu v))^\ell} = \mathrm{cl}(e^{\Lambda_i - t_\mu \cdot \Lambda_i - \ell \Lambda_0 + \ell t_\mu \cdot \Lambda_0}) = \mathrm{cl}(e^{\mathrm{af}(\omega_i) - t_\mu \cdot \mathrm{af}(\omega_i)}) = 1$$

since for any level zero element λ we have $t_\mu(\lambda) = \lambda - \langle \mu, \lambda \rangle \delta$. □

4.4 Ideal sheaf classes

For a reduced word $w = s_{i_1} \cdots s_{i_\ell}$, define $y_w := y_{i_1} \cdots y_{i_\ell} \in \hat{\mathbb{K}}$, which does not depend on the choice of reduced word. By [19, Lemma A.3], we have $y_w = \sum_{v \le w} T_v$. We let $\{\bar{\psi}^w \in K^*_T(\hat{\mathrm{Fl}}) \mid w \in \hat{W}\}$ denote the dual basis to $\{y_w \mid w \in \hat{W}\}$. Thus $\langle y_w, \bar{\psi}^v \rangle = \delta_{w,v}$. The element $\bar{\psi}^w$ is denoted ψ_{KK} in [19].

Remark 4.10 The Schubert basis element $\psi^w \in K^*_T(\hat{\mathrm{Fl}})$ represents the class of the structure sheaf $\mathcal{O}_w$ of a Schubert variety in the thick affine flag variety. The element $\bar{\psi}^w$ represents the ideal sheaf $\mathcal{I}_w$ of the boundary ∂X_w in a Schubert variety X_w. See [19, Appendix A].

Define

$$\bar{G}^w := \varphi^*(\bar{\psi}^w) \in K^*_T(\mathrm{Gr})$$

[1] The conventions here differ by a sign to those in [10]. For example, for us $\psi^{s_i}(s_i) = 1 - e^{\alpha_i}$.

and as usual, we denote by $\bar{G}^w$ the image of this element in $K_T^*(\hat{\text{Fl}})$. Following [17], define $l_u := \sum_{v\in\hat{W}^0: v\leq u} k_v \in \mathbb{L}$ and define $l_u^w \in R(T)$ by

$$l_u := \sum_{w\in\hat{W}} l_u^w y_w. \tag{55}$$

The coefficients l_u^w are related to k_v^x by the formula

$$l_u^w = \sum_{x\leq w}(-1)^{\ell(w)-\ell(x)} \sum_{\substack{v\in\hat{W}^0\\ v\leq u}} k_v^x.$$

We have the following variants of Lemma 4.5, Proposition 4.6, and Theorem 4.7 with identical proofs.

Lemma 4.11 *For $w \in \hat{W}$, we have*

$$\bar{G}^w = \sum_{u\in\hat{W}^0} l_u^w \bar{\psi}_{\text{Gr}}^u \tag{56}$$

where the l_u^w are defined in (55)*, and $\bar{\psi}_{\text{Gr}}^u$ is determined by $\langle l_v, \bar{\psi}_{\text{Gr}}^u\rangle = \delta_{v,u}$.*

Proposition 4.12 *For $w \in \hat{W}$, we have*

$$\Delta(\bar{\psi}^w) = \sum_{w=w_1*w_2} \bar{G}^{w_1} \otimes \bar{\psi}^{w_2}.$$

Theorem 4.13 *For $w \in \hat{W}$, we have*

$$\bar{\psi}^w = \sum_{\substack{(w_1,w_2)\in\hat{W}\times W\\ w_1*w_2=w}} \bar{G}^{w_1} \cup \text{ev}_1^*(\bar{\psi}_{G/B}^{w_2})$$

4.5 •-action on affine Schubert classes

We investigate the behavior of the decomposition in Theorem 4.7 under the •-action of $\hat{\mathbb{K}}$. By Lemma 3.14 and Theorem 3.15, it is enough to separately describe how $\theta(\psi^x)$ for $x \in \hat{W}$, and $\eta(\psi^w)$ for $w \in W$, behave under the •-action. For $\eta(\psi^w)$, Proposition 3.12 gives the following.

Proposition 4.14 *For $a \in \hat{\mathbb{K}}$ and $w \in W$, we have*

$$a \bullet \eta(\psi^w) = \eta(\text{cl}_{\hat{\mathbb{K}}}(a) \bullet \psi^w) \tag{57}$$

and in particular,

$$s_0 \bullet \eta(\psi^w) = \eta(s_\theta \bullet \psi^w) \tag{58}$$

$$y_0 \bullet \eta(\psi^w) = \eta(y_{-\theta} \bullet \psi^w) \tag{59}$$

where $y_{-\theta} := 1 + T_{-\theta}$.

Since $\theta(\psi^x)$ can be expanded in the basis ψ^u for $u \in \hat{W}^0$, it is enough to consider the $\bullet$ action on ψ^u.

Theorem 4.15 *For $\lambda \in P$, $i \in \hat{I}$, and $u \in \hat{W}^0$, we have*

1. $e^\lambda \bullet \psi^u = [\mathcal{L}_\lambda] \cup \psi^u$,
2. $y_i \bullet \psi^u = \psi^u$ *and* $s_i \bullet \psi^u = \psi^u$ *if* $i \in I$,
3. *For* $u \in \hat{W}^0 \setminus \{\mathrm{id}\}$, *we have*

$$y_0 \bullet \psi^u = \psi^{us_0} = \sum_{\substack{(x_1,x_2)\in\hat{W}\times W \\ x_1 * x_2 = us_0}} (-1)^{\ell(x_1)+\ell(x_2)-\ell(u)-1} \theta(\psi^{x_1}) \cup \eta(\psi^{x_2}). \tag{60}$$

4. *For* $u \in \hat{W}^0 \setminus \{\mathrm{id}\}$, *we have*

$$s_0 \bullet \psi^u = e^{-\theta} \psi^u \tag{61}$$
$$+ \sum_{\substack{(x_1,x_2)\in\hat{W}\times W \\ x_1 * x_2 = us_0}} (-1)^{\ell(x_1)+\ell(x_2)-\ell(u)-1} \theta(\psi^{x_1}) \cup \eta((1 - e^{-\theta}) \bullet \psi^{x_2}).$$

Proof These formulae may be deduced from Proposition 4.2 using $s_0 = e^{-\theta} + (1 - e^{-\theta}) y_0$. □

Remark 4.16 The $\cdot$ and $\bullet$ actions of $\hat{\mathbb{K}}$ make $K_T^*(\mathrm{Gr}) \otimes_{R(T)} K_T^*(G/B)$ into a left $(\hat{\mathbb{K}} \times \hat{\mathbb{K}})$-module such that the map (31) is a left $(\hat{\mathbb{K}} \times \hat{\mathbb{K}})$-module isomorphism.

4.6 Recursion

The affine Schubert classes in the tensor product $K_T^*(\mathrm{Gr}) \otimes_{R(T)} K_T^*(G/B)$ are determined by the following recursion.

1. $\psi^u = \psi^u_{\mathrm{Gr}} \otimes 1$ for $u \in \hat{W}^0$, and
2. For all $i \in \hat{I}$,

$$y_i \bullet \psi^w = \begin{cases} \psi^{ws_i} & \text{if } ws_i < w \\ \psi^w & \text{otherwise.} \end{cases}$$

The operator y_i acts on $K_T^*(\mathrm{Gr}) \otimes_{R(T)} K_T^*(G/B)$ by

$$\Delta(y_i) = (1 - e^{\alpha_i}) y_i \otimes y_i + e^{\alpha_i} (y_i \otimes 1 + 1 \otimes y_i - 1 \otimes 1)$$

which follows from (28).

5 Cohomology

In this section, we indicate the modifications necessary for the preceding results to hold in cohomology.

Remark 5.1 It would be interesting to deduce Theorem 5.11 (and other results in cohomology) directly from Theorem 4.7 (and other K-theoretic results), for example by using the Chern character or by taking "lowest degree" terms. In particular, we do not know the relation between the coefficients j_u^w in Theorem 5.7 and the k_u^w in Theorem 4.3.

5.1 Small-torus affine nilHecke ring

Instead of $R(T)$, we work over $S = \mathrm{Sym}_{\mathbb{Z}}(P) \cong H_T^*(\mathrm{pt})$. The algebra $\hat{\mathbb{K}}$ is replaced by the small-torus affine nilHecke ring $\hat{\mathbb{A}}$, as defined in [18, Chapter 4]. Let $\hat{\mathbb{A}}_0$ be the nilCoxeter algebra, the ring generated by elements A_i for $i \in \hat{I}$ which satisfy the braid relations for $\hat{W}$ and the relation $A_i^2 = 0$. We have $\hat{\mathbb{A}}_0 = \bigoplus_{w \in \hat{W}} \mathbb{Z} A_w$ where $A_w = A_{i_1} A_{i_2} \cdots A_{i_\ell}$ for a reduced decomposition $w = s_{i_1} s_{i_2} \cdots s_{i_\ell}$. Let $Q(S)$ be the fraction field of S and let $\hat{\mathbb{A}}_{Q(S)} = Q(S) \otimes_{\mathbb{Q}} \mathbb{Q}[\hat{W}]$ be the twisted group algebra of $\hat{W}$ with coefficients in $Q(S)$, with product $(q' \otimes w)(q \otimes v) = q' w(q) \otimes wv$ for $q, q' \in Q(S)$ and $w, v \in \hat{W}$. Then $\hat{\mathbb{A}}$ is the subring of $\hat{\mathbb{A}}_{Q(S)}$ generated by S and $\hat{\mathbb{A}}_0$. We have the following analogue of (6): $\hat{\mathbb{A}} = \bigoplus_{w \in \hat{W}} S A_w$. Instead of the Demazure product, we will make use of length-additive products. Write $w \doteq uv$ if $w = uv$ and $\ell(w) = \ell(u) + \ell(v)$. Note that $A_u A_v = A_w$ if and only if $w \doteq uv$. This notation naturally extends to longer products.

5.2 $\hat{\mathbb{A}}$-$\hat{\mathbb{A}}$-bimodule structure on cohomology of affine flag variety

Localization identifies $H_T^*(\hat{\mathrm{Fl}})$ with a S-subalgebra of $\mathrm{Fun}(\hat{W}, S)$. We identify a cohomology class $\xi \in H_T^*(\hat{\mathrm{Fl}})$ with a function $\xi \in \mathrm{Fun}(\hat{W}, S)$ taking values $\xi(v)$, $v \in \hat{W}$. For the small torus affine GKM condition see [18, Section 4.2].

There is a S-bilinear perfect pairing $\langle \cdot, \cdot \rangle : \hat{\mathbb{A}} \times H_T^*(\hat{\mathrm{Fl}})$ characterized by $\langle w, \xi \rangle = \xi(w)$.

There is a left action $\xi \mapsto a \cdot \xi$ of $\hat{\mathbb{A}}$ on $H_T^*(\hat{\mathrm{Fl}})$ given by the formulae [18, Chapter 4, Proposition 3.16]

$$(q \cdot \xi)(a) = q\, \xi(a) \tag{62}$$

$$(A_i \cdot \xi)(a) = A_i \cdot \xi(s_i a) + \xi(A_i a) \tag{63}$$

$$(w \cdot \xi)(a) = w\, \xi(w^{-1} a) \tag{64}$$

for $a \in \hat{\mathbb{A}}$, $\xi \in H_T^*(\hat{\mathrm{Fl}})$, $q \in S$, $i \in \hat{I}$, and $w \in \hat{W}$. Here, A_i acts on S via

$$A_i(\lambda) = \langle \alpha_i^\vee, \lambda \rangle \mathrm{id} \tag{65}$$

$$A_i(qq') = A_i(q)q' + (s_i \cdot q)A_i(q'). \tag{66}$$

There is another left action $\xi \mapsto a \bullet \xi$ of $\hat{\mathbb{A}}$ on $H_T^*(\hat{\mathrm{Fl}})$ given by [18, Chapter 4, Section 3.3]

$$(a \bullet \xi)(b) = \xi(ba) \tag{67}$$

for $a, b \in \hat{\mathbb{A}}$ and $\xi \in H_T^*(\hat{\mathrm{Fl}})$.

Let $c_1(\mathcal{L}_\lambda) \in H_T^*(\hat{\mathrm{Fl}})$ denote the first Chern class of the T-equivariant line bundle with weight $\lambda \in X$ on $\hat{\mathrm{Fl}}$. Explicitly [18, Chapter 4, Section 3]

$$\langle t_\mu v, c_1(\mathcal{L}_\lambda) \rangle = v \cdot \lambda \qquad \mu \in Q^\vee, v \in W \tag{68}$$

Lemma 5.2 *For any $\lambda \in X$ and $\xi \in H_T^*(\hat{\mathrm{Fl}})$, we have $\lambda \bullet \xi = c_1(\mathcal{L}_\lambda) \cup \xi$.*

5.3 Endomorphisms

Let $\mathbb{P} = Z_{\hat{\mathbb{A}}}(S)$ be the centralizer of S in $\hat{\mathbb{A}}$, called the Peterson subalgebra. We have the cohomological wrong way map $\varphi^* : H_T^*(\hat{\mathrm{Fl}}) \to H_T^*(\mathrm{Gr})$.

Theorem 5.3 ([26] [15] [18, Chapter 4, Theorem 4.9]) *There is an isomorphism $j : H_*^T(\mathrm{Gr}) \to \mathbb{P}$ making the following commutative diagram of ring and left $R(T)$-module homomorphisms:*

$$\begin{array}{ccc} H_*^T(\mathrm{Gr}) & \xrightarrow{j} & \mathbb{P} \\ \Big\downarrow{\scriptstyle \varphi_*} & & \Big\downarrow \\ H_*^T(\hat{\mathrm{Fl}}) & \longrightarrow & \hat{\mathbb{A}} \end{array}$$

The maps

$$\begin{aligned} p^* &: H_T^*(\mathrm{Gr}) \to H_T^*(\hat{\mathrm{Fl}}) \\ \theta &: H_T^*(\hat{\mathrm{Fl}}) \to H_T^*(\hat{\mathrm{Fl}}) \\ \Delta &: H_T^*(\hat{\mathrm{Fl}}) \to H_T^*(\mathrm{Gr}) \otimes_S H_T^*(\hat{\mathrm{Fl}}) \\ \mathrm{ev}_1^* &: H_T^*(G/B) \to H_T^*(\hat{\mathrm{Fl}}) \\ \eta &: H_T^*(\hat{\mathrm{Fl}}) \to H_T^*(\hat{\mathrm{Fl}}) \\ \kappa &: H_T^*(\hat{\mathrm{Fl}}) \to H_T^*(\hat{\mathrm{Fl}}) \end{aligned}$$

are defined as for K-theory. Lemma 3.3, Proposition 3.4, Lemma 3.5, Lemma 3.6 hold in cohomology with the obvious modifications. Lemma 3.7 holds with $c_1(\mathcal{L}_\lambda)$ replacing $[\mathcal{L}_\lambda]$.

Proposition 5.4 *For $\xi \in H_T^*(\hat{\mathrm{Fl}})$ and $a \in \hat{\mathbb{A}}$, we have*

$$a \bullet \xi = \sum_{(\xi)} \xi_{(1)} \cup \eta(a \bullet \xi_{(2)})$$

where $\Delta(\xi) = \sum_{(\xi)} \xi_{(1)} \otimes \xi_{(2)}$. In particular, taking $a = 1$, we have the identity

$$\cup \circ (1 \otimes \eta) \circ \Delta = 1$$

in $\mathrm{End}_S(H_T^*(\hat{\mathrm{Fl}}))$.

Lemmas 3.9, 3.10, and Propositions 3.11, 3.12, 3.13 hold in cohomology.

5.4 Action of $\hat{\mathbb{A}}$ on tensor products

Equation (28) is replaced by

$$\Delta(A_i) = A_i \otimes 1 + s_i \otimes A_i = 1 \otimes A_i + A_i \otimes s_i \tag{69}$$

Lemma 3.14 holds with no change in cohomology.

5.5 Tensor product decomposition of $H_T^*(\hat{\mathrm{Fl}})$

The left $\hat{\mathbb{A}}$-module structures via $\cdot$ on $H_T^*(\mathrm{Gr})$ and $H_T^*(G/B)$ give a left $\hat{\mathbb{A}}$-module structure on $H_T^*(\mathrm{Gr}) \otimes_S H_T^*(G/B)$.

Theorem 5.5 *There is an S-algebra isomorphism*

$$H_T^*(\mathrm{Gr}) \otimes_S H_T^*(G/B) \cong H_T^*(\hat{\mathrm{Fl}}) \tag{70}$$

$$a \otimes b \mapsto p^*(a) \cup \mathrm{ev}_1^*(b) \tag{71}$$

with componentwise multiplication on the tensor product. This map is also an isomorphism of left $\hat{\mathbb{A}}$-modules under the $\cdot$ action.

5.6 Schubert bases

The S-algebras $H_T^*(\hat{\mathrm{Fl}})$, $H_T^*(\mathrm{Gr})$, and $H_T^*(G/B)$ have equivariant Schubert bases $\{\xi^x \mid x \in \hat{W}\}$, $\{\xi_{\mathrm{Gr}}^u \mid u \in \hat{W}^0\}$, and $\{\xi_{G/B}^w \mid w \in W\}$ respectively. Equations (34) and (35) hold for cohomology Schubert classes.

The analogue of Proposition 4.2 is as follows.

Proposition 5.6 *[18, Chapter 4, Section 3.3] For $\lambda \in X \subset S$ and A_i for $i \in \hat{I}$, on the Schubert basis element $\xi^w \in H_T^*(\hat{\mathrm{Fl}})$ for $w \in \hat{W}$, we have:*

$$A_i \cdot \xi^w = \begin{cases} \xi^{s_i w} & \text{if } s_i w < s \\ 0 & \text{otherwise.} \end{cases} \tag{72}$$

$$A_i \bullet \xi^w = \begin{cases} \xi^{w s_i} & \text{if } w s_i < w \\ 0 & \text{otherwise.} \end{cases} \tag{73}$$

$$\lambda \cdot \xi^w = \lambda \xi^w \tag{74}$$

$$\lambda \bullet \xi^w = c_1(\mathcal{L}_\lambda) \cup \xi^w. \tag{75}$$

5.7 Equivariant affine Stanley classes

Theorem 5.7 ([26] [15] [18]) *For every $w \in \hat{W}^0$, $j_w = k(\xi_w^{\mathrm{Gr}})$ is the unique element of $\mathbb{P}$ of the form*

$$j_w = A_w + \sum_{z \in \hat{W} \setminus \hat{W}^0} j_w^z A_z \tag{76}$$

for some $j_w^z \in S$.

Remark 5.8 By [26] [21] the coefficients j_w^z are equivariant Gromov-Witten invariants for G/B.

For $w \in \hat{W}$ the equivariant affine Stanley class $F^w \in H_T^*(\mathrm{Gr})$ is defined by

$$F^w := \varphi^*(\xi^w) \tag{77}$$

and as usual we also consider F^w an element of $H_T^*(\hat{\mathrm{Fl}})$.

Lemma 5.9 *For $w \in \hat{W}$, we have*

$$F^w = \sum_{u \in \hat{W}^0} j_u^w \xi_{\mathrm{Gr}}^u$$

where the j_u^w are defined in (76).

Proposition 5.10 *For $w \in \hat{W}$, we have $\Delta(\xi^w) = \sum_{w \doteq w_1 w_2} F^{w_1} \otimes \xi^{w_2}$.*

5.8 Coproduct formula for affine Schubert classes

Theorem 5.11 *For $w \in \hat{W}$, we have*

$$\xi^w = \sum_{\substack{(w_1, w_2) \in \hat{W} \times W \\ w_1 w_2 \doteq w}} F^{w_1} \cup \mathrm{ev}_1^*(\xi_{G/B}^{w_2}).$$

5.9 Formulae for Schubert divisors

Corollary 5.12 *For $i \in \hat{I}$ we have*

$$\xi^{s_i} = \begin{cases} F^{s_0} & \text{if } i = 0 \\ F^{s_i} + \mathrm{ev}_1^*(\xi^{s_i}_{G/B}) & \text{otherwise.} \end{cases} \tag{78}$$

Proposition 5.13 *For all $i \in \hat{I}$ and $\lambda \in Q^\vee$ we have*

$$F^{s_i} = \mathrm{level}(\Lambda_i)\, F^{s_0}. \tag{79}$$

5.10 •-action on affine Schubert classes

Proposition 4.14 holds with A_0 replacing y_0 and $A_{-\theta} := -\theta^{-1}(1 - s_\theta)$ replacing $y_{-\theta}$.

Theorem 5.14 *For $\lambda \in X \subset S$, $i \in \hat{I}$, and $u \in \hat{W}^0$, we have*

1. $\lambda \bullet \xi^u = c_1(\mathcal{L}_\lambda) \cup \xi^u$,
2. $A_i \bullet \xi^u = 0$ *and* $s_i \bullet \xi^u = \xi^u$ *if* $i \in I$,
3. *For* $u \in \hat{W}^0 \setminus \{\mathrm{id}\}$

$$A_0 \bullet \xi^u = \xi^{us_0} = \sum_{\substack{(x_1,x_2)\in \hat{W}\times W \\ x_1x_2 \doteq us_0}} \theta(\xi^{x_1}) \cup \eta(\xi^{x_2}).$$

4. *For* $u \in \hat{W}^0 \setminus \{\mathrm{id}\}$

$$s_0 \bullet \xi^u = \xi^u + \sum_{\substack{(x_1,x_2)\in \hat{W}\times W \\ x_1x_2 \doteq us_0}} \theta(\xi^{x_1}) \cup \eta(-\mathrm{cl}(\alpha_0) \bullet \xi^{x_2}).$$

Remark 5.15 The $\cdot$ and $\bullet$ actions of $\hat{\mathbb{A}}$ make $H_T^*(\mathrm{Gr}) \otimes_S H_T^*(G/B)$ into a left $(\hat{\mathbb{A}} \times \hat{\mathbb{A}})$-module such that the map (70) is a left $(\hat{\mathbb{A}} \times \hat{\mathbb{A}})$-module isomorphism.

5.11 Recursion

The affine Schubert classes in the tensor product $H_T^*(\mathrm{Gr}) \otimes_S H_T^*(G/B)$ are determined by the following recursion.

1. $\xi^u = \xi^u_{\mathrm{Gr}} \otimes 1$ for $u \in \hat{W}^0$, and
2. For all $i \in \hat{I}$

$$A_i \bullet \xi^w = \begin{cases} \xi^{ws_i} & \text{if } ws_i < w \\ 0 & \text{otherwise.} \end{cases}$$

Here, the operator A_i acts on $H_T^*(\mathrm{Gr}) \otimes_S H_T^*(G/B)$ by $A_i \bullet (\zeta \otimes \psi) = \zeta \otimes (A_i \bullet \psi)$ if $i \neq 0$ and $A_0 \bullet (\zeta \otimes \psi)$ is computed via (69) and Theorem 5.14.

6 Examples

6.1 Type A in cohomology

Letting $G = \mathrm{SL}(n)$, we now consider the affine Schubert polynomials [24]. Recall the isomorphism $H^*(\hat{\mathrm{Fl}}_G) \cong H^*(\mathrm{Gr}_G) \otimes_{H^*(\mathrm{pt})} H^*(G/B)$. By [15], the cohomology $H^*(\mathrm{Gr}_G)$ is isomorphic to Λ/I_n where Λ is the ring of symmetric functions and I_n is the ideal $\langle m_\lambda \mid \lambda_1 \geq n\rangle$ in Λ. Also, we have the classical Borel isomorphism $H^*(G/B) = \mathbb{Z}[x_1, \ldots, x_n]/\langle e_j(x_1, \ldots, x_n) \mid j \geq 1\rangle$ where $e_j(x_1, \ldots, x_n)$'s are elementary symmetric functions. Hence we have

$$H^*(\hat{\mathrm{Fl}}_G) \cong \Lambda/I_n \otimes_{\mathbb{Z}} \mathbb{Z}[x_1, \ldots, x_n]/\langle e_j(x_1, \ldots, x_n) \mid j \geq 1\rangle.$$

We list some affine Schubert polynomials for $n = 3$, indexed by $w \in \tilde{S}_n$, the affine symmetric group.

w	1	s_0	s_1	s_2	s_1s_0	s_2s_1	$s_2s_1s_0$
$\tilde{\mathfrak{S}}_w$	1	h_1	h_1+x_1	$h_1+x_1+x_2$	h_2	$h_2+h_1x_1+x_1^2$	$m_{2,1}+m_{1,1,1}$

The polynomial $\tilde{\mathfrak{S}}_{s_2s_1}$ can be computed in a number of different ways. First, we can start from $\tilde{\mathfrak{S}}_{s_2s_1s_0}$ which is the same as the affine Schur function indexed by $s_2s_1s_0$, and use the monomial expansion of the affine Schur functions [14]. Then one can act with the divided difference operator $A_i\bullet$ to obtain $\tilde{\mathfrak{S}}_{s_2s_1}$. The action of $A_i\bullet$ is explicitly given in [24, Definition 1.1].

On the other hand, using the coproduct formula (Theorem 5.11) directly give $\tilde{\mathfrak{S}}_{s_2s_1}$:

$$\tilde{\mathfrak{S}}_{s_2s_1} = F_{s_2s_1} + F_{s_2}\mathfrak{S}_{s_1} + \mathfrak{S}_{s_2s_1} = h_2 + h_1x_1 + x_1^2$$

where F_w is the affine Stanley symmetric function, the non-equivariant version of F^w in Section 5, and $\mathfrak{S}_v(x)$ is the Schubert polynomial. Using the coproduct formula together with monomial expansions of F_w [14] and $\mathfrak{S}_v(x)$ [2] provides the following theorem:

Theorem 6.1 *Affine Schubert polynomials are monomial-positive.*

The same coproduct formulae hold in equivariant cohomology, with the affine double Stanley symmetric function F^w [22] replacing F_w, and the double Schubert polynomial $\mathfrak{S}_v(x,y)$ [23] replacing $\mathfrak{S}_v(x)$. However, there is no combinatorially explicit formula for the equivariant affine Stanley classes F^w, see [22, Remark 23].

6.2 Back stable limit

We explain how to obtain the coproduct formula [16, Theorem 3.16] for backstable Schubert polynomials from Theorem 5.11. Let $\overleftarrow{\mathfrak{S}}_w(x) \in \Lambda \otimes_{\mathbb{Z}} \mathbb{Z}[x_i \mid i \in \mathbb{Z}]$ denote the back stable Schubert polynomial from [16], where $w \in S_{\mathbb{Z}} = \langle s_i \mid i \in \mathbb{Z}\rangle$ is an infinite permutation. Let

$$\phi_n : \Lambda \otimes_{\mathbb{Z}} \mathbb{Z}[x_i \mid i \in \mathbb{Z}] \to \Lambda/I_n \otimes_{\mathbb{Z}} \mathbb{Z}[x_1,\dots,x_n]/\langle e_j(x_1,\dots,x_n) \mid j \geq 1\rangle$$

denote the natural quotient ring homomorphism where in the second factor x_i is send to $x_{i \bmod n}$. The back stable Schubert polynomial has a unique expansion $\overleftarrow{\mathfrak{S}}_w(x) = \sum_v a_v(x) \otimes \mathfrak{S}_v(x)$ where $a_v \in \Lambda$ and $\mathfrak{S}_v(x)$ is a finite Schubert polynomial, which may possibly involve negative letters. By shifting w, we may assume that $v \in S_{>0} = \langle s_i \mid i > 0\rangle$, so that $\mathfrak{S}_v(x)$ is a usual Schubert polynomial. We will show that

$$\overleftarrow{\mathfrak{S}}_w(x) = \sum_{\substack{w \doteq uv \\ v \in S_{\neq 0}}} F_u(x) \otimes \mathfrak{S}_v(x) \tag{80}$$

where $F_u(x)$ denotes the Stanley symmetric function and $S_{\neq 0} = \langle s_i \mid i \neq 0\rangle$.

For sufficiently large n, the permutation $w \in S_{\mathbb{Z}}$ gives a well-defined element of the affine symmetric group $\tilde{S}_n$, by sending $s_i, i \in \mathbb{Z}$ to $s_i, i \in \mathbb{Z}/n\mathbb{Z}$. Abusing notation, we denote this element by $w \in \tilde{S}_n$ as well. According to [16, Theorem 10.9], for sufficiently large n, the image $\phi_n(\overleftarrow{\mathfrak{S}}_w(x))$ is equal to the affine Schubert polynomial $\tilde{\mathfrak{S}}_w(x)$, and it is also known that the image $\phi_n(F_u(x))$ is equal to the affine Stanley symmetric function, also denoted F_u.

Given any nonzero $f(x) \in \Lambda \otimes_{\mathbb{Z}} \mathbb{Z}[x_i \mid i \in \mathbb{Z}]$, it is straightforward to see that for sufficiently large n, we must have $\phi_n(f(x)) \neq 0$. Applying this to the difference of the two sides of (80), and using our Theorem 5.11, we see that equality must hold in (80).

6.3 Type A in K-theory

Let $G = \mathrm{SL}(n)$. We now consider affine versions of the Grothendieck polynomials. We have the isomorphism $K^*(\hat{\mathrm{Fl}}_G) \cong K^*(\mathrm{Gr}_G) \otimes_{K^*(\mathrm{pt})} K^*(G/B)$ and identifications $K^*(G/B) = \mathbb{Z}[x_1,\dots,x_n]/\langle e_j(x_1,\dots,x_n) \mid j \geq 1\rangle$ [19]

and $K^*(\mathrm{Gr}_G) \cong \widehat{\Lambda/I_n}$, where $\widehat{\Lambda/I_n}$ denotes the graded completion. By Theorem 4.7, we have the formula

$$\tilde{\mathfrak{G}}_w = \sum_{\substack{(w_1,w_2)\in\hat{W}\times W\\ w_1 * w_2 = w}} (-1)^{\ell(w_1)+\ell(w_2)-\ell(w)} G^{w_1}\mathfrak{G}_{w_2}$$

in $\widehat{\Lambda/I_n} \otimes_{\mathbb{Z}} \mathbb{Z}[x_1,\ldots,x_n]/\langle e_j(x_1,\ldots,x_n) \mid j \geq 1\rangle$, where $\tilde{\mathfrak{G}}_w$ is the affine Grothendieck polynomial, $G^{w_1} \in \widehat{\Lambda/I_n}$ denotes the affine stable Grothendieck polynomial [19], and $\mathfrak{G}_{w_2}$ is the Grothendieck polynomial of Lascoux and Schützenberger. For example, let $n = 3$ and $w = s_2 s_1$. We have

$$\begin{aligned}
\tilde{\mathfrak{G}}_{s_2s_1} &= G^{s_2s_1} - G^{s_2s_1}\mathfrak{G}_{s_1} + G^{s_2}\mathfrak{G}_{s_1} - G^{s_2}\mathfrak{G}_{s_2s_1} + \mathfrak{G}_{s_1s_1}\\
&= G^{s_2s_1}(1-\mathfrak{G}_{s_1}) + G^{s_2}(\mathfrak{G}_{s_1} - \mathfrak{G}_{s_2s_1}) + \mathfrak{G}_{s_2s_1}\\
&= G^{s_2s_1}(1-x_1) + G^{s_2}(x_1 - x_1^2) + x_1^2.
\end{aligned}$$

From [19, A.3.6], we have expansions in terms of Schur functions

$$\begin{aligned}
G^{s_2} &= G_1^{(2)} = s_1 - s_{11} + s_{111} - s_{1111} + \cdots,\\
G^{s_2s_1} &= G_2^{(2)} = s_2 - s_{21} + s_{211} - s_{2111} + \cdots.
\end{aligned}$$

Note that the lowest degree term is $s_2 + s_1x_1 + x_1^2 = \tilde{\mathfrak{S}}_{s_2s_1}$. We plan to compare these formulae with the affine Grothendieck polynomials of Kashiwara and Shimozono [10] in future work.

6.4 Classical type in cohomology

The affine coproduct formula in cohomology can be applied to obtain formulas for Schubert classes in finite-dimensional flag varieties G/B. For classical type we compare these formulas with the Schubert class formulas of Billey and Haiman [3] for nonequivariant cohomology and those of Ikeda, Mihalcea, and Naruse [6] for equivariant cohomology, providing retrospective insight into these formulas.

The affine coproduct formula writes an affine flag variety Schubert class as a sum of products of affine Grassmannian Schubert classes and G/B Schubert classes. The formulas of [3] and [6] write a G/B class of type C_n or D_n as a sum of products of cominuscule Grassmannian Schubert classes and type A flag Schubert classes. To compare our formulae with those in [3, 6], we use the fact that at the bottom of the affine Grassmannian of type $C_n^{(1)}$ or $D_n^{(1)}$ there is a copy of a cominuscule Grassmannian. To perform this comparison it is necessary to use an automorphism of the affine Dynkin diagram.

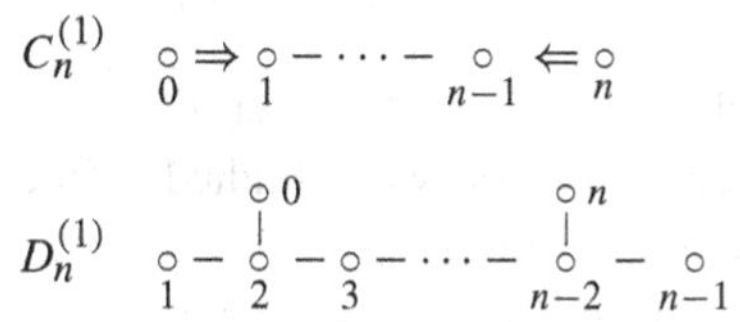

$$C_n^{(1)} \quad \underset{0}{\circ} \Rightarrow \underset{1}{\circ} - \cdots - \underset{n-1}{\circ} \Leftarrow \underset{n}{\circ}$$

$$D_n^{(1)} \quad \underset{1}{\circ} - \underset{2}{\circ} - \underset{3}{\circ} - \cdots - \underset{n-2}{\circ} - \underset{n-1}{\circ} \quad (\text{with } \circ\, 0 \text{ attached to } 2 \text{ and } \circ\, n \text{ attached to } n-2)$$

Figure 1 Affine Dynkin diagrams.

Consider the affine Dynkin diagrams of types $C_n^{(1)}$ and $D_n^{(1)}$ in Figure 1. Let τ be the affine Dynkin automorphism for type $C_n^{(1)}$ or $D_n^{(1)}$ given by $\tau(i) = n - i$ for $i \in \hat{I}$. There are two copies of the classical Weyl group W in $\hat{W}$: the usual one W and $W' = W'_n = \tau(W)$, which is generated by s_j for $j \in \hat{I} \setminus \{n\}$ Let G and G' denote the subgroups of the corresponding loop group (or affine Kac-Moody group) with Weyl groups W and W' respectively, and let G/B and G'/B' be the two finite-dimensional flag varieties (either the symplectic flag variety or the orthogonal flag variety). Finally, note that the subgroup of $\hat{W}$ generated by s_j for $j \in \hat{I} \setminus \{0, n\}$ is isomorphic to the type A_{n-1} Weyl group $W_{A_{n-1}}$.

For $w \in W'$, if we have $w \doteq w_1 w_2$ for $(w_1, w_2) \in \hat{W} \times W$, then $(w_1, w_2) \in W' \times W_{A_{n-1}}$. Applying the affine coproduct formula (Theorem 5.11) and pulling back to $H_T^*(G'/B')$, we have in $H_T^*(G'/B')$ (with $G' = \mathrm{Sp}(2n)$ or $G' = \mathrm{SO}(2n)$) the equality, for $w \in W'$,

$$\xi'^{w} = \sum_{\substack{(w_1, w_2) \in W' \times W_{A_{n-1}} \\ w_1 w_2 \doteq w}} F^{w_1} \cup \tau^* \xi^{w_2}. \tag{81}$$

Here, F^{w_1} is the pullback to $H_T^*(G'/B')$ (under the natural projection from the flag variety to a Grassmannian) of an element of the torus-equivariant cohomology $H_T^*(\mathrm{LG}(n, 2n))$ of the Lagrangian Grassmannian in the $C_n^{(1)}$ case, or an element of the torus-equivariant cohomology $H_T^*(\mathrm{OG}(n, 2n))$ of the orthogonal Grassmannian in the $D_n^{(1)}$ case. Also, ξ^{w_2} denotes a Schubert class in $H_T^*(G/B)$ and τ^* is the composition of pullback maps $H_T^*(G/B) \to H_T^*(\hat{\mathrm{Fl}}_G) \to H_T^*(G'/B')$ (the first one being ev_1^*).

Let us compare (81) to the results of [3, 6] following an argument similar to the one in Section 6.2. For concreteness, let us consider the Schubert polynomial $\mathfrak{S}_w^C \in \mathbb{Z}[z_1, z_2, \ldots] \otimes \Gamma$ of type C [3, Theorem 2.5] (type D is similar), where $\Gamma \subset \Lambda$ is the subring spanned by Q-Schur functions (over $\mathbb{Q}$, the ring Γ is generated by the odd power sum symmetric functions). There is a ring homomorphism

$$\phi_n : \mathbb{Z}[z_1, z_2, \ldots] \otimes \Gamma \to H^*(G'/B')$$

taking $\mathfrak{S}_w^C$ to the Schubert class ξ'^w. Now let $F_v^C \in \Gamma$ denote the type C Stanley symmetric functions, which were studied in the classical setting in [3] [12] [13] and in the affine setting in [20] (see also [27]). By [20], under ϕ_n we have that F_v^C is sent to (the non-equivariant class) $F^v \in H^*(G'/B')$ and the usual Schubert polynomial $\mathfrak{S}_u(z)$ in z-variables is sent to $\tau^*(\xi^u)$. According to [3], the ring $\mathbb{Z}[z_1, z_2, \ldots] \otimes \Gamma$ injects into the projective limit $\varprojlim_n H_T^*(G'/B')$. It follows from (81) that we must have the expansion $\mathfrak{S}_w^C = \sum_{\substack{(w_1, w_2) \in W' \times W_{A_{n-1}} \\ w_1 w_2 \doteq w}} F_{w_1}^C \mathfrak{S}_{w_2}(z)$, which is the Billey–Haiman formula for the type C Schubert polynomials.

A similar formula holds in equivariant cohomology. Equivariantly, our F^{w_1} is a double analogue of the type C or D Stanley symmetric function. Our (81) gives a formula for the double Schubert polynomials of type C or D as a sum of products of double type C or D Stanley symmetric functions and type A double Schubert polynomials. Since the coproduct expansion of a Schubert class is unique, our formula must equal to that in [6]. Our definition of (equivariant) affine Stanley class gives a precise geometric description of the Grassmannian components of the formulas in [6]. We do not obtain a new proof of their formula, because we do not separately know that our F^{w_1} can be compared to the combinatorics in [6]. See also [1, 29].

6.5 Classical type in K-theory

Our coproduct formula in equivariant K-theory should be compared with the classical type double Grothendieck polynomials of A. N. Kirillov and H. Naruse [5] [9] just as our cohomological formula relates to the work of Billey and Haiman. There is a Pieri formula [28] in the K-homology of the type A affine Grassmannian, which gives some coproduct structure constants for K-cohomology Schubert classes.

References

[1] D. Anderson and W. Fulton. Degeneracy loci, Pfaffians, and vexillary signed permutations in types B, C, and D, preprint, 2012, arXiv:1210.2066.

[2] S. Billey, W. Jockusch, and R. Stanley. Some combinatorial properties of Schubert polynomials. J. Algebraic Combin. 2 (1993), no. 4, 345–374.

[3] S. Billey and M. Haiman. Schubert polynomials for the classical groups. J. Amer. Math. Soc. 8 (1995), no. 2, 443–482.

[4] V. Ginzburg. Geometric methods in the representation theory of Hecke algebras and quantum groups. Notes by Vladimir Baranovsky. NATO Adv. Sci. Inst. Ser.

C Math. Phys. Sci., 514, Representation theories and algebraic geometry (Montreal, PQ, 1997), 127–183, Kluwer Acad. Publ., Dordrecht, 1998.

[5] T. Hudson, T. Ikeda, T. Matsumura, and H. Naruse. Double Grothendieck polynomials for symplectic and odd orthogonal Grassmannians. J. Algebra 546 (2020), 294–314.

[6] T. Ikeda, L. C. Mihalcea, and H. Naruse. Double Schubert polynomials for the classical groups. Adv. in Math. 226 (2011) no. 1, 840–886.

[7] V. Kac. Infinite dimensional Lie algebras. Third edition. Cambridge University Press, Cambridge, 1990.

[8] B. Kostant and S. Kumar. T-equivariant K-theory of generalized flag varieties. J. Differential Geom. 32 (1990) no. 2, 549–603.

[9] A. N. Kirillov and H. Naruse. Construction of double Grothendieck polynomials ofclassical types using idCoxeter algebras. Tokyo J. Math. 39, 3 (2017), 695–728.

[10] M. Kashiwara and M. Shimozono. Equivariant K-theory of affine flag manifolds and affine Grothendieck polynomials. Duke Math. J. Volume 148, Number 3 (2009), 501–538.

[11] S. Kumar, Positivity in T-equivariant K-theory of flag varieties associated to Kac-Moody groups. With an appendix by Masaki Kashiwara. J. Eur. Math. Soc. (JEMS) 19 (2017), no. 8, 2469–2519.

[12] T. K. Lam. B and D analogues of stable Schubert polynomials and related insertion algorithms. Ph. D. Thesis, Massachusetts Institute of Technology. 1995.

[13] T. K. Lam. B_n Stanley symmetric functions. Proceedings of the 6th Conference on Formal Power Series and Algebraic Combinatorics (New Brunswick, NJ, 1994). Discrete Math. 157 (1996), no. 1–3, 241–270.

[14] T. Lam. Affine Stanley symmetric functions. Amer. J. Math. 128 (2006) no. 6, 1553–1586.

[15] T. Lam. Schubert polynomials for the affine Grassmannian. J. Amer. Math. Soc. 21 (2008), no. 1, 259–281.

[16] T. Lam, S.-J. Lee, and M. Shimozono. Back stable Schubert calculus, preprint, 2018; arXiv:arXiv:1806.11233.

[17] T. Lam, C. Li, L. C. Mihalcea, M. Shimozono. A conjectural Peterson isomorphism in K-theory. J. Algebra 513 (2018), 326–343.

[18] T. Lam, L. Lapointe, J. Morse, A. Schilling, M. Shimozono, and M. Zabrocki. k-Schur functions and affine Schubert calculus. Fields Institute Monographs, 33. Springer, New York; Fields Institute for Research in Mathematical Sciences, Toronto, ON, 2014. viii+219 pp.

[19] T. Lam, A. Schilling and M. Shimozono. K-theory Schubert calculus of the affine Grassmannian. Comp. Math. 146 (2010), no. 4, 811–852.

[20] T. Lam, A. Schilling and M. Shimozono. Schubert polynomials for the affine Grassmannian of the symplectic group. Math. Z. 264 (2010), no. 4, 765–811.

[21] T. Lam and M. Shimozono. Quantum cohomology of G/P and homology of affine Grassmannian. Acta Math. 204 (2010), no. 1, 49–90.

[22] T. Lam and M. Shimozono. k-double Schur functions and equivariant (co)homology of the affine Grassmannian. Math. Ann. 356 (2013), no. 4, 1379–1404.

[23] A. Lascoux and M.-P. Schützenberger. Polynômes de Schubert. C. R. Acad. Sci. Paris Sér. I Math. 294 (1982), 447–450.

[24] S. J. Lee. Combinatorial description of the cohomology of the affine flag variety. Trans. Amer. Math. Soc. 371 (2019), 4029–4057.
[25] S. A. Mitchell. The Bott filtration of a loop group. Algebraic topology, Barcelona, 1986, 215–226, Lecture Notes in Math., 1298, Springer, Berlin, 1987.
[26] D. Peterson. Quantum cohomology of G/P, Lecture notes, M.I.T., Spring 1997.
[27] S. Pon. Affine Stanley symmetric functions for classical types. J. Algebraic Combin. 36 (2012), no. 4, 595–622.
[28] M. Takigiku. A Pieri-type formula and a factorization formula for sums of K-k-Schur functions, arXiv:1802.06335.
[29] H. Tamvakis. Schubert polynomials and degeneracy locus formulas in "Schubert Varieties, Equivariant Cohomology and Characteristic Classes", 261–314, European Mathematical Society Series of Congress Reports, Zürich, 2018.

19

Bost–Connes Systems and $\mathbb{F}_1$-structures in Grothendieck Rings, Spectra, and Nori Motives

Joshua F. Lieber[a], Yuri I. Manin[b], and Matilde Marcolli[c]

Abstract. We construct geometric lifts of the Bost–Connes algebra to Grothendieck rings and to the associated assembler categories and spectra, as well as to certain categories of Nori motives. These categorifications are related to the integral Bost–Connes algebra via suitable Euler characteristic type maps and zeta functions, and in the motivic case via fiber functors. We also discuss aspects of $\mathbb{F}_1$-geometry, in the framework of torifications, that fit into this general setting.

1 Introduction and summary

This survey/research paper interweaves many different strands that recently became visible in the fabric of algebraic geometry, arithmetics, (higher) category theory, quantum statistics, homotopical "brave new algebra" etc., see especially A. Connes and C. Consani [25] [26]; A. Huber, St. Müller-Stach [43], etc.

In this sense, our present paper can be considered as a continuation and further extension of [54], and we will be relying on much of the work in that paper for details and examples. The motivational starting point in [54] was coming from the interpretation given in [27] of the Bost–Connes quantum statistical mechanical system, and in particular the integral Bost–Connes algebra, as a form of $\mathbb{F}_1$-structure, or "geometry below Spec($\mathbb{Z}$)". The main theme of [54] is an exploration of how this structure manifests itself beyond the usual constructions of $\mathbb{F}_1$-structures on certain classes of varieties over $\mathbb{Z}$. In particular, the results of [54] focus on lifts of the integral Bost–Connes algebra to certain Grothendieck rings and to associated homotopy-theoretic

[a] California Institute of Technology
[b] Max Planck Institute for Mathematics
[c] California Institute of Technology

spectra obtained via assembler categories, and also on another form of $\mathbb{F}_1$-structures arising through quasi-unipotent Morse–Smale dynamical systems.

The main difference between the present paper and [54] consists in a change of the categorical environment: the unifying vision we already considered in [54] was provided by I. Zakharevich's notions of assemblers and scissors congruences: cf. [70], [71], [72], and [21]. In this paper, we continue to use the formalism of assemblers and the associated spectra, but we complement it with categories of Nori motives, [43].

As in [54], we focus primarily on various geometrizations of the Bost–Connes algebra. Some of these constructions take place in Grothendieck rings, like the previous cases considered in [54], and are aimed at lifting the Bost–Connes endomorphisms to the level of homotopy theoretic spectra through the use of Zakharevich's formalism of assembler categories. We focus on the case of relative Grothendieck rings, endowed with appropriate equivariant Euler characteristic maps. For varieties that admits torifications, we introduce zeta functions based on the counting of points over $\mathbb{F}_1$ and over extensions $\mathbb{F}_{1^m}$. We present a more general construction of Bost–Connes type systems associated to exponentiable motivic measures and the associated zeta functions with values in Witt rings, obtained using a lift of the Bost–Connes algebra to Witt rings via Frobenius and Verschiebung maps.

We then consider lifts of the Bost–Connes algebra to Nori motives, where we use a (slightly generalized) version of Nori motives, which may be of independent interest in view of possible versions of equivariant periods. In this categorical setting we show that the fiber functor from Nori motives maps to a categorification of the Bost–Connes algebra previously constructed by Tabuada and the third author, compatibly with the functors realizing the Bost–Connes structure.

1.1 Structure of the paper and main results

Below we will briefly describe the content of the subsequent Sections, and the main results of the paper, with pointers to the specific statements where these are proved.

1.1.1 Bost–Connes systems and relative equivariant Grothendieck rings

In §2, we show the existence of a lift of the Bost–Connes structure to the relative equivariant Grothendieck ring $K_0^{\hat{\mathbb{Z}}}(\mathcal{V}_S)$, extending similar results previously obtained in [54] for the equivariant Grothendieck ring $K_0^{\hat{\mathbb{Z}}}(\mathcal{V})$. The main result in this part of the paper is Theorem 2.11, where the existence of this lift is proved. The rest of the section covers the preliminary results needed for this main result.

In particular, we first introduce the integral Bost–Connes algebra in §2.1, in the form in which it was introduced in [27]. We recall in §2.2 and 2.3 the relative and the equivariant relative Grothendieck rings, and in §2.4 the associated equivariant Euler characteristic map.

In §2.5 we recall from [54] the geometric form of the Verschiebung map that is used in the lifting of the Bost–Connes structure to varieties with suitable $\hat{\mathbb{Z}}$-actions. In §2.6 we introduce the Bost–Connes maps σ_n and $\tilde{\rho}_n$ on classes in $K_0^{\hat{\mathbb{Z}}}(\mathcal{V}_S)$ and Proposition 2.6 shows the way they transform the varieties and the base scheme with their respective $\hat{\mathbb{Z}}$-action.

In §2.7 we recall from [27] the prime decomposition of the integral Bost–Connes algebra, which for a finite set of primes F separates out an F-part and an F-coprime part of the algebra. We then show in §2.8, and in particular Proposition 2.8, that, given a scheme S with a good effectively finite action of $\hat{\mathbb{Z}}$, there is an associated finite set of primes F such that the F-coprime part of the Bost–Connes algebra lifts to endomorphisms of $K_0^{\hat{\mathbb{Z}}}(\mathcal{V}_S)$.

Finally in §2.9 we show how to lift the full Bost–Connes algebra to homomorphisms between Grothendieck rings $K_0^{\hat{\mathbb{Z}}}(\mathcal{V}_{(S,\alpha)})$ where the scheme S and the action α are also transformed by the Bost–Connes map. By an analysis of the structure of periodic points in Lemma 2.9 we show the compatibility with the equivariant Euler characteristic, so we can then prove the main result in Theorem 2.11, showing that the equivariant Euler characteristic intertwines the Bost–Connes maps σ_n and $\tilde{\rho}_n$ on the $K_0^{\hat{\mathbb{Z}}}(\mathcal{V}_{(S,\alpha)})$ rings with the original σ_n and $\tilde{\rho}_n$ maps of the integral Bost–Connes algebra.

1.1.2 Bost–Connes systems on assembler categories and spectra

In §3 we further lift the Bost–Connes structure obtained at the level of Grothendieck rings $K_0^{\hat{\mathbb{Z}}}(\mathcal{V}_S)$ to assembler categories underlying these Grothendieck rings and to the homotopy-theoretic spectra defined by these categories. Again this extends to the equivariant relative case results that were obtained in [54] for the non-relative setting. The main result in this part of the paper is Theorem 3.15, where it is shown that the maps σ_n and $\tilde{\rho}_n$ on the Grothendieck rings $K_0^{\hat{\mathbb{Z}}}(\mathcal{V}_{(S,\alpha)})$ constructed in the previous section lift to functors of the underlying assembler categories, that induce these maps on K_0.

In §3.1 we recall the formalism of assembler categories of [70], underlying scissor congruence relations and Grothendieck rings. In §3.2 we review Segal's Γ-spaces formalism and how one obtains the homotopy-theoretic spectrum associated to an assembler. In §3.3 and §3.4 we then lift this formalism by endowing the main relevant objects with an action of a finite cyclic group, with appropriate compatibility conditions. It is this further structure that provides a framework for the respective lifts of the Bost–Connes algebras, as in the cases discussed in [54] and in the ones we will be discussing in the following sections. We give here a very general definition of Bost–Connes systems in categories, based on endofunctors of subcategories of the automorphism category. In the applications considered in this paper we will be using only the special case where the automorphisms are determined by an effectively finite action of $\hat{\mathbb{Z}}$, but we introduce the more general framework in anticipation of other possible applications.

In §3.5 we construct the assembler underlying the equivariant relative Grothendieck ring $K_0^{\hat{\mathbb{Z}}}(\mathcal{V}_S)$ and we prove the main result in Theorem 3.15 on the lift of the Bost–Connes structure to functors of these assemblers.

1.1.3 Bost–Connes systems on Grothendieck rings and assemblers of torified varieties

In §4 we consider the approach to $\mathbb{F}_1$-geometry via torifications of varieties over $\mathbb{Z}$, introduced in [51]. The main results of this part of the paper are Proposition 4.2 and Proposition 4.5 where we construct assembler categories of torified varieties and we show the existence of a lift of the Bost–Connes algebra to these categories.

In §4.1 we recall the notion of torified varieties from [51] and the different versions of morphisms of torified varieties from [53], and we construct Grothendieck rings of torified varieties for each flavor of morphisms In §4.1.1 we introduce a relative version of these Grothendieck rings of torified varieties. In §4.2 we describe $\mathbb{Q}/\mathbb{Z}$ and $\hat{\mathbb{Z}}$-actions on torifications.

In §4.3 we construct the assembler categories underlying these relative Grothendieck rings and in §4.4 we prove the first main result of this section by constructing the lift of the Bost–Connes structure.

1.1.4 Torified varieties, $\mathbb{F}_1$-points, and zeta functions

This section continues the theme of torified varieties from the previous section but with main focus on some associated zeta functions. We consider two different kinds of zeta function: $\mathbb{F}_1$-zeta functions that count $\mathbb{F}_1$-points of torified varieties, in an appropriates sense that it discussed in §5.1, and dynamical zeta functions associated to endomorphisms of torified varieties that are compatible with the torification. The use of dynamical zeta functions is motivated by a proposal made in [54] for a notion of $\mathbb{F}_1$-structures based on dynamical systems that induce quasi-unipotent maps in homology.

The two main results of this section are Proposition 5.4 and Proposition 5.8 where we show that the $\mathbb{F}_1$-zeta function, respectively the dynamical zeta function, determine exponentiable motivic measures from the Grothendieck rings of torified varieties introduced in the previous section to the ring $W(\mathbb{Z})$ of Witt vectors.

We introduce in §5.1 and §5.2 the counting of $\mathbb{F}_1$-points of a torified variety and its relation to the Grothendieck class. We in show in §5.2 how the Bialynicki–Birula decomposition can be used to determine torifications and we give in §5.3 some explicit examples of computations of Grothendieck classes in simple cases that have physical significance in the context of BPS counting in string theory.

In §5.4 we introduce the $\mathbb{F}_1$-zeta function and we prove Proposition 5.4. In §5.5 we explain how the $\mathbb{F}_1$-zeta function can be obtained from the Hasse–Weil zeta function.

In §5.6 we consider torified varieties with dynamical systems compatible with the torification and the associated Lefschetz and Artin–Mazur dynamical zeta functions. We recall the definition and main properties of these zeta functions in §5.6.1 and we prove in Proposition 5.8 in §5.6.2.

1.1.5 Spectrification of Witt vectors and lifts of zeta functions

In the constructions described in §§ 3 and 4 of [54] and in §§ 2–6 of the present paper we obtain lifts of the integral Bost–Connes algebra to various assembler categories and associated spectra, starting from a ring homomorphism (motivic measure) from the relevant Grothendieck ring to the group ring $\mathbb{Z}[\mathbb{Q}/\mathbb{Z}]$ of the integral Bost–Connes algebra, that is equivariant with respect to the maps σ_n and $\tilde{\rho}_n$ of (4) and (5) of the Bost–Connes algebra and the maps (also denoted by σ_n and $\tilde{\rho}_n$) on the Grothendieck ring induced by a Bost–Connes system on the corresponding assembler category. The motivic measure provides in this way a map that lifts the Bost–Connes structure.

This part of the paper considers then a more general class of zeta functions ζ_μ associated to exponentiable motivic measures $\mu : K_0(\mathcal{V}) \to R$ with values in a commutative ring R, that admit a factorization into linear factors in the subring $W_0(R)$ of the Witt ring $W(R)$.

Our main results in this section are Proposition 6.9, showing that these zeta functions lift to the level of assemblers and spectra, and Proposition 6.14, which shows that the Frobenius and Verschiebung maps on the endomorphism category lift, through the lift of the zeta function, to a Bost–Connes system on the assembler category of the Grothendieck ring of varieties $K_0(\mathcal{V})$.

The main step toward establishing the main results of this section is the construction in §6.2 and §6.3 of a spectrification of the ring $W_0(R)$. This is obtained using its description in terms of the K_0 of the endomorphism category $\mathcal{E}_R$ and of R, and the formalism of Segal Gamma-spaces. The spectrification we use here is not the same as the spectrification of the ring of Witt vectors introduced in [40]. The lifting of Bost–Connes systems via motivic measures in discussed in §6.4, where Proposition 6.14 is proved.

We also consider again in §6.5 the setting on dynamical $\mathbb{F}_1$-structures proposed in [54], with a pair (X, f) of a variety and an endomorphism that induces a quasi-unipotent map in homology, and we associate to these data the operator-theoretic spectrum of the quasi-unipotent map, seen as an element in $\mathbb{Z}[\mathbb{Q}/\mathbb{Z}]$. This determines a spectral map $\sigma : K_0^{\mathbb{Z}}(\mathcal{V}_{\mathbb{C}}) \to \mathbb{Z}[\mathbb{Q}/\mathbb{Z}]$ with the properties of an Euler characteristic.

Another main result in this section is Proposition 6.16, showing that this spectral Euler characteristic lifts to a functor from the assembler category underlying $K_0^{\mathbb{Z}}(\mathcal{V}_{\mathbb{C}})$ to the Tannakian category $\mathrm{Aut}^{\bar{\mathbb{Q}}}_{\mathbb{Q}/\mathbb{Z}}(\mathbb{Q})$ that categorifies the ring $\mathbb{Z}[\mathbb{Q}/\mathbb{Z}]$, and that the resulting functor lifts the Bost–Connes structure on $\mathrm{Aut}^{\bar{\mathbb{Q}}}_{\mathbb{Q}/\mathbb{Z}}(\mathbb{Q})$ described in [57] to a Bost–Connes structure on the assembler of $K_0^{\mathbb{Z}}(\mathcal{V}_{\mathbb{C}})$.

1.1.6 Bost–Connes systems in categories of Nori motives

When we replace the formalism of assembler categories and homotopy theoretic spectra underlying the Grothendieck rings with geometric diagrams and associated Tannakian categories of Nori motives, with the same notion of categorical Bost–Connes systems introduced in Definitions 3.9 and 3.11, we can lift the Euler characteristic type motivic measures to the level of categorifications, where, as in the previous section, the categorification of the Bost–Connes algebra is the one introduced in [57], given by a Tannakian category of $\mathbb{Q}/\mathbb{Z}$-graded vector spaces endowed with Frobenius and Verschiebung endofunctors.

In §7 in this paper we construct Bost–Connes systems in categories of Nori motives. The main result of this part of the paper is Theorem 7.7, which shows that there is a categorical Bost–Connes system on a category of equivariant Nori motives, and that the fiber functor to the categorification of the Bost–Connes algebra constructed in [57] intertwines the respective Bost–Connes endofunctors.

In §7.1 and §7.2 we review the construction of Nori motives from diagrams and their representations. In §7.3 we construct a category of equivariant Nori motives. In §7.4 we describe the endofunctors of this category that implement the Bost–Connes structure and we prove the main result in Theorem 7.7. In §7.6 we generalize this result to the relative case, using Arapura's motivic sheaves version of Nori motives.

Finally, in §7.6 we consider Nori diagrams associated to assemblers and we formulate the question of their "universal cohomological representations". This is a contemporary embodiment of the primordial Grothendieck's dream that motives constitute a universal cohomology theory of algebraic varieties.

2 Bost-Connes systems in Grothendieck rings

In [54] it was shown that the integral Bost–Connes algebra of [27] admits lifts to certain Grothendieck rings, via corresponding equivariant Euler characteristic maps. The cases analyzed in [54] included the cases of the equivariant Grothendieck ring $K_0^{\hat{\mathbb{Z}}}(\mathcal{V})$ and the equivariant Konsevich–Tschinkel Burnside ring $\mathrm{Burn}^{\hat{\mathbb{Z}}}(\mathbb{K})$. We treat here, in a similar way, the case of the relative equivariant Grothendieck ring $K_0^{\hat{\mathbb{Z}}}(\mathcal{V}_S)$. This case is more delicate than the other cases considered in [54], because when the Bost–Connes maps act on the classes in $K_0^{\hat{\mathbb{Z}}}(\mathcal{V}_S)$ they also change the base scheme S with its $\hat{\mathbb{Z}}$-action.

The main result in this section is the existence of a lifting of the Bost–Connes structure to $K_0^{\hat{\mathbb{Z}}}(\mathcal{V}_S)$, proved in Theorem 2.11.

We first review the definition of the integral Bost–Connes algebra in §2.1 and the equivariant relative Grothendieck ring in §2.3. The rest of the section then develops the intermediate steps leading to the proof of the main results of Theorem 2.11.

2.1 Bost-Connes algebra

The Bost–Connes algebra was introduced in [15] as a quantum statistical mechanical system that exhibit the Riemann zeta function as partition function, the generators of the cyclotomic extensions of $\mathbb{Q}$ as values of zero-temperature KMS equilibrium states on arithmetic elements in the algebra, and the abelianized Galois group $\hat{\mathbb{Z}}^* \simeq \mathrm{Gal}(\bar{\mathbb{Q}}/\mathbb{Q})^{ab}$ as group of quantum symmetries. In particular, the arithmetic subalgebra of the Bost–Connes system is given by the semigroup crossed product

$$\mathbb{Q}[\mathbb{Q}/\mathbb{Z}] \rtimes \mathbb{N} \tag{1}$$

of the multiplicative semigroup $\mathbb{N}$ of positive integers acting on the group algebra of the group $\mathbb{Q}/\mathbb{Z}$.

The additive group $\mathbb{Q}/\mathbb{Z}$ can be identified with the multiplicative group ν^* of *roots of unity embedded into* $\mathbb{C}^*$: namely, $r \in \mathbb{Q}/\mathbb{Z}$ corresponds to $e(r) := \exp(2\pi i\, r)$. More generally, the choice of the embedding can be modified by an arbitrary choice of an element in $\hat{\mathbb{Z}}^* = \mathrm{Hom}(\mathbb{Q}/\mathbb{Z}, \mathbb{Q}/\mathbb{Z})$, as is usually done in representations of the Bost–Connes algebra, see [15]. Thus, we will use here interchangeably the notation ζ or r for elements of $\mathbb{Q}/\mathbb{Z}$ assuming a choice of embedding as above. The group algebra $\mathbb{Q}[\nu^*]$ consists of formal finite linear combinations $\sum_{a_\zeta \in \mathbb{Q}} a_\zeta \zeta$ of roots of unity $\zeta \in \nu^*$. Formality means here that the sum is *not* related to the additive structure of $\mathbb{C}$.

The action of the semigroup $\mathbb{N}$ on $\mathbb{Q}[\mathbb{Q}/\mathbb{Z}]$ that defines the crossed product (1) is given by the endomorphisms

$$\rho_n\left(\sum a_\zeta \zeta\right) := \sum a_\zeta \, \frac{1}{n} \sum_{\zeta'^n = \zeta} \zeta'. \tag{2}$$

Equivalently, the algebra (1) is generated by elements $e(r)$ with the relations $e(0) = 1, e(r+r') = e(r)e(r')$, and elements μ_n and μ_n^* satisfying the relations

$$\begin{aligned} &\mu_n^*\mu_n = 1,\ \forall n; && \\ &\mu_n\mu_n^* = \pi_n,\ \forall n && \text{with } \pi_n = \tfrac{1}{n}\textstyle\sum_{nr=0} e(r); \\ &\mu_{nm} = \mu_n\mu_m,\ \forall n,m; && \mu_{nm}^* = \mu_n^*\mu_m^*,\ \forall n,m; \\ &\mu_n^*\mu_m = \mu_m\mu_n^* && \text{if } (n,m) = 1. \end{aligned} \tag{3}$$

The semigroup action (2) is then equivalently written as $\rho_n(a) = \mu_n\, a\, \mu_n^*$, for all $a = \sum a_\zeta \zeta$ in $\mathbb{Q}[\mathbb{Q}/\mathbb{Z}]$. The element $\pi_n \in \mathbb{Q}[\mathbb{Q}/\mathbb{Z}]$ is an idempotent, hence the generators μ_n are isometries but not unitaries. See [15] and §3 of [28] for a detailed discussion of the Bost–Connes system and the role of the arithmetic subalgebra (1).

In [27] an integral model of the Bost–Connes algebra was constructed in order to develop a model of $\mathbb{F}_1$-geometry in which the Bost–Connes system encodes the extensions $\mathbb{F}_{1^m}$, in the sense of [45], of the "field with one element" $\mathbb{F}_1$.

The integral Bost–Connes algebra is obtained by considering the group ring $\mathbb{Z}[\mathbb{Q}/\mathbb{Z}]$, which we can again implicitly identify with $\mathbb{Z}[\nu^*]$ for a choice of embedding $\mathbb{Q}/\mathbb{Z} \hookrightarrow \mathbb{C}$ as roots of unity.

Define its *ring endomorphisms* σ_n:

$$\sigma_n\left(\sum a_\zeta \zeta\right) := \sum a_\zeta \zeta^n. \tag{4}$$

Define *additive maps* $\tilde{\rho}_n\colon \mathbb{Z}[\nu^*] \to \mathbb{Z}[\nu^*]$:

$$\tilde{\rho}_n\left(\sum a_\zeta \zeta\right) := \sum a_\zeta \sum_{\zeta'^n=\zeta} \zeta'. \tag{5}$$

The maps σ_n and $\tilde{\rho}_n$ satisfy the relations

$$\sigma_n \circ \tilde{\rho}_n = n\,\mathrm{id}, \quad \tilde{\rho}_n \circ \sigma_n = n\,\pi_n. \tag{6}$$

The integral Bost–Connes algebra is then defined as the algebra generated by the group ring $\mathbb{Z}[\mathbb{Q}/\mathbb{Z}]$ and generators $\tilde{\mu}_n$ and μ_n^* with the relations

$$\begin{aligned}
&\tilde{\mu}_n\, a\, \mu_n^* = \tilde{\rho}_n(a),\ \forall n; \\
&\mu_n^*\, a = \sigma_n(a)\, \mu_n^*,\ \forall n; \quad a\, \tilde{\mu}_n = \tilde{\mu}_n\, \sigma_n(a),\ \forall n; \\
&\tilde{\mu}_{nm} = \tilde{\mu}_n \tilde{\mu}_m,\ \forall n, m; \quad \mu_{nm}^* = \mu_n^* \mu_m^*,\ \forall n, m; \\
&\tilde{\mu}_n \mu_m^* = \mu_m^* \tilde{\mu}_n \qquad \text{if } (n, m) = 1.
\end{aligned} \tag{7}$$

where the relations in the first two lines hold for all $a = \sum a_\zeta \zeta \in \mathbb{Z}[\mathbb{Q}/\mathbb{Z}]$, with σ_n and $\tilde{\rho}_n$ as in (4) and (5).

The maps $\tilde{\rho}_n$ of the integral Bost–Connes algebra and the semigroup action ρ_n in the rational Bost–Connes algebra (1) are related by

$$\rho_n = \frac{1}{n}\tilde{\rho}_n$$

with $\tilde{\rho}_n$ defined as in (5).

2.2 Relative Grothendieck ring

We describe here a variant of construction of [54], where we work with relative Grothendieck rings and with an Euler characteristic with values in a Grothendieck ring of locally constant sheaves. We show that this relative setting provides ways of lifting to the level of Grothendieck classes certain subalgebras of the integral Bost–Connes algebras associated to the choice of a finite set of non-archimedean places.

Definition 2.1 The relative Grothendieck ring $K_0(\mathcal{V}_S)$ of varieties over a base variety S over a field $\mathbb{K}$ is generated by the isomorphism classes of data $f : X \to S$ of a variety X over S with the relations

$$[f : X \to S] = [f|_Y : Y \to S] + [f|_{X \smallsetminus Y} : X \smallsetminus Y \to S] \tag{8}$$

as in (8) for a closed embedding $Y \hookrightarrow X$ of varieties over S. The product is given by the fibered product $X \times_S Y$. We will write $[X]_S$ as shorthand notation for the class $[f : X \to S]$ in $K_0(\mathcal{V}_S)$.

A morphism $\phi : S \to S'$ induces a base change ring homomorphism $\phi^* : K_0(\mathcal{V}_{S'}) \to K_0(\mathcal{V}_S)$ and a direct image map $\phi_* : K_0(\mathcal{V}_S) \to K_0(\mathcal{V}_{S'})$ which is a group homomorphisms and a morphism of $K_0(\mathcal{V}_{S'})$-modules, but not a ring homomorphism. The class $[\phi : S \to S']$ as an element in $K_0(\mathcal{V}_{S'})$ is the image of $1 \in K_0(\mathcal{V}_S)$ under ϕ_*.

When $S = \mathrm{Spec}(\mathbb{K})$ one recovers the ordinary Grothendieck ring $K_0(\mathcal{V}_\mathbb{K})$.

2.3 Equivariant relative Grothendieck ring

Let X be a variety with a good action $\alpha : G \times X \to X$ by a finite group G and X' a variety with a good action α' by G'. As morphisms we then consider pairs (ϕ, φ) of a morphism $\phi : X \to X'$ and a group homomorphism $\varphi : G \to G'$ such that $\phi(\alpha(g,x)) = \alpha'(\varphi(g), \phi(x))$, for all $g \in G$ and $x \in X$. Thus, isomorphisms of varieties with good G-actions are pairs of an isomorphism $\phi : X \to X'$ of varieties and a group automorphism $\varphi \in \mathrm{Aut}(G)$ with the compatibility condition as above.

Given a base variety (or scheme) S with a given good action α_S of a finite group G, and varieties X, X' over S, with good G-actions $\alpha_X, \alpha_{X'}$ and G-equivariant maps $f : X \to S$ and $f' : X' \to S$, we consider morphisms given by a triple (ϕ, φ, ϕ_S) of a morphism $\phi : X \to X'$, a group homomorphism $\varphi : G \to G$ with the compatibility as above, and an endomorphism $\phi_S : S \to S$ such that $f' \circ \phi = \phi_S \circ f$. Then these maps also satisfy $\phi_S(\alpha_S(g, f(x)) = \alpha_S(\varphi(g), \phi_S(f(x)))$.

Definition 2.2 The relative equivariant Grothendieck ring $K_0^G(\mathcal{V}_S)$ is obtained as follows. Consider the abelian group generated by isomorphism classes $[f : X \to S]$ of varieties over S with compatible good G-actions, with respect to isomorphisms (ϕ, φ, ϕ_S) as above, with the inclusion-exclusion relations generated by equivariant embeddings with compatible G-equivariant maps

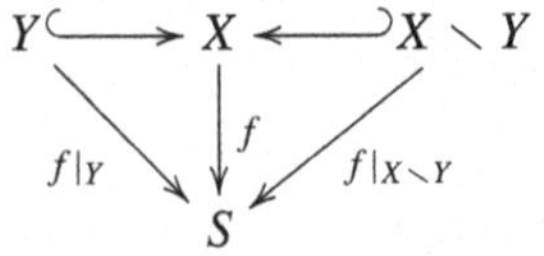

$$\begin{array}{ccccc} Y & \hookrightarrow & X & \hookleftarrow & X \smallsetminus Y \\ & {\scriptstyle f|_Y}\searrow & \downarrow{\scriptstyle f} & \swarrow{\scriptstyle f|_{X \smallsetminus Y}} & \\ & & S & & \end{array} \tag{9}$$

and isomorphisms. This means that we have $[f : X \to S] = [f_Y : Y \to S] + [f_{X \smallsetminus Y} : X \smallsetminus Y \to S]$ if there are isomorphisms $(\phi_Y, \varphi_Y, \phi_{S,Y})$ and $(\phi_{X \smallsetminus Y}, \varphi_{X \smallsetminus Y}, \phi_{S, X \smallsetminus Y})$, such that the diagram commutes.

$$\begin{array}{ccccccccc} Y & \xrightarrow{\phi_Y} & Y & \hookrightarrow & X & \hookleftarrow & X \smallsetminus Y & \xleftarrow{\phi'_{X\smallsetminus Y}} & X \smallsetminus Y \\ \downarrow f_Y & & & \searrow f|_Y & \downarrow f & \swarrow f|_{X\smallsetminus Y} & & & \downarrow f_{X\smallsetminus Y} \\ S & & \xrightarrow{\phi_{S,Y}} & & S & & \xleftarrow{\phi_{S,X\smallsetminus Y}} & & S \end{array} \tag{10}$$

The product $[f : X \to S] \cdot [f' : X' \to S]$ given by $[\tilde{f} : X \times_S X' \to S]$ with $\tilde{f} = f \circ \pi_X = f' \circ \pi_{X'}$ is well defined on isomorphism classes, with the diagonal action $\tilde{\alpha}(g,(x,x')) = (\alpha_X(g,x), \alpha_{X'}(g,x'))$ satisfying $f(\alpha_X(g,x)) = \alpha_S(g, f(x)) = \alpha_S(g, f'(x')) = f'(\alpha_{X'}(g,x'))$.

We will use the following terminology for the $\hat{\mathbb{Z}}$-actions we consider.

Definition 2.3 A good effectively finite action of $\hat{\mathbb{Z}}$ on a variety X is a good action that factors through an action of some quotient $\mathbb{Z}/N\mathbb{Z}$. We will write $\mathbb{Z}/N\mathbb{Z}$-effectively finite when we need to explicitly keep track of the level N.

In the case of the equivariant Grothendieck ring $K_0^{\hat{\mathbb{Z}}}(\mathcal{V})$ considered in [54], we can then also consider a relative version $K_0^{\hat{\mathbb{Z}}}(\mathcal{V}_S)$, with S a variety with a good effectively finite $\hat{\mathbb{Z}}$-action as above. We consider the Grothendieck ring $K_0^{\hat{\mathbb{Z}}}(\mathcal{V}_S)$ given by the isomorphism classes of S-varieties $f : X \to S$ with good effectively finite $\hat{\mathbb{Z}}$-actions with respect to which f is equivariant, with the notion of isomorphism described above. The product is given by the fibered product over S with the diagonal $\hat{\mathbb{Z}}$-action. The inclusion-exclusion relations are as in (8) where $Y \hookrightarrow X$ and $X \smallsetminus Y \hookrightarrow X$ are equivariant embeddings with compatible $\hat{\mathbb{Z}}$-equivariant maps as in (10).

2.4 Equivariant Euler characteristic

There is an Euler characteristic map given by a ring homomorphism

$$\chi_S^{\hat{\mathbb{Z}}} : K_0^{\hat{\mathbb{Z}}}(\mathcal{V}_S) \to K_0^{\hat{\mathbb{Z}}}(\mathbb{Q}_S) \tag{11}$$

to the Grothendieck ring of constructible sheaves over S with $\hat{\mathbb{Z}}$-action, [39], [50], [59], [69].

Lemma 2.4 *Let S be a variety with a good $\mathbb{Z}/N\mathbb{Z}$-effectively finite $\hat{\mathbb{Z}}$-action. Given a constructible sheaf $[\mathcal{F}]$ in $K_0^{\hat{\mathbb{Z}}}(\mathbb{Q}_S)$, let $\mathcal{F}|_{S^g}$ denote the restrictions to the fixed point sets S^g, for $g \in \mathbb{Z}/N\mathbb{Z}$. These determine classes in $K_0(\mathbb{Q}_{S^g}) \otimes \mathbb{Z}[\mathbb{Q}/\mathbb{Z}]$. One obtains in this way a map*

$$\chi : K_0^{\hat{\mathbb{Z}}}(\mathcal{V}_S) \to \bigoplus_{g \in \mathbb{Z}/N\mathbb{Z}} K_0(\mathbb{Q}_{S^g}) \otimes \mathbb{Z}[\mathbb{Q}/\mathbb{Z}]. \tag{12}$$

Proof The $\hat{\mathbb{Z}}$ action on S factors through some $\mathbb{Z}/N\mathbb{Z}$, hence the fixed point sets are given by S^g for $g \in \mathbb{Z}/N\mathbb{Z}$. Given a constructible sheaves $\mathcal{F}$ over S with $\hat{\mathbb{Z}}$-action, consider the restrictions $\mathcal{F}|_{S^g}$. The subgroup $\langle g \rangle$ generated by g acts trivially on S^g, hence for each $s \in S^g$ it acts on the stalk $\mathcal{F}_s$. Thus, these restrictions define classes $[\mathcal{F}|_{S^g}] \in K_0(\mathbb{Q}_{S^g}) \otimes R(\langle g \rangle) \subset K_0(\mathbb{Q}_{S^g}) \otimes \mathbb{Z}[\mathbb{Q}/\mathbb{Z}]$. By precomposing with the Euler characteristic (11) one then obtains the map (12). □

We will also consider the map $K_0^{\hat{\mathbb{Z}}}(\mathcal{V}_S) \to K_0(\mathbb{Q}_{S^G}) \otimes \mathbb{Z}[\mathbb{Q}/\mathbb{Z}]$ given by the Euler characteristic followed by restriction of sheaves to the fixed point set S^G of the group action.

2.4.1 Fixed points and delocalized homology

Equivariant characteristic classes from constructible sheaves to delocalized homology are constructed in [59].

For a variety S with a good action by a finite group G, and a (generalized) homology theory H, the associated delocalized equivariant theory is given by

$$H^G(S) = (\oplus_{g \in G} H(S^g))^G$$

where the disjoint union $\sqcup_g S^g$ of the fixed point sets S^g has an induced G-action $h: S^g \to S^{hgh^{-1}}$. In the case of an abelian group we have $H^G(S) = (\oplus_{g \in G} H(S^g))^G$.

As an observation, we can see explicitly the relation of delocalized homology to the integral Bost–Connes algebra, by considering the following cases (see Remark 2.12). Let S be a variety with a good $\mathbb{Z}/N\mathbb{Z}$-effectively finite $\hat{\mathbb{Z}}$-action. If S has the trivial $\mathbb{Z}/N\mathbb{Z}$-action we have $H^{\mathbb{Z}/N\mathbb{Z}}(S) = H(S) \otimes \mathbb{Z}[\mathbb{Z}/N\mathbb{Z}]$. In particular, if S is just a point, then this is $\mathbb{Z}[\mathbb{Z}/N\mathbb{Z}]$. More generally, there is a morphism

$$\mathbb{Z}[\mathbb{Z}/N\mathbb{Z}] \times H^{\mathbb{Z}/N\mathbb{Z}}(S) \to H^{\mathbb{Z}/N\mathbb{Z}}(S)$$

induced by $H^{\mathbb{Z}/N\mathbb{Z}}(pt) \times H^{\mathbb{Z}/N\mathbb{Z}}(S) \to H^{\mathbb{Z}/N\mathbb{Z} \times \mathbb{Z}/N\mathbb{Z}}(pt \times S) \to H^{\mathbb{Z}/N\mathbb{Z}}(S)$ with the restriction to the diagonal subgroup as the last map.

2.5 The geometric Verschiebung action

We recall here how to construct the geometric Verschiebung action used in [54] to lift the Bost–Connes maps to the level of Grothendieck rings. This has the effect of transforming an action of $\hat{\mathbb{Z}}$ on X that factors through some $\mathbb{Z}/N\mathbb{Z}$ into an action of $\hat{\mathbb{Z}}$ on $X \times Z_n$, with $Z_n = \{1, \ldots, n\}$, that factors through $\mathbb{Z}/Nn\mathbb{Z}$. For $x \in X$, let $\underline{x} = (x, a_i)_{a_i \in Z_n} = (x_i)_{i=1}^n$ be the subset $\{x\} \times Z_n$. For ζ_N a primmitive N-th root of unity, we write in matrix form

$$V_n(\zeta_{Nn}) = \begin{pmatrix} 0 & 0 & \cdots & 0 & \alpha(\zeta_N) \\ 1 & 0 & \cdots & 0 & 0 \\ 0 & 1 & \cdots & 0 & 0 \\ \vdots & & \vdots & & \vdots \\ 0 & 0 & \cdots & 1 & 0 \end{pmatrix}$$

so that we can write

$$V_n(\zeta_{Nn}) \cdot \underline{x} = \begin{cases} (x, a_{i+1}) & i = 1, \ldots, n-1 \\ (\alpha(\zeta_N) \cdot x, a_1) & i = n \end{cases} \tag{13}$$

which satisfies $V_n(\zeta_{Nn})^n = \alpha(\zeta_N) \times \mathrm{Id}_{Z_n}$. The resulting action $\Phi_n(\alpha)$ of $\hat{\mathbb{Z}}$ on $X \times Z_n$ that factors through $\mathbb{Z}/Nn\mathbb{Z}$ is specified by setting

$$\Phi_n(\alpha)(\zeta_{Nn}) \cdot (x, a) = (V_n(\alpha(\zeta_N)) \cdot \underline{x})_a. \tag{14}$$

2.6 Lifting the Bost–Connes endomorphisms

Consider a base scheme S with a good effectively finite action of $\hat{\mathbb{Z}}$. Let $f : X \to S$ be a variety over S with a good effectively finite $\hat{\mathbb{Z}}$ action such that the map is $\hat{\mathbb{Z}}$-equivariant. We denote by $\alpha_S : \hat{\mathbb{Z}} \times S \to S$ the action on S and by $\alpha_X : \hat{\mathbb{Z}} \times X \to X$ the action on X. We write the equivariant relative Grothendieck ring as $K_0^{\hat{\mathbb{Z}}}(\mathcal{V}_{(S,\alpha_S)})$ to explicitly remember the fixed (up to isomorphisms as in §2.3) action on S.

Definition 2.5 Let (S, α_S) be a scheme with a good effectively finite action of $\hat{\mathbb{Z}}$. Let $Z_n = \mathrm{Spec}(\mathbb{Q}^n)$ and let $\Phi_n(\alpha_S)$ be the action of $\hat{\mathbb{Z}}$ on $S \times Z_n$ as in (13) and (14). Given a class $[f : (X, \alpha_X) \to (S, \alpha_S)]$ in $K_0^{\hat{\mathbb{Z}}}(\mathcal{V}_{(S,\alpha_S)})$, with α_X the compatible $\hat{\mathbb{Z}}$-action on X, let

$$\sigma_n[f : (X, \alpha_X) \to (S, \alpha_S)] = [f : (X, \alpha_X \circ \sigma_n) \to (S, \alpha_S \circ \sigma_n)] \tag{15}$$

$$\tilde{\rho}_n[f : (X, \alpha_X) \to (S, \alpha_S)] = [f \times \mathrm{id} : (X \times Z_n, \Phi_n(\alpha_X)) \tag{16}$$

$$\to (S \times Z_n, \Phi_n(\alpha_S))].$$

Proposition 2.6 *For all $n \in \mathbb{N}$ the σ_n defined in* (15) *are ring homomorphisms*

$$\sigma_n : K_0^{\hat{\mathbb{Z}}}(\mathcal{V}_{(S,\alpha_S)}) \to K_0^{\hat{\mathbb{Z}}}(\mathcal{V}_{(S,\alpha_S \circ \sigma_n)}) \tag{17}$$

and the $\tilde{\rho}_n$ defined in (16) *are group homomorphisms*

$$\tilde{\rho}_n : K_0^{\hat{\mathbb{Z}}}(\mathcal{V}_{(S,\alpha_S)}) \to K_0^{\hat{\mathbb{Z}}}(\mathcal{V}_{(S \times Z_n, \Phi_n(\alpha_S))}), \tag{18}$$

with compositions satisfying

$$\tilde{\rho}_n \circ \sigma_n : K_0^{\hat{\mathbb{Z}}}(\mathcal{V}_{(S,\alpha_S)}) \to K_0^{\hat{\mathbb{Z}}}(\mathcal{V}_{(S\times Z_n,\alpha_S\times\alpha_n)}) \to K_0^{\hat{\mathbb{Z}}}(\mathcal{V}_{(S,\alpha_S)})$$

$$\sigma_n \circ \tilde{\rho}_n : K_0^{\hat{\mathbb{Z}}}(\mathcal{V}_{(S,\alpha_S)}) \to K_0^{\hat{\mathbb{Z}}}(\mathcal{V}_{(S,\alpha_S)^{\oplus n}}) \to K_0^{\hat{\mathbb{Z}}}(\mathcal{V}_{(S,\alpha_S)}),$$

with $\sigma_n \circ \tilde{\rho}_n = n\, id$ *and* $\tilde{\rho}_n \circ \sigma_n$ *is the product by* (Z_n, α_n).

Proof Consider the σ_n defined in (15). Since the group $\hat{\mathbb{Z}}$ is commutative and so is its endomorphism ring, these transformations σ_n respect isomorphism classes since for an isomorphism (ϕ, φ, ϕ_S) the actions satisfy

$$\phi_X(\alpha_X(\sigma_n(g), x)) = \alpha'_X(\varphi(\sigma_n(g)), \phi(x)) = \alpha'_X(\sigma_n(\varphi(g)), \phi(x)),$$

and similarly for the actions α_S, α'_S, so that (ϕ, φ, ϕ_S) is also an isomorphism of the images under σ_n. Similarly, the $\tilde{\rho}_n$ defined in (16) are well defined on the isomorphism classes.

As in [54] we see that $\sigma_n \circ \tilde{\rho}_n[f : (X, \alpha_X) \to (S, \alpha_S)] = [f : (X, \alpha_X) \to (S, \alpha_S)]^{\oplus n}$ and $\tilde{\rho}_n \circ \sigma_n[f : (X, \alpha_X) \to (S, \alpha_S)] = [f \times \mathrm{id} : (X \times Z_n, \alpha_X \times \alpha_n) \to (S \times Z_n, \alpha_S \times \alpha_n)]$. The Grothendieck groups $K_0^{\hat{\mathbb{Z}}}(\mathcal{V}_{(S\times Z_n,\alpha_S\times\alpha_n)})$ and $K_0^{\hat{\mathbb{Z}}}(\mathcal{V}_{(S,\alpha_S)^{\oplus n}})$ map to $K_0^{\hat{\mathbb{Z}}}(\mathcal{V}_{(S,\alpha_S)})$ via the morphism induced by composition with the natural maps of the respective base varieties to (S, α_S). □

The fact that the ring homomorphisms (17) and (18) determine a lift of the ring endomorphism $\sigma_n : \mathbb{Z}[\mathbb{Q}/\mathbb{Z}] \to \mathbb{Z}[\mathbb{Q}/\mathbb{Z}]$ and group homomorphisms $\tilde{\rho}_n : \mathbb{Z}[\mathbb{Q}/\mathbb{Z}] \to \mathbb{Z}[\mathbb{Q}/\mathbb{Z}]$ of the integral Bost–Connes algebra is discussed in Proposition 2.8 and §2.9.

We know from [54] that the integral Bost–Connes algebra lifts to the equivariant Grothendieck ring $K^{\hat{\mathbb{Z}}}(\mathcal{V}_{\mathbb{Q}})$ via maps σ_n and $\tilde{\rho}_n$ that, respectively, precompose the action with the Bost–Connes endomorphism σ_n and apply a geometric form of the Verschiebung map. The main difference with the relative case considered here lies in the fact that the lifts to the equivariant relative Grothendieck rings given by the maps (17) and (18) need to transform in a compatible way the actions on both X and S.

Remark 2.7 Because the maps σ_n and $\tilde{\rho}_n$ of (17) and (18) simultaneously modify the action on the varieties and on the base scheme S, they do not give endomorphisms of the same $K_0^{\hat{\mathbb{Z}}}(\mathcal{V}_{(S,\alpha_S)})$. However, given (S, α_S), it is possible to identify a subalgebra of the integral Bost–Connes algebra that lift to endomorphisms of a corresponding subring of $K_0^{\hat{\mathbb{Z}}}(\mathcal{V}_{(S,\alpha_S)})$, using the notion of "prime decomposition" of the Bost–Connes algebra. We discuss this more carefully in §2.7, §2.8 and §2.9.

2.7 Prime decomposition of the Bost–Connes algebra

As in [27], for each prime p, we can decompose the group $\mathbb{Q}/\mathbb{Z}$ into a product $\mathbb{Q}_p/\mathbb{Z}_p \times (\mathbb{Q}/\mathbb{Z})^{(p)}$, where $\mathbb{Q}_p/\mathbb{Z}_p$ is the Prüfer group, namely the subgroup of elements of $\mathbb{Q}/\mathbb{Z}$ where the denominator is a power of p, isomorphic to $\mathbb{Z}[\frac{1}{p}]/\mathbb{Z}$, while $(\mathbb{Q}/\mathbb{Z})^{(p)}$ consists of the elements with denominator prime to p.

Similarly, given a finite set F of primes, we can decompose $\mathbb{Q}/\mathbb{Z} = (\mathbb{Q}/\mathbb{Z})_F \times (\mathbb{Q}/\mathbb{Z})^F$, where the first term $(\mathbb{Q}/\mathbb{Z})_F$ is identified with fractions in $\mathbb{Q}/\mathbb{Z}$ whose denominator has prime factor decomposition consisting only of primes in F, while elements in $(\mathbb{Q}/\mathbb{Z})^F$ have denominators prime to all $p \in F$. The group ring decomposes accordingly as $\mathbb{Z}[(\mathbb{Q}/\mathbb{Z})_F] \otimes \mathbb{Z}[(\mathbb{Q}/\mathbb{Z})^F]$.

The subsemigroup $\mathbb{N}_F \subset \mathbb{N}$ generated multiplicatively by the primes $p \in F$ acts on $\mathbb{Z}[(\mathbb{Q}/\mathbb{Z})_F] \otimes_{\mathbb{Z}} \mathbb{Q} = \mathbb{Q}[(\mathbb{Q}/\mathbb{Z})_F]$ by endomorphisms

$$\rho_n(e(r)) = \frac{1}{n} \sum_{nr'=r} e(r'), \quad n \in \mathbb{N}_F,\ r \in (\mathbb{Q}/\mathbb{Z})_F.$$

The corresponding morphisms $\sigma_n(e(r)) = e(nr)$ and maps $\tilde{\rho}_n(e(r)) = \sum_{nr'=r} e(r')$ act on $\mathbb{Z}[(\mathbb{Q}/\mathbb{Z})_F]$ and we can consider the associated algebra $\mathcal{A}_{\mathbb{Z},F}$ generated by $\mathbb{Z}[(\mathbb{Q}/\mathbb{Z})_F]$ and $\tilde{\mu}_n$, μ_n^* with $n \in \mathbb{N}_F$, with the relations

$$\tilde{\mu}_{nm} = \tilde{\mu}_n \tilde{\mu}_m, \quad \mu_{nm}^* = \mu_n^* \mu_m^*, \quad \mu_n^* \tilde{\mu}_n = n, \quad \tilde{\mu}_n \mu_m^* = \mu_m^* \tilde{\mu}_n, \tag{19}$$

where the first two relations hold for arbitrary $n, m \in \mathbb{N}$, the third for arbitrary $n \in \mathbb{N}$ and the forth for $n, m \in \mathbb{N}$ satisfying $(n, m) = 1$, and the relations

$$x\tilde{\mu}_n = \tilde{\mu}_n \sigma_n(x) \quad \mu_n^* x = \sigma_n(x) \mu_n^*, \quad \tilde{\mu}_n x \mu_n^* = \tilde{\rho}_n(x), \tag{20}$$

for any $x \in \mathbb{Z}[\mathbb{Q}/\mathbb{Z}]$, where $\tilde{\rho}_n(e(r)) = \sum_{nr'=r} e(r')$, and with

$$\mathcal{A}_{\mathbb{Z},F} \otimes_{\mathbb{Z}} \mathbb{Q} = \mathbb{Q}[(\mathbb{Q}/\mathbb{Z})_F] \rtimes \mathbb{N}_F.$$

We refer to $\mathcal{A}_{\mathbb{Z},F}$ as the F-part of the integral Bost–Connes algebra.

The decomposition $\mathbb{N} = \mathbb{N}_F \times \mathbb{N}^{(F)}$, where $\mathbb{N}^{(F)}$ is generated by all primes $p \notin F$, gives also an algebra $\mathcal{A}_{\mathbb{Z}}^{(F)}$ generated by $\mathbb{Z}[(\mathbb{Q}/\mathbb{Z})^F]$ and the $\tilde{\mu}_n$ and μ_n^* as in (20) with $p \notin F$ with

$$\mathcal{A}_{\mathbb{Z}}^{(F)} \otimes_{\mathbb{Z}} \mathbb{Q} = \mathbb{Q}[(\mathbb{Q}/\mathbb{Z})^F] \rtimes \mathbb{N}^{(F)}.$$

We refer to $\mathcal{A}_{\mathbb{Z}}^{(F)}$ as the F-coprime part of the integral Bost–Connes algebra.

2.8 Lifting the F_N-coprime Bost–Connes algebra

Let $F = F_N$ be the set of prime factors of N and let $\mathbb{Z}[(\mathbb{Q}/\mathbb{Z})^F]$ denote, as before, the part of the group ring of $\mathbb{Q}/\mathbb{Z}$ involving only denominators relatively prime to N. The semigroup $\mathbb{N}^{(F)}$ is generated by primes $p \nmid N$ and we consider the morphisms $\sigma_n(e(r)) = e(nr)$ and maps $\tilde{\rho}_n(e(r)) = \sum_{nr'=r} e(r')$ with $n \in \mathbb{N}^{(F)}$ and $r \in (\mathbb{Q}/\mathbb{Z})^F$ as discussed above.

Proposition 2.8 *Let S be a base scheme endowed with a good $\mathbb{Z}/N\mathbb{Z}$-effectively finite action of $\hat{\mathbb{Z}}$. Let $\mathcal{Z}_{n,S}$ be defined as $\mathcal{Z}_{n,S} = S \times Z_n$, with $Z_n = \mathrm{Spec}(\mathbb{Q}^n)$, with the action $\Phi_n(\alpha_S)$ obtained as in* (13) *and* (14)*. The endomorphisms $\sigma_n : \mathbb{Z}[(\mathbb{Q}/\mathbb{Z})^{F_N}] \to \mathbb{Z}[(\mathbb{Q}/\mathbb{Z})^{F_N}]$ with $n \in \mathbb{N}^{(F_N)}$ of the F_N-coprime part of the integral Bost–Connes algebra lift to endomorphisms $\sigma_n : K_0^{\hat{\mathbb{Z}}}(\mathcal{V}_S) \to K_0^{\hat{\mathbb{Z}}}(\mathcal{V}_S)$, as in* (15)*, which define a semigroup action of the multiplicative group $\mathbb{N}^{(F_N)}$ on the Grothendieck ring $K_0^{\hat{\mathbb{Z}}}(\mathcal{V}_S)$. The maps $\tilde{\rho}_n$, for $n \in \mathbb{N}^{(F_N)}$, lift to group homomorphisms $\tilde{\rho}_n : K_0^{\hat{\mathbb{Z}}}(\mathcal{V}_S) \to K_0^{\hat{\mathbb{Z}}}(\mathcal{V}_S)$, as in* (16)*, so that $\sigma_n \circ \tilde{\rho}_n[f : X \to S] = [f : X \to S]^{\oplus n}$ and $\tilde{\rho}_n \circ \sigma_n[f : X \to S] = [f : X \to S] \cdot \mathcal{Z}_{n,S}$.*

Proof Given the base variety S with a good $\mathbb{Z}/N\mathbb{Z}$-effectively finite $\hat{\mathbb{Z}}$-action, let $F = F_N$ denote the set of prime factors of N. Let X be a variety over S, with a $\hat{\mathbb{Z}}$-equivariant map $f : (X,\alpha_X) \to (S,\alpha_S)$, where we explicitly write the actions, satisfying $f(\alpha_X(\zeta,x)) = \alpha_S(\zeta, f(x))$. For $(N,n) = 1$, the maps $\sigma_n : [f : (X,\alpha_X) \to (S,\alpha_S)] = [f : (X,\alpha_X \circ \sigma_n) \to (S,\alpha_S \circ \sigma_n)]$, as in (15), satisfy $(S,\alpha_S \circ \sigma_n) \simeq (S,\alpha_S)$ with the notion of isomorphism discussed in §2.3, since $\zeta \mapsto \sigma_n(\zeta)$ is an automorphism of $\mathbb{Z}/N\mathbb{Z}$. Thus, the maps σ_n, for $n \in \mathbb{N}^{(F_N)}$ determine a semigroup action of $\mathbb{N}^{(F_N)}$ by endomorphisms of $K_0^{\hat{\mathbb{Z}}}(\mathcal{V}_S)$.

Consider then $(\mathcal{Z}_{n,N}, \Phi_n(\alpha_S))$ as above, which we write equivalently as $\tilde{\rho}_n(S,\alpha_S)$ where $\tilde{\rho}_n$ is the lift of the Bost–Connes map to $K^{\hat{\mathbb{Z}}}(\mathcal{V})$ as in Proposition 3.5 of [54]. We know that $\tilde{\rho}_n \circ \sigma_n[S,\alpha_S] = [S,\alpha_S] \cdot [Z_n,\alpha_n]$ in $K^{\hat{\mathbb{Z}}}(\mathcal{V})$. Since for $(n,N) = 1$ we have $(S,\alpha_S \circ \sigma_n) \simeq (S,\alpha_S)$, this gives $(\mathcal{Z}_{n,N}, \Phi_n(\alpha_S)) \simeq (S \times Z_n, \alpha_S \times \gamma_n)$. Then setting $\tilde{\rho}_n(f : X \to S) = (\tilde{f} : X \times_S \mathcal{Z}_{n,S} \to S)$ with $\tilde{f} = f \circ \pi_X$ gives $X \times_S \mathcal{Z}_{n,S} \simeq X \times Z_n$, and the composition properties for $\tilde{\rho}_n \circ \sigma_n$ and $\sigma_n \circ \tilde{\rho}_n$ are satisfied.

Given a class $[f : X \to S]$, let $[\mathcal{F}_{X,S}]$ be the class in $K_0^{\hat{\mathbb{Z}}}(\mathcal{Q}_S)$ of the constructible sheaf given by the Euler characteristic (11) of $[f : X \to S]$. Let $[\mathcal{F}_{X,S}|_{S^{\mathbb{Z}/N\mathbb{Z}}}]$ be the resulting class in $K_0(S^{\mathbb{Z}/N\mathbb{Z}}) \otimes \mathbb{Z}[\mathbb{Q}/\mathbb{Z}]$ obtained by restriction to the fixed point set $S^{\mathbb{Z}/N\mathbb{Z}}$ with the element in $\mathbb{Z}[\mathbb{Q}/\mathbb{Z}]$ specifying the representation of $\hat{\mathbb{Z}}$ on the stalks of the sheaf $\mathcal{F}_{X,S}|_{S^{\mathbb{Z}/N\mathbb{Z}}}$. For $(N,n) = 1$, the action of σ_n by automorphisms of $\mathbb{Z}/N\mathbb{Z}$ wih the resulting action by isomorphisms of S induces an action by isomorphisms on the $K_0(S^{\mathbb{Z}/N\mathbb{Z}})$ part and the usual Bost–Connes action on $\mathbb{Z}[\mathbb{Q}/\mathbb{Z}]$. The restriction of the semigroup action of $\mathbb{N}^{(F_N)}$ to the subring $\mathbb{Z}[(\mathbb{Q}/\mathbb{Z})^{F_N}]$ is then the image of the action of the maps σ_n and $\tilde{\rho}_n$ on the preimage of this subring under the morphism $K_0^{\hat{\mathbb{Z}}}(\mathcal{V}_S) \to K_0(\mathbb{Q}_{SG}) \otimes \mathbb{Z}[\mathbb{Q}/\mathbb{Z}]$. □

While this construction captures a lift of the $\mathbb{Z}[(\mathbb{Q}/\mathbb{Z})^{F_N}]$ part of the Bost–Connes algebra with the semigroup action of $\mathbb{N}^{(F_N)}$, the fact that the endomorphisms σ_n acting on the roots of unity in $\mathbb{Z}/N\mathbb{Z}$ are automorphisms when $(N,n) = 1$ loses some of the interesting structure of the Bost–Connes algebra, which stems from the partial invertibility of these morphisms. Thus, one also wants to recover the structure of the complementary part of the Bost–Connes algebra with the group ring $\mathbb{Z}[(\mathbb{Q}/\mathbb{Z})_{F_N}]$ and the semigroup $\mathbb{N}_{F_N}$.

2.9 Lifting the full Bost–Connes algebra

Unlike the $\mathbb{Z}[(\mathbb{Q}/\mathbb{Z})^{F_N}]$ part of the Bost–Connes algebra described above, when one considers the full Bost–Connes algebra, including the F_N-part, the lift to the Grothendieck ring no longer consists of endomorphisms of a fixed $K_0^{\hat{\mathbb{Z}}}(\mathcal{V}_{(S,\alpha)})$, but is given as in Proposition 2.6 by homomorphisms as in (15), (17) and (16), (18),

$$\sigma_n : K_0^{\hat{\mathbb{Z}}}(\mathcal{V}_{(S,\alpha_S)}) \to K_0^{\hat{\mathbb{Z}}}(\mathcal{V}_{(S,\alpha_S \circ \sigma_n)}),$$
$$\tilde{\rho}_n : K_0^{\hat{\mathbb{Z}}}(\mathcal{V}_{(S,\alpha_S)}) \to K_0^{\hat{\mathbb{Z}}}(\mathcal{V}_{(S\times Z_n, \Phi_n(\alpha_S))}).$$

For G a finite abelian group with a good action $\alpha : G \times S \to S$ on a variety S, let $(S,\alpha)_k^G = \{s \in S : \alpha(g^k, s) = s, \forall g \in G\}$ denote the set of periodic points of period k, with $(S,\alpha)_1^G = (S,\alpha)^G$ the set of fixed points. We always have $(S,\alpha)_k^G \subseteq (S,\alpha)_{km}^G$ for all $m \in \mathbb{N}$, hence in particular a copy of the fixed point set $(S,\alpha)^G$ is contained in all $(S,\alpha)_k^G$. For $G = \mathbb{Z}/N\mathbb{Z}$, with ζ_N a primitive N-th root of unity generator, the set of k-periodic points is given by $(S,\alpha)_k^{\mathbb{Z}/N\mathbb{Z}} = \{s \in S : \alpha(\zeta_N^k, s) = s\}$.

Lemma 2.9 *The sets of periodic points satisfy $(S,\alpha \circ \sigma_n)_k^G = (S,\alpha)_{nk}^G$. The sets $(S \times Z_n, \Phi_n(\alpha))_k^G$ can be non-empty only when $n|k$ with*

$$(S \times Z_n, \Phi_n(\alpha))_k^G = ((S,\alpha)_{k/n}^G)^n.$$

Proof Under the action $\alpha \circ \sigma_n$ the periodicity condition means $\alpha \circ \sigma_n(\zeta^k, s) = \alpha(\zeta^{nk}, s) = s$ for all $\zeta \in G$ hence the identification

$$(S,\alpha \circ \sigma_n)_k^G = (S,\alpha)_{nk}^G.$$

In the case of the geometric Verschiebung action $\Phi_n(\alpha)$ on $S \times Z_n$, the k-periodicity condition $\Phi_n(\alpha)(\zeta^k, (s,z)) = (s,z)$ implies that $n|k$ for the k-periodicity in the $z \in Z_n$ variable and that $\alpha(\zeta^{k/n}, s) = s$. □

The identification $(S,\alpha\circ\sigma_n)^G_k = (S,\alpha)^G_{nk}$ implies the inclusion

$$(S,\alpha)^G_k \subseteq (S,\alpha\circ\sigma_n)^G_k$$

and in particular the inclusion of the fixed point sets

$$(S,\alpha)^G \subseteq (S,\alpha\circ\sigma_n)^G.$$

Similarly, $(S \times Z_n, \Phi_n(\alpha))^G \subseteq ((S,\alpha)^G)^n$. Since these inclusions will in general be strict, due to the fact that the endomorphisms σ_n are not automorphisms, one cannot simply use the map given by the equivariant Euler characteristic followed by the restriction to the fixed point set

$$K_0^{\hat{\mathbb{Z}}}(\mathcal{V}_S) \to K_0(\mathbb{Q}_{S^{\hat{\mathbb{Z}}}}) \otimes \mathbb{Z}[\mathbb{Q}/\mathbb{Z}]$$

to lift the Bost–Connes endomorphisms to the maps (17) and (18) of Proposition 2.6. However, a simple variant of the same idea, where we consider sets of periodic points, gives the lift of the full Bost–Connes algebra to the equivariant relative Grothendieck rings $K_0^G(\mathcal{V}_{(S,\alpha)})$.

Consider the equivariant Euler characteristic map followed by the restrictions to the sets of periodic points

$$K_0^G(\mathcal{V}_{(S,\alpha)}) \xrightarrow{\chi_S^G} K_0^G(\mathbb{Q}_{(S,\alpha)}) \to \bigoplus_{k\geq 1} K_0^G(\mathbb{Q}_{(S,\alpha)^G_k}). \tag{21}$$

Also, for a given $n \in \mathbb{N}$, consider the same map composed with the projection to the summands with $n|k$

$$\chi^G_{S,n} : K_0^G(\mathcal{V}_{(S,\alpha)}) \xrightarrow{\chi_S^G} K_0^G(\mathbb{Q}_{(S,\alpha)}) \to \bigoplus_{k\geq 1\,:\, n|k} K_0^G(\mathbb{Q}_{(S,\alpha)^G_k}). \tag{22}$$

For simplicity we consider the case where the fixed point set and periodic points sets of the action (S,α) are all finite sets.

Definition 2.10 Let (S,α) be a variety with a good effectively finite $\hat{\mathbb{Z}}$-action. Consider data $(A_{(S,\alpha),n}, f_{(S,\alpha),n})$ and $(B_{(S,\alpha)}, h_{(S,\alpha)})$ of a family of rings $A_{(S,\alpha),n}$ with $n \in \mathbb{N}$ and $B_{(S,\alpha)}$ and ring homomorphisms $f_{(S,\alpha),n} : K_0^G(\mathcal{V}_{(S,\alpha)}) \to A_{(S,\alpha),n} \otimes \mathbb{Z}[\mathbb{Q}/\mathbb{Z}]$ and $h_{(S,\alpha)} : K_0^G(\mathcal{V}_{(S,\alpha)}) \to B_{(S,\alpha)} \otimes \mathbb{Z}[\mathbb{Q}/\mathbb{Z}]$. The maps $f_{(S,\alpha),n}$ and $h_{(S,\alpha)}$ are said to intertwine the Bost–Connes structure if there are ring isomorphisms

$$J_n : A_{(S,\alpha),n} \to B_{(S,\alpha\circ\sigma_n)}$$

and isomorphisms of abelian groups

$$\tilde{J}_n : B_{(S,\alpha)} \to A_{(S\times Z_n,\Phi_n(\alpha))},$$

such that the following holds.

1. There is a commutative diagram of ring homomorphisms

$$\begin{array}{ccc} K_0^{\hat{\mathbb{Z}}}(\mathcal{V}_{(S,\alpha)}) & \xrightarrow{f_{(S,\alpha),n}} & A_{(S,\alpha),n} \otimes \mathbb{Z}[\mathbb{Q}/\mathbb{Z}] \\ \Big\downarrow \sigma_n & & \Big\downarrow J_n \otimes \sigma_n \\ K_0^{\hat{\mathbb{Z}}}(\mathcal{V}_{(S,\alpha\circ\sigma_n)}) & \xrightarrow{h_{(S,\alpha)}} & B_{(S,\alpha\circ\sigma_n)} \otimes \mathbb{Z}[\mathbb{Q}/\mathbb{Z}] \end{array}$$

   where the maps $\sigma_n : K_0^{\hat{\mathbb{Z}}}(\mathcal{V}_{(S,\alpha)}) \to K_0^{\hat{\mathbb{Z}}}(\mathcal{V}_{(S,\alpha\circ\sigma_n)})$ are as in (17) and the maps $\sigma_n : \mathbb{Z}[\mathbb{Q}/\mathbb{Z}] \to \mathbb{Z}[\mathbb{Q}/\mathbb{Z}]$ are the endomorphisms of the integral Bost–Connes algebra.
2. There is a commutative diagram of group homomorphisms

$$\begin{array}{ccc} K_0^{\hat{\mathbb{Z}}}(\mathcal{V}_{(S,\alpha)}) & \xrightarrow{h_{(S,\alpha)}} & B_{(S,\alpha)} \otimes \mathbb{Z}[\mathbb{Q}/\mathbb{Z}] \\ \Big\downarrow \tilde{\rho}_n & & \Big\downarrow \tilde{J}_n \otimes \tilde{\rho}_n \\ K_0^{\hat{\mathbb{Z}}}(\mathcal{V}_{(S\times Z_n,\Phi_n(\alpha))}) & \xrightarrow{f_{(S\times Z_n,\Phi_n(\alpha)),n}} & A_{(S\times Z_n,\Phi_n(\alpha)} \otimes \mathbb{Z}[\mathbb{Q}/\mathbb{Z}] \end{array}$$

   where the maps $\tilde{\rho}_n : K_0^{\hat{\mathbb{Z}}}(\mathcal{V}_{(S,\alpha)}) \to K_0^{\hat{\mathbb{Z}}}(\mathcal{V}_{(S\times Z_n,\Phi_n(\alpha))})$ are as in (18) and the $\tilde{\rho}_n : \mathbb{Z}[\mathbb{Q}/\mathbb{Z}] \to \mathbb{Z}[\mathbb{Q}/\mathbb{Z}]$ are the maps (5) of the integral Bost–Connes algebra.

Theorem 2.11 *Let (S,α) be a variety with a good effectively finite $\hat{\mathbb{Z}}$-action, such that the set $(S,\alpha)_k^{\hat{\mathbb{Z}}}$ of k-periodic points for this action is finite, for all $k \geq 1$. Then the maps* (21) *and* (22) *intertwine the Bost–Connes structure in the sense of Definition 2.10.*

Proof Under the assumptions that all the $(S,\alpha)_k^G$ for $k \geq 0$ are finite sets, we can identify the target of the map with $\oplus_k K_0(\mathbb{Q}_{(S,\alpha)_k^G}) \otimes R(G)$. In the case of varieties with good effectively finite $\hat{\mathbb{Z}}$ actions, we obtain in this way ring homomorphisms

$$\chi^{\hat{\mathbb{Z}}}_{(S,\alpha)} : K_0^{\hat{\mathbb{Z}}}(\mathcal{V}_{(S,\alpha)}) \to \bigoplus_{k\geq 1} K_0(\mathbb{Q}_{(S,\alpha)_k^{\hat{\mathbb{Z}}}}) \otimes \mathbb{Z}[\mathbb{Q}/\mathbb{Z}]$$

$$\chi^{\hat{\mathbb{Z}}}_{(S,\alpha),n} : K_0^{\hat{\mathbb{Z}}}(\mathcal{V}_{(S,\alpha)}) \to \bigoplus_{k\geq 1 : n|k} K_0(\mathbb{Q}_{(S,\alpha)_k^{\hat{\mathbb{Z}}}}) \otimes \mathbb{Z}[\mathbb{Q}/\mathbb{Z}].$$

These maps fit in the following commutative diagrams of ring homomorphisms

$$\begin{CD} K_0^{\hat{\mathbb{Z}}}(\mathcal{V}_{(S,\alpha)}) @>{\chi^{\hat{\mathbb{Z}}}_{(S,\alpha),n}}>> \bigoplus_{n|k} K_0(\mathbb{Q}_{(S,\alpha)^{\hat{\mathbb{Z}}}_k}) \otimes \mathbb{Z}[\mathbb{Q}/\mathbb{Z}] \\ @V{\sigma_n}VV @VV{J_n \otimes \sigma_n}V \\ K_0^{\hat{\mathbb{Z}}}(\mathcal{V}_{(S,\alpha\circ\sigma_n)}) @>>{\bar{\chi}^{\hat{\mathbb{Z}}}_{(S,\alpha\circ\sigma_n)}}> \bigoplus_{\ell} K_0(\mathbb{Q}_{(S,\alpha\circ\sigma_n)^{\hat{\mathbb{Z}}}_\ell}) \otimes \mathbb{Z}[\mathbb{Q}/\mathbb{Z}] \end{CD}$$

where the map $(J_n)_{k,\ell}$ is non-trivial for $k = \ell n$ and identifies $K_0(\mathbb{Q}_{(S,\alpha)^{\hat{\mathbb{Z}}}_\ell})$ with $K_0(\mathbb{Q}_{(S,\alpha\circ\sigma_n)^{\hat{\mathbb{Z}}}_k})$, while the maps $\sigma_n : \mathbb{Z}[\mathbb{Q}/\mathbb{Z}] \to \mathbb{Z}[\mathbb{Q}/\mathbb{Z}]$ are the Bost–Connes endomorphisms. Similarly, we obtain commutative diagrams of group homomorphisms

$$\begin{CD} K_0^{\hat{\mathbb{Z}}}(\mathcal{V}_{(S,\alpha)}) @>{\chi^{\hat{\mathbb{Z}}}_{(S,\alpha)}}>> \bigoplus_{\ell} K_0(\mathbb{Q}_{(S,\alpha)^{\hat{\mathbb{Z}}}_\ell}) \otimes \mathbb{Z}[\mathbb{Q}/\mathbb{Z}] \\ @V{\tilde{\rho}_n}VV @VV{\tilde{J}_n \otimes \tilde{\rho}_n}V \\ K_0^{\hat{\mathbb{Z}}}(\mathcal{V}_{(S\times Z_n,\Phi_n(\alpha))}) @>>{\chi^{\hat{\mathbb{Z}}}_{(S\times Z_n,\Phi_n(\alpha)),n}}> \bigoplus_{n|k} K_0(\mathbb{Q}_{(S\times Z_n,\Phi_n(\alpha))^{\hat{\mathbb{Z}}}_k}) \otimes \mathbb{Z}[\mathbb{Q}/\mathbb{Z}] \end{CD}$$

where $(\tilde{J}_n)_{\ell,k}$ is non-trivial for $k = \ell n$ and maps the Grothendieck ring of constructible sheaves $K_0(\mathbb{Q}_{(S,\alpha)^{\hat{\mathbb{Z}}}_k})$ to $K_0(\mathbb{Q}_{(S,\alpha)^{\hat{\mathbb{Z}}}_k})^{\oplus n}$ and identifies the latter with the Grothendieck ring $K_0(\mathbb{Q}_{(S\times Z_n,\Phi_n(\alpha))^{\hat{\mathbb{Z}}}_\ell})$. □

Remark 2.12 A similar argument can be given using a map obtained by composing the equivariant Euler characteristic considered here with values in $K_0^{\hat{\mathbb{Z}}}(\mathbb{Q}_S)$ with equivariant characteristic classes from constructible sheaves to delocalized equivariant homology as in [59], see §2.4.1.

3 From Grothendieck Rings to Spectra

In this section we show that the Bost–Connes structure can be lifted further from the level of the relative Grothendieck ring to the level of spectra, using the assembler category construction of [70].

The results of this section are a natural continuation of the results in [54]. The general theme considered there consisted of the following steps:

- Appropriate equivariant Euler characteristic maps from certain $\hat{\mathbb{Z}}$-equivariant Grothendieck rings to the group ring $\mathbb{Z}[\mathbb{Q}/\mathbb{Z}]$ are constructed.

- These Euler characteristic maps are then used to lift the Bost–Connes operations σ_n and $\tilde{\rho}_n$ from $\mathbb{Z}[\mathbb{Q}/\mathbb{Z}]$ to corresponding operations in the equivariant Grothendieck ring.
- Assembler categories with K_0 given by the equivariant Grothendieck ring are constructed.
- Endofunctors σ_n and $\tilde{\rho}_n$ of these assembler categories are constructed so that they induce the Bost–Connes structure in the Grothendieck ring.
- Induced maps of spectra are obtained from these endofunctors through the Gamma-space construction that associated a spectrum to an assembler category.

The construction of Bost–Connes operations σ_n and $\tilde{\rho}_n$ on the equivariant Grothendieck rings was generalized in the previous section to the case of relative Grothendieck rings. This section deals with the corresponding generalization of the remaining steps.

We start this section by a brief survey in §3.1 of Zakharevich's formalism of *assemblers* which axiomatizes the "scissors congruence" relations (8).

A general framework for categorical Bost–Connes systems is introduced in §3.3 and §3.4 in terms of subcategories of the automorphism category (in our examples encoding the $\hat{\mathbb{Z}}$-actions) and endofunctors σ_n and $\tilde{\rho}_n$ implementing the Bost–Connes structure.

In §3.5 we construct an assembler category for the equivariant relative Grothendieck ring, and we prove the main result of this section, Theorem 3.15, on the lifting of the Bost–Connes structure to this assembler category.

3.1 Assemblers

Below we will recall the basics of a general formalism for scissors congruence relations applicable in algebraic geometric contexts defined by I. Zakharevich in [70] and [71]. The abstract form of scissors congruences consists of categorical data called *assemblers*, which in turn determine a homotopy–theoretic *spectrum*, whose homotopy groups embody scissors congruence relations. This formalism is applied in [72] in the framework producing an assembler and a spectrum whose π_0 recovers the Grothendieck ring of varieties. This is used to obtain a characterisation of the kernel of multiplication by the Lefschetz motive, which provides a general explanation for the observations of [14], [58] on the fact that the Lefschetz motive is a zero divisor in the Grothendieck ring of varieties.

Consider a (small) category $\mathcal{C}$ and an object X in $\mathcal{C}$.

Definition 3.1 A *sieve* $\mathcal{S}$ over X in $\mathcal{C}$ is a family of morphisms $f_i : X_i' \to X$ (also called "objects over X") satisfying the following conditions:

a) Any isomorphism with target X belongs to $\mathcal{S}$ (as a family with one element).
b) If a morphism $X' \to X$ belongs to $\mathcal{S}$, then its precomposition with any other morphism in $\mathcal{C}$ with target X'

$$X'' \to X' \to X$$

also belongs to $\mathcal{S}$.

It follows that composition of any two morphisms in $\mathcal{S}$ composable in $\mathcal{C}$ itself belongs to $\mathcal{S}$ so that any sieve is a category in its own right.

Definition 3.2 A Grothendieck topology on a category $\mathcal{C}$ consists of the assignment of a collection of sieves $\mathcal{J}(X)$ given for all objects X in $\mathcal{C}$, with the following properties:

a) the total overcategory $\mathcal{C}/X$ of morphisms with target X is a member of the collection $\mathcal{J}(X)$.
b) The pullback of any sieve in $\mathcal{J}(X)$ under a morphism $f : Y \to X$ exists and is a sieve in $\mathcal{J}(Y)$. Here pullback of a sieve is defined as the family of pullbacks of its objects, $X' \to X$, whereas pullback of such an object with respect to $Y \to X$ is defined as $pr_Y : Y \times_X X' \to Y$.
c) given $\mathcal{C}' \in \mathcal{J}(X)$ and a sieve $\mathcal{T}$ in $\mathcal{C}/X$, if for every $f : Y \to X$ in $\mathcal{C}'$ the pullback $f^*\mathcal{T}$ is in $\mathcal{J}(Y)$ then $\mathcal{T}$ is in $\mathcal{J}(X)$.

For more details, see [47], Chapters 16 and 17, or [43], pp. 20–22.

Let $\mathcal{C}$ be a category with a Grothendieck topology. Zakharevich's notion of an assembler category is then defined as follows.

Definition 3.3 A collection of morphisms $\{f_i : X_i \to X\}_{i \in I}$ in $\mathcal{C}$ is a *covering family* if the full subcategory of $\mathcal{C}/X$ that contains all the morphisms of $\mathcal{C}$ that factor through the f_i,

$$\{g : Y \to X \mid \exists i \in I \ h : Y \to X_i \ \text{ such that } f_i \circ h = g\},$$

belongs to the sieve collection $\mathcal{J}(X)$.

In a category $\mathcal{C}$ with an initial object $\emptyset$ two morphisms $f : Y \to X$ and $g : W \to X$ are called *disjoint* if the pullback $Y \times_X W$ exists and is equal to $\emptyset$. A collection $\{f_i : X_i \to X\}_{i \in I}$ in $\mathcal{C}$ is disjoint if f_i and f_j are disjoint for all $i \neq j \in I$.

Definition 3.4 An assembler category $\mathcal{C}$ is a small category endowed with a Grothendieck topology, which has an initial object $\emptyset$ (with the empty family as covering family), and where all morphisms are monomorphisms, with the property that any two finite disjoint covering families of X in $\mathcal{C}$ have a common refinement that is also a finite disjoint covering family.

A morphism of assemblers is a functor $F : \mathcal{C} \to \mathcal{C}'$ that is continuous for the Grothendieck topologies and preserves the initial object and the disjointness property, that is, if two morphisms are disjoint in $\mathcal{C}$ their images are disjoint in $\mathcal{C}'$.

For X a finite set, the coproduct of assemblers $\bigvee_{x\in X} \mathcal{C}_x$ is a category whose objects are the initial object $\emptyset$ and all the non–initial objects of the assemblers $\mathcal{C}_x$. Morphisms of non–initial objects are induced by those of $\mathcal{C}_x$.

Consider a pair $(\mathcal{C}, \mathcal{D})$ where $\mathcal{C}$ is an assembler category, and $\mathcal{D}$ is a sieve in $\mathcal{C}$.

One has then an associated assembler $\mathcal{C} \smallsetminus \mathcal{D}$ defined as the full subcategory of $\mathcal{C}$ containing all the objects that are not initial objects of $\mathcal{D}$. The assembler structure on $\mathcal{C} \smallsetminus \mathcal{D}$ is determined by taking as covering families in $\mathcal{C} \smallsetminus \mathcal{D}$ those collections $\{f_i : X_i \to X\}_{i\in I}$ with X_i, X objects in $\mathcal{C}\smallsetminus\mathcal{D}$ that can be completed to a covering family in $\mathcal{C}$, namely such that there exists $\{f_j : X_j \to X\}_{j\in J}$ with X_j in $\mathcal{D}$ such that $\{f_i : X_i \to X\}_{i\in I} \cup \{f_j : X_j \to X\}_{j\in J}$ is a covering family in $\mathcal{C}$.

Moreover, there is a morphism of assemblers $\mathcal{C} \to \mathcal{C} \smallsetminus \mathcal{D}$ that maps objects of $\mathcal{D}$ to $\emptyset$ and objects of $\mathcal{C} \smallsetminus \mathcal{D}$ to themselves and morphisms with source in $\mathcal{C} \smallsetminus \mathcal{D}$ to themselves and morphisms with source in $\mathcal{D}$ to the unique morphism to the same target with source $\emptyset$. The data $(\mathcal{C}, \mathcal{D}, \mathcal{C} \smallsetminus \mathcal{D})$ are called the abstract scissors congruences.

The construction of Γ-spaces, which we review more in detail in §3.2, then provides the homotopy theoretic spectra associated to assembler categories as in [70]. This construction of assembler categories and spectra provides the formalism we use here and in the previous paper [54] to lift Bost–Connes type algebras to the level of Grothendieck rings and spectra.

3.2 From categories to Γ-spaces and spectra

The Segal construction [65] associates a Γ-space (hence a spectrum) to a category $\mathcal{C}$ with a zero object and a categorical sum. Let Γ^0 be the category of finite pointed sets with objects $\underline{n} = \{0, 1, \ldots, n\}$ and morphisms $f : \underline{n} \to \underline{m}$ the functions with $f(0) = 0$. Let Δ_* denote the category of pointed simplicial sets. The construction can be generalized to symmetric monoidal categories,

[68]. The associated Γ-space $F_{\mathcal{C}} : \Gamma^0 \to \Delta_*$ is constructed as follows. First assign to a finite pointed set X the category $P(X)$ with objects all the pointed subsets of X with morphisms given by inclusions. A functor $\Phi_X : P(X) \to \mathcal{C}$ is summing if it maps $\emptyset \in P(X)$ to the zero object of $\mathcal{C}$ and given $S, S' \in P(X)$ with $S \cap S' = \{\star\}$ the base point of X, the morphism $\Phi_X(S) \oplus \Phi_X(S') \to \Phi_X(S \cup S')$ is an isomorphism. Let $\Sigma_{\mathcal{C}}(X)$ be the category whose objects are the summing functors Φ_X with morphisms the natural transformations that are isomorphisms on all objects of $P(X)$. Setting

$$\Sigma_{\mathcal{C}}(f)(\Phi_X)(S) = \Phi_X(f^{-1}(S)),$$

for a morphisms $f : X \to Y$ of pointed sets and $S \in P(Y)$ gives a functor $\Sigma_{\mathcal{C}} : \Gamma^0 \to \text{Cat}$ to the category of small categories. Composing with the nerve $\mathcal{N}$ gives a functor

$$F_{\mathcal{C}} = \mathcal{N} \circ \Sigma_{\mathcal{C}} : \Gamma^0 \to \Delta_*$$

which is the Γ-space associated to the category $\mathcal{C}$. The functor $F_{\mathcal{C}} : \Gamma^0 \to \Delta_*$ obtained in this way is extended to an endofunctor $F_{\mathcal{C}} : \Delta_* \to \Delta_*$ via the coend

$$F_{\mathcal{C}}(K) = \int^{\underline{n}} K^n \wedge F_{\mathcal{C}}(\underline{n}).$$

One obtains the spectrum $\mathbb{X} = F_{\mathcal{C}}(\mathbb{S})$ associated to the Γ-space $F_{\mathcal{C}}$ by setting $\mathbb{X}_n = F_{\mathcal{C}}(S^n)$ with maps $S^1 \wedge F_{\mathcal{C}}(S^n) \to F_{\mathcal{C}}(S^{n+1})$. The construction is functorial in $\mathcal{C}$, with respect to functors preserving sums and the zero object.

When $\mathcal{C}$ is the category of finite sets, $F_{\mathcal{C}}(\mathbb{S})$ is the sphere spectrum $\mathbb{S}$, and when $\mathcal{C} = \mathcal{P}_R$ is the category of finite projective modules over a commutative ring R, the spectrum $F_{\mathcal{P}_R}(\mathbb{S}) = K(R)$ is the K-theory spectrum of R.

The Segal construction determines a functor from the category of small symmetric monoidal categories to the category of -1-connected spectra. It is shown in [68] that this functor determines an equivalence of categories between the stable homotopy category of -1-connected spectra and a localization of the category of small symmetric monoidal categories, obtained by inverting morphisms sent to weak homotopy equivalences by the functor.

Given an assembler category $\mathcal{C}$, one considers a category $\mathcal{W}(\mathcal{C})$ with objects $\{A_i\}_{i\in I}$ given by collections of non-initial objects A_i in $\mathcal{C}$ indexed by finite sets and morphisms $\phi : \{A_i\}_{i\in I} \to \{B_j\}_{j\in J}$ consisting of a map of finite sets $f : I \to J$ and morphisms $\phi_i : A_i \to B_{f(i)}$ that form disjoint covering families $\{\phi_i \mid i \in f^{-1}(j)\}$, for all $j \in J$. One then obtains a Γ-space as the functor that assigns to a finite pointed set (X, x_0) the simplicial set $\mathcal{N}\mathcal{W}(X \wedge \mathcal{C})$, the nerve of the category $\mathcal{W}(X \wedge \mathcal{C})$ where $X \wedge \mathcal{C}$ is the assembler $X \wedge \mathcal{C} = \bigvee_{x\in X\smallsetminus\{x_0\}} \mathcal{C}$.

The spectrum associated to the assembler $\mathcal{C}$ is the spectrum defined by this Γ space, namely $X_n = \mathcal{NW}(S^n \wedge \mathcal{C})$.

For another occurrence of Γ-spaces in the context of $\mathbb{F}_1$-geometry, see [26].

3.3 Automorphism category and enhanced assemblers

We describe in this and the next subsection a general formalism of "enhanced assemblers" underlying all the explicit cases of Bost–Connes structures in Grothendieck rings discussed in [54] and in some of the later sections of this paper.

We first recall the definition of the automorphism category.

Definition 3.5 The automorphism category $\text{Aut}(C)$ of a category C is given by:

(i) Objects of $\text{Aut}(C)$ are pairs $\hat{X} = (X, v_X)$ where $X \in \text{Obj}(C)$ and $v_X : X \to X$ is an automorphism of X.
(ii) Morphisms $\hat{f} : (X, v_X) \to (Y, v_Y)$ in $\text{Aut}(C)$ are morphisms $f : X \to Y$ such that $f \circ v_X = v_Y \circ f : X \to Y$ in C.
(iii) The forgetful functor sends $\hat{X}$ to X and $\hat{f}$ to f.

We use here a standard categorical notation according to which, say, $f \circ v_X$ is the precomposition of f with v_X.

Thus, we can make the following general definition. In the following we will be especially interested in the case where the chosen subcategory is determined by a group action, see Remark 3.7.

Definition 3.6 Let C be a category. We will call here an *enhancement* of C a pair consisting of a choice of a subcategory $\hat{C}$ of the automorphism category $\text{Aut}(C)$ and the forgetful functor $\hat{C} \to C$, where objects (X, v_X) of $\hat{C}$ have automorphisms $v_X : X \to X$ of finite order.

The main idea here is that a subcategory $\hat{C}$ of the automorphism category of C is where the endofunctors defining the lifts of the Bost–Connes structure are defined, as we make more precise in Definitions 3.9 and 3.11.

Remark 3.7 In the cases considered in [54] and in this paper, the subcategory $\hat{C}$ of $\text{Aut}(C)$ is usually determined by a finite group action, so that elements of $\hat{C}$ are of the form $(X, \alpha_X(g))$ with $\alpha_X : G \times X \to X$ the group action. However, one expects other interesting examples that are not necessarily given by group actions, hence it is worth considering this more general formulation.

Remark 3.8 Assume that C is endowed with a structure of assembler. Then a series of constructions presented in §§ 3 and 4 of [54] and in §§ 2–6 of this paper, and restricted there to various categories of schemes, show in fact how this structure of assembler can be lifted from C to $\hat{C}$.

In particular the Bost–Connes type structures we are investigating can be formulated broadly in this setting of enhanced assemblers as follows.

3.4 Bost–Connes systems on categories

Let $\hat{C}$ be an enhancement of a category C, in the sense of Definition 3.6.

Definition 3.9 We assume here that C is an additive (symmetric) monoidal category and that the enhancement $\hat{C}$ is compatible with this structure. A Bost–Connes system in an enhancement $\hat{C}$ of C consists of two families of endofunctors $\{\sigma_n\}_{n\in\mathbb{N}}$ and $\{\tilde{\rho}_n\}_{n\in\mathbb{N}}$ of $\hat{C}$ with the following properties:

1. The functors σ_n are compatible with both the additive and the (symmetric) monoidal structure, while the functors $\tilde{\rho}_n$ are functors of additive categories.
2. For all $n, m \in \mathbb{N}$ these endofunctors satisfy
$$\sigma_{nm} = \sigma_n \circ \sigma_m, \quad \tilde{\rho}_{nm} = \tilde{\rho}_n \circ \tilde{\rho}_m.$$
3. The compositions satisfy
$$\begin{aligned} \sigma_n \circ \tilde{\rho}_n(X, \upsilon_X) &= (X, \upsilon_X)^{\oplus n} \\ \tilde{\rho}_n \circ \sigma_n(X, \upsilon_X) &= (X, \upsilon_X) \otimes (Z_n, \upsilon_n), \end{aligned} \tag{23}$$
for an object (Z_n, υ_n) in $\hat{C}$ that depends on n but not on (X, υ_X), and similarly on morphisms.

Here $\oplus$ refers to the additive structure of C and $\otimes$ to the monoidal structure.

Remark 3.10 In all the explicit cases considered in [54] and in this paper, the endofunctors σ_n and $\tilde{\rho}_n$ of Definition 3.9 have the form

$$\sigma_n(X, \upsilon_X) = (X, \upsilon_X \circ \sigma_n) \quad \text{and} \quad \tilde{\rho}_n(X, \upsilon_X) = (X \times Z_n, \Phi_n(\upsilon_X)),$$

where the endomorphism υ_X is the action of a generator of some finite cyclic group $\mathbb{Z}/N\mathbb{Z}$ quotient of $\hat{\mathbb{Z}}$ and the action satisfies $\upsilon_X \circ \sigma_n(\zeta, x) = \upsilon_X(\sigma_n(\zeta), x)$, where $\sigma_n(\zeta) = \zeta^n$ is the Bost–Connes map of (4), while the action $\Phi_n(\upsilon_X)$ on $X \times Z_n$ is a geometric form of the Verschiebung, as will be discussed more explicitly in §2.5. The object (Z_n, υ_n) in Definition 3.9

plays the role of the element $n\pi_n$ in the integral Bost–Connes algebra and the relations (23) play the role of the relations (6).

This definition covers the main examples considered in §§ 3 and 4 of [54] obtained using the assembler categories associated to the equivariant Grothendieck ring $K_0^{\hat{\mathbb{Z}}}(\mathcal{V})$ of varieties with a good $\hat{\mathbb{Z}}$-action factoring through some finite cyclic quotient and to the equivariant version $\mathrm{Burn}^{\hat{\mathbb{Z}}}$ of the Kontsevich–Tschinkel Burnside ring. This same definition also accounts for the construction we will discuss in §4 of this paper, based on assembler categories associated to torified varieties (see Remark 4.6).

The more general formulation given in Definition 3.9 is motivated by the fact that one expects other significant examples of categorical Bost–Connes structures where the choice of the subcategory $\hat{C}$ of the automorphism category $\mathrm{Aut}(C)$ is not determined by the action of a cyclic group as in the cases discussed here. Such more general classes of categorical Bost–Connes systems are not discussed in the present paper, but they are a motivation for future work, for which we just set the general framework in this section.

A generalization of Definition 3.9 is needed when considering relative cases, in particular the lift to assemblers of the construction presented in §2 for relative equivariant Grothendieck rings $K_0^{\hat{\mathbb{Z}}}(\mathcal{V}_S)$. The reason why we need the following modification of Definition 3.9 is the fact that, in the relative setting, the base scheme S itself has its enhancement structure (the group action, in the specific examples) modified by the endofunctors implementing the Bost–Connes structure and this needs to be taken into account. We will see this additional structure more explicitly applied in §3.5, in the specific case where the automorphisms are determined by a group action (see Remark 3.16).

Definition 3.11 Let $\hat{\mathcal{I}}$ be an enhancement of an additive (symmetric) monoidal category $\mathcal{I}$ as above, endowed with a Bost–Connes system given by endofunctors $\{\sigma_n^{\mathcal{I}}\}$ and $\{\tilde{\rho}_n^{\mathcal{I}}\}$ of $\hat{\mathcal{I}}$ as in Definition 3.9, with α_n the object in $\hat{\mathcal{I}}$ with $\tilde{\rho}_n \circ \sigma_n(\alpha) = \alpha \otimes \alpha_n$. Let $\{\hat{C}_\alpha\}_{\alpha \in \hat{\mathcal{I}}}$ be a collection of enhancements of additive (symmetric) monoidal categories C_α, indexed by the objects of the auxiliary category $\hat{\mathcal{I}}$, endowed with functors $f_n : \hat{C}_{\alpha^{\oplus n}} \to \hat{C}_\alpha$ and $h_n : \hat{C}_{\alpha\times\beta} \to \hat{C}_\alpha$. Let $\{\sigma_n\}_{n\in\mathbb{N}}$ and $\{\tilde{\rho}_n\}_{n\in\mathbb{N}}$ be two collections of functors

$$\sigma_n : \hat{C}_\alpha \to \hat{C}_{\sigma_n^{\mathcal{I}}(\alpha)} \quad \text{and} \quad \tilde{\rho}_n : \hat{C}_\alpha \to \hat{C}_{\tilde{\rho}_n^{\mathcal{I}}(\alpha)}$$

satisfying the properties:

1. The functors σ_n are compatible with both the additive and the (symmetric) monoidal structure, while the functors $\tilde{\rho}_n$ are functors of additive categories.

2. For all $n, m \in \mathbb{N}$ these functors satisfy

$$\sigma_{nm} = \sigma_n \circ \sigma_m, \quad \tilde{\rho}_{nm} = \tilde{\rho}_n \circ \tilde{\rho}_m.$$

3. The compositions

$$\sigma_n \circ \tilde{\rho}_n : \hat{\mathcal{C}}_\alpha \to \hat{\mathcal{C}}_{\alpha^{\oplus n}} \quad \text{and} \quad \tilde{\rho}_n \circ \sigma_n : \hat{\mathcal{C}}_\alpha \to \hat{\mathcal{C}}_{\alpha \otimes \alpha_n}$$

satisfy

$$\begin{aligned} f_n \circ \sigma_n \circ \tilde{\rho}_n(X, v_X)_\alpha &= (X, v_X)_\alpha^{\oplus n} \quad \text{and} \\ h_n \circ \tilde{\rho}_n \circ \sigma_n(X, v_X)_\alpha &= (X, v_X)_\alpha \otimes (Z_n, v_n)_\alpha, \end{aligned} \tag{24}$$

for an object $(Z_n, v_n)_\alpha$ in $\hat{\mathcal{C}}_\alpha$ that depends on n and α, but not on (X, v_X).

We will first focus on the case of assembler categories, as those were at the basis of our constructions of Bost–Connes systems in [54], but we will also consider in §7 a different categorical setting that will allow us to identify analogous structures at a motivic level, following the formalism of geometric diagrams and Nori motives.

3.5 Assemblers for the relative Grothendieck ring

We consider the relative Grothendieck ring $K_0(\mathcal{V}_S)$ of varieties over a base variety S over a field $\mathbb{K}$, as in Definition 2.1.

An assembler $\mathcal{C}_S$ such that the associated spectrum $K(\mathcal{C}_S)$ has $K_0(\mathcal{C}_S) = \pi_0 K(\mathcal{C}_S)$ given by the relative Grothendieck ring $K_0(\mathcal{V}_S)$ can be obtained as a slight modification of the construction given in [72] for the ordinary Grothendieck ring $K_0(\mathcal{V}_{\mathbb{K}})$.

Definition 3.12 The assembler $\mathcal{C}_S$ for the relative Grothendieck ring $K_0(\mathcal{V}_S)$ has objects $f : X \to S$ that are varieties over S and morphisms that are locally closed embeddings of varieties over S.

Lemma 3.13 *The category $\mathcal{C}_S$ of Definition 3.12 is indeed as assembler, with the Grothendieck topology on $\mathcal{C}_S$ is generated by the covering families*

$$\{Y \hookrightarrow X, X \smallsetminus Y \hookrightarrow X\}$$

with compatible maps (9)

$$\begin{array}{ccccc} Y & \hookrightarrow & X & \hookleftarrow & X \smallsetminus Y \\ & {\scriptstyle f|_Y} \searrow & \downarrow {\scriptstyle f} & \swarrow {\scriptstyle f|_{X \smallsetminus Y}} & \\ & & S & & \end{array} \tag{25}$$

Proof The argument is the same as in [70], [72] and in [54]. In this setting finite disjoint covering families are maps

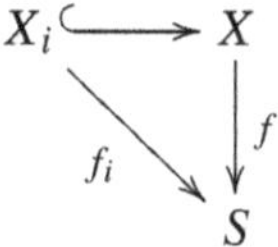

where $X_i = Y_i \smallsetminus Y_{i-1}$ with commutative diagrams

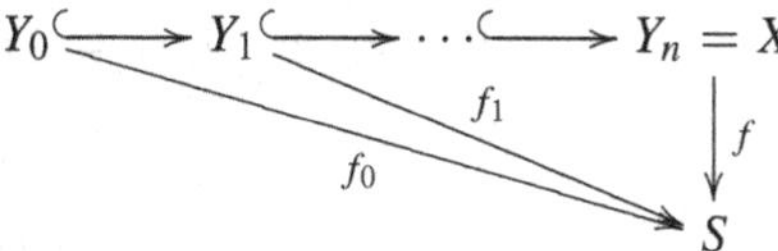

The category has pullbacks, hence as shown in [70] (Remark after Definition 2.4) this suffices to obtain that any two finite disjoint covering families have a common refinement. Morphisms are embeddings compatible with the structure maps as in (25) hence in particular monomorphisms. Theorem 2.3 of [70] then shows that the spectrum $K(\mathcal{C}_S)$ associated to this assembler category has $\pi_0 K(\mathcal{C}_S) = K_0(\mathcal{V}_S)$. □

In a similar way we obtain an assembler category and spectrum for the equivariant version $K_0^{\hat{\mathbb{Z}}}(\mathcal{V}_S)$. The argument is as in the previous case and in Lemma 4.5.1 of [54], using the inclusion-exclusion relations (10).

Corollary 3.14 *An assembler category $\mathcal{C}^{\hat{\mathbb{Z}}}_{(S,\alpha)}$ for $K_0^{\hat{\mathbb{Z}}}(\mathcal{V}_{(S,\alpha)})$ is constructed as in Lemma 3.13 with objects the $\hat{\mathbb{Z}}$-equivariant $f : X \to S$, morphisms given by $\hat{\mathbb{Z}}$-equivariant locally closed embeddings of varieties over S and with Grothendieck topology generated by the covering families given by $\hat{\mathbb{Z}}$-equivariant maps as in* (9) *and* (10).

As in Proposition 4.2 of [54], we show that the lifting of the integral Bost–Connes algebra obtained in Proposition 2.6 and Theorem 2.11 further lifts to functors of the associated assembler categories, with the σ_n compatible with the monoidal structure, but not the $\tilde{\rho}_n$.

Theorem 3.15 *The maps $\sigma_n : (f : (X,\alpha_X) \to (S,\alpha)) \mapsto (f : (X,\alpha_X \circ \sigma_n) \to (S,\alpha \circ \sigma_n))$ and $\tilde{\rho}_n : (f : (X,\alpha_X) \to (S,\alpha)) \mapsto (f \times \mathrm{id} : (X \times Z_n, \Phi_n(\alpha_X)) \to (S \times Z_n, \Phi_n(\alpha)))$ define functors of the assembler categories $\sigma_n : \mathcal{C}^{\hat{\mathbb{Z}}}_{(S,\alpha)} \to \mathcal{C}^{\hat{\mathbb{Z}}}_{(S,\alpha\circ\sigma_n)}$ and $\tilde{\rho}_n : \mathcal{C}^{\hat{\mathbb{Z}}}_{(S,\alpha)} \to \mathcal{C}^{\hat{\mathbb{Z}}}_{(S\times Z_n,\Phi_n(\alpha))}$. The functors σ_n are compatible with the monoidal structure.*

Proof The functors σ_n defined as above on objects are compatibly defined on morphisms by assigning to a locally closed embedding

$$\begin{array}{ccc} (Y,\alpha_Y) \xrightarrow{j} (X,\alpha_X) & \overset{\sigma_n}{\longmapsto} & (Y,\alpha_Y \circ \sigma_n) \xrightarrow{j} (X,\alpha_X \circ \sigma_n) \\ {\scriptstyle f_Y} \searrow \quad \downarrow {\scriptstyle f_X} & & {\scriptstyle f_Y} \searrow \quad \downarrow {\scriptstyle f_X} \\ (S,\alpha) & & (S,\alpha \circ \sigma_n) \end{array}$$

Similarly, we define the $\tilde{\rho}_n$ on morphisms by

$$\begin{array}{ccc} (Y,\alpha_Y) \xrightarrow{j} (X,\alpha_X) & \overset{\rho_n}{\longmapsto} & (Y \times Z_n, \Phi_n(\alpha_Y)) \xrightarrow{j} (X \times Z_n, \Phi_n(\alpha_X)) \\ {\scriptstyle f_Y} \searrow \quad \downarrow {\scriptstyle f_X} & & {\scriptstyle f_Y} \searrow \quad \downarrow {\scriptstyle f_X} \\ (S,\alpha) & & (S \times Z_n, \Phi_n(\alpha)) \end{array}$$

The functors σ_n are compatible with the monoidal structure since

$$\begin{aligned} \sigma_n(X,\alpha_X) \times \sigma_n(X',\alpha_{X'}) &= (X \times X', (\alpha \times \alpha') \circ \Delta \circ \sigma_n) \\ &= \sigma_n((X,\alpha_X) \times (X',\alpha_{X'})). \end{aligned}$$ □

The functor of assembler categories determines an induced map of spectra and in turn an induced map of homotopy groups. By construction the induced maps on the π_0 homotopy agree with the maps (17) and (18) of Proposition 2.6. □

Remark 3.16 We can associate to the assembler category $\mathcal{C}^{\hat{\mathbb{Z}}}_{(S,\alpha)}$ of Corollary 3.14 with the endofunctors σ_n and $\tilde{\rho}_n$ a categorical Bost–Connes structure in the sence of Definition 3.11, where the objects are $f : X \to S$ as above with the automorphisms given by elements $g \in \hat{\mathbb{Z}}$ acting on $f : X \to S$ through the action by $\alpha_X(g)$ on X and by $\alpha_S(g)$ on S, intertwined by the equivariant map f.

4 Torifications, $\mathbb{F}_1$-points, zeta functions, and spectra

In this section we relate the point of view developed in [54], with lifts of the Bost–Connes system to Grothendieck rings and spectra, to the approach to $\mathbb{F}_1$-geometry based on torifications. This was first introduced in [51]. Weaker forms of torification were also considered in [53], which allow for the development of a form of $\mathbb{F}_1$-geometry suitable for the treatment of certain classical moduli spaces.

The approach we follow here, in order the relate the case of torified geometry with the Bost–Connes systems on Grothendieck rings, assemblers, and spectra discussed in [54], is based on the following simple setting. Instead of working with the equivariant Grothendieck rings $K_0^{\hat{\mathbb{Z}}}(\mathcal{V})$ and $K_0^{\hat{\mathbb{Z}}}(\mathcal{V}_S)$, where one assumes the varieties have a good effectively finite $\hat{\mathbb{Z}}$-action, we consider here a variant that connects to the torifications point of view on $\mathbb{F}_1$-geometry of [51]. We replace varieties with $\mathbb{Z}/N\mathbb{Z}$-effectively finite $\hat{\mathbb{Z}}$-actions with varieties with a $\mathbb{Q}/\mathbb{Z}$-action induced by a torification, where the group schemes $\mathfrak{m}_n$ of n-th roots of unity, given by the kernels

$$1 \to \mathfrak{m}_n \to \mathbb{G}_m \xrightarrow{\lambda \mapsto \lambda^n} \mathbb{G}_m \to 1$$

determine a diagonal embedding in each torus and an action by multiplication. This is a very restrictive class of varieties, because the existence of a torification on a variety implies that the Grothendieck class is a sum of classes of tori with non-negative coefficients. The resulting construction will be more restrictive than the one considered in [54]. We will see, however, that one can still see in this context several interesting phenomena, especially in connection with the "dynamical" approach to $\mathbb{F}_1$-geometry proposed in [54].

4.1 Torifications

A torification of an algebraic variety X defined over $\mathbb{Z}$ is a decomposition $X = \sqcup_{i\in\mathcal{I}} T_i$ into algebraic tori $T_i = \mathbb{G}_m^{d_i}$. Weaker to stronger forms of torification [53] include

1. *torification of the Grothendieck class*: $[X] = \sum_{i\in\mathcal{I}}(\mathbb{L}-1)^{d_i}$ with $\mathbb{L}$ the Lefschetz motive;
2. *geometric torification*: $X = \sqcup_{i\in\mathcal{I}} T_i$ with $T_i = \mathbb{G}_m^{d_i}$;
3. *affine torification*: the existence of an affine covering compatible with the geometric torification, [51];
4. *regular torification*: the closure of each torus in the geometric torification is also a union of tori of the torification, [51].

Similarly, there are different possibilities when one considers morphisms of torified varieties, see [53]. In view of describing associated Grothendieck rings, we review the different notions of morphisms. The Grothendieck classes are then defined with respect to the corresponding type of isomorphism.

A torified morphism of geometric torifications in the sense of [51] between torified varieties $f : (X, T) \to (Y, T')$ is a morphism $f : X \to Y$ of varieties together with a map $h : I \to J$ of the indexing sets of the torifications

$X = \sqcup_{i \in I} T_i$ and $Y = \sqcup_{j \in J} T'_j$ such that the restriction of f to tori T_i is a morphism of algebraic groups $f_i : T_i \to T'_{h(i)}$. There are then three different classes of morphisms of torified varieties that were introduced in [53]: strong, ordinary, and weak morphisms. To describe them, one first defines strong, ordinary, and weak equivalences of torifications, and one then uses these to define the respective class of morphisms.

Let T and T' be two geometric torifications of a variety X.

1. The torifications (X,T) and (X,T') are *strongly equivalent* if the identity map on X is a torified morphism as above.
2. The torifications (X,T) and (X,T') are *ordinarily equivalent* if there exists an automorphism $\phi : X \to X$ that is a torified morphism.
3. The torifications (X,T) and (X,T') are *weakly equivalent* if X has two decompositions $X = \cup_i X_i$ and $X = \cup_j X'_j$ into a disjoint union of subvarieties, compatible with the torifications, such that there are isomorphisms of varieties $\phi_i : X_i \to X'_{j(i)}$ that are torified.

In the weak case a "decomposition compatible with torifications" means that the intersections $T_i \cap X_j$ of the tori of T with the pieces of the decomposition (when non-empty) are tori of the torification of X_j, and similarly for $T'_i \cap X'_j$. In general weakly equivalent torification are not ordinarily equivalent because the maps ϕ_i need not glue together to define a single map ϕ on all of X.

We then have the following classes of morphisms of torified varieties from [53]:

1. *strong morphisms*: these are torified morphisms in the sense of [51], namely morphisms that restrict to morphisms of tori of the respective torifications.
2. *ordinary morphisms*: an ordinary morphism of torified varieties (X,T) and (Y,T') is a morphism $f : X \to Y$ such that becomes a torified morphism after composing with strong isomorphisms, that is, $\phi_Y \circ f \circ \phi_X : (X,T) \to (Y,T')$ is a strong morphism of torified varieties, for some isomorphisms $\phi_X : X \to X$ and $\phi_Y : Y \to Y$. In other words, if we denote by T_ϕ and T'_ϕ the torifications such that $\phi_X : (X,T) \to (X,T_\phi)$ and $\phi_Y : (Y,T'_\phi) \to (Y,T')$ are torified, then $f : (X,T_\phi) \to (Y,T'_\phi)$ is torified.
3. *weak morphisms*: the torified varieties (X,T) and (Y,T') admit decompositions $X = \sqcup_i X_i$ and $Y = \sqcup_j Y_i$, compatible with the torifications, such that there exist ordinary morphisms $f_i : (X_i,T_i) \to (Y_{f(i)}, T'_{f(i)})$ of these subvarieties.

Note that the strong isomorphisms $\phi_X : (X,T) \to (X,T_\phi)$ and $\phi_Y : (Y,T'_\phi) \to (Y,T')$ used in the definition of ordinary morphisms are ordinary equivalences of the torifications T and T_ϕ, respectively T' and T'_ϕ.

Given these notions of morphisms, we can correspondingly construct Grothendieck rings $K_0(\mathcal{T})^s$, $K_0(\mathcal{T})^o$, and $K_0(\mathcal{T})^w$ in the following way.

As an abelian group $K_0(\mathcal{T})^s$ is generated by isomorphism classes $[X,T]_s$ of pairs of a torifiable variety X and a torification T modulo strong isomorphisms, with the inclusion-exclusion relations $[X,T]_s = [Y,T_Y]_s + [X \smallsetminus Y, T_{X\smallsetminus Y}]_s$ whenever $(Y,T_Y) \hookrightarrow (X,T)$ is a strong morphism (that is, the inclusion of Y in X is compatible with the torification: Y is a union of tori of the torification of X) and (Y,T_Y) is a *complemented subvariety* in (X,T), which means that the complement $X \smallsetminus Y$ is also a union of tori of the torification so that the inclusion of $(X \smallsetminus Y, T_{X\smallsetminus Y})$ in (X,T) is also a strong morphism. This complemented condition is very strong. Indeed, one can see that, for example, there are in general very few complemented points in a torified variety. The product operation is $[X,T]_s \cdot [Y,T']_s = [X \times Y, T \times T']_s$ with the torification $T \times T'$ given by the product tori $T_{ij} = T_i \times T'_j = \mathbb{G}_m^{d_i+d_j}$.

The abelian group $K_0(\mathcal{T})^o$ is generated by isomorphism classes $[X]_o$ varieties that admit a torification with respect to ordinary isomorphisms, with the inclusion-exclusion relations $[X]_o = [Y]_o + [X \smallsetminus Y]_o$ whenever the inclusions $Y \hookrightarrow X$ and $X \smallsetminus Y \hookrightarrow X$ are ordinary morphisms. The product is the class of the Cartesian product $[X]_o \cdot [Y]_o = [X \times Y]_o$.

The abelian group $K_0(\mathcal{T})^w$ is generated by the isomorphism classes $[X]_w$ of torifiable varieties X with respect to weak morphisms, with the inclusion-exclusion relations $[X]_w = [Y]_w + [X \smallsetminus Y]_w$ whenever the inclusions $Y \hookrightarrow X$ and $X \smallsetminus Y \hookrightarrow X$ are weak morphisms. The product structure is again given by $[X]_w \cdot [Y]_w = [X \times Y]_w$.

The reader can consult the explicit examples given in [53] to see how these notions (and the resulting Grothendieck rings) can be different. For example, as mentioned in §2.2 of [53], consider the variety $X = \mathbb{P}^1 \times \mathbb{P}^1$ and consider on it two torifications T and T', where T is the standard torification given by the decomposition of each $\mathbb{P}^1$ into cells $\mathbb{A}^0 \cup \mathbb{A}^1$, with the cell $\mathbb{A}^1$ torified as $\mathbb{A}^0 \cup \mathbb{G}_m$, while T' is the torification where in the big cell $\mathbb{A}^2$ of $\mathbb{P}^1 \times \mathbb{P}^1$ we take a torification of the diagonal $\mathbb{A}^1$ and a torification of the complement of the diagonal, and we use the same torification of the lower dimensional cells as in T. These two torifications are related by a weak isomorphism, hence the elements $(\mathbb{P}^1 \times \mathbb{P}^1, T)$ and $(\mathbb{P}^1 \times \mathbb{P}^1, T')$ define the same class in $K_0(\mathcal{T})^w$, but they are not related by an ordinary isomorphism so they define different classes in $K_0(\mathcal{T})^o$.

Note however that, in all these cases, the Grothendieck classes $[X]_a$ with $a = s, o, w$ have the form $[X]_a = \sum_{n\geq 0} a_n \mathbb{T}^n$ with $a_n \in \mathbb{Z}_+$ and $\mathbb{T}^n = [\mathbb{G}_m^n]$.

In the following, whenever we simply write $a = s, o, w$ without specifying one of the three choices of morphisms, it means that the stated property holds for all of these choices.

4.1.1 Relative case

In a similar way, we can construct relative Grothendieck rings $K_S(\mathcal{T})^a$ with $a = s, o, w$ where in the strong case $S = (S, T_S)$ is a choice of a variety with an assigned torification, with $K_S(\mathcal{T})^s$ generated as an abelian group by isomorphisms classes $[f : (X,T) \to (S,T_S)]$ where f is a strong morphism of torified varieties and the isomorphism class is taken with respect to strong isomorphisms ϕ, ϕ_S such that the diagram commutes

$$\begin{array}{ccc} (X,T) & \xrightarrow{\phi} & (X',T') \\ {\scriptstyle f}\downarrow & & \downarrow{\scriptstyle f'} \\ (S,T_S) & \xrightarrow{\phi_S} & (S,T_S) \end{array}$$

with inclusion-exclusion relations

$$[f : (X,T) \to (S,T_S)] =$$

$$[f|_{(Y,T_Y)} : (Y,T_Y) \to (S,T_S)] + [f|_{(X\smallsetminus Y, T_{X\smallsetminus Y})} : (X \smallsetminus Y, T_{X\smallsetminus Y}) \to (S,T_S)]$$

where $\iota_Y : (Y,T_Y) \hookrightarrow (X,T)$ is a strong morphism and (Y,T_Y) is complemented with $\iota_{X\smallsetminus Y} : (X \smallsetminus Y, T_{X\smallsetminus Y}) \hookrightarrow (X,T)$ also a strong morphism and both these inclusions are compatible with the map $f : (X,T) \to (S,T_S)$, so that $f_Y = f \circ \iota_Y$ and $f|_{(X\smallsetminus Y, T_{X\smallsetminus Y})} = f \circ \iota_{X\smallsetminus Y}$ are strong morphisms. The construction for ordinary and weak morphism is similar, with the appropriate changes in the definition.

4.2 Group actions

In order to operate on Grothendieck classes with Bost–Connes type endomorphisms, we introduce appropriate group actions.

Torified varieties carry natural $\mathbb{Q}/\mathbb{Z}$ actions, since the roots of unity embed diagonally in each torus of the torification and act on it by multiplication. However, we will also be interested in considering good effectively finite $\hat{\mathbb{Z}}$-actions, in the sense already discussed in [54], that is, actions of $\hat{\mathbb{Z}}$ as in Definition 2.3.

Remark 4.1 The main reason for working with $\hat{\mathbb{Z}}$-actions rather than with $\mathbb{Q}/\mathbb{Z}$ actions lies in the fact that, in the construction of the geometric Vershiebung action discussed in §2.5 we need to be able to describe the cyclic permutation action of $\mathbb{Z}/n\mathbb{Z}$ on the finite set Z_n as an action factoring through $\mathbb{Z}/n\mathbb{Z}$. This cannot be done in the case of $\mathbb{Q}/\mathbb{Z}$-actions because there are no nontrivial group homomorphisms $\mathbb{Q}/\mathbb{Z} \to \mathbb{Z}/n\mathbb{Z}$ since $\mathbb{Q}/\mathbb{Z}$ is infinitely divisible.

In the case of the natural $\mathbb{Q}/\mathbb{Z}$-actions on torifications, we consider objects of the form (X, T, α) where X is a torifiable variety, T a choice of a torification, and $\alpha : \mathbb{Q}/\mathbb{Z} \times X \to X$ an action of $\mathbb{Q}/\mathbb{Z}$ determined by an embedding of $\mathbb{Q}/\mathbb{Z}$ as roots of unity in $\mathbb{G}_m(\mathbb{C}) = \mathbb{C}^*$, which act on each torus $T_i = \mathbb{G}_m^{k_i}$ diagonally by multiplication. An embedding of $\mathbb{Q}/\mathbb{Z}$ in $\mathbb{G}_m$ is determined by an invertible element in $\mathrm{Hom}(\mathbb{Q}/\mathbb{Z}, \mathbb{G}_m) = \hat{\mathbb{Z}}$, hence the action α is uniquely determined by the torification T and by the choice of an element in $\hat{\mathbb{Z}}^*$.

The corresponding morphisms are, respectively, strong, ordinary, or weak morphisms of torified varieties compatible with the $\mathbb{Q}/\mathbb{Z}$-actions, in the sense that the resulting torified morphism (after composing with isomorphisms or with local isomorphisms in the ordinary and weak case) are $\mathbb{Q}/\mathbb{Z}$-equivariant. We can then proceed as above and obtain equivariant Grothendieck rings $K_0^{\mathbb{Q}/\mathbb{Z}}(\mathcal{T})^s$, $K_0^{\mathbb{Q}/\mathbb{Z}}(\mathcal{T})^o$, and $K_0^{\mathbb{Q}/\mathbb{Z}}(\mathcal{T})^w$ of torified varieties.

In the case of good $\mathbb{Z}/N\mathbb{Z}$-effectively finite $\hat{\mathbb{Z}}$-actions, the setting is essentially the same. We consider objects of the form (X, T, α) where X is a torifiable variety, T a choice of a torification, and $\alpha : \mathbb{Z}/N\mathbb{Z} \times X \to X$ is given by the action of the N-th roots of unity on the tori $T_i = \mathbb{G}_m^{k_i}$ by multiplication. Thus, a good $\hat{\mathbb{Z}}$-action is determined by T, by the choice of an embedding of roots of unity in $\mathbb{G}_m$ (an element of $\hat{\mathbb{Z}}^*$) as above, and by the choice of $N \in \mathbb{N}$ that determines which subgroup of roots of unity is acting.

This choice of good $\mathbb{Z}/N\mathbb{Z}$-effectively finite $\hat{\mathbb{Z}}$-actions, with strong, ordinary, or weak morphisms whose associated torified morphisms are $\mathbb{Z}/N\mathbb{Z}$-equivariant, determine equivariant Grothendieck rings $K_0^{\hat{\mathbb{Z}}}(\mathcal{T})^s$, $K_0^{\hat{\mathbb{Z}}}(\mathcal{T})^o$, and $K_0^{\hat{\mathbb{Z}}}(\mathcal{T})^w$ of torified varieties with good effectively finite $\hat{\mathbb{Z}}$-actions.

4.3 Assembler and spectrum of torified varieties

As in the previous cases of $K_0^{\hat{\mathbb{Z}}}(\mathcal{V})$ of [54] and in the case of $K_0^{\hat{\mathbb{Z}}}(\mathcal{V}_S)$ discussed above, we consider the Grothendieck rings $K_0(\mathcal{T})^s$, $K_0(\mathcal{T})^o$, and $K_0(\mathcal{T})^w$ and their corresponding equivariant versions $K_0^{\mathbb{Q}/\mathbb{Z}}(\mathcal{T})^s$, $K_0^{\mathbb{Q}/\mathbb{Z}}(\mathcal{T})^o$, $K_0^{\mathbb{Q}/\mathbb{Z}}(\mathcal{T})^w$, and $K_0^{\hat{\mathbb{Z}}}(\mathcal{T})^s$, $K_0^{\hat{\mathbb{Z}}}(\mathcal{T})^o$, $K_0^{\hat{\mathbb{Z}}}(\mathcal{T})^w$ from the point of view of assemblers and spectra developed in [70], [71], [72].

Proposition 4.2 *For $a = s, o, w$, the category $\mathcal{C}^a_{\mathcal{T}}$ has objects that are pairs (X,T) of a torifiable variety and a torification, with morphisms the locally closed embeddings that are, respectively, strong, ordinary, or weak morphisms of torified varieties. The Grothendieck topology is generated by the covering families*

$$\{(Y,T_Y) \hookrightarrow (X,T_X), (X \smallsetminus Y, T_{X\smallsetminus Y}) \hookrightarrow (X,T_X)\} \tag{26}$$

where both embeddings are strong, ordinary, or weak morphisms, respectively. The category $\mathcal{C}^a_{\mathcal{T}}$ is an assembler with spectrum $K(\mathcal{C}^a_{\mathcal{T}})$ satisfying $\pi_0 K(\mathcal{C}^a_{\mathcal{T}}) = K_0(\mathcal{T})^a$. Similarly, for $G = \mathbb{Q}/\mathbb{Z}$ or $G = \hat{\mathbb{Z}}$ let $\mathcal{C}^{G,a}_{\mathcal{T}}$ be the category with objects (X,T,α) given by a torifiable variety X with a torification T and a G-action α of the kind discussed in §4.2 and morphisms the locally closed embeddings that are G-equivariant strong, ordinary, or weak morphisms. The Grothendieck topology is generated by covering families (26) with G-equivariant embeddings. The category $\mathcal{C}^{G,a}_{\mathcal{T}}$ is also an assembler, whose associated spectrum $K(\mathcal{C}^{G,a}_{\mathcal{T}})$ satisfies $\pi_0 K(\mathcal{C}^{G,a}_{\mathcal{T}}) = K_0^G(\mathcal{T})^a$.

Proof The argument is again as in [70], see Lemma 3.13. We check that the category admits pullbacks. In the strong case, if (Y,T_Y) and $(Y',T_{Y'})$ are objects with morphisms $f : (Y,T_Y) \hookrightarrow (X,T_X)$ and $f' : (Y',T_{Y'}) \hookrightarrow (X,T_X)$ given by embeddings that are strong morphisms of torified varieties. This means that the tori of the torification T_Y are restrictions to Y of tori of the torification T_X of X. Thus, both Y and Y' are unions of subcollections of tori of T_X. Their intersection $Y \cap Y'$ will then also inherit a torification consisting of a subcollection of tori of T_X and the resulting embedding $(Y \cap Y', T_{Y\cap Y'}) \hookrightarrow (X,T_X)$ is a strong morphism of torified varieties. In the ordinary case, we consider embeddings $f : Y \hookrightarrow X$ and $f' : Y' \hookrightarrow X$ that are ordinary morphisms of torified varieties, which means that, for isomorphisms ϕ_X, ϕ'_X, ϕ_Y, $\phi_{Y'}$, the compositions

$$\phi_X \circ f \circ \phi_Y : (Y,T_Y) \hookrightarrow (X,T_X), \quad \phi'_X \circ f' \circ \phi_{Y'} : (Y',T_{Y'}) \hookrightarrow (X,T_X)$$

are (strong) torified morphisms. Thus, the tori of the torifications T_Y and $T_{Y'}$ are subcollections of tori of X, under the embeddings $\phi_X \circ f \circ \phi_Y$ and $\phi'_X \circ f' \circ \phi_{Y'}$. The intersection $\phi_X \circ f \circ \phi_Y(Y) \cap \phi'_X \circ f' \circ \phi_{Y'}(Y') \subset X$ is isomorphic to a copy of $Y \cap Y'$ and has an induced torification $T_{Y\cap Y'}$ by a subcollection of tori of T_X. The embedding of $Y \cap Y'$ in X with this image is an ordinary morphism with respect to this torification. The weak case is constructed similarly to the ordinary case on the pieces of the decomposition. The equivariant cases are constructed analogously, as discussed in the case of equivariant Grothendieck rings of varieties in [54]. □

Corollary 4.3 *There are inclusions of assemblers $\mathcal{C}^s_{\mathcal{T}} \hookrightarrow \mathcal{C}^o_{\mathcal{T}} \hookrightarrow \mathcal{C}^w_{\mathcal{T}}$ that induce maps of K-theory, in particular $K_0(\mathcal{T})^s \to K_0(\mathcal{T})^o$ and $K_0(\mathcal{T})^o \to K_0(\mathcal{T})^w$. Similarly, for the G-equivariant cases of Proposition 4.2.*

Proof Since for morphisms strong implies ordinary and ordinary implies weak, one obtains inclusions of assemblers as stated. □

4.4 Lifting of the Bost–Connes system for torifications

We consider here lifts of the integral Bost–Connes algebra to the Grothendieck rings $K_0^{\hat{\mathbb{Z}}}(\mathcal{T})^s$, $K_0^{\hat{\mathbb{Z}}}(\mathcal{T})^o$, and $K_0^{\hat{\mathbb{Z}}}(\mathcal{T})^w$ and to the assemblers and spectra $K^{\hat{\mathbb{Z}}}(\mathcal{C}^s_{\mathcal{T}})$, $K^{\hat{\mathbb{Z}}}(\mathcal{C}^o_{\mathcal{T}})$, and $K^{\hat{\mathbb{Z}}}(\mathcal{C}^w_{\mathcal{T}})$.

Definition 4.4 We regard the zero-dimensional variety Z_n as a torified variety with the torification consisting of n zero dimensional tori and with a good $\hat{\mathbb{Z}}$ action factoring through $\mathbb{Z}/n\mathbb{Z}$ that cyclically permutes the points of Z_n. We write (Z_n, T_0, γ) for this object. For (X, T, α) a triple of a torifiable variety X, a given torification T, and an effectively finite action α of $\hat{\mathbb{Z}}$, we then set, for all $n \in \mathbb{N}$,

$$\begin{aligned} \sigma_n(X,T,\alpha) &= (X,T,\alpha \circ \sigma_n) \\ \tilde{\rho}_n(X,T,\alpha) &= (X \times Z_n, \sqcup_{a \in Z_n} T, \Phi_n(\alpha)). \end{aligned} \tag{27}$$

Proposition 4.5 *The σ_n and $\tilde{\rho}_n$ defined as in* (27) *determine endofunctors of the assembler categories $\mathcal{C}^{\hat{\mathbb{Z}},a}_{\mathcal{T}}$ that induce, respectively, ring homomorphisms $\sigma_n : K^{\hat{\mathbb{Z}}}(\mathcal{C}^a_{\mathcal{T}}) \to K^{\hat{\mathbb{Z}}}(\mathcal{C}^a_{\mathcal{T}})$ and group homomorphisms $\tilde{\rho}_n : K^{\hat{\mathbb{Z}}}(\mathcal{C}^a_{\mathcal{T}}) \to K^{\hat{\mathbb{Z}}}(\mathcal{C}^a_{\mathcal{T}})$ with the Bost–Connes relations*

$$\tilde{\rho}_n \circ \sigma_n(X,T,\alpha) = (X,T,\alpha) \times (Z_n, T_0, \gamma)$$

$$\sigma_n \circ \tilde{\rho}_n(X,T,\alpha) = (X,T,\alpha)^{\oplus n}.$$

Proof The proof is completely analogous to the case discussed in Theorem 3.15 and to the similar cases discussed in [54]. □

Remark 4.6 The σ_n and $\tilde{\rho}_n$ defined as in (27) determine a categorical Bost–Connes system as in Definition 3.9, where the objects are pairs (X, T) and the automorphisms are elements $g \in \hat{\mathbb{Z}}$ acting through the effectively finite action $\alpha(g)$.

Remark 4.7 Bost–Connes type quantum statistical mechanical systems associated to individual toric varieties (and more generally to varieties admitting torifications) were constructed in [44]. Here instead of Bost–Connes endomorphisms of individual varieties we are interested in a Bost–Connes system over the entire Grothendieck ring and its associated spectrum.

Remark 4.8 Variants of the construction above can be obtained by considering the multivariable versions of the Bost–Connes system discussed in [55], with actions of subsemigroups of $M_N(\mathbb{Z})^+$ on $\mathbb{Q}[\mathbb{Q}/\mathbb{Z}]^{\otimes N}$, that is, subalgebras of the crossed product algebra

$$\mathbb{Q}[\mathbb{Q}/\mathbb{Z}]^{\otimes N} \rtimes_\rho M_N(\mathbb{Z})^+$$

generated by $e(\underline{r})$ and μ_α, μ_α^* with

$$\rho_\alpha(e(\underline{r})) = \mu_\alpha e(\underline{r})\mu_\alpha^* = \frac{1}{\det \alpha} \sum_{\alpha(\underline{s})=\underline{r}} e(\underline{s})$$

$$\sigma_\alpha(e(\underline{r})) = \mu_\alpha^* e(\underline{r})\mu_\alpha = e(\alpha(\underline{r})).$$

The relevance of this more general setting to $\mathbb{F}_1$-geometries lies in a result of Borger and de Smit [13] showing that every torsion free finite rank Λ-ring embeds in some $\mathbb{Z}[\mathbb{Q}/\mathbb{Z}]^{\otimes N}$ with the action of $\mathbb{N}$ determined by the Λ-ring structure compatible with the diagonal subsemigroup of $M_N(\mathbb{Z})^+$.

5 Torified varieties and zeta functions

We discuss in this section the connection between the dynamical point of view on $\mathbb{F}_1$-geometry proposed in [54] and the point of view based on torifications.

We first discuss in §5.1 and §5.2 the notion of $\mathbb{F}_1$-points of a torified variety and its relation to the torification of the Grothendieck class, with some explicit examples. We then introduce the $\mathbb{F}_1$-zeta function in §5.4 and we show its main properties in Proposition 5.4, while in §5.5 we explain the relation between the $\mathbb{F}_1$-zeta function and the Hasse–Weil zeta function.

In §5.6 we consider the point of view on $\mathbb{F}_1$-structures proposed in [54] based on dynamical systems inducing quasi-uniponent endomorphisms on homology, in the particular case of torified varieties with dynamical systems compatible with the torification. We focus on the associated dynamical zeta functions, the Lefschetz zeta function and the Artin–Mazur zeta function, whose properties we recall in §5.6.1. We then prove in Proposition 5.8 that the resulting dynamical zeta function have similar properties to the $\mathbb{F}_1$-zeta function in the sense that both define exponentiable motivic measures from the Grothendieck rings of torified varieties to the Witt ring.

5.1 Counting $\mathbb{F}_1$-points

Assuming that a variety X over $\mathbb{Z}$ admits an $\mathbb{F}_1$-structure, regarded here as one of several possible forms of torified structure recalled above, [51], [53], the number of points of X over $\mathbb{F}_1$ is computed as the $q \to 1$ limit of the counting function $N_X(q)$ of points over $\mathbb{F}_q$ of the mod p reduction of X, for q a power of p. Any form of torified structure in particular implies that the variety is polynomially countable, hence that the counting function $N_X(q)$ is a polynomial in q with $\mathbb{Z}$-coefficients. The limit $\lim_{q\to 1} N_X(q)$, possibly normalized by a power of $q-1$, is interpreted as the number of $\mathbb{F}_1$-points of X, see [67]. Similarly, one can define "extensions" $\mathbb{F}_{1^m}$ of $\mathbb{F}_1$, in the sense of [45] (see also [27]). These corresponds to actions of the groups $\mathfrak{m}_m$ of m-th roots of unity. In terms of a torified structure, the points over $\mathbb{F}_{1^m}$ count m-th roots of unity in each torus of the decomposition. In terms of the counting function $N_X(q)$ the counting of points of X over the extension $\mathbb{F}_{1^m}$ is obtained as the value $N_X(m+1)$, see Theorem 4.10 of [24] and Theorem 1 of [29]). Summarizing, we have the following.

Lemma 5.1 *Let X be a variety over $\mathbb{Z}$ with torified Grothendieck class*

$$[X] = \sum_{i=0}^{N} a_i \mathbb{T}^i \tag{28}$$

with coefficients $a_i \in \mathbb{Z}_+$ and $\mathbb{T} = [\mathbb{G}_m] = \mathbb{L} - 1$. Then the number of points over $\mathbb{F}_{1^m}$ of X is given by

$$\#X(\mathbb{F}_{1^m}) = \sum_{i=0}^{N} a_i\, m^i . \tag{29}$$

In particular, $\#X(\mathbb{F}_1) = a_0 = \chi(X)$ the Euler characteristic.

5.2 Bialynicki-Birula decompositions and torified geometries

As shown in [5], [18], the motive of a smooth projective variety with action of the multiplicative group admits a decomposition, obtained via the method of Bialynicki-Birula, [9], [10], [11]. We recall the result here, in a particular case which gives rise to examples of torified varieties.

Lemma 5.2 *Let X be a smooth projective k-variety X endowed with a $\mathbb{G}_m$ action such that the fixed point locus $X^{\mathbb{G}_m}$ admits a torification of the Grothendieck class. Then X also admits a torification of the Grothendieck class. Consider the filtration $X = X_n \supset X_{n-1} \supset \cdots \supset X_0 \supset \emptyset$ with affine fibrations $\phi_i : X_i \smallsetminus X_{i-1} \to Z_i$ over the components $X^{\mathbb{G}_m} = \sqcup_i Z_i$, associated* *to the Bialynicki-Birula decomposition. If the fixed point locus $X^{\mathbb{G}_m}$ admits a geometric torification such that the restrictions of the fibrations ϕ_i to the individual tori of the torification of Z_i are trivializable, then X also admits a geometric torification.*

Proof The Bialynicki-Birula decomposition, [9], [10], [11], see also [42], shows that a smooth projective k-variety X endowed with a $\mathbb{G}_m$ action has smooth closed fixed point locus $X^{\mathbb{G}_m}$ which decomposes into a finite union of components $X^{\mathbb{G}_m} = \sqcup_i Z_i$, of dimensions $\dim Z_i$ the dimension of $TX_z^{\mathbb{G}_m}$ at $z \in Z_i$. The variety X has a filtration $X = X_n \supset X_{n-1} \supset \cdots \supset X_0 \supset \emptyset$ with affine fibrations $\phi_i : X_i \smallsetminus X_{i-1} \to Z_i$ of relative dimension d_i equal to the dimension of the positive eigenspace of the $\mathbb{G}_m$-action on the tangent space of X at points of Z_i. The scheme $X_i \smallsetminus X_{i-1}$ is identified with $\{x \in X : \lim_{t\to 0} tx \in Z_i\}$ under the $\mathbb{G}_m$-action $t : x \mapsto tx$, with $\phi_i(x) = \lim_{t\to 0} tx$. As shown in [18], the object $M(X)$ in the category of correspondences Corr_k with integral coefficients (and in the category of Chow motives) decomposes as

$$M(X) = \bigoplus_i M(Z_i)(d_i), \tag{30}$$

where $M(Z_i)$ are the motives of the components of the fixed point set and $M(Z_i)(d_i)$ are Tate twists. The class in the Grothendieck ring $K_0(\mathcal{V}_k)$ decomposes then as

$$[X] = \sum_i [Z_i]\mathbb{L}^{d_i}. \tag{31}$$

It is then immediate that, if the components Z_i admit a geometric torification (respectively, a torification of the Grothendieck class) then the variety X also does. If $Z_i = \cup_{j=1}^{n_i} T_{ij}$ with $T_{ij} = \mathbb{G}_m^{a_{ij}}$ or, respectively $[Z_i] = \sum_{j=1}^{n_i} (\mathbb{L} - 1)^{a_{ij}}$, then $X = \cup_{i=0}^{n} (X_i \smallsetminus X_{i-1}) = \cup_{i=0}^{n} \mathcal{F}^{d_i}(Z_i)$, where $\mathcal{F}^{d_i}(Z_i)$ denotes the total space of the affine fibration $\phi_i : X_i \smallsetminus X_{i-1} \to Z_i$ with fibers $\mathbb{A}^{d_i}$. The Grothendieck class is then torified by

$$[X] = \sum_{i=1}^{n} \sum_{j=1}^{n_i} \mathbb{T}^{a_{ij}} \left(1 + \sum_{k=1}^{d_i} \binom{d_i}{k} \mathbb{T}^k \right),$$

with $\mathbb{T} = \mathbb{L} - 1$ the class of the multiplicative group $\mathbb{T} = [\mathbb{G}_m]$, and where the affine spaces are torified by

$$\mathbb{L}^n - 1 = \sum_{k=1}^{n} \binom{n}{k} \mathbb{T}^k.$$

If the restriction of the fibration $\mathcal{F}^{d_i}(Z_i)$ to the individual tori T_{ij} of the torification of Z_i is trivial, then it can be torified by a products $T_{ij} \times T_k$ of the torus T_{ij} and the tori T_k of a torification of the fiber affine space $\mathbb{A}^{d_i}$. This determines a a geometric torification of the affine fibrations $\mathcal{F}^{d_i}(Z_i)$, hence of X. □

5.3 An example of torified varieties

A physically significant example of torified varieties of the type described in Lemma 5.2 arises in the context of BPS state counting of [22]. Refined BPS state counting computes the multiplicities of BPS particles with charges in a lattice (K-theory changes of even D-branes) for assigned spin quantum numbers of a Spin(4) $= SU(2) \times SU(2)$ representation, see [22], [23], [33].

We mention here the following explicit example from [23], namely the case of the moduli space $\mathcal{M}_{\mathbb{P}^2}(4,1)$ of Gieseker semi-stable shaved on $\mathbb{P}^2$ with Hilbert polynomial equal to $4m + 1$. In this case, it is proved in [23] that $\mathcal{M}_{\mathbb{P}^2}(4,1)$ has a torus action of $\mathbb{G}_M^2$ for which the fixed point locus consists of 180 isolated points and 6 components isomorphic to $\mathbb{P}^1$. The Grothendieck class, obtained through the Bialynicki-Birula decomposition [23] is given by

$$
\begin{aligned}
[\mathcal{M}_{\mathbb{P}^2}(4,1)] = {} & 1 + 2\mathbb{L} + 6\mathbb{L}^2 + 10\mathbb{L}^3 + 14\mathbb{L}^4 + 15\mathbb{L}^5 \\
& + 16\mathbb{L}^6 + 16\mathbb{L}^7 + 16\mathbb{L}^8 + 16\mathbb{L}^9 + 16\mathbb{L}^{10} + 16\mathbb{L}^{11} \\
& + 15\mathbb{L}^{12} + 14\mathbb{L}^{13} + 10\mathbb{L}^{14} + 6\mathbb{L}^{15} + 2\mathbb{L}^{16} + \mathbb{L}^{17}.
\end{aligned}
$$

Note that, for a smooth projective variety with Grothendieck class that is a polynomial in the Lefschetz motive $\mathbb{L}$, the Poincaré polynomial and the Grothendieck class are related by replacing x^2 with $\mathbb{L}$, since the variety is Hodge–Tate. In torified form the above gives

$$
\begin{aligned}
[\mathcal{M}_{\mathbb{P}^2}(4,1)] = {} & \mathbb{T}^{17} + 19\,\mathbb{T}^{16} + 174\,\mathbb{T}^{15} + 1020\,\mathbb{T}^{14} \\
& + 4284\,\mathbb{T}^{13} + 13665\,\mathbb{T}^{12} + 34230\,\mathbb{T}^{11} + 68678\,\mathbb{T}^{10} \\
& + 111606\,\mathbb{T}^{9} + 147653\,\mathbb{T}^{8} + 159082\,\mathbb{T}^{7} + 139008\,\mathbb{T}^{6} \\
& + 97643\,\mathbb{T}^{5} + 54320\,\mathbb{T}^{4} + 23370\,\mathbb{T}^{3} + 7468\,\mathbb{T}^{2} + 1632\,\mathbb{T} + 192,
\end{aligned}
$$

where $192 = \chi(\mathcal{M}_{\mathbb{P}^2}(4,1))$ is the Euler characteristics, which is also the number of points over $\mathbb{F}_1$. The number of points over $\mathbb{F}_{1^m}$ gives 864045 for $m = 1$ (the number of tori in the torification), 383699680 for $m = 2$ (roots of unity of order two), 36177267945 for $m = 3$ (roots of unity of order three), etc.

In this example, the Euler characteristic $\chi(\mathcal{M}_{\mathbb{P}^2}(4,1))$, which can also be seen as the number of $\mathbb{F}_1$-points, is interpreted physically as determining the

BPS counting. It is natural to ask whether the counting of $\mathbb{F}_{1^m}$-points, which corresponds to the counting of roots of unity in the tori of the torification, can also carry physically significant information.

Other examples of torified varieties relevant to physics can be found in the context of quantum field theory, see [6] and [60].

5.3.1 BPS counting and the virtual motive

The formulation of the refined BPS counting given in [22] can be summarized as follows. The virtual motive $[X]_{\rm vir} = \mathbb{L}^{-n/2}[X]$, with $n = \dim(X)$, is a class in the ring of motivic weights $K_0(\mathcal{V})[\mathbb{L}^{-1/2}]$, see [12]. When X admits a $\mathbb{G}_m$ action and a Bialynicki-Birula decomposition as discussed in the previous section, where all the components Z_i of the fixed point locus of the $\mathbb{G}_m$-action have Tate classes $[Z_i] = \sum_j c_{ij}\mathbb{L}^{b_{ij}} \in K_0(\mathcal{V})$, with $c_{ij} \in \mathbb{Z}$ and $b_{ij} \in \mathbb{Z}_+$, the virtual motive $[X]_{\rm vir}$ is a Laurent polynomial in the square root $\mathbb{L}^{1/2}$ of the Lefschetz motive,

$$[X]_{\rm vir} = \sum_{i,j} c_{ij}\, \mathbb{L}^{b_{ij}+d_i-1/2}, \tag{32}$$

where, as before, d_i is the dimension of the positive eigenspace of the $\mathbb{G}_m$-action on the tangent space of X at points of Z_i. In applications to BPS counting, one considers the virtual motive of a moduli space M that admits a perfect obstruction theory, so that it has virtual dimension zero and an associated invariant $\#_{\rm vir}M$ which is computed by a virtual index

$$\#_{\rm vir}M = \chi_{\rm vir}(M, K^{1/2}_{M,\rm vir}) = \chi(M, K^{1/2}_{M,\rm vir} \otimes \mathcal{O}_{M,\rm vir}),$$

where $\mathcal{O}_{M,\rm vir}$ is the virtual structure sheaf and $K^{1/2}_{M,\rm vir}$ is a square root of the virtual canonical bundle, see [35].

5.3.2 The formal square root of the Leftschetz motive

The formal square root $\mathbb{L}^{1/2}$ of the Leftschetz motive that occurs in (32) as Grothendieck class can be introduced, at the level of the category of motives, as shown in §3.4 of [48], using the Tannakian formalism, [30]. Let $\mathcal{C} = \mathrm{Num}^{\dagger}_{\mathbb{Q}}$ be the Tannakian category of pure motives with the numerical equivalence relation and the Koszul sign rule twist $\dagger$ in the tensor structure, with motivic Galois group $G = \mathrm{Gal}(\mathcal{C})$. The inclusion of the Tate motives (with motivic Galois group $\mathbb{G}_m$) determines a group homomorphism $t : G \to \mathbb{G}_m$, which satisfies $t \circ w = 2$ with the weight homomorphism $w : \mathbb{G}_m \to G$ (see §5 of [31]). The category $\mathcal{C}(\mathbb{Q}(\frac{1}{2}))$ obtained by adjoining a square root of the Tate motive to $\mathcal{C}$ is then obtained as the Tannakian category whose Galois group is the fibered product

$$G^{(2)} = \{(g,\lambda) \in G \times \mathbb{G}_m \,:\, t(g) = \lambda^2\}.$$

The construction of square roots of Tate motives described in [48] was generalized in [49] to arbitrary n-th roots of Tate motives, obtained via the same Tannakian construction, with the category $\mathcal{C}(\mathbb{Q}(\frac{1}{n}))$ obtained by adjoining an n-th root of the Tate motive determined by its Tannakian Galois group

$$G^{(n)} = \{(g,\lambda) \in G \times \mathbb{G}_m \,:\, t(g) = \sigma_n(\lambda)\},$$

with $\sigma_n : \mathbb{G}_m \to \mathbb{G}_m$, $\sigma_n(\lambda) = \lambda^n$. The category $\hat{\mathcal{C}}$ obtained by adjoining to $\mathcal{C} = \mathrm{Num}^{\dagger}_{\mathbb{Q}}$ arbitrary roots of the Tate motives is the Tannakian category with Galois group $\hat{G} = \varprojlim_n G^{(n)}$. The category $\hat{\mathcal{C}}$ has an action of $\mathbb{Q}^*_+$ by automorphisms induced by the endomorphisms σ_n of $\mathbb{G}_m$. These roots of Tate motives give rise to a good formalism of $\mathbb{F}_\zeta$-geometry, with ζ a root of unity, lying "below" $\mathbb{F}_1$-geometry and expressed at the motivic level in terms of a Habiro ring type object associated to the Grothendieck ring of orbit categories of $\hat{\mathcal{C}}$, see [49].

5.4 Counting $\mathbb{F}_1$-points and zeta function

For a variety X over $\mathbb{Z}$ that is polynomially countable (that is, the counting functions $N_X(q) = \#X_p(\mathbb{F}_q)$ with X_p the mod p reduction is a polynomial in q with $\mathbb{Z}$ coefficients) the counting of points over the "extensions" $\mathbb{F}_{1^m}$ (in the sense of [45]) can be obtained as the values $N_X(m+1)$ (see Theorem 4.10 of [24] and Theorem 1 of [29]). As we discussed earlier, in the case of a torified variety, with Grothendieck class $[X] = \sum_{i\geq 0} a_i \mathbb{T}^i$ with $a_i \in \mathbb{Z}_+$, this corresponds to the counting given in (29). This is the counting of the number of m-th roots of unity in each torus $\mathbb{T}^i = [\mathbb{G}^i_m]$ of the torification.

For a variety X over a finite field $\mathbb{F}_q$ the Hasse–Weil zeta function is given, in logarithmic form by

$$\log Z_{\mathbb{F}_q}(X,t) = \sum_{m\geq 1} \frac{\#X(\mathbb{F}_{q^m})}{m} t^m. \tag{33}$$

In the case of torified varieties, there is an analogous zeta function over $\mathbb{F}_1$. We think of this $\mathbb{F}_1$-zeta function as defined on torified Grothendieck classes, $Z_{\mathbb{F}_1}([X],t)$. In the case of geometric torifications, we can regard it as a function of the variety and the torification, $Z_{\mathbb{F}_1}((X,T),t)$. For simplicity of notation, we will simply write $Z_{\mathbb{F}_1}(X,t)$ by analogy to the Hasse–Weil zeta function, with

$$\log Z_{\mathbb{F}_1}(X,t) := \sum_{m\geq 1} \frac{\#X(\mathbb{F}_{1^m})}{m} t^m. \tag{34}$$

Lemma 5.3 *Let X be a variety over $\mathbb{Z}$ with a torified Grothendieck class $[X] = \sum_{k\geq 0} a_k \mathbb{T}^k$ with $a_k \in \mathbb{Z}_+$. Then the $\mathbb{F}_1$-zeta function is given by*

$$\log Z_{\mathbb{F}_1}(X,t) = \sum_{k=0}^{N} a_k \operatorname{Li}_{1-k}(t), \tag{35}$$

where $\operatorname{Li}_s(t)$ is the polylogarithm function with $\operatorname{Li}_1(t) = -\log(1-t)$ and for $k \geq 1$

$$\operatorname{Li}_{1-k}(t) = \left(t\frac{d}{dt}\right)^{k-1} \frac{t}{1-t}.$$

Proof For $[X] = \sum_{k\geq 0} a_k \mathbb{T}^k$ with $a_k \in \mathbb{Z}_+$ as above, we can consider a similar zeta function based on the counting of $\mathbb{F}_{1^m}$-points described above. Using (29), we obtain an expression of the form

$$\log Z_{\mathbb{F}_1}(X,t) = \sum_{m\geq 1} \frac{\#X(\mathbb{F}_{1^m})}{m} t^m$$
$$= \sum_{k=0}^{N} a_k \sum_{m\geq 1} m^{k-1} t^m = \sum_{k=0}^{N} a_k \operatorname{Li}_{1-k}(t),$$

given by a linear combination of polylogarithm functions $\operatorname{Li}_s(t)$ at integer values $s \leq 1$. □

Such polylogarithm functions can be expressed explicitly in the form $\operatorname{Li}_1(t) = -\log(1-t)$ and for $k \geq 1$

$$\operatorname{Li}_{1-k}(t) = \left(t\frac{d}{dt}\right)^{k-1} \frac{t}{1-t} = \sum_{\ell=0}^{k-1} \ell!\ S(k,\ell+1) \left(\frac{t}{1-t}\right)^{\ell+1},$$

with $S(k,r)$ the Stirling numbers of the second kind

$$S(k,r) = \frac{1}{r!} \sum_{j=0}^{r} (-1)^{r-j} \binom{r}{j} j^k.$$

As in the case of the Hasse–Weil zeta function over $\mathbb{F}_q$ (see [62]), the $\mathbb{F}_1$-zeta function gives an exponentiable motivic measure.

Proposition 5.4 *The $\mathbb{F}_1$-zeta function is an exponentiable motivic measure, that is, a ring homomorphism $Z_{\mathbb{F}_1} : K_0(\mathcal{T})^a \to W(\mathbb{Z})$ from the Grothendieck ring of torified varieties (with either $a = w, o, s$) to the Witt ring.*

Proof Clearly with respect to addition in the Grothendieck ring of torified varieties we have $[X] + [X'] = \sum_{i\geq 0} a_i \mathbb{T}^i + \sum_{j\geq 0} a'_j \mathbb{T}^j = \sum_{k\geq 0} b_k \mathbb{T}^k$ with $b_k = a_k + a'_k$, hence

$$\log Z_{\mathbb{F}_1}([X]+[X'],t) = \sum_{k=0}^{N} b_k \,\mathrm{Li}_{1-k}(t) = \log Z_{\mathbb{F}_1}([X],t) + \log Z_{\mathbb{F}_1}([X'],t).$$

The behavior with respect to products $[X]\cdot[Y]$ in the Grothendieck ring of torified varieties can be analyzed as in [62] for the Hasse–Weil zeta function. We view the $\mathbb{F}_1$-zeta function

$$Z_{\mathbb{F}_1}(X,t) = \exp\left(\sum_{k=0}^{N} a_k \,\mathrm{Li}_{1-k}(t)\right)$$

as the element in the Witt ring $W(\mathbb{Z})$ with ghost components $\#X(\mathbb{F}_{1^m}) = \sum_{k=0}^{N} m^k$, by writing the ghost map $\mathrm{gh} : W(\mathbb{Z}) \to \mathbb{Z}^{\mathbb{N}}$ as

$$\mathrm{gh} : Z(t) = \exp\left(\sum_{m\geq 1} \frac{N_m}{m} t^m\right) \mapsto t\frac{d}{dt}\log Z(t) = \sum_{m\geq 1} N_m t^m \mapsto (N_m)_{m\geq 1}.$$

The ghost map is an injective ring homomorphism. Thus, it suffices to see that on the ghost components $N_m(X\times Y) = N_m(X)\cdot N_m(Y)$. If $[X] = \sum_{k\geq 0} a_k \mathbb{T}^k$ and $[Y] = \sum_{\ell\geq 0} b_\ell \mathbb{T}^\ell$ then $[X\times Y] = \sum_{n\geq 0}\sum_{k+\ell=n} a_k b_\ell \mathbb{T}^n$ and $N_m(X\times Y) = \sum_{n\geq 0}\sum_{k+\ell=n} a_k b_\ell m^n = N_m(X)\cdot N_m(Y)$. □

5.5 Relation to the Hasse–Weil zeta function

We discuss here the relation between the $\mathbb{F}_1$-zeta function $Z_{\mathbb{F}_1}(X,t)$ introduced in (34) above, for a variety X over $\mathbb{Z}$ with torified Grothendieck class $[X] = \sum_{k\geq 0} a_k \mathbb{T}^k$, and the Hesse–Weil zeta function $Z_{\mathbb{F}_q}(X,t)$, defined as in (33).

Definition 5.5 Consider the following elements in the Witt ring $W(\mathbb{Z})$, for $k\geq 0$:

$$Z_{0,k,q}(t) := \exp\left(\sum_{m\geq 1} (q-1)^k \frac{t^m}{m}\right) = \frac{1}{(1-t)^{(q-1)^k}} \tag{36}$$

$$Z_{1,k,q}(t) := \exp\left(\sum_{m\geq 1} (\#\mathbb{P}^{m-1}(\mathbb{F}_q))^k \frac{t^m}{m}\right) = \exp\left(\sum_{m\geq 1}\left(\sum_{\ell=0}^{m-1} q^\ell\right)^k \frac{t^m}{m}\right). \tag{37}$$

Lemma 5.6 *Let $Z_{\mathbb{F}_q}(\mathbb{T}^k,t)$ be the Hasse-Weil zeta function of a torus $\mathbb{T}^k$. The function $Z_{0,k,q}(t)$ of (36) divides $Z_{\mathbb{F}_q}(\mathbb{T}^k,t)$ in the Witt ring with quotient the function $Z_{1,k,q}(t)$ of (37).*

Proof Given elements $Q = Q(t)$ and $P = P(t)$ in the Witt ring $W(\mathbb{Z})$, we have that Q divides P iff the ghost components q_m of Q in $\mathbb{Z}^{\mathbb{N}}$ divide the corresponding ghost components p_m of P. There is then an element $S = S(t)$ in $W(\mathbb{Z})$, with ghost components $s_m = p_m/q_m$, such that the Witt product gives $S \star_W Q = P$. The m-th ghost components of $Z_{\mathbb{F}_q}(\mathbb{T}^k,t)$ is $(q^m-1)^k = \#\mathbb{T}^k(\mathbb{F}_{q^m})$, and we have $(q^m-1)^k/(q-1)^k = (1+q+\cdots+q^{m-1})^k$. □

Given elements $Q, P \in W(\mathbb{Z})$ such that $Q|P$ as above, we write $S = P/_W Q$ for the resulting element $S \in W(\mathbb{Z})$ with $S \star_W Q = P$.

The $\mathbb{F}_1$-zeta function of (34) is obtained from the Hasse–Weil zeta function of (33) in the following way.

Proposition 5.7 *Let X be a variety X over $\mathbb{Z}$ with torified Grothendieck class $[X] = \sum_{k\geq 0} a_k \mathbb{T}^k$. The $\mathbb{F}_1$-zeta function is given by*

$$\begin{aligned} Z_{\mathbb{F}_1}(X,t) &= \lim_{q\to 1} {}^W\textstyle\sum_{k\geq 0} \left(Z_{\mathbb{F}_q}(\mathbb{T}^k,t) \,/_W\, Z_{0,k,q}(t)\right)^{a_k} \\ &= \lim_{q\to 1} {}^W\textstyle\sum_{k\geq 0} Z_{1,k,q}(t)^{a_k}, \end{aligned} \tag{38}$$

while the Hasse–Weil zeta function is given by

$$Z_{\mathbb{F}_q}(X,t) = {}^W\sum_{k\geq 0} Z_{\mathbb{F}_q}(\mathbb{T}^k,t)^{a_k}, \tag{39}$$

where ${}^W\sum$ denotes the sum in the Witt ring.

Proof For the Hasse–Weil zeta function we have

$$\begin{aligned} Z_{\mathbb{F}_q}(X,t) &= \exp\left(\sum_{m\geq 1} \#X(\mathbb{F}_{q^m})\frac{t^m}{m}\right) = \exp\left(\sum_{k\geq 0} a_k \sum_{m\geq 1} (q^m-1)^k \frac{t^m}{m}\right) \\ &= \prod_{k\geq 0} \exp(a_k \log Z_{\mathbb{F}_q}(\mathbb{T}^k,t)), \end{aligned}$$

hence we get (39). To obtain the $\mathbb{F}_1$-zeta function we then use Lemma 5.6 and the fact that $(q^m-1)^k/(q-1)^k = (1+q+\cdots+q^{m-1})^k$, with $\lim_{q\to 1}(1+q+\cdots+q^{m-1})^k = m^k$. □

5.6 Dynamical zeta functions

The dynamical approach to $\mathbb{F}_1$-structures proposed in [54] is based on the existence of an endomorphism $f : X \to X$ that induces a quasi-unipotent morphism f_* on the homology $H_*(X,\mathbb{Z})$. In particular, this means that the map f_* has eigenvalues that are roots of unity.

In the case of a variety X endowed with a torification $X = \sqcup_i T^{d_i}$, one can consider in particular endomorphisms $f : X \to X$ that preserve the torification and that restrict to endomorphisms of each torus T^{d_i}.

We recall the definition and main properties of the relevant dynamical zeta functions, which we will consider in Proposition 5.8.

5.6.1 Properties of dynamical zeta functions

In general to a self-map $f : X \to X$, one can associate the dynamical Artin–Mazur zeta function and the homological Lefschetz zeta function. A particular class of maps with the property that they induce quasi-unipotent morphisms in homology is given by the Morse–Smale diffeomorphisms of smooth manifolds, see [66]. These are diffeomorphisms characterized by the properties that the set of nonwandering points is finite and hyperbolic, consisting of a finite number of periodic points, and for any pair of these points x, y the stable and unstable manifolds $W^s(x)$ and $W^u(y)$ intersect transversely. Morse–Smale diffeomorphisms are structurally stable among all diffeomorphisms, [36], [66].

The Lefschetz zeta function is given by

$$\zeta_{\mathcal{L},f}(t) = \exp\left(\sum_{m\geq 1} \frac{L(f^m)}{m} t^m\right), \tag{40}$$

with $L(f^m)$ the Lefschetz number of the m-th iterate f^m,

$$L(f^m) = \sum_{k\geq 0} (-1)^k \mathrm{Tr}((f^m)_* \mid H_k(X,\mathbb{Q})).$$

For a function with finite sets of fixed points $\mathrm{Fix}(f^m)$ this is also equal to

$$L(f^m) = \sum_{x\in \mathrm{Fix}(f^m)} \mathcal{I}(f^m, x),$$

with $\mathcal{I}(f^m, x)$ the index of the fixed point. The zeta function can then be written as a rational function of the form

$$\zeta_{\mathcal{L},f}(t) = \prod_k \det(1 - t\, f_* | H_k(X,\mathbb{Q}))^{(-1)^{k+1}}.$$

In the case of a map f with finitely many periodic points, all hyperbolic, the Lefschetz zeta function can be equivalently written (see [36]) as the rational function

$$\zeta_{\mathcal{L},f}(t) = \prod_\gamma (1 - \Delta_\gamma t^{p(\gamma)})^{(-1)^{u(\gamma)+1}}.$$

Here the product is over periodic orbits γ with least period $p(\gamma)$ and $u(\gamma) = \dim E^u_x$, for $x \in \gamma$, is the dimension of the span of eigenvectors of $Df_x^{p(\gamma)} : T_x M \to T_x M$ with eigenvalues λ with $|\lambda| > 1$. One has $\Delta_\gamma = \pm 1$ according to whether $Df_x^{p(\gamma)}$ is orientation preserving or reversing. The relation comes from the identity $\mathcal{I}(f^m, x) = (-1)^{u(\gamma)}\Delta_\gamma$. The Artin–Mazur zeta function is given by

$$\zeta_{AM,f}(t) = \exp\left(\sum_{m\geq 1} \frac{\#\mathrm{Fix}(f^m)}{m} t^m\right). \tag{41}$$

The case of Morse–Smale diffeomorphisms can be treated as in [37] to obtain rationality and a description in terms of the homological zeta functions.

In the setting of real tori $\mathbb{R}^d/\mathbb{Z}^d$, one can considers the case of a toral endomorphism specified by a matrix $M \in M_d(\mathbb{Z})$. In the hyperbolic case, the counting of isolated fixed points of M^m is given by $|\det(1 - M^m)|$ and the dynamical Artin–Mazur zeta function is expressible in terms of the Lefschetz zeta function, associated to the signed counting of fixed points, through the fact that the Lefschetz zeta function agrees with the zeta function

$$\zeta_M(t) = \exp\left(\sum_{n\geq 1} \frac{t^n}{n} a_n\right), \quad \text{with} \quad a_n = \det(1 - M^n), \tag{42}$$

where $a_n = \det(1 - M^n)$ is a signed fixed point counting. The general relation between the zeta functions for the signed $\det(1 - M^n)$ and for $|\det(1 - M^m)|$ is shown in [4] for arbitrary toral endomorphisms, with $M \in M_d(\mathbb{Z})$.

In the case of complex algebraic tori $T^d = \mathbb{G}_m^d(\mathbb{C})$, one can similarly consider the endomorphisms action of the semigroup of matrices $M \in M_d(\mathbb{Z})^+$ by the linear action on $\mathbb{C}^d$ preserving $\mathbb{Z}^d$ and the exponential map $0 \to \mathbb{Z} \to \mathbb{C} \to \mathbb{C}^* \to 1$ so that, for $M = (m_{ab})$ and $\lambda_a = \exp(2\pi i u_a)$, with the action given by

$$\lambda = (\lambda_a) \mapsto M(\lambda) = \exp\left(2\pi i \sum_b m_{ab} u_b\right).$$

The subgroup $\mathrm{SL}_n(\mathbb{Z}) \subset M_n(\mathbb{Z})^+$ acts by automorphisms. These generalize the Bost–Connes endomorphisms $\sigma_n : \mathbb{G}_m \to \mathbb{G}_m$, which correspond to the ring homomorphisms of $\mathbb{Z}[t, t^{-1}]$ given by $\sigma_n : P(t) \mapsto P(t^n)$ and determine multivariable versions of the Bost–Connes algebra, see [55]. We can consider in this way maps of complex algebraic tori $T^d_{\mathbb{C}} = \mathbb{G}_m^d(\mathbb{C})$ that induce maps of the real tori obtained as the subgroup $T^d_{\mathbb{R}} = U(1)^d \subset \mathbb{G}_m^d(\mathbb{C})$, and associate to these maps the Lefschetz and Artin–Mazur zeta functions of the induced map of real tori.

5.6.2 Torifications and dynamical zeta functions

In the case of a variety with a torification, we consider endomorphisms $f : X \to X$ that preserve the tori of the torification and restrict to each torus

T^{d_i} to a diffeomorphism $f_i : T_{\mathbb{R}}^{d_i} \to T_{\mathbb{R}}^{d_i}$. In particular, we consider toral endomorphism with a matrix $M_i \in M_{d_i}(\mathbb{Z})$, we can associate to the pair (X, f) a zeta function of the form

$$\zeta_{\mathcal{L},f}(X,t) = \prod_i \zeta_{\mathcal{L},f_i}(t), \quad \zeta_{AM,f}(X,t) = \prod_i \zeta_{AM,f_i}(t). \tag{43}$$

Proposition 5.8 *The zeta functions* (40) *and* (41) *define exponentiable motivic measures on the Grothendieck ring $K_0^{\mathbb{Z}}(\mathcal{V}_{\mathbb{C}})$ of §6 of [54] with values in the Witt ring $W(\mathbb{Z})$. The zeta functions* (43) *define exponentiable motivic measures on the Grothendieck ring $K_0(\mathcal{T})^a$ of torified varieties with values in $W(\mathbb{Z})$.*

Proof The Grothendieck ring $K_0^{\mathbb{Z}}(\mathcal{V}_{\mathbb{C}})$ considered in §6 of [54] consists of pairs (X, f) of a complex quasi-projective variety and an automorphism $f : X \to X$ that induces a quasi-uniponent map f_* in homology. The addition is simply given by the disjoint union, and both the counting of periodic points $\#\mathrm{Fix}(f^m)$ and the Lefschetz numbers $L(f^m)$ behave additively under disjoint unions. Thus, the zeta functions $\zeta_{\mathcal{L},f}(t)$ and $\zeta_{AM,f}(t)$, seen as elements in the Witt ring $W(\mathbb{Z})$ add

$$\zeta_{\mathcal{L},f_1 \sqcup f_2}(t) = \exp\left(\sum_{m\geq 1} \frac{L((f_1 \sqcup f_2)^m)}{m} t^m\right)$$
$$= \exp\left(\sum_{m\geq 1} \frac{L(f_1^m)}{m} t^m\right) \cdot \exp\left(\sum_{m\geq 1} \frac{L(f_2^m)}{m} t^m\right) = \zeta_{\mathcal{L},f_1}(t) +_W \zeta_{\mathcal{L},f_2}(t)$$

and similarly for $\zeta_{AM,f_1\sqcup f_2}(t) = \zeta_{AM,f_1}(t) +_W \zeta_{AM,f_2}(t)$. The product is given by the Cartesian product $(X_1, f_1) \times (X_2, f_2)$. Since $\mathrm{Fix}((f_1 \times f_2)^m) = \mathrm{Fix}(f_1^m) \times \mathrm{Fix}(f_2^m)$ and the same holds for Lefschetz numbers since

$$L((f_1 \times f_2)^m)$$
$$= \sum_{k\geq 0} \sum_{\ell+r=k} (-1)^{\ell+r} \mathrm{Tr}((f_1^m)_* \otimes (f_2^m)_* \mid H_\ell(X_1, \mathbb{Q}) \otimes H_r(X_2, \mathbb{Q}))$$

which gives $L(f_1^m) \cdot L(f_2^m)$. Thus, we can use as in Proposition 5.4 the fact that the ghost map $\mathrm{gh} : W(\mathbb{Z}) \to \mathbb{Z}^{\mathbb{N}}$

$$\mathrm{gh} : \exp\left(\sum_{m\geq 1} \frac{N_m}{m} t^m\right) \mapsto \sum_{m\geq 1} N_m t^m \mapsto (N_m)_{m\geq 1}$$

is an injective ring homomorphism to obtain the multiplicative property. The case of the torified varieties and the zeta functions (43) is analogous, combining the additive and multiplicative behavior of the fixed point counting and the Lefschetz numbers on each torus and of the decomposition into tori as in Proposition 5.4. □

In the case of quasi-unipotent maps of tori the Lefschetz zeta function can be computed completely explicitly. Indeed, it is shown in [7], [8] that, for a quasi-unipotent self map $f : T^n_{\mathbb{R}} \to T^n_{\mathbb{R}}$, the Lefschetz zeta function has an explicit form that is completely determined by the map on the first homology. Under the quasi-unipotent assumption all the eigenvalues of the induced map on H_1 are roots of unity, hence the characteristic polynomial $\det(1-t\, f_*|H_1(X))$ is a product of cyclotomic polynomials $\Phi_{m_1}(t)\cdots\Phi_{m_N}(t)$ where

$$\Phi_m(t) = \prod_{d|m}(1-t^d)^{\mu(m/d)},$$

with $\mu(n)$ the Möbius function. It is shown in [8] that the Lefschetz zeta function has the form

$$\zeta_{\mathcal{L},f}(t) = \prod_{d|m}(1-t^d)^{-s_d}, \tag{44}$$

where $m = \mathrm{lcm}\{m_1,\ldots,m_N\}$ and

$$s_d = \frac{1}{d}\sum_{k|d} F_k\, \mu(d/k)$$

$$F_k = \prod_{i=1}^{N} (\Phi_{m_i/(k,m_i)}(1))^{\varphi(m_i)/\varphi(m_i/(k,m_i))}$$

where the Euler function

$$\varphi(m) = m \prod_{p|m,\ p \text{ prime}} (1-p^{-1})$$

is the degree of $\Phi_m(t)$.

Remark 5.9 The properties of dynamical Artin–Mazur zeta functions change significantly when, instead of considering varieties over $\mathbb{C}$ one considers varieties in positive characteristic, [16], [19]. The prototype model of this phenomenon is illustrated by considering the Bost–Connes endomorphisms $\sigma_n : \lambda \mapsto \lambda^n$ of $\mathbb{G}_m(\bar{\mathbb{F}}_p)$. In this case, the dynamical zeta function of σ_n is rational or transcendental depending on whether p divides n (Theorem 1.2 and 1.3 and §3 of [16] and Theorem 1 of [17]). Similar phenomena in the more general case of endomorphisms of Abelian varieties in positive characteristic have been investigated in [19]. In the positive characteristic setting, where one is considering the characteristic p version of the Bost–Connes system of [27], one should then replace the dynamical zeta function by the tame zeta function considered in [19].

6 Spectra and zeta functions

We have already discussed in §5.4 and §5.6 zeta functions arising from certain counting functions that define ring homomorphisms from suitable Grothendieck rings to the Witt ring $W(\mathbb{Z})$. We consider here a more general setting of exponentiable motivic measures.

A motivic measure is a ring homomorphism $\mu : K_0(\mathcal{V}) \to R$, from the Grothendieck ring of varieties $K_0(\mathcal{V})$ to a commutative ring R. Examples include the counting measure, for varieties defined over finite fields, which counts the number of algebraic points over $\mathbb{F}_q$, the topological Euler characteristic or the Hodge–Deligne polynomials for complex algebraic varieties.

The Kapranov motivic zeta function [46] is defined as

$$\zeta(X,t) = \sum_{n=0}^{\infty} [S^n(X)] t^n,$$

where $S^n(X) = X^n/S_n$ are the symmetric products of X and $[S^n(X)]$ are the classes in $K_0(\mathcal{V})$. Similarly, the zeta function of a motivic measure is defined as

$$\zeta_\mu(X,t) = \sum_{n=0}^{\infty} \mu(S^n(X))\, t^n. \tag{45}$$

It is viewed as an element in the Witt ring $W(R)$. The addition in $K_0(\mathcal{V})$ is mapped by the zeta function to the addition in $W(R)$, which is the usual product of the power series,

$$\zeta_\mu(X \sqcup Y,t) = \zeta_\mu(X,t) \cdot \zeta_\mu(Y,t) = \zeta_\mu(X,t) +_{W(R)} \zeta_\mu(Y,t). \tag{46}$$

The motivic measure $\mu : K_0(\mathcal{V}) \to R$ is said to be exponentiable (see [62], [63]) if the zeta function (45) defines a ring homomorphism

$$\zeta_\mu : K_0(\mathcal{V}) \to W(R),$$

that is, if in addition to (46) one also has

$$\zeta_\mu(X \times Y,t) = \zeta_\mu(X,t) \star_{W(R)} \zeta_\mu(Y,t). \tag{47}$$

We investigate here how to lift the zeta functions of exponentiable motivic measures to the level of spectra. To this purpose, we first investigate how to construct a spectrum whose π_0 is a dense subring $W_0(R)$ of the Witt ring $W(R)$ and then we consider how to lift the ring homomorphisms given by zeta functions ζ_μ of exponentiable measures with a rationality and a factorization condition.

6.1 The Endomorphism Category

Let R be a commutative ring. We denote by $\mathcal{E}_R$ the endomorphism category of R, which is defined as follows (see [1], [2], [32]).

Definition 6.1 The category $\mathcal{E}_R$ has objects given by the pairs (E, f) of a finite projective module E over R and an endomorphism $f \in \mathrm{End}_R(E)$, and morphisms given by morphisms $\phi : E \to E'$ of finite projective modules that commute with the endomorphisms, $f' \circ \phi = \phi \circ f$. The endomorphism category has direct sum $(E, f) \oplus (E', f') = (E \oplus E', f \oplus f')$ and tensor product $(E, f) \otimes (E', f') = (E \otimes E', f \otimes f')$.

The category of finite projective modules over R is identified with the subcategory corresponding to the objects $(E, 0)$ with trivial endomorphism.

An exact sequence in $\mathcal{E}_R$ is a sequence of objects and morphisms in $\mathcal{E}_R$ which is exact as a sequence of finite projective modules over R (forgetting the endomorphisms). This determines a collection of admissible short exact sequence (and of admissible monomorphisms and epimorphisms). The endomorphism category $\mathcal{E}_R$ is then an exact category, hence it has an associated K-theory defined via the Quillen Q-construction, [61]. This assigns to the exact category $\mathcal{E}_R$ the category $\mathcal{Q}\mathcal{E}_R$ with the same objects and morphisms $\mathrm{Hom}_{\mathcal{Q}\mathcal{E}_R}((E, f), (E', f'))$ given by diagrams

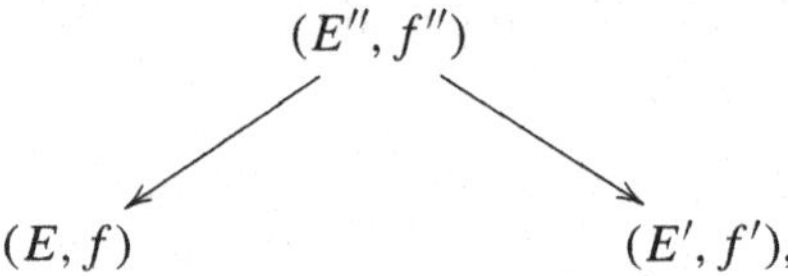

where the first arrow is an admissible epimorphism and the second an admissible monomorphism, with composition given by pullback. By the Quillen construction the K-theory of $\mathcal{E}_R$ is then $K_{n-1}(\mathcal{E}_R) = \pi_n(\mathcal{N}(\mathcal{Q}\mathcal{E}_R))$, with $\mathcal{N}(\mathcal{Q}\mathcal{E}_R)$ the nerve of $\mathcal{Q}\mathcal{E}_R$.

The forgetful functor $(E, f) \mapsto E$ induces a map on K-theory

$$K_n(\mathcal{E}_R) \to K_n(\mathcal{P}_R) = K_n(R),$$

which is a split surjection. Let

$$\mathcal{E}_n(R) := \mathrm{Ker}(K_n(\mathcal{E}_R) \to K_n(R)).$$

In the case of K_0 an explicit description is given by the following, [1], [2]. Let $K_0(\mathcal{E}_R)$ denote the K_0 of the endomorphism category $\mathcal{E}_R$. It is a ring with the product structure induced by the tensor product. It is proved in [1], [2] that the quotient

$$W_0(R) = K_0(\mathcal{E}_R)/K_0(R) \tag{48}$$

embeds as a dense subring of the big Witt ring $W(R)$ via the map

$$L : (E, f) \mapsto \det(1 - t\, M(f))^{-1}, \tag{49}$$

with $M(f)$ the matrix associated to $f \in \mathrm{End}_R(E)$, where $\det(1 - t\, M(f))^{-1}$ is viewed as an element in $\Lambda(R) = 1 + tR[[t]]$. As a subring $W_0(R) \hookrightarrow W(R)$ of the big Witt ring, $W_0(R)$ consists of the rational Witt vectors

$$W_0(R) = \left\{ \frac{1 + a_1 t + \cdots + a_n t^n}{1 + b_1 t + \cdots + b_m t^m} \mid a_i, b_i \in R,\ n, m \geq 0 \right\}.$$

Equivalently, one can consider the ring $\mathcal{R} = (1 + tR[t])^{-1}R[t]$ and identify the above with $1 + t\mathcal{R}$, where the multiplication in $1 + t\mathcal{R}$ corresponds to the addition in the Witt ring, and the Witt product is determined by the identity $(1 - at) \star (1 - bt) = (1 - abt)$.

This description of Witt rings in terms of endomorphism categories was applied to investigate the arithmetic structures of the Bost–Connes quantum statistical mechanical system, see [24], [56], [57].

This relation between the Grothendieck ring and Witt vectors was extended to the higher K-theory in [38], where an explicit description for the kernels $\mathcal{E}_n(R)$ is obtained, by showing that

$$\mathcal{E}_{n-1}(R) = \mathrm{Coker}(K_n(R) \to K_n(\mathcal{R})),$$

where $\mathcal{R} = (1 + tR[t])^{-1}R[t]$ and $K_n(R) \to K_n(\mathcal{R})$ is a split injection. The identification above is obtained in [38] by showing that there is an exact sequence

$$0 \to K_n(R) \to K_n(\mathcal{R}) \to K_{n-1}(\mathcal{E}_R) \to K_{n-1}(R) \to 0. \tag{50}$$

The identification (48) for K_0 is then recovered as the case with $n = 0$ that gives an identification $\mathcal{E}_0(R) \simeq 1 + t\mathcal{R}$.

6.2 Spectrum of the Endomorphism Category and Witt vectors

Let $\mathcal{P}_R$ denote the category of finite projective modules over a commutative ring R with unit. Also let $\mathcal{E}_R$ be the endomorphism category recalled above. By the Segal construction described in §3.2, we obtain associated Γ-spaces $F_{\mathcal{P}_R}$ and $F_{\mathcal{E}_R}$ and spectra $F_{\mathcal{P}_R}(\mathbb{S}) = K(R)$, the K-theory spectrum of R, and $F_{\mathcal{E}_R}(\mathbb{S})$, the spectrum of the endomorphism category.

We obtain in the following way a functorial "spectrification" of the Witt ring $W_0(R)$, namely a spectrum $\mathbb{W}(R)$ with $\pi_0\mathbb{W}(R) = W_0(R)$.

Definition 6.2 For a commutative ring R, with $\mathcal{P}_R$ the category of finite projective modules and $\mathcal{E}_R$ the category of endomorphisms, the spectrum $\mathbb{W}(R)$ is defined as the cofiber $\mathbb{W}(R) := F_{\mathcal{E}_R}(\mathbb{S})/F_{\mathcal{P}_R}(\mathbb{S})$ obtained from the Γ-spaces $F_{\mathcal{P}_R} : \Gamma^0 \to \Delta_*$ and $F_{\mathcal{E}_R} : \Gamma^0 \to \Delta_*$ associated to the categories $\mathcal{P}_R$ and $\mathcal{E}_R$.

Lemma 6.3 *For a commutative ring R, the inclusion of the category $\mathcal{P}_R$ of finite projective modules as the subcategory of the endomorphism category $\mathcal{E}_R$ determines a long exact sequence*

$$\cdots \to \pi_n(F_{\mathcal{P}_R}(\mathbb{S})) \to \pi_n(F_{\mathcal{E}_R}(\mathbb{S})) \to$$

$$\pi_n(F_{\mathcal{E}_R}(\mathbb{S})/F_{\mathcal{P}_R}(\mathbb{S})) \to \pi_{n-1}(F_{\mathcal{P}_R}(\mathbb{S})) \to \cdots$$

$$\cdots \to \pi_0(F_{\mathcal{P}_R}(\mathbb{S})) \to \pi_0(F_{\mathcal{E}_R}(\mathbb{S})) \to \pi_0(F_{\mathcal{E}_R}(\mathbb{S})/F_{\mathcal{P}_R}(\mathbb{S}))$$

of the homotopy groups of the spectra $F_{\mathcal{P}_R}(\mathbb{S})$, $F_{\mathcal{E}_R}(\mathbb{S})$ with cofiber $\mathbb{W}(R)$ as in Definition 6.2. The spectrum $\mathbb{W}(R)$ satisfies $\pi_0\mathbb{W}(R) = W_0(R)$.

Proof The functoriality of the Segal construction implies that the inclusion of $\mathcal{P}_R$ as the subcategory of $\mathcal{E}_R$ given by objects $(E,0)$ with trivial endomorphism determines a map of Γ-spaces $F_{\mathcal{P}_R} \to F_{\mathcal{E}_R}$, which is a natural transformation of the functors $F_{\mathcal{P}_R} : \Gamma^0 \to \Delta_*$ and $F_{\mathcal{E}_R} : \Gamma^0 \to \Delta_*$. After passing to endofunctors $F_{\mathcal{P}_R} : \Delta_* \to \Delta_*$ and $F_{\mathcal{E}_R} : \Delta_* \to \Delta_*$ we obtain a map of spectra $K(R) \to F_{\mathcal{E}_R}(\mathbb{S})$, induced by the inclusion of $\mathcal{P}_R$ as subcategory of $\mathcal{E}_R$. The category Δ_* of simplicial sets has products and equalizers, hence pullbacks. Thus, given two functors $F, F' : \Gamma^0 \to \Delta_*$, a natural transformation $\alpha : F \to F'$ is mono if and only if for all objects $X \in \Gamma^0$ the morphism $\alpha_X : F(X) \to F'(X)$ is a monomorphism in Δ_*. An embedding $\mathcal{C} \hookrightarrow \mathcal{C}'$ determines by composition an embedding $\Sigma_{\mathcal{C}}(X) \hookrightarrow \Sigma_{\mathcal{C}'}(X)$ of the categories of summing functors, for each object $X \in \Gamma^0$. This gives a monomorphism $F_{\mathcal{C}}(X) = \mathcal{N}\Sigma_{\mathcal{C}}(X) \to F_{\mathcal{C}'}(X) = \mathcal{N}\Sigma_{\mathcal{C}'}(X)$, hence a monomorphism $F_{\mathcal{C}} \to F_{\mathcal{C}'}$ of Γ-spaces. Arguing as in Lemma 1.3 of [64] we then obtain from such a map $F_{\mathcal{C}} \to F_{\mathcal{C}'}$ of Γ-spaces a long exact sequence of homotopy groups of the associated spectra

$$\cdots \to \pi_n(F_{\mathcal{C}}(\mathbb{S})) \to \pi_n(F_{\mathcal{C}'}(\mathbb{S})) \to$$

$$\pi_n(F_{\mathcal{C}'}(\mathbb{S})/F_{\mathcal{C}}(\mathbb{S})) \to \pi_{n-1}(F_{\mathcal{C}}(\mathbb{S})) \to \cdots$$

$$\cdots \to \pi_0(F_{\mathcal{C}}(\mathbb{S})) \to \pi_0(F_{\mathcal{C}'}(\mathbb{S})) \to \pi_0(F_{\mathcal{C}'}(\mathbb{S})/F_{\mathcal{C}}(\mathbb{S})),$$

where $F_{\mathcal{C}'}(\mathbb{S})/F_{\mathcal{C}}(\mathbb{S})$ is the cofiber. When applied to the subcategory $\mathcal{P}_R \hookrightarrow \mathcal{E}_R$ this gives the long exact sequence

$$\cdots \to \pi_n(F_{\mathcal{P}_R}(\mathbb{S})) \to \pi_n(F_{\mathcal{E}_R}(\mathbb{S})) \to$$

$$\pi_n(F_{\mathcal{E}_R}(\mathbb{S})/F_{\mathcal{P}_R}(\mathbb{S})) \to \pi_{n-1}(F_{\mathcal{P}_R}(\mathbb{S})) \to \cdots$$

$$\cdots \to \pi_0(F_{\mathcal{P}_R}(\mathbb{S})) \to \pi_0(F_{\mathcal{E}_R}(\mathbb{S})) \to \pi_0(F_{\mathcal{E}_R}(\mathbb{S})/F_{\mathcal{P}_R}(\mathbb{S})).$$

Here we have $\pi_n(F_{\mathcal{P}_R}(\mathbb{S})) = K_n(R)$. Moreover, by construction we have $\pi_0(F_{\mathcal{E}_R}(\mathbb{S})) = K_0(\mathcal{E}_R)$ so that we identify

$$\pi_0(F_{\mathcal{E}_R}(\mathbb{S})/F_{\mathcal{P}_R}(\mathbb{S})) = W_0(R) = K_0(\mathcal{E}_R)/K_0(R).$$

Thus, the spectrum $\mathbb{W}(R) := F_{\mathcal{E}_R}(\mathbb{S})/F_{\mathcal{P}_R}(\mathbb{S})$ given by the cofiber of $F_{\mathcal{P}_R}(\mathbb{S}) \to F_{\mathcal{E}_R}(\mathbb{S})$ provides a spectrum whose zeroth homotopy group is the Witt ring $W_0(R)$. □

The forgetful functor $\mathcal{E}_R \to \mathcal{P}_R$ also induces a corresponding map of Γ-spaces $F_{\mathcal{E}_R} \to F_{\mathcal{P}_R}$. Moreover, one can also construct a spectrum with π_0 equal to $W_0(R)$ using the characterization given in [38], that we recalled above, in terms of the map on K-theory (and on K-theory spectra) $K(R) \to K(\mathcal{R})$ with $\mathcal{R} = (1 + rR[t])^{-1}R[t]$. One can obtain in this way a reformulation in terms of spectra of the result of [38]. However, for our purposes here, it is preferable to work with the spectrum constructed in Lemma 6.3.

We give a variant of Lemma 6.3 that will be useful in the following. We denote by $\mathcal{P}_R^{\pm}$ and $\mathcal{E}_R^{\pm}$, respectively, the categories of $\mathbb{Z}/2\mathbb{Z}$-graded finite projective R-modules and the $\mathbb{Z}/2\mathbb{Z}$-graded endomorphism category with objects given by pairs $\{((E_+, f_+), (E_-, f_-))\}$, which we write simply as $(E_\pm, f_\pm)$ and with morphisms $\phi : E_\pm \to E'_\pm$ of $\mathbb{Z}/2\mathbb{Z}$-graded finite projective modules that commute with $f_\pm$. The sum in $\mathcal{E}_R^{\pm}$ is given by

$$(E_\pm, f_\pm) \oplus (E'_\pm, f'_\pm) = ((E_+ \oplus E'_+, E_- \oplus E'_-), (f_+ \oplus f'_+, f_- \oplus f'_-))$$

while the tensor product $(E_\pm, f_\pm) \otimes (E'_\pm, f'_\pm)$ is given by

$$(E_\pm, f_\pm) \otimes (E'_\pm, f'_\pm) = ((F_+, g_+), (F_-, g_-))$$

where

$$F_+ = E_+ \otimes E'_+ \oplus E_- \otimes E'_-$$

$$g_+ = f_+ \otimes f'_+ \oplus f_- \otimes f'_-$$

$$F_- = E_+ \otimes E'_- \oplus E_- \otimes E'_+$$

$$g_- = f_+ \otimes f'_- \oplus f_- \otimes f'_+ .$$

Again we consider $\mathcal{P}_R^\pm$ as a subcategory of $\mathcal{E}_R^\pm$ with trivial endomorphisms.

Lemma 6.4 *The map $\delta : K_0(\mathcal{E}_R^\pm) \to K_0(\mathcal{E}_R)$ given by $[E_\pm, f_\pm] \mapsto [E_+, f_+] - [E_-, f_-]$ is a ring homomorphism and it descends to a ring homomorphism*

$$K_0(\mathcal{E}_R^\pm)/K_0(\mathcal{P}_R^\pm) \to K_0(\mathcal{E}_R)/K_0(R) \simeq W_0(R).$$

Proof The map is clearly compatible with sums. Compatibility with product also holds since $[E_\pm, f_\pm] \cdot [E'_\pm, f'_\pm] \mapsto ([E_+, f_+] - [E_-, f_-]) \cdot ([E'_+, f'_+] - [E'_-, f'_-])$. Moreover, it maps $K_0(\mathcal{P}_R^\pm)$ to $K_0(\mathcal{P}_R)$. □

As before, the categories $\mathcal{P}_R^\pm$ and $\mathcal{E}_R^\pm$ have associated Γ-spaces $F_{\mathcal{P}_R^\pm} : \Gamma^0 \to \Delta_*$ and $F_{\mathcal{E}_R^\pm} : \Gamma^0 \to \Delta_*$ and spectra $F_{\mathcal{P}_R^\pm}(\mathbb{S})$ and $F_{\mathcal{E}_R^\pm}(\mathbb{S})$. The following result follows as in Lemma 6.3.

Lemma 6.5 *The inclusion of $\mathcal{P}_R^\pm$ as a subcategory of $\mathcal{E}_R^\pm$ induces a long exact sequence*

$$\cdots \to \pi_n(F_{\mathcal{P}_R^\pm}(\mathbb{S})) \to \pi_n(F_{\mathcal{E}_R^\pm}(\mathbb{S})) \to$$

$$\pi_n(F_{\mathcal{E}_R^\pm}(\mathbb{S})/F_{\mathcal{P}_R^\pm}(\mathbb{S})) \to \pi_{n-1}(F_{\mathcal{P}_R^\pm}(\mathbb{S})) \to \cdots$$

$$\cdots \to \pi_0(F_{\mathcal{P}_R^\pm}(\mathbb{S})) \to \pi_0(F_{\mathcal{E}_R^\pm}(\mathbb{S})) \to \pi_0(F_{\mathcal{E}_R^\pm}(\mathbb{S})/F_{\mathcal{P}_R^\pm}(\mathbb{S}))$$

of the homotopy groups of the spectra $F_{\mathcal{P}_R^\pm}(\mathbb{S})$ and $F_{\mathcal{E}_R^\pm}(\mathbb{S})$, which at the level of π_0 gives $K_0(\mathcal{P}_R^\pm) \to K_0(\mathcal{E}_R^\pm) \to K_0(\mathcal{E}_R^\pm)/K_0(\mathcal{P}_R^\pm)$.

We denote by $\mathbb{W}^\pm(R) = F_{\mathcal{E}_R^\pm}(\mathbb{S})/F_{\mathcal{P}_R^\pm}(\mathbb{S})$ the cofiber of

$$F_{\mathcal{P}_R^\pm}(\mathbb{S}) \to F_{\mathcal{E}_R^\pm}(\mathbb{S}).$$

Remark 6.6 It is important to point out that our treatment of Witt vectors and their spectrification, as presented in this section, differs from the one in [40] (see especially Theorem 2.2.9 and equation (2.2.11) in that paper), and in [20]. Nonetheless, the circle action on THH that is used to obtain the spectrum TR is closely related to the Bost–Connes structure investigated in the present paper. A more direct relation between Bost–Connes structures and topological Hochschild and cyclic homology will also relate naturally to the point of view on $\mathbb{F}_1$-geometry developed in [26]. We will leave this topic for future work.

6.3 Exponentiable measures and maps of Γ-spaces

The problem of lifting to the level of spectra the Hasse–Weil zeta function associated to the counting motivic measure for varieties over finite fields was discussed in [21]. We consider here a very similar setting and procedure, where we want to lift a zeta function $\zeta_\mu : K_0(\mathcal{V}) \to W(R)$ associated to an exponentiable motivic measure to the level of spectra. To this purpose, we make some assumptions of rationality and the existence of a factorization for our zeta functions of exponentiable motivic measures. We then consider the spectrum $K(\mathcal{V})$ of [70], [72] with $\pi_0 K(\mathcal{V}) = K_0(\mathcal{V})$ and a spectrum, obtained from a Γ-space, associated to the subring $W_0(R)$ of the big Witt ring $W(R)$.

Definition 6.7 A motivic measure, that is, a ring homomorphism μ : $K_0(\mathcal{V}) \to R$ of the Grothendieck ring of varieties to a commutative ring R, is called factorizable is it satisfies the following three properties:

1. **exponentiability**: the associated zeta function $\zeta_\mu(X,t)$ is a ring homomorphism $\zeta_\mu : K_0(\mathcal{V}) \to W(R)$ to the Witt ring of R;
2. **rationality**: the homomorphism ζ_μ factors through the inclusion of the subring $W_0(R)$ of the Witt ring, $\zeta_\mu : K_0(\mathcal{V}) \to W_0(R) \hookrightarrow W(R)$, and
3. **factorization**: the rational functions $\zeta_\mu(X,t)$ admit a factorization into linear factors
$$\zeta_\mu(X,t) = \frac{\prod_i (1-\alpha_i t)}{\prod_j (1-\beta_j t)} = \zeta_{\mu,+}(X,t) \ -_W \ \zeta_{\mu,-}(X,t)$$
where $\zeta_{\mu,+}(X,t) = \prod_j (1-\beta_j t)^{-1}$ and $\zeta_{\mu,-}(X,t) = \prod_i (1-\alpha_i t)^{-1}$ and $-_W$ is the difference in the Witt ring, that is the ratio of the two polynomials.

Lemma 6.8 *A factorizable motivic measure $\mu : K_0(\mathcal{V}) \to R$, as in Definition 6.7, determines a functor $\Phi_\mu : \mathcal{C}_\mathcal{V} \to \mathcal{E}_R^\pm$ where $\mathcal{C}_\mathcal{V}$ is the assembler category encoding the scissor-congruence relations of the Grothendieck ring $K_0(\mathcal{V})$ and $\mathcal{E}_R^\pm$ is the $\mathbb{Z}/2\mathbb{Z}$-graded endomorphism category.*

Proof The objects of $\mathcal{C}_{\mathcal{V}}$ are varieties X and the morphisms are locally closed embeddings, [70], [72]. To an object X we assign an object of $\mathcal{E}_R$ obtained in the following way. Consider a factorization

$$\zeta_\mu(X,t) = \frac{\prod_{i=1}^n (1-\alpha_i t)}{\prod_{j=1}^m (1-\beta_j t)} = \zeta_{\mu,+}(X,t) \; -_W \; \zeta_{\mu,-}(X,t)$$

as above of the zeta function of X. Let $E_+^{X,\mu} = R^{\oplus m}$ and $E_-^{X,\mu} = R^{\oplus n}$ with endomorphisms $f_\pm^{X,\mu}$ respectively given in matrix form by $M(f_+^{X,\mu}) = \mathrm{diag}(\beta_j)_{j=1}^m$ and $M(f_-^{X,\mu}) = \mathrm{diag}(\alpha_i)_{i=1}^n$.

The pair $(E_\pm^{X,\mu}, f_\pm^{X,\mu})$ is an object of the endomorphism category $\mathcal{E}_R^\pm$. Given an embedding $Y \hookrightarrow X$, the zeta function satisfies

$$\zeta_\mu(X,t) = \zeta_\mu(Y,t)\cdot\zeta_\mu(X \smallsetminus Y,t) = \zeta_\mu(Y,t) \; +_W \; \zeta_\mu(X \smallsetminus Y,t).$$

Using the factorizations of each term, this gives

$$(E_\pm^{X,\mu}, f_\pm^{X,\mu}) = (E_\pm^{Y,\mu}, f_\pm^{Y,\mu}) \oplus (E_\pm^{X\smallsetminus Y,\mu}, f_\pm^{X\smallsetminus Y,\mu}),$$

hence a morphism in $\mathcal{E}_R^\pm$ given by the canonical morphism to the direct sum

$$(E_\pm^{Y,\mu}, f_\pm^{Y,\mu}) \to (E_\pm^{X,\mu}, f_\pm^{X,\mu}). \qquad \square$$

Proposition 6.9 *The functor $\Phi_\mu : \mathcal{C}_{\mathcal{V}} \to \mathcal{E}_R^\pm$ of Lemma 6.8 induces a map of Γ-spaces and of the associated spectra $\Phi_\mu : K(\mathcal{V}) \to F_{\mathcal{E}_R^\pm}(\mathbb{S})$. The induced maps on the homotopy groups has the property that the composition*

$$K_0(\mathcal{V}) \xrightarrow{\Phi_\mu} K_0(\mathcal{E}_R^\pm) \xrightarrow{\delta} K_0(\mathcal{E}_R) \to K_0(\mathcal{E}_R)/K_0(R) = W_0(R) \tag{51}$$

with δ as in Lemma 6.4, is given by the zeta function $\zeta_\mu : K_0(\mathcal{V}) \to W_0(R)$.

Proof The Γ-space associated to the assembler category $\mathcal{C}_{\mathcal{V}}$ is obtained in the following way, [70], [72]. One first associates to the assembler category $\mathcal{C}_{\mathcal{V}}$ another category $\mathcal{W}(\mathcal{C}_{\mathcal{V}})$ whose objects are finite collections $\{X_i\}_{i\in I}$ of non-initial objects of $\mathcal{C}_{\mathcal{V}}$ with morphisms $\varphi = (f, f_i) : \{X_i\}_{i\in I} \to \{X'_j\}_{j\in J}$ given by a map of the indexing sets $f : I \to J$ and morphisms $f_i : X_i \to X'_{f(i)}$ in $\mathcal{C}_{\mathcal{V}}$, such that, for every fixed $j \in J$ the collection $\{f_i : X_i \to X'_j : i \in f^{-1}(j)\}$ is a disjoint covering family of the assembler $\mathcal{C}_{\mathcal{V}}$. This means, in the case of the assembler $\mathcal{C}_{\mathcal{V}}$ underlying the Grothendieck ring of varieties, that the f_i are closed embeddings of the varieties X_i in the given X'_j with disjoint images. We first show that the functor $\Phi_\mu : \mathcal{C}_{\mathcal{V}} \to \mathcal{E}_R^\pm$ of Lemma 6.8 extends to a functor (for which we still use the same notation) $\Phi_\mu : \mathcal{W}(\mathcal{C}_{\mathcal{V}}) \to \mathcal{E}_R^\pm$. We define $\Phi_\mu(\{X_i\}_{i\in I}) = \oplus_{i\in I}\Phi_\mu(X_i) = \oplus_{i\in I}(E_\pm^{X_i,\mu}, f_\pm^{X_i,\mu})$. Given a covering family $\{f_i : X_i \to X'_j : i \in f^{-1}(j)\}$ as above, each morphism $f_i : X_i \to X'_j$ determines a morphism $\Phi_\mu(f_i) : (E_\pm^{X_i,\mu}, f_\pm^{X_i,\mu}) \to (E_\pm^{X'_j,\mu}, f_\pm^{X'_j,\mu})$ given by the canonical morphism to the direct sum $(E_\pm^{X_i,\mu}, f_\pm^{X_i,\mu}) \to (E_\pm^{X_i,\mu}, f_\pm^{X_i,\mu}) \oplus (E_\pm^{X'_j \smallsetminus X_i,\mu}, f_\pm^{X'_j \smallsetminus X_i,\mu})$. This determines a morphism

$$\Phi_\mu(\varphi) : \oplus_{i\in I}(E_\pm^{X_i,\mu}, f_\pm^{X_i,\mu}) \to \oplus_{j\in J}(E_\pm^{X'_j,\mu}, f_\pm^{X'_j,\mu}).$$

We then show that the functor $\Phi_\mu : \mathcal{W}(\mathcal{C}_\mathcal{V}) \to \mathcal{E}_R^\pm$ constructed in this way determines a map of the associated Γ-spaces. The Γ-space associated to $\mathcal{W}(\mathcal{C}_\mathcal{V})$ is constructed in [70], [72] as the functor that assigns to a finite pointed set $S \in \Gamma^0$ the simplicial set given by the nerve $\mathcal{N}\mathcal{W}(S \wedge \mathcal{C}_\mathcal{V})$, where the coproduct of assemblers $S \wedge \mathcal{C}_\mathcal{V} = \bigvee_{s\in S\smallsetminus\{s_0\}} \mathcal{C}_\mathcal{V}$ has an initial object and a copy of the non-initial objects of $\mathcal{C}_\mathcal{V}$ for each point $s \in S \smallsetminus \{s_0\}$ and morphisms induced by those of $\mathcal{C}_\mathcal{V}$. This means that we can regard objects of $\mathcal{W}(S \wedge \mathcal{C}_\mathcal{V})$ as collections $\{X_{s,i}\}_{i\in I}$, for some $s \in S \smallsetminus \{s_0\}$ and morphisms $\varphi_s = (f_s, f_{s,i}) : \{X_{s,i}\}_{i\in I} \to \{X'_{s,j}\}_{j\in J}$ as above. In order to obtain a map of Γ-spaces between $F_\mathcal{V} : S \mapsto \mathcal{N}\mathcal{W}(S \wedge \mathcal{C}_\mathcal{V})$ and $F_{\mathcal{E}_R^\pm} : S \mapsto \mathcal{N}\Sigma_{\mathcal{E}_R^\pm}(S)$, we construct a functor $\mathcal{W}(S \wedge \mathcal{C}_\mathcal{V}) \to \Sigma_{\mathcal{E}_R^\pm}(S)$ from the category $\mathcal{W}(S \wedge \mathcal{C}_\mathcal{V})$ described above to the category of summing functors $\Sigma_{\mathcal{E}_R^\pm}(S)$. To an object $X_{S,I} := \{X_{s,i}\}_{i\in I}$ in $\mathcal{W}(S \wedge \mathcal{C}_\mathcal{V})$ we associate a functor $\Phi_{X_{S,I}} : \mathcal{P}(S) \to \mathcal{E}_R^\pm$ that maps a subset $A_+ = \{s_0\} \sqcup A \in \mathcal{P}(X)$ to $\Phi_{X_{S,I}}(A_+) = \oplus_{a\in A}\Phi_\mu(\{X_{a,i}\}_{i\in I})$ where $\Phi_\mu : \mathcal{W}(\mathcal{C}_\mathcal{V}) \to \mathcal{E}_R^\pm$ is the functor constructed above. It is a summing functor since $\Phi_{X_{S,I}}(A_+ \cup B_+) = \Phi_{X_{S,I}}(A_+) \oplus \Phi_{X_{S,I}}(B_+)$ for $A_+ \cap B_+ = \{s_0\}$. This induces a map of simplicial sets $\mathcal{N}\mathcal{W}(S \wedge \mathcal{C}_\mathcal{V}) \to \mathcal{N}\Sigma_{\mathcal{E}_R^\pm}(S)$ which determines a natural transformation of the functors $F_\mathcal{V} : S \mapsto \mathcal{N}\mathcal{W}(S \wedge \mathcal{C}_\mathcal{V})$ and $F_{\mathcal{E}_R^\pm} : S \mapsto \mathcal{N}\Sigma_{\mathcal{E}_R^\pm}(S)$. This map of Γ-spaces in turn determines a map of the associated spectra and an induced map of their homotopy groups. It remains to check that the induced map at the level of π_0 agrees with the expected map of Grothendieck rings $K_0(\mathcal{V}) \to K_0(\mathcal{E}_R^\pm)$, hence with the zeta function when further mapped to $K_0(\mathcal{E}_R)$ and to the quotient $K_0(\mathcal{E}_R)/K_0(R)$. This is the case since by construction the induced map $\pi_0 K(\mathcal{V}) = K_0(\mathcal{V}) \to K_0(\mathcal{E}_R^\pm) = \pi_0 F_{\mathcal{E}_R^\pm}(\mathbb{S})$ is given by the assignment $[X] \mapsto [E_\pm^{X,\mu}, f_\pm^{X,\mu}]$. □

Corollary 6.10 *The map of Grothendieck rings given by the composition* (51) *also lifts to a map of spectra.*

Proof It is possible to realize the map $\delta : K_0(\mathcal{E}_R^\pm) \to K_0(\mathcal{E}_R)$ described in Lemma 6.4 at the level of spectra. The K-theory spectrum of an abelian category $\mathcal{A}$ is weakly equivalent to the K-theory spectrum of the category of bounded chain complexes over $\mathcal{A}$. In fact, this holds more generally for $\mathcal{A}$ an exact category closed under kernels. Thus, in the case of the category $\mathcal{E}_R$, there is a weak equivalence $K(\mathrm{Ch}^\flat(\mathcal{E}_R)) \tilde{\to} K(\mathcal{E}_R)$ which descends on the level π_0 to the map $K_0(\mathrm{Ch}^\flat(\mathcal{E}_R)) \tilde{\to} K_0(\mathcal{E}_R)$ given by $[E^\cdot, f^\cdot] \mapsto \sum_k (-1)^k [E^k, f^k]$. To an object $(E^\pm, f^\pm)$ of $\mathcal{E}_R^\pm$ we can assign a chain complex in $\mathrm{Ch}^\flat(\mathcal{E}_R)$ of the form $0 \to (E^-, f^-) \overset{0}{\to} (E^+, f^+) \to 0$, where (E^+, f^+) sits in degree 0. This descends on the level of K-theory to a map $K(\mathcal{E}_R^\pm) \to K(\mathrm{Ch}^\flat(\mathcal{E}_R))$, which at the level of π_0 gives the map $[E^\pm, f^\pm] \mapsto [E^+, f^+] - [E^-, f^-]$. The functor $\mathcal{E}_R^\pm \to \mathrm{Ch}^\flat(\mathcal{E}_R)$ used here does not respect tensor products, although the induced map $\delta : K_0(\mathcal{E}_R^\pm) \to K_0(\mathcal{E}_R)$ at the level of K_0 is compatible with products. Thus, the composition (51) can also be lifted at the level of spectra. □

It should be noted that the construction of a derived motivic zeta function outlined above is not the first to appear in the literature. In [21], the authors describe a derived motivic measure $\zeta : K(\mathcal{V}_k) \to K(\mathrm{Rep}_{cts}(\mathrm{Gal}(k^s/k); \mathbb{Z}_\ell))$ from the Grothendieck spectrum of varieties to the K-theory spectrum of the category of continuous ℓ-adic Galois representations. This map corresponds to the assignment $X \mapsto H^*_{\mathrm{et},c}(X \times_k k^s, \mathbb{Z}_\ell)$. In particular, they show that when $k = \mathbb{F}_q$ for ℓ coprime to q, on the level of π_0, ζ corresponds to the Hasse-Weil zeta function. They then use ζ to prove that $K_1(\mathcal{V}_{\mathbb{F}_q})$ is not only nontrivial, but contains interesting algebro-geometric data.

Essentially, the approach in [21] was to start with a Weil Cohomology theory (in this case, ℓ-adic cohomology) and then to construct a derived motivic measure realizing on the level of K-theory the assignment to a variety X of its corresponding cohomology groups. The methods used in the case of ℓ-adic cohomology may not immediately generalize to other Weil cohomology theories. This method has yielded deep insight into the world of algebraic geometry. Our approach here, in contrast, is to take an interesting class of motivic measures, namely Kapranov motivic zeta functions (exponentiable motivic measures, [46], [62], [63]), and to determine reasonable conditions under which such a motivic measure can be derived directly. This method still needs to be studied further to yield additional insights into what it captures about the geometry of varieties.

6.4 Bost–Connes type systems via motivic measures

The lifting of the integral Bost–Connes algebra to various Grothendieck rings, their assembler categories, and the associated spectra, that we discussed in [54] and in the earlier sections of this paper, can be viewed as an instance of a more general kind of operation. As discussed in [25], there is a close relation between the endomorphisms σ_n and the maps $\tilde{\rho}_n$ of the integral Bost–Connes algebra and the operation of Frobenius and Verschiebung in the Witt ring. Thus, we can formulate a more general form of the question investigated above, of lifting of the integral Bost–Connes algebra to a Grothendieck ring through an Euler characteristic map, in terms of lifting the Frobenius and Verschiebung operations of a Witt ring to a Grothendieck ring through the zeta function ζ_μ of an exponentiable motivic measure. A prototype example of this more general setting is provided by the Hasse–Weil zeta function $Z : K_0(\mathcal{V}_{\mathbb{F}_q}) \to W(\mathbb{Z})$, which has the properties that the action of the Frobenius F_n on the Witt ring $W(\mathbb{Z})$ corresponds to passing to a field extension, $F_n Z(X_{\mathbb{F}_q}, t) = Z(X_{\mathbb{F}_{q^n}}, t)$ and the action of the Verschiebung V_n on the Witt ring $W(\mathbb{Z})$ is related to the Weil restriction of scalars from $\mathbb{F}_{q^n}$ to $\mathbb{F}_q$ (see [62] for a precise statement).

Recall that, if one denotes by $[a]$ the elements $[a] = (1 - at)^{-1}$ in the Witt ring $W(R)$, for $a \in R$, then the Frobenius ring homomorphisms $F_n : W(R) \to W(R)$ of the Witt ring are determined by $F_n([a]) = [a^n]$ and the Verschiebung group homomorphisms $V_n : W(R) \to W(R)$ are defined on an arbitrary $P(t) \in W(R)$ as $V_n : P(t) \mapsto P(t^n)$. These operations satisfy an analog of the Bost–Connes relations

$$\begin{aligned} &F_n \circ F_m = F_{nm}, \qquad V_n \circ V_m = V_{nm}, \\ &F_n \circ V_n = n \cdot \mathrm{id}, \\ &F_n \circ V_m = V_m F_n \quad \text{if } (n,m) = 1. \end{aligned} \tag{52}$$

These correspond, respectively, to the semigroup structure of the σ_n and $\tilde{\rho}_n$ of the integral Bost–Connes algebra and the relations $\sigma_n \circ \tilde{\rho}_n = n \cdot \mathrm{id}$, while the last relation is determined in the Bost–Connes case by the commutation of the generators $\tilde{\mu}_n$ and μ_m^* for $(n,m) = 1$.

Definition 6.11 A factorizable motivic measure $\mu : K_0(\mathcal{V}) \to R$, in the sense of Definition 6.7, is of *Bost–Connes type* if there is a lift to $K_0(\mathcal{V})$ of the Frobenius F_n and Verschiebung V_n of the Witt ring $W(R)$ to $K_0(\mathcal{V})$ such that the diagrams commute:

$$\begin{array}{ccc} K_0(\mathcal{V}) & \xrightarrow{\zeta_\mu} & W(R) \\ \downarrow{\sigma_n} & & \downarrow{F_n} \\ K_0(\mathcal{V}) & \xrightarrow{\zeta_\mu} & W(R) \end{array} \qquad \begin{array}{ccc} K_0(\mathcal{V}) & \xrightarrow{\zeta_\mu} & W(R) \\ \downarrow{\tilde{\rho}_n} & & \downarrow{V_n} \\ K_0(\mathcal{V}) & \xrightarrow{\zeta_\mu} & W(R) \end{array}$$

Such a motivic measure $\mu : K_0(\mathcal{V}) \to R$ is of *homotopic Bost–Connes type* if the maps σ_n and $\tilde{\rho}_n$ in the diagrams above also lift to endofunctors of the assembler category $\mathcal{C}_{\mathcal{V}}$ of the Grothendieck ring $K_0(\mathcal{V})$ with the endofunctors σ_n compatible with the monoidal structure.

Definition 6.12 The Frobenius and Verschiebung on the category $\mathcal{E}_R^{\pm}$ are defined as the endofunctors $F_n(E, f) = (E, f^n)$ and $V_n(E_{\pm}, f_{\pm}) = (E_{\pm}^{\oplus n}, V_n(f_{\pm}))$ with $V_n(f)$ defined by

$$V_n : (E, f) \mapsto (E^{\oplus n}, V_n(f)), \quad V_n(f) = \begin{pmatrix} 0 & 0 & \cdots & 0 & f \\ 1 & 0 & \cdots & 0 & 0 \\ 0 & 1 & \cdots & 0 & 0 \\ \vdots & \vdots & \cdots & \vdots & \vdots \\ 0 & 0 & \cdots & 1 & 0 \end{pmatrix}. \tag{53}$$

It is worth noting that the endofunctors of Definition 6.12 are akin to those used in the definitions of topological cyclic and topological restriction homology, [41].

Lemma 6.13 *The Frobenius and Verschiebung F_n and V_n of Definition 6.12 are endofunctors of the category $\mathcal{E}_R^{\pm}$ with the property that the maps they induce on $W_0(R) = K_0(\mathcal{E}_R)/K_0(R)$ agree with the restrictions to $W_0(R) \subset W(R)$ of the Frobenius and Verschiebung maps. These endofunctors determine natural transformations (still denoted F_n and V_n) of the Γ-space $F_{\mathcal{E}_R^{\pm}} : \Gamma^0 \to \Delta_*$.*

Proof The homomorphism $K_0(\mathcal{E}_R) \to W_0(R)$ given by

$$(E, f) \mapsto L(E, f) = \det(1 - tM(f))^{-1}$$

sends the pair (R, f_a) with f_a acting on R as multiplication by $a \in R$ to the element $[a] = (1 - at)^{-1}$ in the Witt ring. The action of the Frobenius $F_n([a]) = [a^n]$ is induced from the Frobenius $F_n(E, f) = (E, f^n)$ which is an endofunctor of $\mathcal{E}_R$. This extends to a compatible endofunctor of $\mathcal{E}_R^{\pm}$ by $F_n(E_{\pm}, f_{\pm}) = (E_{\pm}, f_{\pm}^n)$. Similarly, the Verschiebung map that sends $\det(1 - tM(f))^{-1} \mapsto \det(1 - t^n M(f))^{-1}$ is induced from the Verschiebung on $\mathcal{E}_R$ given by (53), since we have $L(E^{\oplus n}, V_n(f)) = \det(1 - t^n M(f))^{-1}$, with compatible endofunctors $V_n(E_{\pm}, f_{\pm}) = (E_{\pm}^{\oplus n}, V_n(f_{\pm}))$ on $\mathcal{E}_R^{\pm}$. The Frobeniius and Verschiebung on $\mathcal{E}_R^{\pm}$ induce natural transformations of the Γ-space $F_{\mathcal{E}_R^{\pm}} : \Gamma^0 \to \Delta_*$ by composition of the summing functors $\Phi : \mathcal{P}(X) \to \mathcal{E}_R^{\pm}$ in $\Sigma_{\mathcal{E}_R^{\pm}}(X)$ with the endofunctors F_n and V_n of $\mathcal{E}_R^{\pm}$. □

Proposition 6.14 *Let $\mu : K_0(\mathcal{V}) \to R$ be a factorizable motivic measure, as in Definition 6.7, that is of homotopical Bost–Connes type. Then the endofunctors σ_n and $\tilde{\rho}_n$ of the assembler category $\mathcal{C}_{\mathcal{V}}$ determine natural* *transformations (still denoted by σ_n and $\tilde{\rho}_n$) of the associated Γ-space $F_{\mathcal{V}} : \Gamma^0 \to \Delta_*$ that fit in the commutative diagrams*

$$\begin{array}{ccc} F_{\mathcal{V}} & \xrightarrow{\Phi_\mu} & F_{\mathcal{E}_R^\pm} \\ \downarrow{\scriptstyle \sigma_n} & & \downarrow{\scriptstyle F_n} \\ F_{\mathcal{V}} & \xrightarrow{\Phi_\mu} & F_{\mathcal{E}_R^\pm} \end{array} \qquad \begin{array}{ccc} F_{\mathcal{V}} & \xrightarrow{\Phi_\mu} & F_{\mathcal{E}_R^\pm} \\ \downarrow{\scriptstyle \tilde{\rho}_n} & & \downarrow{\scriptstyle V_n} \\ F_{\mathcal{V}} & \xrightarrow{\Phi_\mu} & F_{\mathcal{E}_R^\pm} \end{array}$$

where $\Phi_\mu : F_{\mathcal{V}} \to F_{\mathcal{E}_R^\pm}$ is the natural transformation of Γ-spaces of (6.9) *and F_n and V_n are the natural transformations of Lemma 6.13.*

Proof The natural transformation Φ_ν is determined as in Proposition 6.9 by the functor $\Phi_\mu : \mathcal{C}_{\mathcal{V}} \to \mathcal{E}_R^\pm$ that assigns $\Phi_\mu : X \mapsto (E_\pm^X, f_\pm^X)$ constructed as in Lemma 6.8. Suppose we have endofunctors σ_n and $\tilde{\rho}_n$ of the assembler category $\mathcal{C}_{\mathcal{V}}$ that induce maps σ_n and $\tilde{\rho}_n$ on $K_0(\mathcal{V})$ that lift the Frobenius and Verschiebung maps of $W(R)$ through the zeta function $\zeta_\mu : K_0(\mathcal{V}) \to W(R)$. This means that $\zeta_\mu(\sigma_n(X),t) = F_n\zeta_\mu(X,t)$ and $\zeta_\mu(\tilde{\rho}_n(X),t) = V_n\zeta_\mu(X,t) = \zeta_\mu(X,t^n)$. By Lemma 6.13, we have $F_n\zeta_\mu(X,t) = L(F_n(E_\pm^X, f_\pm^X)) = L(E_\pm^X, (f_\pm^X)^n)$ and $V_n\zeta_\mu(X,t) = L(V_n(E_\pm^X, f_\pm^X)) = L((E_\pm^X)^{\oplus n}, V_n(f_\pm^X))$. This shows the compatibilities of the natural transformations in the diagrams above. □

6.5 Spectra and spectra

We apply a construction similar to the one discussed in the previous subsections to the case of the map $(X, f) \mapsto \sum_{\lambda \in \mathrm{Spec}(f_*)} m_\lambda \lambda$ that assigns to a variety over $\mathbb{C}$ with a quasi-unipotent map the spectrum of the induced map f_* in homology, seen as an element in $\mathbb{Z}[\mathbb{Q}/\mathbb{Z}]$, as in §6 of [54].

In this section the term spectrum will appear both in its homotopy theoretic sense and in its operator sense. Indeed, we consider here a lift to the level of spectra (in the homotopy theoretic sense) of the construction described in §6 of [54], based on the spectrum (in the operator sense) Euler characteristic.

We consider here a setting as in [34], [39], where (X, f) is a pair of a variety over $\mathbb{C}$ and an endomorphism $f : X \to X$ such that the induced map f_* on $H_*(X, \mathbb{Z})$ has spectrum consisting of roots of unity. As discussed in [54] and in a related form in [34] the spectrum determines a ring homomorphism (an Euler characteristic)

$$\sigma : K_0^{\mathbb{Z}}(\mathcal{V}_{\mathbb{C}}) \to \mathbb{Z}[\mathbb{Q}/\mathbb{Z}] \tag{54}$$

where $K_0^{\mathbb{Z}}(\mathcal{V}_{\mathbb{C}})$ denotes the Grothendieck ring of pairs (X, f) with the operations defined by the disjoint union and the Cartesian product. It is shown in [54] that one can lift the operations σ_n and $\tilde{\rho}_n$ of the integral Bost–Connes algebra from $\mathbb{Z}[\mathbb{Q}/\mathbb{Z}]$ to $K_0^{\mathbb{Z}}(\mathcal{V}_{\mathbb{C}})$ via the "spectral Euler characteristic" (54), and that the operations can further be lifted from $K_0^{\mathbb{Z}}(\mathcal{V}_{\mathbb{C}})$ to a (homotopy theoretic) spectrum with π_0 equal to $K_0^{\mathbb{Z}}(\mathcal{V}_{\mathbb{C}})$ via the assembler category construction of [70].

In the next sub section we discuss how to lift the right hand side of (54), namely the original Bost–Connes algebra $\mathbb{Z}[\mathbb{Q}/\mathbb{Z}]$ with the operations σ_n and $\tilde{\rho}_n$ to the level of a homotopy theoretic spectrum, so that the spectral Euler characteristic (54) becomes induced by a map of spectra.

6.5.1 Bost–Connes Tannakian categorification and lifting of the spectral Euler characteristic

To construct a categorification of the map (54) compatible with the Bost–Connes structure, we use the lift of the left-hand-side of (54) to an assembler category, as in Proposition 6.6 of [54], while for the right-hand-side of (54) we use the categorification of Bost–Connes system constructed in [57].

We begin by recalling the categorification of the Bost–Connes algebra of [57]. Let $\mathrm{Vect}^{\bar{\mathbb{Q}}}_{\mathbb{Q}/\mathbb{Z}}(\mathbb{Q})$ be the category of pairs $(W, \oplus_{r\in\mathbb{Q}/\mathbb{Z}} \bar{W}_r)$ with W a finite dimensional $\mathbb{Q}$-vector space and $\oplus_r \bar{W}_r$ a $\mathbb{Q}/\mathbb{Z}$-graded vector space with $\bar{W} = W \otimes \bar{\mathbb{Q}}$. This is a neutral Tannakian category with fiber functor the forgetful functor $\omega : \mathrm{Vect}^{\bar{\mathbb{Q}}}_{\mathbb{Q}/\mathbb{Z}}(\mathbb{Q}) \to \mathrm{Vect}(\mathbb{Q})$ and with $\mathrm{Aut}^{\otimes}(\omega) = \mathrm{Spec}(\bar{\mathbb{Q}}[\mathbb{Q}/\mathbb{Z}]^G)$ and $G = \mathrm{Gal}(\bar{\mathbb{Q}}/\mathbb{Q})$, see Theorem 3.2 of [57]. The category $\mathrm{Vect}^{\bar{\mathbb{Q}}}_{\mathbb{Q}/\mathbb{Z}}(\mathbb{Q})$ is endowed with additive symmetric monoidal functors $\sigma_n(W) = W$ and $\overline{\sigma_n(W)_r} = \oplus_{r':\sigma_n(r')=r} \bar{W}_{r'}$ if r is in the range of σ_n and zero otherwise and additive functors $\tilde{\rho}_n(W) = W^{\oplus n}$ and $\overline{\tilde{\rho}_n(W)_r} = \bar{W}_{\sigma_n(r)}$ satisfying $\sigma_n \circ \tilde{\rho}_n = n \cdot \mathrm{id}$ that induce the Bost–Connes maps on $\mathbb{Q}[\mathbb{Q}/\mathbb{Z}]$.

As shown in Theorem 3.18 of [57], this category can be equivalently described as a category of automorphisms $\mathrm{Aut}^{\bar{\mathbb{Q}}}_{\mathbb{Q}/\mathbb{Z}}(\mathbb{Q})$ with objects pairs (W, ϕ) of a $\mathbb{Q}$-vector space V and a G-equivariant diagonalizable automorphism of $\bar{W}$ with eigenvalues that are roots of unity (seen as elements in $\mathbb{Q}/\mathbb{Z}$). There is an equivalence of categories between $\mathrm{Vect}^{\bar{\mathbb{Q}}}_{\mathbb{Q}/\mathbb{Z}}(\mathbb{Q})$ and $\mathrm{Aut}^{\bar{\mathbb{Q}}}_{\mathbb{Q}/\mathbb{Z}}(\mathbb{Q})$ under which the functors σ_n and $\tilde{\rho}_n$ correspond, respectively, to the Frobenius and Verschiebung on $\mathrm{Aut}^{\bar{\mathbb{Q}}}_{\mathbb{Q}/\mathbb{Z}}(\mathbb{Q})$, given by

$$F_n : (W, \phi) \mapsto (W, \phi^n), \quad V_n : (W, \phi) \mapsto (W^{\oplus n}, V_n(\phi)), \tag{55}$$

with

$$V_n(\phi) = \begin{pmatrix} 0 & 0 & \cdots & 0 & \phi \\ 1 & 0 & \cdots & 0 & 0 \\ 0 & 1 & \cdots & 0 & 0 \\ \vdots & & \vdots & & \vdots \\ 0 & 0 & \cdots & 1 & 0 \end{pmatrix}. \tag{56}$$

The equivalence is realized by mapping $(W, \phi) \mapsto (W, \oplus_r \bar{W}_r)$ where $\bar{W}_r$ are the eigenspaces of ϕ with eigenvalue $r \in \mathbb{Q}/\mathbb{Z}$.

Remark 6.15 Conceptually, the first description of the categorification in terms of the Tannakian category $\mathrm{Vect}^{\bar{\mathbb{Q}}}_{\mathbb{Q}/\mathbb{Z}}(\mathbb{Q})$ is closer to the integral Bost–Connes algebra as introduced in [27], while its equivalent description in terms of $\mathrm{Aut}^{\bar{\mathbb{Q}}}_{\mathbb{Q}/\mathbb{Z}}(\mathbb{Q})$ is closer to the reinterpretation of the Bost–Connes algebra in terms of Frobenius and Verschiebung operators, as in [25]. Since we have introduced here the Bost–Connes algebra in the form of [27], we are recalling both of these descriptions of the categorification, even though in the following we will be using only the one in terms of $\mathrm{Aut}^{\bar{\mathbb{Q}}}_{\mathbb{Q}/\mathbb{Z}}(\mathbb{Q})$.

Proposition 6.16 *Let $\mathcal{C}^{\mathbb{Z}}_{\mathbb{C}}$ be the assembler category underlying the Grothendieck ring $K_0^{\mathbb{Z}}(\mathcal{V}_{\mathbb{C}})$, as in Proposition 6.6 of [54]. The assignment*

$$\Phi(X, f) = (H_*(X, \mathbb{Q}), \oplus_r E_r(f_*)),$$

where $E_r(f_*)$ is the eigenspace with eigenvalue $r \in \mathbb{Q}/\mathbb{Z}$, determines a functor $\Phi : \mathcal{C}^{\mathbb{Z}}_{\mathbb{C}} \to \mathrm{Aut}^{\bar{\mathbb{Q}}}_{\mathbb{Q}/\mathbb{Z}}(\mathbb{Q})$ that lifts the Frobenius and Vershiebung functors on $\mathrm{Aut}^{\bar{\mathbb{Q}}}_{\mathbb{Q}/\mathbb{Z}}(\mathbb{Q})$ to the endofunctors σ_n and $\tilde{\rho}_n$ of $\mathcal{C}^{\mathbb{Z}}_{\mathbb{C}}$ implementing the Bost–Connes structure.

Proof We can construct the functor from the assembler category $\mathcal{C}^{\mathbb{Z}}_{\mathbb{C}}$ of §6 of [54], underlying $K_0^{\mathbb{Z}}(\mathcal{V}_{\mathbb{C}})$ to $\mathrm{Aut}^{\bar{\mathbb{Q}}}_{\mathbb{Q}/\mathbb{Z}}(\mathbb{Q})$ by following along the lines of Lemma 6.8 and Proposition 6.9, where we assign $\Phi(X, f) = (H_*(X, \mathbb{Q}), \oplus_r E_r(f_*))$ where $E_r(f_*)$ is the eigenspace with eigenvalue $r \in \mathbb{Q}/\mathbb{Z}$. The Bost–Connes algebra then lifts to the Frobenius and Vershiebung functors on $\mathrm{Aut}^{\bar{\mathbb{Q}}}_{\mathbb{Q}/\mathbb{Z}}(\mathbb{Q})$ and the latter lift to geometric Frobenius and Verschiebung operations on the pairs (X, f) mapping to (X, f^n) and to $(X \times Z_n, \Phi_n(f))$. □

This point of view, that replaces the Bost–Connes algebra with its categorification in terms of the Tannakian category $\mathrm{Aut}^{\bar{\mathbb{Q}}}_{\mathbb{Q}/\mathbb{Z}}(\mathbb{Q})$ as in [57] will also be useful in Section 7, where we reformulate our categorical setting, by passing from Grothendieck rings, assemblers and spectra, to Tannakian categories of Nori motives, and we compare in Lemma 7.4 and Theorem 7.7 the categorification of the Bost–Connes algebra obtained via Nori motives with the one of [57] recalled here.

7 Bost–Connes systems in categories of Nori motives

We introduce in this section a motivic framework, with Bost–Connes type systems on Tannakian categories of motives. The main result in this part of the paper will be Theorem 7.7, showing the existence of a fiber functor from the Tannakian category of Nori motives with good effectively finite $\hat{\mathbb{Z}}$-action to the Tannakian category $\mathrm{Aut}^{\bar{\mathbb{Q}}}_{\mathbb{Q}/\mathbb{Z}}(\mathbb{Q})$ that lifts the Bost–Connes system given by Frobenius and Verschiebung on the target category to a Bost–Connes system on Nori motives. Proposition 7.9 then extends this Bost–Connes structure to the relative case of motivic sheaves.

This is a natural generalization of the approach to Grothendieck rings via assemblers, which can be extended in an interesting way to the domain of motives, namely, Nori motives.

Roughly speaking, the theory of Nori motives starts with lifting the relations

$$[f : X \to S] = [f|_Y : Y \to S] + [f|_{X \smallsetminus Y} : X \smallsetminus Y \to S] \qquad (57)$$

of (relative) Grothendieck rings $K_0(\mathcal{V}_S)$ to the level of "diagrams", which intuitively can be imagined as "categories without multiplication of morphisms."

7.1 Nori diagrams

More precisely, (cf. Definition 7.1.1 of [43], p. 137), we have the following definitions.

Definition 7.1 A diagram (also called a quiver) D is a family consisting of a set of vertices $V(D)$ and a set of oriented edges, $E(D)$. Each edge e either connects two different vertices, going, say, from a vertex $\partial_{\mathrm{out}} e = v_1$ to a vertex $\partial_{\mathrm{in}} e = v_2$, or else is "an identity", starting and ending with one and the same vertex v. We will consider only diagrams with one identity for each vertex.

Diagrams can be considered as objects of a category, with obvious morphisms.

Definition 7.2 Each small category $\mathcal{C}$ can be considered as a diagram $D(\mathcal{C})$, with $V(D(\mathcal{C})) = \mathrm{Ob}\,\mathcal{C}$, $E(D(\mathcal{C})) = \mathrm{Mor}\,\mathcal{C}$, so that each morphism $X \to Y$ "is" an oriented edge from X to Y. More generally, *a representation T*

of a diagram D in a (small) category $\mathcal{C}$ is a morphism of directed graphs $T : D \to D(\mathcal{C})$.

Notice that a considerably more general treatment of graphs with markings, including diagrams etc. in the operadic environment, can be found in [52]. We do not use it here, although it might be highly relevant.

7.2 From geometric diagrams to Nori motives

We recall the main idea in the construction of Nori motives from geometric diagrams. For more details, see [43], pp. 140–144.

1. *Start* with the following data:
   a) a diagram D;
   b) a noetherian commutative ring with unit R and the category of finitely generated R–modules R-Mod;
   c) a representation T of D in R-Mod, in the sense of Definition 7.2.
2. *Produce* from them the category $C(D,T)$ defined in the following way:
   d1) If D is finite, then $C(D,T)$ is the category of finitely generated R–modules equipped with an R–linear action of $End(T)$.
   d2) If D is infinite, first consider all its finite subdiagrams F.
   d3) For each F construct $C(F,T|_F)$ as in d1). Then apply the following limiting procedure:

$$C(D,T) := \mathrm{colim}_{F \subseteq D\,\mathrm{finite}}\ C(F,T|_F)$$

   Thus, the category $C(D,T)$ has the following structure:
   - Objects of $C(D,T)$ will be all objects of the categories $C(F,T|_F)$. If $F \subset F'$, then each object X_F of $C(F,T|_F)$ can be canonically extended to an object of $C(F',T|_{F'})$.
   - Morphisms from X to Y in $C(D,T)$ will be defined as colimits over F of morphisms from X_F to Y_F with respect to these extensions.

   d4) The fact that $C(D,T)$ has a functor to R-Mod follows directly from the definition and the finite case.
3. *The result* is called *the diagram category* $C(D,T)$.
   It is an R–linear abelian category which is endowed with R–linear faithful exact forgetful functor

$$f_T : C(D,T) \to R\text{-Mod}.$$

7.2.1 Universal diagram category

The following results explain why abstract diagram categories play a central role in the formalism of Nori motives: they formalise the Grothendieck intuition of motives as objects of a universal cohomology theory.

Theorem 7.3 [43]

(i) Any representation $T : D \to R$-Mod can be presented as post-composition of the forgetful functor f_T with an appropriate representation $\tilde{T} : D \to C(D,T)$:

$$T = f_T \circ \tilde{T}.$$

with the following universal property:

Given any R–linear abelian category A with a representation $F : D \to A$ and R–linear faithful exact functor $f : A \to R$-Mod with $T = f \circ F$, it factorizes through a faithful exact functor $L(F) : C(D,T) \to A$ compatibly with the decomposition

$$T = f_T \circ \tilde{T}.$$

(ii) The functor $L(F)$ is unique up to unique isomorphism of exact additive functors.

For proofs, cf. [43], pp. 140–141 and p. 167.

7.2.2 Nori geometric diagrams

If we start not with an abstract category but with a "geometric" category $\mathcal{C}$ (in the sense that its objects are spaces/varieties/schemes, possibly endowed with additional structures), in which one can define morphisms of closed embeddings $Y \hookrightarrow X$ (or $Y \subset X$) and morphisms of complements to closed embeddings $X \setminus Y \to X$, we can define the Nori diagram of *effective pairs* $D(\mathcal{C})$ in the following way (see [43], pp. 207–208).

a) One vertex of $D(\mathcal{C})$ is a triple (X, Y, i) where $Y \hookrightarrow X$ is a closed embedding, and i is an integer.
b) Besides obvious identities, there are edges of two types.
b1) Let (X, Y) and (X', Y') be two pairs of closed embeddings. Every morphism $f : X \to X'$ such that $f(Y) \subset Y'$ produces functoriality edges f^* (or rather (f^*, i)) going from (X', Y', i) to (X, Y, i).
b2) Let $(Z \subset Y \subset X)$ be a stair of closed embeddings. Then it defines coboundary edges ∂ from (Y, Z, i) to $(X, Y, i+1)$.

7.2.3 (Co)homological representatons of Nori geometric diagrams

If we start not just from the initial category of spaces $\mathcal{C}$, but rather from a pair $(\mathcal{C}, H)$ where H is a cohomology theory, then assuming reasonable properties of this pair, we can define the respective representation T_H of $D(\mathcal{C})$ that we will call a *(co)homological representation of* $D(\mathcal{C})$.

For a survey of such pairs $(\mathcal{C}, H)$ that were studied in the context of Grothendieck's motives, see [43], pp. 31–133. The relevant cohomology theories include, in particular, singular cohomology, and algebraic and holomorphic de Rham cohomologies.

Below we will consider the basic example of cohomological representations of Nori diagrams that leads to Nori motives.

7.2.4 Effective Nori motives

We follow [43], pp. 207–208. Take as a category $\mathcal{C}$, the starting object in the definition of Nori geometric diagrams above, the category $\mathcal{V}_k$ of varieties X defined over a subfield $k \subset \mathbb{C}$.

We can then define the Nori diagram $D(\mathcal{C})$ as above. This diagram will be denoted Pairs^{eff} from now on,

$$\mathrm{Pairs}^{eff} = D(\mathcal{V}_k).$$

The category of effective mixed Nori motives is the diagram category $C(\mathrm{Pairs}^{eff}, H^*)$ where $H^i(X, \mathbb{Z})$ is the respective singular cohomology of the analytic space X^{an} (cf. [43], pp. 31–34 and further on).

It turns out (see [43], Proposition 9.1.2. p. 208) that the map

$$H^* : \mathrm{Pairs}^{eff} \to \mathbb{Z}\text{-Mod}$$

sending (X, Y, i) to the relative singular cohomology $H^i(X(\mathbb{C}), Y(\mathbb{C}); \mathbb{Z})$, naturally extends to a representation of the respective Nori diagram in the category of finitely generated abelian groups $\mathbb{Z}$-Mod.

7.3 Category of equivariant Nori motives

We now introduce the specific category of Nori motives that we will be using for the construction of the associated Bost–Connes system.

Let $D(\mathcal{V})$ the Nori geometric diagrams associated to the category $\mathcal{V}$ of varieties over $\mathbb{Q}$, constructed as described in §7.2.

As in [54] and in § 2 of this paper, we consider here the category $\mathcal{V}^{\hat{\mathbb{Z}}}$ of varieties X with a good effectively finite action of $\hat{\mathbb{Z}}$. We can view $\mathcal{V}^{\hat{\mathbb{Z}}}$ as an enhancement $\hat{\mathcal{V}}$ of the category $\mathcal{V}$, in the sense described in §3.3.

Define the Nori diagram of *effective pairs* $D(\mathcal{V}^{\hat{\mathbb{Z}}})$ as we recalled earlier in §7.2:

a) One vertex of $D(\mathcal{V}^{\hat{\mathbb{Z}}})$ is a triple $((X,\alpha_X),(Y,\alpha_Y),i)$, of varieties X and Y with good effectively finite $\hat{\mathbb{Z}}$ actions, $\alpha_X : \hat{\mathbb{Z}} \times X \to X$ and $\alpha_Y : \hat{\mathbb{Z}} \times Y \to Y$, and an integer i, together with a closed embedding $j : Y \hookrightarrow X$ that is equivariant with respect to the $\hat{\mathbb{Z}}$ actions. For brevity, we will denote such a triple $(\hat{X},\hat{Y},i)$ and call it a closed embedding in the enhancement $\hat{\mathcal{V}}$.
b) Identity edges, functoriality edges, and coboundary edges are obvious enhancements of the respective edges defined in §7.2, with the requirement that all these maps are $\hat{\mathbb{Z}}$-equivariant.
b1) Let $(\hat{X},\hat{Y})$ and $(\hat{X}',\hat{Y}')$ be two pairs of closed embeddings in $\hat{\mathcal{V}}$. Every morphism $f : X \to X'$ such that $f(Y) \subset Y'$ and $f \circ \alpha_X = \alpha_{X'} \circ f$ produces functoriality edges f^* (or rather (f^*,i)) going from $((X',\alpha_{X'}),(Y',\alpha_{Y'}),i)$ to $((X,\alpha_X),(Y,\alpha_Y),i)$.
b2) Let $(Z \subset Y \subset X)$ be a stair of closed embeddings compatible with enhancements (equivariant with respect tot the $\hat{\mathbb{Z}}$-actions). Then it defines coboundary edges ∂

$$((Y,\alpha_Y),(Z,\alpha_Z),i) \to ((X,\alpha_X),(Y,\alpha_Y),i+1).$$

We have thus defined he Nori geometric diagram of enhanced effective pairs, which we denote equivalently by $D(\hat{\mathcal{V}})$ or $D(\mathcal{V}^{\hat{\mathbb{Z}}})$.

Notice that forgetting in this diagram all enhancements, we obtain the map $D(\hat{\mathcal{V}}) \to D(\mathcal{V})$ which is *injective* both on vertices and edges.

7.4 Bost–Connes system on Nori motives

We now construct a Bost–Connes system on a category of Nori motives obtained from the diagram $D(\mathcal{V}^{\hat{\mathbb{Z}}})$ described above, which lifts to the level of motives the categorification of the Bost–Connes algebra constructed in [57].

As we recalled in §6.5.1, we can describe the categorification of [57] of the Bost–Connes algebra in terms of the Tannakian category $\mathrm{Vec}^{\bar{\mathbb{Q}}}_{\mathbb{Q}/\mathbb{Z}}(\mathbb{Q})$ with suitable functors σ_n and $\tilde{\rho}_n$ constructed as in Theorem 3.7 of [57] or in terms of an equivalent Tannakian category $\mathrm{Aut}^{\bar{\mathbb{Q}}}_{\mathbb{Q}/\mathbb{Z}}(\mathbb{Q})$ endowed with Frobenius and Verschiebung functors. We are going to use here the second description.

Lemma 7.4 *The assignment*

$$T : ((X,\alpha_X),(Y,\alpha_Y),i) \mapsto H^i(X(\mathbb{C}),Y(\mathbb{C}),\mathbb{Q})$$

determines a representation $T : D(\mathcal{V}^{\hat{\mathbb{Z}}}) \to \mathrm{Aut}^{\bar{\mathbb{Q}}}_{\mathbb{Q}/\mathbb{Z}}(\mathbb{Q})$ *of the diagram* $D(\mathcal{V}^{\hat{\mathbb{Z}}})$ *constructed above.*

Proof As discussed in the previous subsection, we view the elements $((X,\alpha_X),(Y,\alpha_Y),i)$ of $D(\mathcal{V}^{\hat{\mathbb{Z}}})$ in terms of an enhancement $\hat{\mathcal{V}}$ of the category $\mathcal{V}$ defined as in §3.3, by choosing a primitive root of unity that generates the cyclic group $\mathbb{Z}/N\mathbb{Z}$, so that the actions α_X and α_Y are determined by self maps v_X and v_Y as in §3.3. We identify the element above with $((X,v_X),(Y,v_Y),i)$, which we also denoted by $(\hat{X},\hat{Y},i)$ in the previous subsection. Since the embedding $Y \hookrightarrow X$ is $\hat{\mathbb{Z}}$-equivariant, the map v_Y is the restriction to Y of the map v_X under this embedding. We denote by ϕ^i the induced map on the cohomology $H^i(X(\mathbb{C}),Y(\mathbb{C}),\mathbb{Q})$. The eigenspaces of ϕ^i are the subspaces of the decomposition of $H^i(X(\mathbb{C}),Y(\mathbb{C}),\mathbb{Q})$ according to characters of $\hat{\mathbb{Z}}$, that is, elements in $\mathrm{Hom}(\hat{\mathbb{Z}},\mathbb{C}^*) = \nu^* \simeq \mathbb{Q}/\mathbb{Z}$. Thus, we obtain an object $(H^i(X(\mathbb{C}),Y(\mathbb{C}),\mathbb{Q}),\phi^i)$ in the category $\mathrm{Aut}^{\bar{\mathbb{Q}}}_{\mathbb{Q}/\mathbb{Z}}(\mathbb{Q})$. Edges in the diagram are $\hat{\mathbb{Z}}$-equivariant maps so they induce morphisms between the corresponding objects in the category $\mathrm{Aut}^{\bar{\mathbb{Q}}}_{\mathbb{Q}/\mathbb{Z}}(\mathbb{Q})$. □

One can also see in a similar way that the fiber functor

$$T : ((X,\alpha_X),(Y,\alpha_Y),i) \mapsto H^i(X(\mathbb{C}),Y(\mathbb{C}),\mathbb{Q})$$

determines an object in the category $\mathrm{Vec}^{\bar{\mathbb{Q}}}_{\mathbb{Q}/\mathbb{Z}}(\mathbb{Q})$. Indeed, the pair (X,Y) with $Y \subset X$ is endowed with compatible good effectively finite $\hat{\mathbb{Z}}$-actions α_X and α_Y, hence the singular cohomology $H^i(X(\mathbb{C}),Y(\mathbb{C}),\mathbb{Q})$ carries a resulting $\hat{\mathbb{Z}}$-representation. Thus, the vector space $H^i(X(\mathbb{C}),Y(\mathbb{C}),\bar{\mathbb{Q}})$ can be decomposed into eigenspaces of this representations according to characters $\chi \in \mathrm{Hom}(\hat{\mathbb{Z}},\mathbb{G}_m) = \mathbb{Q}/\mathbb{Z}$. Thus, we obtain a decomposition of $H^i(X(\mathbb{C}),Y(\mathbb{C}),\bar{\mathbb{Q}}) = \oplus_{r\in\mathbb{Q}/\mathbb{Z}}\bar{V}_r$ as a $\mathbb{Q}/\mathbb{Z}$-graded vector space. We choose to work with the category $\mathrm{Aut}^{\bar{\mathbb{Q}}}_{\mathbb{Q}/\mathbb{Z}}(\mathbb{Q})$ because the Bost–Connes structure is more directly expressed in terms of Frobenius and Verschiebung, which will make the lifting of this structure to the resulting category of Nori motives more immediately transparent, as we discuss below.

The representation $T : D(\mathcal{V}^{\hat{\mathbb{Z}}}) \to \mathrm{Aut}^{\bar{\mathbb{Q}}}_{\mathbb{Q}/\mathbb{Z}}(\mathbb{Q})$ replaces, at this motivic level, our previous use in [54] of the equivariant Euler characteristics $K_0^{\hat{\mathbb{Z}}}(\mathcal{V}) \to \mathbb{Z}[\mathbb{Q}/\mathbb{Z}]$ (see [50]) as a way to lift the Bost–Connes algebra. We proceed in the following way to obtain the Bost–Connes structure in this setting.

Definition 7.5 Let D be a diagram, endowed with a representation $T : D \to \mathrm{Aut}^{\bar{\mathbb{Q}}}_{\mathbb{Q}/\mathbb{Z}}(\mathbb{Q})$, and let $\mathcal{C}(D,T)$ be the associated diagram category, obtained as in §7.2, with the induced functor $\tilde{T} : \mathcal{C}(D,T) \to \mathrm{Aut}^{\bar{\mathbb{Q}}}_{\mathbb{Q}/\mathbb{Z}}(\mathbb{Q})$. We say that the functor $\tilde{T}$ intertwines the Bost–Connes structure, if there are endofunctors σ_n and $\tilde{\rho}_n$ of $\mathcal{C}(D,T)$ (where the σ_n but not the $\tilde{\rho}_n$ are compatible with the tensor product structure) such that the following diagrams commute,

$$\begin{array}{ccc} \mathcal{C}(D,T) & \xrightarrow{\tilde{T}} & \mathrm{Aut}^{\bar{\mathbb{Q}}}_{\mathbb{Q}/\mathbb{Z}}(\mathbb{Q}) \\ \Big\downarrow{\scriptstyle \sigma_n} & & \Big\downarrow{\scriptstyle F_n} \\ \mathcal{C}(D,T) & \xrightarrow{\tilde{T}} & \mathrm{Aut}^{\bar{\mathbb{Q}}}_{\mathbb{Q}/\mathbb{Z}}(\mathbb{Q}) \end{array} \qquad \begin{array}{ccc} \mathcal{C}(D,T) & \xrightarrow{\tilde{T}} & \mathrm{Aut}^{\bar{\mathbb{Q}}}_{\mathbb{Q}/\mathbb{Z}}(\mathbb{Q}) \\ {\scriptstyle \tilde{\rho}_n}\Big\uparrow & & \Big\uparrow{\scriptstyle V_n} \\ \mathcal{C}(D,T) & \xrightarrow{\tilde{T}} & \mathrm{Aut}^{\bar{\mathbb{Q}}}_{\mathbb{Q}/\mathbb{Z}}(\mathbb{Q}) \end{array}$$

where on the right-hand-side of the diagrams, the F_n and V_n are the Frobenius and Verschiebung on $\mathrm{Aut}^{\bar{\mathbb{Q}}}_{\mathbb{Q}/\mathbb{Z}}(\mathbb{Q})$, defined as in (55) and (56).

Definition 7.6 For $((X,\alpha_X),(Y,\alpha_Y),i)$ in the category $\mathcal{C}(D(\mathcal{V}^{\hat{\mathbb{Z}}}),T)$ of Nori motives associated to the diagram $D(\mathcal{V}^{\hat{\mathbb{Z}}})$ define

$$\sigma_n : ((X,\alpha_X),(Y,\alpha_Y),i) \mapsto ((X,\alpha_X \circ \sigma_n),(Y,\alpha_Y \circ \sigma_n),i) \tag{58}$$

$$\tilde{\rho}_n : ((X,\alpha_X),(Y,\alpha_Y),i) \mapsto (X \times Z_n, \Phi_n(\alpha_X),(Y \times Z_n, \Phi_n(\alpha_Y)),i), \tag{59}$$

where $Z_n = \mathrm{Spec}(\mathbb{Q}^n)$ and $\Phi_n(\alpha)$ is the geometric Verschiebung defined as in §2.5.

Theorem 7.7 *The σ_n and $\tilde{\rho}_n$ of* (58) *and* (59) *determine a Bost–Connes system on the category $\mathcal{C}(D(\mathcal{V}^{\hat{\mathbb{Z}}}),T)$ of Nori motives associated to the diagram $D(\mathcal{V}^{\hat{\mathbb{Z}}})$. The representation $T : D(\mathcal{V}^{\hat{\mathbb{Z}}}) \to \mathrm{Aut}^{\bar{\mathbb{Q}}}_{\mathbb{Q}/\mathbb{Z}}(\mathbb{Q})$ constructed above has the property that the induced functor*

$$\mathcal{C}(D(\mathcal{V}^{\hat{\mathbb{Z}}}),T) \to \mathrm{Aut}^{\bar{\mathbb{Q}}}_{\mathbb{Q}/\mathbb{Z}}(\mathbb{Q})$$

intertwines the endofunctors σ_n and $\tilde{\rho}_n$ of the Bost–Connes system on $\mathcal{C}(D(\mathcal{V}^{\hat{\mathbb{Z}}}),T)$ and the Frobenius F_n and Verschiebung V_n of the Bost–Connes structure on $\mathrm{Aut}^{\bar{\mathbb{Q}}}_{\mathbb{Q}/\mathbb{Z}}(\mathbb{Q})$.

Proof Consider the mappings σ_n and $\tilde{\rho}_n$ defined in (58) and (59), The effect of the transformation σ_n, when written in terms of the data $((X,\upsilon_X),(Y,\upsilon_Y),i)$ is to send $\upsilon_X \mapsto \upsilon_X^n$ and $\upsilon_Y \mapsto \upsilon_Y^n$, hence it induces the Frobenius map F_n acting on $(H^i(X(\mathbb{C}),Y(\mathbb{C}),\mathbb{Q}),\phi^i)$ in $\mathrm{Aut}^{\bar{\mathbb{Q}}}_{\mathbb{Q}/\mathbb{Z}}(\mathbb{Q})$. Similarly, we have $T(X\times Z_n, \Phi_n(\alpha_X),(Y\times Z_n,\Phi_n(\alpha_Y)),i) = H^i(X\times Z_n, Y\times Z_n, \mathbb{Q})$ where by the relative version of the Künneth formula we have $(H^i(X(\mathbb{C}) \times Z_n(\mathbb{C}), Y(\mathbb{C}) \times Z_n(\mathbb{C}),\mathbb{Q}) \simeq H^i(X(\mathbb{C}),Y(\mathbb{C}),\mathbb{Q})^{\oplus n}$ with the induced map $V_n(\phi^i)$. The maps

σ_n and $\tilde{\rho}_n$ defined as above determine self maps of the diagram $D(\mathcal{V}^{\hat{\mathbb{Z}}})$. By Lemma 7.2.6 of [43] given a map $F : D_1 \to D_2$ of diagrams and a representation $T : D_2 \to R$-Mod, there is an R-linear exact functor $\mathcal{F} : C(D_1, T \circ F) \to C(D_2, T)$ such that the following diagram commutes:

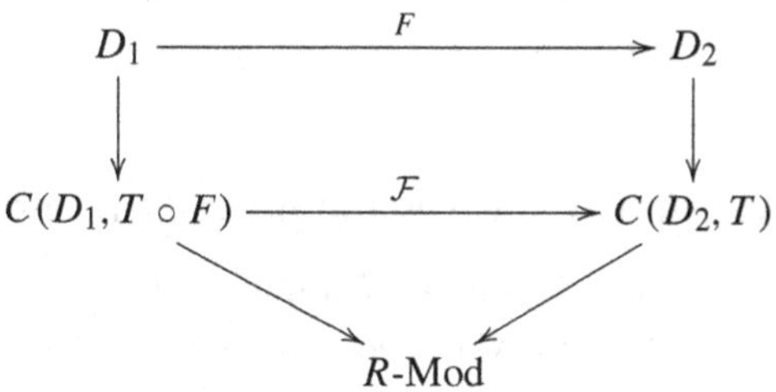

We still denote by σ_n and $\tilde{\rho}_n$ the endofunctors induced in this way on $C(D(\mathcal{V}^{\hat{\mathbb{Z}}}), T)$. To check the compatibility of the σ_n functors with the monoidal structure, we use the fact that for Nori motives the product structure is constructed using "good pairs" (see §9.2.1 of [43]), that is, elements (X, Y, i) with the property that $H^j(X, Y, \mathbb{Z}) = 0$ for $j \neq i$. For such elements the product is given by $(X, Y, i) \times (X', Y', j) = (X \times X', X \times Y' \cup Y \times X', i + j)$. The diagram category $C(\text{Good}^{eff}, T)$ obtained by replacing effective pairs Pairs^{eff} with good effective pairs Good^{eff} is equivalent to $C(\text{Pairs}^{eff}, T)$ (Theorem 9.2.22 of [43]), hence the tensor structure defined in this way on $C(\text{Good}^{eff}, T)$ determines the tensor structure of $C(\text{Pairs}^{eff}, T)$ and on the resulting category of Nori motives, see §9.3 of [43]. Thus, to check the compatibility of the functors σ_n with the tensor structure it suffices to see that on a product of good pairs, where indeed we have

$$\sigma_n((X, \alpha_X), (Y, \alpha_Y), i) \times \sigma_n((X', \alpha'_X), (Y', \alpha'_Y), j) = (\mathcal{X}, \mathcal{Y}, i + j),$$

where

$$\mathcal{X} := (X \times X', (\alpha_X \times \alpha'_X) \circ \Delta \circ \sigma_n)$$

$$\mathcal{Y} := (X \times Y', (\alpha_X \times \alpha'_Y) \circ \Delta \circ \sigma_n) \cup (Y \times X', (\alpha_Y \times \alpha'_X) \circ \Delta \circ \sigma_n)$$

and the latter $(\mathcal{X}, \mathcal{Y}, i + j)$ is equal to

$$\sigma_n(((X, \alpha_X), (Y, \alpha_Y), i) \times ((X', \alpha'_X), (Y', \alpha'_Y), j)).$$

The functors $\tilde{\rho}_n$ are not compatible with the tensor product structure, as expected. □

Remark 7.8 In [57] a motivic interpretation of the categorification of the Bost–Connes algebra is given by identifying the Tannakian category

$\mathrm{Vec}^{\bar{\mathbb{Q}}}_{\mathbb{Q}/\mathbb{Z}}(\mathbb{Q})$ with a limit of orbit categories of Tate motives. Here we presented a different motivic categorification of the Bost–Connes algebra by lifting the Bost–Connes structure to the level of the category of Nori motives. In [57] a motivic Bost–Connes structure was also constructed using the category of motives over finite fields and the larger class of Weil numbers replacing the roots of unity of the Bost–Connes system.

7.5 Motivic sheaves and the relative case

The argument presented in Theorem 7.7 lifting the Bost–Connes structure to the category of Nori motives, which provides a Tannakian category version of the lift to Grothendieck rings via the equivariant Euler characteristics $K_0^{\hat{\mathbb{Z}}}(\mathcal{V}) \to \mathbb{Z}[\mathbb{Q}/\mathbb{Z}]$, can also be generalized to the relative setting, where we considered the Euler characteristic

$$\chi_S^{\hat{\mathbb{Z}}} : K_0^{\hat{\mathbb{Z}}}(\mathcal{V}_S) \to K_0^{\hat{\mathbb{Z}}}(\mathbb{Q}_S)$$

with values in the Grothendieck ring of constructible sheaves, discussed in §2 of this paper. The categorical setting of Nori motives that is appropriate for this relative case is the Nori category of motivic sheaves introduced in [3].

We recall here briefly the construction of the category of motivic sheaves of [3] and we show that the Bost–Connes structure on the category of Nori motives described in Theorem 7.7 extends to this relative setting.

Consider pairs $(X \to S, Y)$ of varieties over a base S with $Y \subset X$ endowed with the restriction $f_Y : Y \to S$. Morphisms $f : (X \to S, Y) \to (X' \to S, Y')$ are morphisms of varieties $h : X \to X'$ satisfying the commutativity of

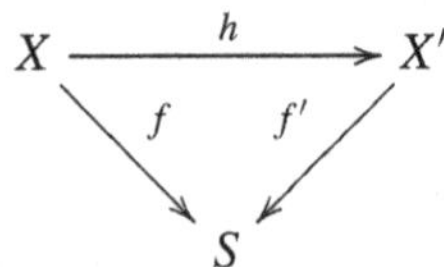

and such that $h(Y) \subset Y'$. As before, we consider varieties endowed with good effectively finite $\hat{\mathbb{Z}}$-action. We denote by (S, α) the base with its good effectively finite $\hat{\mathbb{Z}}$-action and by $((X\alpha_X) \to (S, \alpha), (Y, \alpha_Y))$ the pairs as above where we assume that the map $f : X \to S$ and the inclusion $Y \hookrightarrow X$ are $\hat{\mathbb{Z}}$-equivariant.

Following [3], a diagram $D(\mathcal{V}_S)$ is obtained by considering as vertices elements of the form $(X \to S, Y, i, w)$ with $(X \to S, Y)$ a pair as above, $i \in \mathbb{N}$ and $w \in \mathbb{Z}$. The edges are given by the three types of edges

1. geometric morphisms $h : (X \to S, Y) \to (X' \to S, Y')$ as above determine edges $h^* : (X' \to S, Y', i, w) \to (X \to S, Y, i, w)$;
2. connecting morphisms $\partial : (Y \to S, Z, i, w) \to (X \to S, Y, i+1, w)$ for a chain of inclusions $Z \subset Y \subset X$;
3. twisted projections:

$$(X, Y, i, w) \to (X \times \mathbb{P}^1, Y \times \mathbb{P}^1 \cup X \times \{0\}, i+2, w+1).$$

For consistency with our previous notation we have here written the morphisms in the contravariant (cohomological) way rather than in the covariant (homological) way used in §3.3 of [3].

Note that in the previous section, following [43] we described the effective Nori motives as $\mathcal{MN}^{eff} = C(\text{Pairs}^{eff}, T)$, with the category of Nori motives $\mathcal{MN}$ being then obtained as the localization of $\mathcal{MN}^{eff}$ at $(\mathbb{G}_m, \{1\}, 1)$ (inverting the Lefschetz motive). Here in the setting of [3] the Tate motives are accounted for in the diagram construction by the presence of the twist w and the last class of edges.

Given $f : X \to S$ and a sheaf $\mathcal{F}$ on X one has $H^i_S(X; \mathcal{F}) = R^i f_* \mathcal{F}$. In the case of a pair $(f : X \to S, Y)$, let $j : X \smallsetminus Y \hookrightarrow X$ be the inclusion and consider $H^i_S(X, Y; \mathcal{F}) = R^i f_* j_! \mathcal{F}|_{X \smallsetminus Y}$. The diagram representation T in this case maps $T(X \to S, Y, i, w) = H^i_S(X, Y, \mathcal{F})(w)$ to the (Tate twisted) constructible sheaf $H^i_S(X, Y; \mathcal{F})$. It is shown in [3] that the Nori formalism of geometric diagrams applies to this setting and gives rise to a Tannakian category of motivic sheaves $\mathcal{MN}_S$. In particular one considers the case where $\mathcal{F}$ is constant with $\mathcal{F} = \mathbb{Q}$, so that the diagram representation $T : D(\mathcal{V}_S) \to \mathbb{Q}_S$ and the induced functor on $\mathcal{MN}_S$ replace at the motivic level the Euler characteristic map on the relative Grothendieck ring $K_0(\mathcal{V}_S) \to K_0(\mathbb{Q}_S)$.

As in the previous cases, we consider an enhancement of this category of motivic sheaves, in the sense of §3.3, by introducing good effectively finite $\hat{\mathbb{Z}}$-actions. We modify the construction of [3] in the following way.

We consider a diagram $D(\mathcal{V}^{\hat{\mathbb{Z}}}_{(S,\alpha)})$ where the vertices are elements

$$((X, \alpha_X) \to (S, \alpha), (Y, \alpha_Y), i, w)$$

so that the maps $f : X \to S$ and the inclusion $Y \hookrightarrow X$ are $\hat{\mathbb{Z}}$-equivariant, and with morphisms as above, where all the maps are required to be compatible with the $\hat{\mathbb{Z}}$-actions. One obtains by the same procedure as in [3] a category of equivariant motivic sheaves $\mathcal{MN}^{\hat{\mathbb{Z}}}_S$. The representation above maps $D(\mathcal{V}^{\hat{\mathbb{Z}}}_{(S,\alpha)})$ to $\hat{\mathbb{Z}}$-equivariant constructible sheaves over (S, α). Then the same argument we used §2 at the level of Grothendieck rings, assemblers and spectra applies to this setting and gives the following result.

Proposition 7.9 *The maps of diagrams*

$$\sigma_n : D(\mathcal{V}^{\hat{\mathbb{Z}}}_{(S,\alpha)}) \to D(\mathcal{V}^{\hat{\mathbb{Z}}}_{(S,\alpha\circ\sigma_n)})$$

$$\tilde{\rho}_n : D(\mathcal{V}^{\hat{\mathbb{Z}}}_{(S,\alpha)}) \to D(\mathcal{V}^{\hat{\mathbb{Z}}}_{(S\times Z_n,\Phi_n(\alpha))})$$

given by

$$\sigma_n((X,\alpha_X) \to (S,\alpha),(Y,\alpha_Y),i,w) =$$

$$((X,\alpha_X \circ \sigma_n) \to (S,\alpha \circ \sigma_n),(Y,\alpha_Y \circ \sigma_n),i,w)$$

$$\tilde{\rho}_n((X,\alpha_X) \to (S,\alpha),(Y,\alpha_Y),i,w) =$$

$$((X \times Z_n,\Phi_n(\alpha_X)) \to (S \times Z_n,\Phi_n(\alpha)),(Y \times Z_n,\Phi_n(\alpha_Y)),i,w)$$

determine functors of the resulting category of motivic sheaves $\mathcal{MN}^{\hat{\mathbb{Z}}}_S$ such that $\sigma_n \circ \tilde{\rho}_n = n\,\mathrm{id}$ and $\tilde{\rho}_n \circ \sigma_n$ is a product with (Z_n,α_n). Thus, one obtains on the category $\mathcal{MN}^{\hat{\mathbb{Z}}}_S$ a Bost–Connes system as in Definition 3.11.

Proof The argument is as in Proposition 2.6, using again, as in Theorem 7.7 the fact that maps of diagrams induce functors of the resulting categories of Nori motives. □

7.6 Nori geometric diagrams for assemblers, and a challenge

We conclude this section on Bost–Connes systems and Nori motives by formulating a question about Nori diagrams and assembler categories.

According to the Nori formalism as it is presented in [43], we must start with a "geometric" category C of spaces/varieties/schemes, possibly endowed with additional structures, in which one can define morphisms of closed embeddings $Y \hookrightarrow X$ (or $Y \subset X$) and morphisms of complements to closed embeddings $X \setminus Y \to X$, Then the Nori diagram of *effective pairs* $D(C)$ is defined as in [43], pp. 207–208, see §7.2.2.

In the current context, *objects* of our category C will be *assemblers* $\mathcal{C}$ (of course, described in terms of a category of lower level). In particular, each such $\mathcal{C}$ is endowed with a Grothendieck topology.

A *vertex* of the Nori diagram $D(C)$ will be a triple $(\mathcal{C},\mathcal{C} \setminus \mathcal{D},i)$ where its first two terms are taken from an abstract scissors congruence in C, and i is an integer. Intuitively, this means that we we are considering the canonical embedding $\mathcal{C} \setminus \mathcal{D} \hookrightarrow \mathcal{C}$ as an analog of closed embedding. This intuition makes translation of the remaining components of Nori's diagrams obvious, except for one: *what is the geometric meaning of the integer* i *in* $(\mathcal{C},\mathcal{C}\setminus\mathcal{D},i)$?

The answer in the general context of assemblers, seemingly, was not yet suggested, and already in the algebraic–geometric contexts is non–obvious and non–trivial. Briefly, i translates to the level of Nori geometric diagrams the *weight filtration* of various cohomology theories (cf. [43], 10.2.2, pp. 238–241), and the existence of such translation and its structure are encoded in several versions of *Nori's Basic Lemma* independently and earlier discovered by A. Beilinson and K. Vilonen (cf. [43], 2.5, pp. 45–59).

The most transparent and least technical version of the Basic Lemma ([43], Theorem 2.5.2 , p. 46) shows that in algebraic geometry the existence of weight filtration is based upon special properties of *affine schemes.* Lifts of Bost–Connes algebras to the level of cohomology based upon the techniques of *enhancement* also require a definition of *affine* assemblers. Since we do not know its combinatorial version, the enhancements that we can study now, force us to return to algebraic geometry.

This challenge suggests to think about other possible geometric contexts in which dimensions/weights of the relevant objects may take, say, p–adic values (as in the theory of p–adic weights of automorphic forms inaugurated by J. P. Serre), or rational values (as it happens in some corners of "geometries below Spec $\mathbb{Z}$"), or even real values (as in various fractal geometries).

Can one transfer the scissors congruences imagery there?

See, for example, the formalism of Farey semi–intervals as base of ∞–adic topology.

Acknowledgment

We thank the referee for many very detailed and useful comments and suggestions on how to improve the structure and presentation of the paper. The first and third authors were supported in part by the Perimeter Institute for Theoretical Physics. The third author is also partially supported by NSF grants DMS-1707882 and DMS-2104330 and by NSERC Discovery Grant RGPIN-2018-04937 and Accelerator Supplement grant RGPAS-2018-522593.

References

[1] G. Almkvist, *Endomorphisms of finitely generated projective modules over a commutative ring*, Ark. Mat. 11 (1973) 263–301.

[2] G. Almkvist, *The Grothendieck ring of the category of endomorphisms*, J. Algebra 28 (1974) 375–388.

[3] D. Arapura, *An abelian category of motivic sheaves*, Adv. Math. 233 (2013), 135–195. [arXiv:0801.0261]

[4] M. Baake, E. Lau, V. Paskunas, *A note on the dynamical zeta function of general toral endomorphisms*, Monatsh. Math., Vol.161 (2010) 33–42. [arXiv:0810.1855]

[5] S. del Baño, *On the Chow motive of some moduli spaces*, J. Reine Angew. Math. 532 (2001) 105–132

[6] D. Bejleri, M. Marcolli, *Quantum field theory over* $\mathbb{F}_1$, J. Geom. Phys. 69 (2013) 40–59. [arXiv:1209.4837]

[7] P. Berrizbeitia, V.F. Sirvent, *On the Lefschetz zeta function for quasi-unipotent maps on the n-dimensional torus*, J. Difference Equ. Appl. 20 (2014), no. 7, 961–972.

[8] P. Berrizbeitia, M.J. González, A. Mendoza, V.F. Sirvent, *On the Lefschetz zeta function for quasi-unipotent maps on the n-dimensional torus. II: The general case*, Topology Appl. 210 (2016), 246–262.

[9] A. Bialynicki-Birula, *Some theorems on actions of algebraic groups*, Ann. Math. (2) 98 (1973) 480–497.

[10] A. Bialynicki-Birula, *Some properties of the decompositions of algebraic varieties determined by actions of a torus*, Bull. Acad. Polon. Sci. Sér. Sci. Math. Astronom. Phys. 24 (1976) 667–674

[11] A. Bialynicki-Birula, J.B. Carrell, W.M. McGovern, *Algebraic quotients. Torus actions and cohomology. The adjoint representation and the adjoint action*, Vol.131 of Encyclopaedia of Mathematical Sciences. Invariant Theory and Algebraic Transformation Groups, II, Springer 2002.

[12] Kai Behrend, Jim Bryan, Balázs Szendröi, *Motivic degree zero Donaldson–Thomas invariants*, Invent. Math. 192 (2013) 111–160. [arXiv:0909.5088]

[13] J. Borger, B. de Smit, *Galois theory and integral models of* Λ*-rings*, Bull. Lond. Math. Soc. 40 (2008), no. 3, 439–446. [arXiv:0801.2352]

[14] L. Borisov, *The class of the affine line is a zero divisor in the Grothendieck ring*, arXiv:1412.6194.

[15] J.B. Bost, A. Connes. *Hecke algebras, type III factors and phase transitions with spontaneous symmetry breaking in number theory*. Selecta Math. (N.S.) 1 (1995) no. 3, pp. 411–457.

[16] A. Bridy, *The Artin-Mazur zeta function of a dynamically affine rational map in positive characteristic*, Journal de Théorie des Nombres de Bordeaux, Vol.28 (2016) 301–324. [arXiv:1306.5267]

[17] A. Bridy, *Transcendence of the Artin-Mazur zeta function for polynomial maps of* $\mathbb{A}^1(\mathbb{F}_p)$, Acta Arith. 156 (2012), no. 3, 293–300. [arXiv:1202.0362]

[18] P. Brosnan, *On motivic decompositions arising from the method of Bialynicki-Birula*, Invent. Math. 161 (2005) 91–111. [arXiv:math/0407305]

[19] J. Byszewski, G. Cornelissen, *Dynamics on Abelian varieties in positive characteristic*, arXiv:1802.07662.

[20] J.A. Campbell, *Facets of the Witt Vectors*, arXiv:1910.10206.

[21] J.A. Campbell, J. Wolfson, I. Zakharevich, *Derived* ℓ*-adic zeta functions*, arXiv:1703.09855.

[22] J. Choi, S. Katz, A. Klemm, *The refined BPS index from stable pair invariants*, Commun. Math. Phys. 328 (2014) 903–954. [arXiv:1210.4403]

[23] J. Choi, M. Maican, *Torus action on the moduli spaces of torsion plane sheaves of multiplicity four*, Journal of Geometry and Physics 83 (2014) 18–35. [arXiv:1304.4871]

[24] A. Connes, C. Consani, *Schemes over $\mathbb{F}_1$ and zeta functions*, Compos. Math. 146 (2010) 1383–1415. [arXiv:0903.2024]

[25] A. Connes, C. Consani, *On the arithmetic of the BC-system*, J. Noncommut. Geom. 8 (2014) no. 3, 873–945. [arXiv:1103.4672]

[26] A. Connes, C. Consani, *Absolute algebra and Segal's Γ-rings: au dessous de* Spec($\mathbb{Z}$), J. Number Theory 162 (2016), 518–551 [arXiv:1502.05585]

[27] A. Connes, C. Consani, M. Marcolli, *Fun with $\mathbb{F}_1$*, J. Number Theory 129 (2009) 1532–1561. [arXiv:0806.2401]

[28] A. Connes, M. Marcolli, *Noncommutative geometry, quantum fields and motives*, Colloquium Publications, Vol.55, American Mathematical Society, 2008.

[29] A. Deitmar, *Remarks on zeta functions and K-theory over $\mathbb{F}_1$*, Proc. Japan Acad. Ser. A Math. Sci. 82 (2006) 141–146. [arXiv:math/0605429]

[30] P. Deligne, *Catégories tensorielles*, Mosc. Math. J. 2 (2002), no. 2, 227–248.

[31] P. Deligne, J.S. Milne, *Tannakian categories, in Hodge Cycles, Motives, and Shimura Varieties*, Lecture Notes in Mathematics, Vol. 900, Springer 1982, pp. 101–228

[32] A.W.M. Dress, C. Siebeneicher, *The Burnside ring of profinite groups and the Witt vector construction*, Advances in Mathematics Vol.70 (1988) N.1, 87–132.

[33] J.M. Drézet, M Maican, *On the geometry of the moduli spaces of semi-stable sheaves supported on plane quartics*, Geom. Dedicata 152 (2011) 17–49. [arXiv:0910.5327]

[34] W. Ebeling, S.M. Gusein-Zade, *Higher-order spectra, equivariant Hodge–Deligne polynomials, and Macdonald-type equations*, in "Singularities and computer algebra", pp. 97–108, Springer, 2017. [arXiv:1507.08088]

[35] B. Fantechi, L. Göttsche, *Riemann-Roch theorems and elliptic genus for virtually smooth schemes*, Geom. Topol. 14 (2010) no. 1, 83–115. [arXiv:0706.0988]

[36] J.M. Franks, *Some smooth maps with infinitely many hyperbolic periodic points*, Trans. Amer. Math. Soc. 226 (1977), 175–179.

[37] J.M. Franks, *Homology and the zeta function for diffeomorphisms*, International Conference on Dynamical Systems in Mathematical Physics (Rennes, 1975), pp. 79–88. Astérisque, No. 40, Soc. Math. France, 1976.

[38] D.R. Grayson, *The K-theory of endomorphisms*, J. Algebra 48 (1977) no. 2, 439–446.

[39] S.M. Gusein-Zade, *Equivariant analogues of the Euler characteristic and Macdonald type equations*, Russian Math. Surveys 72 (2017) 1, 1–32.

[40] L. Hesselholt, *Witt vectors of non-commutative rings and topological cyclic homology*, Acta Math., Vol. 178 (1997) N.1, 109–141.

[41] L. Hesselholt, I. Madsen, *On the K-theory of finite algebras over Witt vectors of perfect fields*, Topology, 36 (1997) N.1, 29–101.

[42] W.H. Hesselink, *Concentration under actions of algebraic groups*, Lect. Notes Math., vol. 867 (1981) 55–89.

[43] A. Huber, St. Müller–Stach. *Periods and Nori motives.* With contributions by Benjamin Friedrich and Jonas von Wangenheim. Erg der Math und ihrer Grenzgebiete, vol. 65, Springer 2017.

[44] Z. Jin, M. Marcolli, *Endomotives of toric varieties*, J. Geom. Phys. 77 (2014), 48–71. [arXiv:1309.4101]

[45] M. Kapranov, A. Smirnov, *Cohomology determinants and reciprocity laws: number field case*, Unpublished manuscript.

[46] M. Kapranov, *The elliptic curve in the S-duality theory and Eisenstein series for Kac-Moody groups*, arXiv:math/0001005.

[47] M. Kashiwara, P. Shapira, *Categories and Sheaves*, Springer, 2005.

[48] M. Kontsevich, Y. Soibelman, *Cohomological Hall algebra, exponential Hodge structures and motivic Donaldson-Thomas invariants*, Commun. Number Theory Phys. Vol. 5 (2011) N.2, 231–352. [arXiv:1006.2706]

[49] C. Lo, M. Marcolli, *$\mathbb{F}_\zeta$-geometry, Tate motives, and the Habiro ring*, International Journal of Number Theory, 11 (2015), no. 2, 311–339. [arXiv:1310.2261]

[50] E. Looijenga, *Motivic measures*, Séminaire N. Bourbaki, 1999–2000, exp. no 874, 267–297.

[51] J. López Peña, O. Lorscheid, *Torified varieties and their geometries over $\mathbb{F}_1$*, Math. Z. 267 (2011) 605–643. [arXiv:0903.2173]

[52] Yu. Manin, D. Borisov. *Generalized operads and their inner cohomomorphisms.* In: Geometry and Dynamics of Groups and Spaces (In memory of Aleksander Reznikov). Ed. by M. Kapranov et al. Progress in Math., vol. 265, Birkhäuser, Boston, pp. 247–308. Preprint math.CT/0609748.

[53] Yuri I. Manin, Matilde Marcolli, *Moduli operad over $\mathbb{F}_1$*, in "Absolute Arithmetic and $\mathbb{F}_1$-Geometry", 331–361, Eur. Math. Soc., 2016. [arXiv:1302.6526]

[54] Yuri I. Manin, Matilde Marcolli, *Homotopy types and geometries below* $\mathrm{Spec}(\mathbb{Z})$, in "Dynamics: topology and numbers", pp. 27–56, Contemp. Math., 744, Amer. Math. Soc., 2020. [arXiv:1806.10801]

[55] M. Marcolli, *Cyclotomy and endomotives*, p-Adic Numbers Ultrametric Anal. Appl. 1 (2009), no. 3, 217–263. [arXiv:0901.3167]

[56] M. Marcolli, Z. Ren, *q-Deformations of statistical mechanical systems and motives over finite fields*, p-Adic Numbers Ultrametric Anal. Appl. 9 (2017) no. 3, 204–227. [arXiv:1704.06367]

[57] M. Marcolli, G. Tabuada, *Bost-Connes systems, categorification, quantum statistical mechanics, and Weil numbers*, J. Noncommut. Geom. 11 (2017) no. 1, 1–49. [arXiv:1411.3223]

[58] N. Martin, *The class of the affine line is a zero divisor in the Grothendieck ring: an improvement*, C. R. Math. Acad. Sci. Paris 354 (2016), no. 9, pp. 936–939. [arXiv:1604.06703]

[59] L. Maxim, J. Schürmann, *Equivariant characteristic classes of external and symmetric products of varieties*, Geom. Topol. 22 (2018) no. 1, 471–515. [arXiv:1508.04356]

[60] S. Müller-Stach, B. Westrich, *Motives of graph hypersurfaces with torus operations*, Transform. Groups 20 (2015), no. 1, 167–182. [arXiv:1301.5221]

[61] D. Quillen, *Higher algebraic K-theory. I*, in "Algebraic K-theory, I: Higher K-theories", Lecture Notes in Math., Vol. 341 (1973) 85–147.

[62] N. Ramachandran, *Zeta functions, Grothendieck groups, and the Witt ring*, Bull. Sci. Math. Soc. Math. Fr. 139 (2015) N.6, 599–627 [arXiv:1407.1813]

[63] N. Ramachandran, G. Tabuada, *Exponentiable motivic measures*, J. Ramanujan Math. Soc. 30 (2015), no. 4, 349–360. [arXiv:1412.1795]

[64] S. Schwede, *Stable homotopical algebra and Γ-spaces*, Math. Proc. Phil. Soc. Vol.126 (1999) 329–356.

[65] G. Segal, *Categories and cohomology theories*, Topology, Vol.13 (1974) 293–312.

[66] M. Shub, D. Sullivan, *Homology theory and dynamical systems*, Topology 14 (1975) 109–132.

[67] C. Soulé, *Les variétés sur le corps à un élément*, Mosc. Math. J. 4 (2004), 217–244.

[68] R.W. Thomason, *Symmetric monoidal categories model all connective spectra*, Theory and Applications of Categories, Vol.1 (1995) N.5, 78–118.

[69] J.L. Verdier, *Caractéristique d'Euler-Poincaré*, Bull. Soc. Math. France 101 (1973) 441–445.

[70] I. Zakharevich, *The K-theory of assemblers*, Adv. Math. 304 (2017), 1176–1218. [arXiv:1401.3712]

[71] I. Zakharevich, *On K_1 of an assembler*, J. Pure Appl. Algebra 221 (2017), no. 7, 1867–1898. [arXiv:1506.06197]

[72] I. Zakharevich, *The annihilator of the Lefschetz motive*, Duke Math. J. 166 (2017), no. 11, 1989–2022. [arXiv:1506.06200]

20

Nef Cycles on Some Hyperkähler Fourfolds

John Christian Ottem[a]

For Bill Fulton on his 80th birthday

Abstract. We study the cones of surfaces on varieties of lines on cubic fourfolds and Hilbert schemes of points on K3 surfaces. From this we obtain new examples of nef cycles which fail to be pseudoeffective.

1 Introduction

The cones of curves and effective divisors are essential tools in the study of algebraic varieties. Here the intersection pairing between curves and divisors allows one to interpret these cones geometrically in terms of their duals; the cones of nef divisors and movable curves respectively. In intermediate dimensions however, very little is known about the behaviour of the cones of effective cycles and their dual cones, the cones of *nef cycles*. Here a surprising feature is that nef cycles may fail to be pseudoeffective, showing that the usual geometric intuition for 'positivity' does not extend so easily to higher codimension. In the paper [7], Debarre–Ein–Lazarsfeld–Voisin presented examples of such cycles of codimension two on abelian fourfolds.

In this paper, we show that similar examples can also be found on certain hyperkähler manifolds (or 'irreducible holomorphic symplectic varieties'). For these varieties, the cones of curves and divisors are already well-understood thanks to several recent advances in hyperkähler geometry [2, 3, 5, 4, 12]. Our main result is the following:

Theorem 1.1 *Let X be the variety of lines of a very general cubic fourfold. Then the cone of pseudoeffective 2-cycles on X is strictly contained in the cone of nef 2-cycles.*

[a] University of Oslo

On such a hyperkähler fourfold, the most interesting cohomology class is the second Chern class $c_2(X)$, which represents a positive rational multiple of the Beauville–Bogomolov form on $H^2(X,\mathbb{Z})$. This fact already implies that the class carries a certain amount of positivity, because it intersects products of effective divisors non-negatively (see (1)). To prove Theorem 1.1 we show that this class is in the interior of the cone of nef cycles, but not in the cone of effective cycles. The main idea of the proof is to deform X to the Hilbert square of an elliptic K3 surface. Here the claim for $c_2(X)$ follows from the fact that it has intersection number zero with the fibers of a Lagrangian fibration.

Thanks to B. Bakker, R. Lazarsfeld, B. Lehmann, U. Rieß, C. Vial, C. Voisin and L. Zhang for useful discussions. I am grateful to D. Huybrechts for his helpful comments and for inviting me to Bonn where this paper was written.

2 Preliminaries

We work over the complex numbers. For a smooth projective variety X, let $N_k(X)$ (resp. $N^k(X)$) denote the $\mathbb{R}$-vector space of dimension (resp. codimension) k cycles on X modulo numerical equivalence. In $N_k(X)$ we define the pseudoeffective cone $\overline{\mathrm{Eff}}_k(X)$ to be the closure of the cone spanned by classes of k-dimensional subvarieties. A class $\alpha \in N_k(X)$ is said to be *big* if it lies in the interior of $\overline{\mathrm{Eff}}_k(X)$. This is equivalent to having $\alpha = \epsilon h^{\dim X - k} + e$ for h the class of a very ample divisor; $\epsilon > 0$; and e an effective 2-cycle with $\mathbb{R}$-coefficients.

A codimension k-cycle is said to be *nef* if it has non-negative intersection number with any k-dimensional subvariety. We let $\mathrm{Nef}^k(X) \subset N^k(X)$ denote the cone spanned by nef cycles; by the definition this is the dual cone of $\overline{\mathrm{Eff}}_k(X)$. For the varieties in this paper, it is known that numerical and homological equivalence coincide, so we may consider these as cones in $H_{2k}(X,\mathbb{R})$ and $H^{2k}(X,\mathbb{R})$ respectively. If $Y \subset X$ is a subvariety, we let $[Y] \in H^*(X,\mathbb{R})$ denote its corresponding cohomology class.

For a variety X, we denote by $X^{[n]}$ the Hilbert scheme parameterizing length n subschemes of X.

2.1 Hyperkähler fourfolds

We will study effective 2-cycles on certain hyperkähler varieties (or holomorphic symplectic varieties). By definition, such a variety is a smooth, simply connected algebraic variety admitting a non-degenerate holomorphic two-form

ω generating $H^{2,0}(X)$. In dimension four there are currently two known examples of hyperkähler manifolds up to deformation: Hilbert schemes of two points on a K3 surface and generalized Kummer varieties.

A hyperkähler manifold X carries an integral, primitive quadratic form q on the cohomology group $H^2(X,\mathbb{Z})$ called the Beauville–Bogomolov form. The signature of this form is $(3, b_2(X) - 3)$, and $(1, \rho - 1)$ when restricted to the Picard lattice $\mathrm{Pic}(X) = H^{1,1}(X) \cap H^2(X,\mathbb{Z})$. For the hyperkähler fourfolds considered in this paper it is known that the second Chern class $c_2(X) = c_2(T_X)$ represents a positive rational multiple of the Beauville–Bogomolov form [17]. It follows from this and standard properties of the Beauville–Bogomolov form that

$$c_2(X) \cdot D_1 \cdot D_2 \geq 0 \tag{1}$$

for any two distinct prime divisors D_1, D_2 on X.

2.2 Specialization of effective cycles

In the proof of Theorem 1.1, we will need a certain semi-continuity result for effective cycles. This result is likely well-known to experts, but we include it here for the convenience of the reader and for future reference.

Proposition 2.1 *Let $f : \mathcal{X} \to T$ be a smooth projective morphism over a smooth variety T and suppose that $\alpha \in H^{k,k}(\mathcal{X},\mathbb{Z})$ is a class such that $\alpha|_{\mathcal{X}_t}$ is effective on the general fiber. Then $\alpha|_{\mathcal{X}_t}$ is effective for every $t \in T$.*

Proof This follows essentially from the theory of relative Hilbert schemes. The main point is that there are at most finitely many components $\rho_i : H_i \to T$ of the Hilbert scheme parameterizing cycles supported in the fibers of f with cohomology class α. For such a component H_i, let $\pi_i : \mathcal{Z}_i \to H_i$ denote the universal family of H_i. This fits into the diagram

$$\begin{array}{ccc} \mathcal{Z}_i & \longrightarrow & \mathcal{X} \\ \downarrow{\scriptstyle \pi_i} & & \downarrow{\scriptstyle f} \\ H_i & \xrightarrow{\rho_i} & T \end{array}$$

where π_i is flat. Let $T' = \bigcup_i \rho_i(H_i)$, where the union is taken over all indices i such that ρ_i is not surjective. T' is a proper closed subset of T.

Let $t_0 \in T$ be any point. By assumption the class $\alpha|_{\mathcal{X}_t}$ is represented by an effective cycle Z_t on $\mathcal{X}_t$ for some $t \in T - T'$. By construction, there exists a component H of the above Hilbert scheme, a universal family $\pi : \mathcal{Z} \to H$, a point $h \in H$ so that $Z_t = \mathcal{Z}_h$, and so that $\rho(H) = T$. We have $\mathcal{Z} \subset \mathcal{X}'$, where $\mathcal{X}' = \mathcal{X} \times_T H$. Note that we have two sections, $[\mathcal{Z}]$ and $\rho^*\alpha$, of the local system $R^{2n-2k}\pi_*\mathbb{Z}$. These agree over $\rho^{-1}(t)$ so they agree everywhere. It follows that the cycle $\mathcal{Z}|_{\mathcal{X}'_{h_0}} \in CH^{n-k}(\mathcal{X}_{t_0})$ is an effective cycle on $\mathcal{X}_{t_0}$ with class $\alpha|_{t_0}$, as desired. □

By a limit argument, we get the following result which says that just like in the case of divisors, the effective cones can only become larger after specialization:

Corollary 2.2 *With the notation of Proposition 2.1, if $\alpha|_t$ is big on a very general fiber, it is big on every fiber.*

3 The variety of lines of a cubic fourfold

Let $Y \subset \mathbb{P}^5$ be a smooth cubic fourfold. The variety of lines $X = F(Y)$ on Y is a smooth 4-dimensional subvariety of the Grassmannian $Gr(2,6)$. A fundamental result due to Beauville and Donagi [6] says that X is a holomorphic symplectic variety. Moreover, X is deformation-equivalent to the Hilbert square $S^{[2]}$ of a K3 surface.

From the embedding of X in the Grassmannian we obtain natural cycle classes on X. In particular, if U denotes the universal subbundle on $Gr(2,6)$, we can consider the Chern classes $g = c_1(U^\vee)|_X$ and $c = c_2(U^\vee)|_X$. Note that g coincides with the polarization of X given by the Plücker embedding of X in $\mathbb{P}^{14}$. When Y is very general, a Noether–Lefschetz type argument shows that the vector space of degree four Hodge classes $H^{2,2}(X) \cap H^4(X,\mathbb{Q})$ is two-dimensional, generated by g^2 and c. We have the following intersection numbers:

$$g^4 = 108, \quad g^2 \cdot c = 45, \quad c^2 = 27. \tag{2}$$

The cubic polynomial defining Y shows that X is the zero-set of a section of the vector bundle $S^3U^\vee$ on $Gr(2,6)$. Using this description, a standard Chern class computation shows that

$$c_2(X) = 5g^2 - 8c.$$

See for example [1] or [17] for detailed proofs of these statements.

3.1 Surfaces in the variety of lines

There are several interesting surfaces on the fourfold X. The most basic of these are the restrictions of codimension two Schubert cycles from the ambient

Grassmannian $Gr(2,6)$. In terms of g^2 and c, these cycles are given by $g^2 - c$ and c. Moreover, since on $Gr(2,6)$ every effective cycle is nef (this is true on any homogeneous variety), also their classes remain nef and effective when restricted to X.

There are two natural surfaces on X with class proportional to $g^2 - c$. Fixing a general line $l \subset Y$, the surface parameterizing lines meeting l represents the class $\frac{1}{3}(g^2 - c)$. Also, the variety of lines of 'second type' (that is, lines with normal bundle $\mathcal{O}(1)^2 \oplus \mathcal{O}(-1)$ in Y) is a surface with corresponding class $5(g^2 - c)$ (see [6]).

The class c is also represented by an irreducible surface. Indeed, for a general hyperplane $H = \mathbb{P}^4 \subset \mathbb{P}^5$ the Fano surface of lines in the cubic threefold $H \cap Y$,

$$\Sigma_H = \{[l] \in X \mid l \subset H\},$$

is a smooth surface of general type with $[\Sigma_H] = c$.

We also have the following less obvious example. We can view Y as the general hyperplane section of some cubic fivefold $V \subset \mathbb{P}^6$. The variety of planes in V is a smooth surface $F_2(V)$. As Y is general, there is an embedding $F_2(V) \to F(Y)$ given by associating a plane to its intersection with the hyperplane section. The class of the image is a surface of general type with class $63c$ (see [11]).

By the following result of Voisin [20], the class $c = c_2(U^\vee)$ lies the boundary in the cone of effective 2-cycles on X:

Lemma 3.1 (Voisin) *Let X be the variety of lines on a very general cubic fourfold. Then the class c is extremal in the effective cone of 2-cycles.*

This result deserves a few comments. First of all, we note that even though the class c is the restriction of an extremal Schubert cycle on $Gr(2,6)$, there is a priori no reason to expect that it should remain extremal when restricted to X. To see how subtle this is, we note that Voisin [20, Theorem 2.9] showed that in the case where X is instead the variety of lines of a cubic *fivefold* Y, the corresponding class c of lines in a hyperplane section of Y is in fact big on X (in the sense that it lies in the interior of the effective cone).

It is in fact quite surprising that the class c is extremal in $\overline{\mathrm{Eff}}_2(X)$, as the surface $\Sigma_H \subset X$ behaves in many ways like a complete intersection. For instance, varying the hyperplane H, it is clear that it deforms in a large family covering X, and $\Sigma_H \cdot V > 0$ for any other surface $V \subset X$. In fact, Voisin showed that the surface above gives a counterexample to a question of Peternell [15], who asked whether a smooth subvariety with ample normal bundle is big. This question has a positive answer for curves and divisors [14]. In the current setting, the normal bundle of Σ_H coincides with the restriction of $U^\vee$, which is ample on Σ_H.

Remark 3.2 Even though the subvariety Σ_H is very positively embedded in X, it is not an 'ample subscheme' in the sense of [13]. Essentially, [13] defines a smooth subvariety $Y \subset X$ to be ample if large powers of the ideal sheaf kill cohomology of every coherent sheaf $\mathscr{F}$ in degrees $< \dim Y$ (that is, $H^k(X, I_Y^m \otimes \mathscr{F}) = 0$ for $k < \dim Y$ and $m \gg 0$). This is a generalization of the notion of an ample divisor; see [13] for basic results on such subschemes. To see why Σ_H is not ample, we can use the fact that ample subschemes satisfy Lefschetz hyperplane theorems on rational homology[13], so that their Betti numbers agree $b_i(X) = b_i(Y)$ for $i < \dim Y$. However, in our case X is simply connected, whereas both Σ_H and $F_2(V)$ have non-zero first Betti number. As ample subschemes also have ample normal bundles, this raises the question whether Peternell's question has a positive answer when restricted to ample subschemes, that is, whether smooth ample subvarieties have big cycle classes. As usual, the answer is affirmative for curves and divisors (see [14]).

3.2 Lagrangian submanifolds

Voisin's proof of Lemma 3.1 is quite interesting: it uses the fact that Σ_H is a Lagrangian submanifold of X. Indeed, the class of the 2-form ω is the image of a primitive cohomology class $\sigma^{3,1} \in H^4(X, \mathbb{Q})_{prim}$ under the incidence correspondence $P \subset F \times X$; from this it follows that ω restricts to 0 on any surface with class proportional to c (see [20, Lemma 1.1]).

Now, the (2, 2)-form $\omega \wedge \overline{\omega}$ either vanishes on a surface V (in which case V is Lagrangian), or restricts to a multiple of the volume form on V. In the latter case, the integral $\int_V \omega \wedge \overline{\omega}$ is strictly positive. Hence any surface V so that $\omega|_Y$ vanishes cannot be homologous to a cycle of the form $\epsilon h^2 + e$ where h is ample and e is effective. Consequently, since numerical and homological equivalence coincide on X, we see that the class of $\omega \wedge \overline{\omega}$ defines a supporting hyperplane of the effective cone of surfaces.

This argument works in greater generality (using the Hodge–Riemann relations) and one obtains the following:

Proposition 3.3 *Let X be a smooth projective variety of dimension n and let ω be a closed p-form representing a non-zero class $[\omega] \in H^{p,0}(X)$. If $V \subset X$ is a p-dimensional subvariety so that $\omega|_V = 0$, then V is not homologous to a big cycle.*

In particular, if X is an holomorphic symplectic variety of dimension four, and $V \subset X$ is a Lagrangian surface, then $[V]$ is in the boundary of $\overline{\mathrm{Eff}}_2(X)$.

This result can be used to prove extremality of other cycle classes on hyperkähler varieties. Two standard examples of Lagrangian subvarieties are: (i) Any surface $V \subset X$ with $H^{2,0}(V) = 0$ (e.g., a rational surface); and (ii) $C^{[2]} \subset S^{[2]}$ for a curve C in a K3 surface S. For instance, if $Y \subset \mathbb{P}^5$ is a cubic fourfold containing a plane P, then the dual plane $P^\vee \subset X = F(Y)$ parameterizing lines in P is extremal. (See also [16, §3.2]).

3.3 Proof of Theorem 1.1

Voisin's result gives one face of the pseudoeffective cone of X. To bound the other half of $\overline{\mathrm{Eff}}_2(X)$, we will consider the class $c_2(X) = 5g^2 - 8c$. Note that this class is already quite positive, since it intersects products of divisors D_1D_2 non-negatively (cf. equation (1)). In fact, this class will be shown to lie in the interior of the cone of nef 2-cycles, and is *strictly nef*, in the sense that $c_2(X) \cdot Z > 0$ for every irreducible surface $Z \subset X$. We will first show that it is not big. We first consider a special case:

Proposition 3.4 *Let X be a hyperkähler manifold of dimension $2n$, admitting a Lagrangian fibration. Then $c_k(X)$ is not big, for $1 \leq k \leq n$.*

Proof Let $f : X \to B$ denote this fibration. By a theorem of Matsushita, $\dim B = n$ and the general fiber A of f of is an n-dimensional abelian variety. In particular, $\Omega_A^1 \simeq \mathcal{O}_A^n$. Let $Y = A \cap H_1 \cap \cdots \cap H_{n-k}$, where $H_i \in |h|$, and h is a very ample line bundle on X. The restriction of the normal bundle sequence

$$0 \to T_A|_Y = \mathcal{O}_Y^n \to T_X|_Y \to N_A|_Y = \mathcal{O}_Y^n \to 0$$

to Y shows that $c_k(X) \cdot A \cdot h^{n-k} = 0$. Now if c_k were a big cycle, then it would be numerically equivalent to $\epsilon h^k + e$, for some $\epsilon > 0$ and e an effective cycle with $\mathbb{R}$−coefficients. However, the cycle $A \cdot h^{n-k}$ is nef (and non-zero), and so it has strictly positive intersection number with h^k. Hence $c_k(X)$ cannot be big. □

Lemma 3.5 *Let $X = S^{[n]}$ for a very general K3 surface S. Then the Chern classes $c_k(X)$ are not big for $1 \leq k \leq n$.*

Proof By Proposition 2.1, it suffices to prove this for a special K3 surface S. Choosing an elliptic K3 surface $S \to \mathbb{P}^1$, we obtain a Lagrangian fibration $S^{[n]} \to (\mathbb{P}^1)^{[n]} = \mathbb{P}^n$, and so the claim follows from Proposition 3.4. □

Corollary 3.6 *Let $X = F(Y)$ for a very general cubic fourfold Y. Then the second Chern class $c_2(X) = 5g^2 - 8c$ is not big.*

Proof As in the previous lemma, it suffices to exhibit one cubic fourfold for which the statement is true. We specialize to a Pfaffian cubic fourfold Y so that $F(Y) = S^{[2]}$ for a degree 14 K3 surface S (provided that Y does not contain any planes [6]). Choosing Y so that S contains an elliptic fibration, we obtain the required abelian surface fibration $S^{[2]} \to (\mathbb{P}^1)^{[2]} = \mathbb{P}^2$ on $F(Y)$, and we conclude using Proposition 3.4. □

Combining this with Voisin's result, we obtain the following bound for the effective cone $\overline{\mathrm{Eff}}_2(X)$:

Corollary 3.7 *Let X be the variety of lines on a very general cubic fourfold. Then*

$$\mathbb{R}_{\geq 0}c + \mathbb{R}_{\geq 0}(g^2 - c) \subseteq \overline{\mathrm{Eff}}_2(X) \subseteq \mathbb{R}_{\geq 0}c + \mathbb{R}_{\geq 0}(g^2 - \tfrac{8}{5}c). \tag{3}$$

Finally, we deduce the statement in Theorem 1.1.

Proof of Theorem 1.1 By the previous corollary, we have that $\overline{\mathrm{Eff}}_2(X) \subseteq \mathbb{R}_{\geq 0}c + \mathbb{R}_{\geq 0}c_2(X)$. Using the intersection numbers (2), we find by duality:

$$\begin{aligned}\mathrm{Nef}^2(X) &\supseteq \left(\mathbb{R}_{\geq 0}c + \mathbb{R}_{\geq 0}(5g^2 - 8c)\right)^{\vee}\\ &= \mathbb{R}_{\geq 0}(20c - g^2) + \mathbb{R}_{\geq 0}(3g^2 - 5c)\\ &\not\supseteq \overline{\mathrm{Eff}}_2(X).\end{aligned}$$

□

In the above proof, the class $c_2(X)$ plays an important role. The theorem implies that $c_2(X)$ is in the interior of the nef cone, but not big. We do not know if $c_2(X)$ is not pseudoeffective (i.e., a limit of effective cycles) or if there are special cubic fourfolds for which $c_2(X)$ is effective.

The role of $c_2(X)$ is also interesting from the viewpoint of generic *non-projective* deformations X of the fourfold above, since in that case we have $H^{2,2}(X, \mathbb{C}) \cap H^4(X, \mathbb{Q}) = \mathbb{Q}c_2(X)$. In this case, Verbitsky [18] proved that X contains no analytic subvarieties of positive dimension at all.

Remark 3.8 In his thesis [16], Max Rempel also considered examples of pseudoeffective cycles on hyperkähler fourfolds, in particular also on the variety of lines of a cubic fourfold. Among his many results, Rempel shows that the class $3g^2 - 5c$ (which is proportional to $c_2(X) - \frac{1}{5}g^2$) has no effective multiple. The proof involves an interesting geometric argument using Voisin's rational self-map $\varphi : X \dashrightarrow X$.

Remark 3.9 (An alternative definition of nefness) In the Nakai–Moishezon criterion, a line bundle L is ample if and only if for every $r = 1, \ldots, \dim X$ we have $L^r \cdot Z > 0$ for every subvariety of dimension r. This suggests the following naïve fix for the definition of nefness for any cycle class: $\gamma \in N^k(X)$ could be defined to be 'nef' if the restriction cycle $i^*\gamma \in N^k(Y)$ is pseudoeffective for every subvariety $i : Y \hookrightarrow X$. In particular, taking i to be the identity map, γ would itself be pseudoeffective on X. The geometric meaning of this condition is on the other hand not so clear.

This definition is related to Fulger and Lehmann's notion of *universal pseudoeffectivity* [8] where the $f^*\gamma$ is required to be pseudoeffective for any morphism of projective varieties $f : Y \to X$. It is an interesting question whether these notions are equivalent.

4 Generalized Kummer varieties

Let X denote a generalized Kummer variety of dimension 4 of an abelian surface A. Recall that these are defined as the fiber over 0 of the addition map $\Sigma : A^{[3]} \to A$. In this case, the group $H^2(X,\mathbb{Z})$ can be identified with $H^2(A,\mathbb{Z}) \oplus \mathbb{Z}e$, where the class e is one half of the divisor corresponding to non-reduced subschemes in $A^{[3]}$. As before, $c_2(X)$ represents a positive multiple of the Beauville–Bogomolov form [10].

There are 81 distinguished Kummer surfaces Z_τ on X, whose classes are linearly independent in $H^4(X,\mathbb{Q})$: Z_0 is the closure of the locus of points $(0, a, -a)$ with $a \in A - 0$, and the other Z_τ are the translates of Z_0 via the 3-torsion points $A[3]$. Hassett–Tschinkel [10, Proposition 5.1] showed that

$$c_2(X) = \frac{1}{3} \sum_{\tau \in A[3]} [Z_\tau].$$

In particular, $3c_2(X)$ represents an effective cycle. Moreover, using the same argument as in Lemma 3.5 we obtain

Proposition 4.1 *Let X be a generic projective generalized Kummer fourfold. Then the second Chern class $c_2(X)$ is effective, but not big.*

References

[1] Ekaterina Amerik. A computation of invariants of a rational selfmap, Ann. Fac. Sci. Toulouse 18 (2009),

[2] Ekaterina Amerik and Misha Verbitsky. Morrison-Kawamata cone conjecture for hyperkahler manifolds. Ann. Sci. Éc. Norm. Supér.(4) 50.4 (2017): 973–993.

[3] Benjamin Bakker. A classification of Lagrangian planes in holomorphic symplectic varieties. Central European Journal of Mathematics 12 (2014) 952–975.

[4] Arend Bayer, Emanuele Macrì. MMP for moduli of sheaves on K3s via wall-crossing: nef and movable cones, Lagrangian fibrations. Inventiones mathematicae 198 (2014) 505–590.

[5] Arend Bayer, Brendan Hassett and Yuri Tschinkel: Mori cones of holomorphic symplectic varieties of K3 type. Annales scientifiques de l'ENS 48, fascicule 4 (2015), 941–950.

[6] Arnaud Beauville and Ron Donagi. La variété des droites d'une hypersurface cubique de dimension 4. *C. R. Acad. Sci. Paris Sér. I Math.*, 301 (1985) 703–706,

[7] Olivier Debarre, Lawrence Ein, Robert Lazarsfeld, and Claire Voisin. Pseudoeffective and nef classes on abelian varieties. *Composito Mathematica*, 147 (2011) 1793–1818.

[8] Mihai Fulger and Brian Lehmann. Positive cones of dual cycle classes. Alg. Geom. 4 (2017), no. 1, 1–28.

[9] Mihai Fulger and Brian Lehmann. Zariski decompositions of numerical cycle classes. J. Alg. Geom. 26 (2017), no. 1, 43–106.

[10] Brendan Hassett, Yuri Tschinkel. Hodge theory and Lagrangian planes on generalized Kummer fourfolds. Mosc. Math. J., 13 (2013), 33–56.

[11] Atanas Iliev, Laurent Manivel. Cubic hypersurfaces and integrable systems. Amer. J. Math. 130 (2008), no. 6, 1445–1475.

[12] Eyal Markman. A survey of Torelli and monodromy results for holomorphic-symplectic varieties. In Complex and differential geometry, volume 8 of Springer Proc. Math. Springer, Heidelberg, 2011. 257–322.

[13] John Christian Ottem. Ample subvarieties and q-ample divisors. Adv. Math. 229 (2012), 2868–2887.

[14] John Christian Ottem. On subvarieties with ample normal bundle, J. of the Eur. Math. Soc. 18 (11) (2016) 2459–2468.

[15] Thomas Peternell. Submanifolds with ample normal bundles and a conjecture of Hartshorne. *Interactions of classical and numerical algebraic geometry*, Contemp. Math., 496, Amer. Math. Soc., Providence, RI, (2009) 317–330.

[16] Max Rempel. Positivité des cycles dans les varietés algebraiques. PhD thesis (2012).

[17] Mingmin Shen, Charles Vial. The Fourier transform for certain hyperkähler fourfolds. Memoirs of the American Mathematical Society 240 (2016).

[18] Misha Verbitsky. Trianalytic subvarieties of hyperkaehler manifolds, GAFA vol. 5 no. 1 (1995) 92–104.

[19] Claire Voisin. A counterexample to the Hodge conjecture extended to Kähler varieties. International Mathematics Research Notices 2002.20 (2002) 1057–1075.

[20] Claire Voisin. Coniveau 2 complete intersections and effective cones. *Geom. Funct. Anal.* Vol. 19 (2010) 1494–1513.

[21] Claire Voisin. Chow Rings, Decomposition of the Diagonal, and the Topology of Families (AM-187). Princeton University Press, 2014.

21

Higher Order Polar and Reciprocal Polar Loci

Ragni Piene[a]

To Bill Fulton on the occasion of his 80th birthday.

Abstract. In this note we introduce higher order polar loci as natural generalizations of the classical polar loci, replacing the role of tangent spaces by that of higher order osculating spaces. The close connection between polar loci and dual varieties carries over to a connection between higher order polar loci and higher order dual varieties. We generalize the duality between the degrees of polar classes of a variety and those of its dual variety to varieties that are reflexive to a higher order. In particular, the degree of the "top" (highest codimension) polar class of order k is equal to the degree of the kth dual variety. We define higher order Euclidean normal bundles and use them to define higher order reciprocal polar loci and classes. We give examples of how to compute the degrees of the higher order polar and reciprocal polar classes in some special cases: curves, scrolls, and toric varieties.

1 Introduction

The theory of polar varieties, or polar loci, has a long and rich history. The terminology *pole* and *polar* goes back at least to Servois (1811) and Gergonne (1813). Poncelet's more systematic treatment of *reciprocal polars* was presented in his "Mémoire sur la théorie générale des polaires réciproques" in 1824, though it did not appear in print until 1829. For a discussion of the – at times heated – debate between Poncelet and Gergonne concerning the principles of "reciprocity" versus that of "duality", see [10]. Apparently, in the end, every geometer adopted both terms and used them interchangeably. Less known, perhaps, is the work of Bobillier, who (in 1828) was the first to replace the conic as "directrice" by a curve of arbitrary degree d, so that the polar curve of a point is a curve of degree $d - 1$, and the polar points of a line

[a] University of Oslo

is the intersection of the polar curves of the points on the line, hence equal to $(d-1)^2$ points, the number of base points of the corresponding pencil. For a summary of this early history, see [10] and [13].

In this note we introduce higher order polar loci as natural generalizations of the classical polar loci. The close connection between polar loci and dual varieties carries over to a connection between higher order polar loci and higher order dual varieties. In particular, the degree of the "top" (highest codimension) polar class of order k is equal to the degree of the kth dual variety.

In a series of papers, Bank, Giusti, Heintz, Mbakop, and Pardo introduced what they called "dual polar varieties" and used them to find real solutions to polynomial equations. These loci were studied further in [19] for real plane curves, and more generally in [26], under the name of "reciprocal polar varieties." These variants of polar loci are defined with respect to a quadric, in order to get a notion of orthogonality, and sometimes with respect also to a hyperplane at infinity. The orthogonality enables one to define *Euclidean normal bundles*, as studied in [4], [7], and [26]. For example, the (generic) Euclidean distance degree introduced in [7] is the degree of the "top" reciprocal polar class (see [26]). Note that the definition of reciprocal polar loci given here differs slightly from the one given in [26], but the degrees of the classes are the same.

In the next section we recall the definition of the classical polar loci and their classes, and their relation to the Mather Chern classes. In the third section we define the higher order polar loci and their classes and discuss how the latter can be computed. In the case of a smooth, k-regular variety, the kth order polar classes can be expressed in terms of its Chern classes and the hyperplane class, and there are also other cases when it is possible to compute these classes. In the fourth section, we introduce the higher order Euclidean normal bundles and use them, in the following section, to define higher order reciprocal polar loci and classes. The three last sections give examples of how to compute the degrees of the higher order polar and reciprocal polar classes in some special cases: curves, scrolls, and toric varieties.

Acknowledgments. I am grateful to Patrick Popescu-Pampu for alerting me to the work of Bobillier and to the thesis [10]. I thank the referee for asking a "natural question," which is answered in Theorem 3.5.

2 Polar loci and Mather Chern classes

Let V be a vector space of dimension $n+1$ over an algebraically closed field of characteristic 0. The polar loci of an m-dimensional projective variety

$X \subset \mathbb{P}(V)$ are defined as follows: Let $L_i \subset \mathbb{P}(V)$ be a linear subspace of codimension $m - i + 2$. The *ith polar locus* of X with respect to L_i is

$$M_i(L_i) := \overline{\{P \in X_{\mathrm{sm}} \mid \dim(T_{X,P} \cap L_i) \geq i - 1\}},$$

where $T_{X,P}$ denotes the projective tangent space to X at the (smooth) point P. Note that $M_0(L_0) = X$. The rational equivalence class $[M_i(L_i)]$ is independent of L_i for general L_i, and will be denoted $[M_i]$ and called the *ith polar class* of X [23, Prop. (1.2), p. 253]. The classes $[M_i]$ are *projective invariants* of X: the ith polar class of a (general) linear projection of X is the projection of the ith polar class of X, and the ith polar class of a (general) linear section is the linear section of the ith polar class (see [23, Thm. (4.1), p. 269; Thm. (4.2), p. 270].

Let us recall the definition of the Mather Chern class $c^{\mathrm{M}}(X)$ of an m-dimensional variety X. Let $\widetilde{X} \subseteq \mathrm{Grass}_m(\Omega^1_X)$ denote the Nash transform of X, i.e., $\widetilde{X}$ is the closure of the image of the rational section $X \dashrightarrow \mathrm{Grass}_m(\Omega^1_X)$ given by the locally free rank m sheaf $\Omega^1_X|_{X_{\mathrm{sm}}}$. The Mather Chern class of X is $c^{\mathrm{M}}(X) := \nu_*(c(\Omega^\vee) \cap [\widetilde{X}])$, where Ω is the tautological sheaf on $\mathrm{Grass}_m(\Omega^1_X)$ and $\nu\colon \widetilde{X} \to X$.

In 1978 we showed the following, generalizing the classical Todd–Eger formulas to the case of singular varieties:

Theorem 2.1 *[25, Thm. 3] The polar classes of X are given by*

$$[M_i] = \sum_{j=0}^{i} (-1)^j \binom{m-j+1}{m-i+1} h^{i-j} \cap c_j^M(X),$$

where $h := c_1(\mathcal{O}_{\mathbb{P}^n}(1))$ is the class of a hyperplane. Vice versa, the Mather Chern classes of X are given by

$$c_i^{\mathrm{M}}(X) = \sum_{j=0}^{i} (-1)^j \binom{m-j+1}{m-i+1} h^{i-j} \cap [M_j].$$

3 Higher order polar loci

Let $X \subset \mathbb{P}(V)$ be a projective variety of dimension m, and $P \in X$ a general point. There is a sequence of osculating spaces to X at P:

$$\{P\} \subseteq T_{X,P} = \mathrm{Osc}^1_{X,P} \subseteq \mathrm{Osc}^2_{X,P} \subseteq \mathrm{Osc}^3_{X,P} \subseteq \cdots \subseteq \mathbb{P}(V),$$

defined via the sheaves of principal parts of $\mathcal{O}_X(1)$ as follows. Let $V_X := V \otimes \mathcal{O}_X$ denote the trivial $(n+1)$-bundle on X, and consider the k-jet map (see e.g. [22, p. 492])

$$j_k\colon V_X \to \mathcal{P}^k_X(1).$$

Let $X_{k-\text{cst}} \subseteq X$ denote the open dense where the rank of j_k is constant, say equal to $m_k + 1$. Then for $P \in X_{k-\text{cst}}$, $\text{Osc}^k_{X,P} = \mathbb{P}((j_k)_P(V)) \subseteq \mathbb{P}(V)$. Note that $m_0 = 0$, $m_1 = m$, and $\dim \text{Osc}^k_{X,P} = m_k$.

Assume $m_k < n$. Let $L_{k,i} \subset \mathbb{P}(V)$ be a linear subspace of codimension $m_k - i + 2$. The *ith polar locus of order k* of X with respect to $L_{k,i}$ is

$$M_{k,i}(L_{k,i}) := \overline{\{P \in X_{k-\text{cst}} \mid \dim(\text{Osc}^k_{X,P} \cap L_{k,i}) \geq i-1\}}.$$

Note that $M_{1,i}(L_{1,i}) = M_i(L_{1,i})$ is the classical ith polar locus with respect to $L_{1,i}$, and that $M_{k,0}(L_{k,0}) = X$ for all k.

The $(m_k + 1)$-quotient $V_{X_{k-\text{cst}}} \to j_k(V_{X_{k-\text{cst}}})$ gives a rational section

$$X \dashrightarrow \text{Grass}_{m_k+1}(V_X) = X \times \text{Grass}_{m_k+1}(V).$$

Its closure $\widetilde{X}^k \subseteq X \times \text{Grass}_{m_k+1}(V)$, together with the projection map $\nu^k \colon \widetilde{X}^k \to X$, is called the *kth Nash transform* of $X \subset \mathbb{P}(V)$. Note that $\nu^1 = \nu \colon \widetilde{X}^1 = \widetilde{X} \to X$ is the usual Nash transform. Let $V_{\widetilde{X}^k} \to \mathcal{P}^k$ denote the induced $(m_k + 1)$-quotient. We call $\mathcal{P}^k$ *the kth order osculating bundle* of X. The projection map $\varphi_k \colon \widetilde{X}^k \to \text{Grass}_{m_k+1}(V)$ is the *kth associated map* of $X \subset \mathbb{P}(V)$ (see [28] and [21]).

Theorem 3.1 *The class of $M_{k,i}(L_{k,i})$ is independent of $L_{k,i}$, for general $L_{k,i}$, and is given by*

$$[M_{k,i}] = \nu^k_*(c_i(\mathcal{P}^k) \cap [\widetilde{X}^k]).$$

Proof The proof is similar to that of [23, Prop. (1.2), p. 253]. □

We call $[M_{k,i}]$ the *ith polar class of order k* of X.

Proposition 3.2 *Let $V' \subseteq V$ be a general subspace, with* $\dim V' > m_k + 1$, *and let $f \colon X \to \mathbb{P}(V')$ denote the corresponding linear projection. Then the image of the ith polar class of order k of $X \subset \mathbb{P}(V)$ is the same as the ith polar class of order k of $f(X) \subset \mathbb{P}(V')$.*

Proof The proof is similar to that of [23, Thm. (4.1), p. 269]. □

In the case $k = 1$, we proved [23, Thm. (4.2), p. 270] the following result, which does not have an obvious generalization to the case $k \geq 2$.

Proposition 3.3 *Let $\mathbb{P}(W) \subset \mathbb{P}(V)$ be a (general) linear subspace of codimension s. Set $Y := X \cap \mathbb{P}(W)$. For $0 \leq i \leq m - s$, the ith polar class of $Y \subset \mathbb{P}(W)$ is equal to the intersection of the ith polar class of X with $\mathbb{P}(W)$.*

Recall the definition of higher order dual varieties (introduced in [24]). The points of the dual projective space $\mathbb{P}(V)^\vee := \mathbb{P}(V^\vee)$ are the hyperplanes $H \subset \mathbb{P}(V)$. The *kth order dual variety of X* is

$$X^{(k)} := \overline{\{H \in \mathbb{P}(V)^\vee \mid H \supseteq \mathrm{Osc}^k_{X,P}, \text{ for some } P \in X_{k-\mathrm{cst}}\}}.$$

In particular, $X^{(1)} = X^\vee$ is the dual variety of X.

Set $\mathcal{K}^k := \mathrm{Ker}(V_{\widetilde{X}^k} \to \mathcal{P}^k)$; it is a $(n - m_k)$-bundle. Then $X^{(k)} \subset \mathbb{P}(V)^\vee$ is equal to the image of $\mathbb{P}((\mathcal{K}^k)^\vee) \subset \widetilde{X}^k \times \mathbb{P}(V)^\vee$ via the projection on the second factor. Let $p : \mathbb{P}((\mathcal{K}^k)^\vee) \to X$ and $q : \mathbb{P}((\mathcal{K}^k)^\vee) \to X^{(k)}$ denote the projections.

Proposition 3.4 *Let $L \subset \mathbb{P}(V)^\vee$ be a linear subspace of codimension $n - m_k + i - 1$ and set $L_{k,i} := L^\vee \subset \mathbb{P}(V)^{\vee\vee} = \mathbb{P}(V)$. Then*

$$p(q^{-1}(X^{(k)} \cap L)) = M_{k,i}(L_{k,i}).$$

Proof If L is general, then so is $L_{k,i}$. The dimension of $L_{k,i}$ is $n - 1 - \dim L = n - 1 - (m_k - i + 1) = n - m_k + i - 2$, hence the codimension of $L_{k,i}$ is $m_k - i + 2$. If $H \in X^{(k)} \cap L$, then there is a $P \in X_{k-\mathrm{cst}}$ such that $(P, H) \in \mathbb{P}((\mathcal{K}^k)^\vee)$. Hence $\mathrm{Osc}^k_{X,P} \subseteq H$ and $L_{k,i} \subseteq H$, so the intersection $\mathrm{Osc}^k_{X,P} \cap L_{k,i}$ has dimension $\geq i - 1$. Therefore $P \in M_{k,i}(L_{k,i})$. Conversely, if $P \in M_{k,i}(L_{k,i})$, then $\dim \mathrm{Osc}^k_{X,P} \cap L_{k,i} \geq i - 1$, so that $\mathrm{Osc}^k_{X,P}$ and $L_{k,i}$ do not span $\mathbb{P}(V)$. Hence there is a hyperplane H that contains both these spaces, and so $H \in X^{(k)} \cap L$. □

The "expected dimension" of $X^{(k)}$ is equal to the dimension of $\mathbb{P}((\mathcal{K}^k)^\vee)$, which is $m + n - m_k - 1$. Let $\delta_k := m + n - m_k - 1 - \dim X^{(k)}$ denote the *kth dual defect* of X. Let $\overline{X} \subset \mathbb{P}(V) \times \mathbb{P}(V)^\vee$ denote the image of $\mathbb{P}((\mathcal{K}^k)^\vee)$, and let $\overline{X^{(k)}} \subset \mathbb{P}(V)^\vee \times \mathbb{P}(V) \cong \mathbb{P}(V) \times \mathbb{P}(V)^\vee$ denote the corresponding variety constructed for $X^{(k)}$. It was shown in [24, Prop. 1, p. 336] that $\overline{X} \subseteq \overline{X^{(k)}}$, so that equality holds iff their dimensions are equal. In this case, we say that X is *k-reflexive*, and we then have $X = (X^{(k)})^{(k)}$. For example, a non-degenerate *curve* $X \subset \mathbb{P}(V)$ is $(n - 1)$-reflexive (see Section 6).

Theorem 3.5 *Assume X is k-reflexive. Then the degree of the ith polar class of order k of $X^{(k)}$ is given by*

$$\deg[M^\vee_{k,i}] = \deg[M_{k,m-\delta_k-i}].$$

In particular, the degree of $X^{(k)}$ is equal to the degree of the $(m - \delta_k)$th polar class $[M_{k,m-\delta_k}]$ of order k of X.

Proof For $k = 1$, this is [13, Thm. (4), p. 189], where it follows immediately from the definition of the ranks (corresponding to the degrees of the polar classes). Here we use Proposition 3.4: if h and $h^\vee$ denote the hyperplane classes of $\mathbb{P}(V)$ and $\mathbb{P}(V)^\vee$ respectively, then the class $[M_{k,m-\delta_k-i}]$ is the pushdown to X of the class $(h^\vee)^{n-m_k+m-\delta_k-i-1} \cap [\overline{X}]$. By definition, $\delta_k = m + n - 1 - m_k - m^\vee$, where $m^\vee := \dim X^{(k)}$, so this is the same as $(h^\vee)^{m^\vee - i} \cap [\overline{X}]$.

Hence its degree is the degree of $h^{n-1-m_k^\vee+i}(h^\vee)^{m^\vee-i}\cap[\overline{X}]$, where $m_k^\vee$ denotes the dimension of a general kth osculating space of $X^{(k)}$. Similarly, the class $[M_{k,i}^\vee]$ is the pushdown to $X^{(k)}$ of the class $h^{n-m_k^\vee+i-1}\cap[\overline{X^{(k)}}] = h^{n-m_k^\vee+i-1}\cap[\overline{X}]$ and has degree equal to the degree of $(h^\vee)^{m^\vee-i}h^{n-1-m_k^\vee+i}\cap[\overline{X}]$.

Note that $i = m - \delta_k$ is the largest $i \leq m$ such that $[M_{k,i}] \neq 0$. □

Assume $X \subset \mathbb{P}(V)$ is generically k-regular, i.e., the map $j_k\colon V_X \to \mathcal{P}_X^k(1)$ is generically surjective. The *kth Jacobian ideal* $\mathcal{J}_k$ is the $\binom{m+k}{k}$th Fitting ideal of $\mathcal{P}_X^k(1)$ [21, (2.9)]. Note that $\mathcal{J}_1 = F^{m+1}(\mathcal{P}_X^1(1)) = F^{m+1}(\mathcal{P}_X^1) = F^m(\Omega_X^1)$ is the ordinary Jacobian ideal. Let $\pi_k\colon X^k \to X$ denote the blow-up of $\mathcal{J}_k$. Then, by [29, 5.4.3], setting

$$\mathcal{A}_k := \operatorname{Ann}_{\pi_k^*\mathcal{P}_X^k(1)}(F^{\binom{m+k}{k}}(\pi_k^*\mathcal{P}_X^k(1))),$$

$\pi_k^*\mathcal{P}_X^k(1)/\mathcal{A}_k$ is a $\binom{m+k}{k}$-bundle. Let $\mathcal{I}_k$ denote the 0th Fitting ideal of the cokernel of the map $V_{X^k} \to \pi_k^*\mathcal{P}_X^k(1)/\mathcal{A}_k$ and $\overline{\pi}_k\colon \overline{X}^k \to X^k$ the blow-up of $\mathcal{I}_k$. Then [23, Lemma (1.1), p.252] the image $\overline{\mathcal{P}}^k$ of $\overline{V}_{\overline{X}^k}$ in $\overline{\pi}_k\pi_k^*\mathcal{P}_X^k(1)/\mathcal{A}_k$ is a $\binom{m+k}{k}$-bundle. Hence we get a $\binom{m+k}{k}$-quotient $V_{\overline{X}^k} \to \overline{\mathcal{P}}^k$ which agrees with $V_X \to \mathcal{P}_X^k(1)$ above $X_{k-\mathrm{cst}}$. Note that, as discussed in the case $k = 1$ in [23, p. 255], the map $\pi_k \circ \overline{\pi}_k\colon \overline{X}^k \to X$ factors via the kth Nash transform $\nu^k\colon \widetilde{X}^k \to X$. In particular, we have

$$[M_{k,i}] = \pi_{k*}\overline{\pi}_{k*}(c_i(\overline{\mathcal{P}}^k)\cap[\overline{X}^k]).$$

In some cases, the Chern classes of $\overline{\mathcal{P}}^k$ can be computed in terms of the Chern classes of $\mathcal{P}_X^k(1)$ and the invertible sheaves $\mathcal{J}_k\mathcal{O}_{\overline{X}^k}$ and $\mathcal{I}_k\mathcal{O}_{\overline{X}^k}$. We shall see in Section 6 that this is the case when X is a curve. Another case is the following.

Proposition 3.6 *Assume $X \subset \mathbb{P}(V)$ is smooth and generically k-regular, and that $m_k = n - 1$. Then*

$$[M_{k,i}] = c_1(\mathcal{P}_X^k(1))^i \cap [X] - \sum_{j=0}^{i-1}\binom{i}{j}c_1(\mathcal{P}_X^k(1))^j \cap s_{m-i+j}(I_k, X),$$

where $s_{m-i+j}(I_k, X) = -\overline{\pi}_{k*}(c_1(\mathcal{I}_k\mathcal{O}_{\overline{X}^k})^{i-j}\cap[\overline{X}^k])$ denote the Segre classes of the subscheme $I_k \subset X$ defined by the ideal $\mathcal{I}^k := F^0(\mathrm{Coker}\ j^k)$.

Proof Since $m_k = n-1$, the (locally free) kernel $\overline{\mathcal{K}}^k := \mathrm{Ker}(V_{\overline{X}^k} \to \overline{\mathcal{P}}^k)$ has rank 1, hence $c_i(\overline{\mathcal{P}}^k) = c_1(\overline{\mathcal{P}}^k)^i$ holds. We also have $\bigwedge^n\overline{\mathcal{P}}^k \cong \bigwedge^n\overline{\pi}_k^*\mathcal{P}_X^k(1)\otimes\mathcal{I}^k\mathcal{O}_{\overline{X}^k}$, hence $c_1(\overline{\mathcal{P}}^k) = c_1(\overline{\pi}_k^*\mathcal{P}_X^k(1)) + c_1(\mathcal{I}^k\mathcal{O}_{\overline{X}^k})$, so the result follows by applying the projection formula. □

4 Higher order Euclidean normal bundles

Let V and V' be vector spaces of dimensions $n+1$ and n, and $V \to V'$ a surjection. Set $H_\infty := \mathbb{P}(V') \subset \mathbb{P}(V)$, and call it the *hyperplane at infinity*. A non-degenerate quadratic form on V' defines an isomorphism $V' \cong (V')^\vee$ and a non-singular quadric $Q_\infty \subset H_\infty$.

Let $L' := \mathbb{P}(W) \subset \mathbb{P}(V')$ be a linear space, and set

$$K := \mathrm{Ker}(V'^\vee \cong V' \to W) \subset V'^\vee.$$

The *polar* of L' with respect to Q_∞ is the linear space $L'^\perp := \mathbb{P}(K^\vee) \subset \mathbb{P}(V')$.

Given a linear space $L \subset \mathbb{P}(V)$ and a point $P \in L$, define the *orthogonal space to L at P* by

$$L_P^\perp := \langle P, (L \cap H_\infty)^\perp \rangle.$$

If $L \nsubseteq H_\infty$, the dimension of $(L \cap H_\infty)^\perp$ is equal to $n-1-(\dim L - 1) - 1 = n - \dim L - 1$, and the dimension of $L_P^\perp$ is $n - \dim L$ if $P \notin (L \cap H_\infty)^\perp$ and $n - \dim L - 1$ if $P \in (L \cap H_\infty)^\perp$. Note that if $P \in (L \cap H_\infty)^\perp$, then $P \in L \cap H_\infty$ and hence $P \in P^\perp$. This implies that $P \in Q_\infty$ and $L \cap H_\infty \subseteq T_{Q_\infty, P}$.

Let $X \subset \mathbb{P}(V)$ be a variety of dimension m. Assume $X \nsubseteq H_\infty$. Let $P \in X$ be a non-singular point. The *Euclidean normal space* to X at P with respect to H_∞ and Q_∞ is the linear space $N_{X,P} := T_{X,P}^\perp$. Let $X_1 \subseteq X$ denote the set of non-singular points P such that $T_{X,P} \nsubseteq H_\infty$, and such that $T_{X,P} \cap H_\infty \nsubseteq T_{Q_\infty, P}$ if $P \in Q_\infty$. Then $\dim N_{X,P} = n - m$ for $P \in X_1$.

By replacing tangent spaces by higher order osculating spaces we may define higher order Euclidian normal spaces. Let $k \geq 1$ be such that the dimension m_k of a general kth order osculating space to X is less than n. For $P \in X_{k-\mathrm{cst}}$, set

$$N_{X,P}^k := (\mathrm{Osc}_{X,P}^k)^\perp.$$

Let $X_k \subseteq X_{k-\mathrm{cst}}$ denote the set of points P such that $\mathrm{Osc}_{X,P}^k \nsubseteq H_\infty$, and such that $\mathrm{Osc}_{X,P}^k \cap H_\infty \nsubseteq T_{Q_\infty, P}$ if $P \in Q_\infty$. Then $\dim N_{X,P}^k = n - m_k$ for $P \in X_k$.

With notations as in Section 3, consider the exact sequence

$$0 \to \mathcal{K}^k \to V_{\tilde{X}^k} \to \mathcal{P}^k \to 0.$$

Assume the hyperplane $H_\infty = \mathbb{P}(V') \subset \mathbb{P}(V)$ is *general*. Then it follows from [22, Lemma (4.1), p. 483] that the dual $V'^\vee_{\tilde{X}^k} \to (\mathcal{K}^k)^\vee$ of the composed map $\mathcal{K}^k \to V_{\tilde{X}^k} \to V'_{\tilde{X}^k}$ is surjective.

Set

$$\mathcal{X}_k := \overline{\{(P, P')|P \in X_k, P' \in N^k_{X,P}\}} \subset X \times \mathbb{P}(V).$$

Let $p_k : \mathcal{X}_k \to X$ and $q_k : \mathcal{X}_k \to \mathbb{P}(V)$ denote the projections. The dimension of $\mathcal{X}_k$ is $n - m_k + m$. The sheaf $p_{k*}q_k^*\mathcal{O}_{\mathbb{P}(V)}(1)$ is generically locally free, with rank $n - m_k + 1$. Let $\nu_k : \widetilde{X}_k \to X$ be the "Nash transform" such that $\nu_k^* p_{k*}q_k^*\mathcal{O}_{\mathbb{P}(V)}(1)$ admits a locally free rank $n - m_k + 1$ quotient bundle $\mathcal{E}^k$. By the definition of the normal spaces, it follows that $\mathcal{E}^k$ restricted to X_k splits as $(\mathcal{K}^k)^\vee|_{X_k} \oplus \mathcal{O}_{X_k}(1)$, and that, by replacing if necessary $\nu^k : \widetilde{X}^k \to X$ and $\nu_k : \widetilde{X}_k \to X$ by a further Nash transform (by abuse of notation, also denoted $\nu^k : \widetilde{X}^k \to X$), the map $V_{\widetilde{X}^k} \to (\mathcal{K}^k)^\vee \oplus \nu^{k*}\mathcal{O}_X(1)$ factors via the surjection $V_{\widetilde{X}^k} \to \mathcal{E}^k$. We call $\mathcal{E}^k$ the *kth Euclidean normal bundle* to X. The difference between $\mathcal{E}^k$ and $(\mathcal{K}^k)^\vee \oplus \nu^{k*}\mathcal{O}_X(1)$ measures where $V_{\widetilde{X}_k} \to (\mathcal{K}^k)^\vee \oplus \nu^{k*}\mathcal{O}_X(1)$ is not surjective.

Proposition 4.1 *For a generic choice of a hyperplane H_∞ and a quadric $Q_\infty \subset H_\infty$, the kth Euclidean normal bundle $\mathcal{E}^k$ splits as $(\mathcal{K}^k)^\vee \oplus \nu^{k*}\mathcal{O}_X(1)$ on an appropriate Nash modification $\nu^k : \widetilde{X}^k \to X$.*

Proof We need to identify the points $P \in \widetilde{X}^k$ such that $\alpha : V_{\widetilde{X}^k} \to (\mathcal{K}^k)^\vee \oplus \nu^{k*}\mathcal{O}_X(1)$ is not surjective at P. First of all, by genericity of H_∞, we may assume that H_∞ intersects X transversally, i.e., H_∞ intersects each Whitney stratum of X transversally. This means that H_∞ does not contain a tangent space, nor a limit tangent space, to X at any point of $X \cap H_\infty$. (This is the content of the surjectivity of $V'_{\widetilde{X}^1} \to (\mathcal{K}^1)^\vee$.) This implies that H_∞ does not contain a kth order osculating space, nor a limit of such a space, at ay point of $X \cap H_\infty$. Hence α is surjective at all points of $\widetilde{X}^k$ above $X \setminus H_\infty$. Now consider $P \in \widetilde{X}^k$ mapping to $X \cap H_\infty$. The map α is not surjective at P only if $\nu^k(P) \in (\widetilde{O}_P \cap H_\infty)^\perp$, where $\widetilde{O}_P$ is the limit kth osculating space to X at $\nu^k(P)$ corresponding to P. But this implies that $\nu^k(P) \in \nu^k(P)^\perp$, hence $\nu^k(P) \in Q_\infty$. Therefore $\nu^k(P)^\perp = T_{Q_\infty, \nu^k(P)}$, and hence $T_{Q_\infty, \nu^k(P)} \supseteq \widetilde{O}_P \cap H_\infty \supseteq \widetilde{T}_P \cap H_\infty$. But we may assume this does not happen: by [12, Cor. 4, p. 291], a general quadric $Q_\infty \subset H_\infty$ intersects each of the Whitney strata of $X \cap H_\infty$ transversally. □

The morphism $\varphi_k \colon \widetilde{X}^k \to \mathrm{Grass}_{m_k+1}(V)$ corresponding to the quotient $V_{\widetilde{X}^k} \to \mathcal{P}^k$ is the kth order associated map of $X \subset \mathbb{P}(V)$. The *kth order associated normal map* is the morphism

$$\psi_k \colon \widetilde{X}^k \to \mathrm{Grass}_{n-m_k+1}(V),$$

defined by the quotient $V_{\widetilde{X}^k} \to \mathcal{E}^k$.

5 Higher order reciprocal polar loci

Instead of imposing conditions on the osculating spaces of a variety, one can similarly impose conditions on the higher order Euclidean normal spaces. In what follows, we assume that H_∞ and Q_∞ are general with respect to X.

For each $i = 0, \ldots, m$ and $k \geq 1$, let $L^{k,i} \subset \mathbb{P}(V)$ be a general linear space of codimension $n - m_k + i$ and define the *ith reciprocal polar locus of order* k with respect to $L^{k,i}$ to be

$$M_{k,i}^{\perp}(L^{k,i}) := \overline{\{P \in X_k \mid N_{X,P}^k \cap L^{k,i} \neq \emptyset\}}.$$

Note that, with $p_k : \mathcal{X}_k \to X$ and $q_k : \mathcal{X}_k \to \mathbb{P}(V)$ as in the previous section, $M_{k,i}^{\perp}(L^{k,i}) = p_k q_k^{-1}(L^{k,i})$, and $M_{k,0}^{\perp}(L^{k,0}) = X$ for all $k \geq 1$.

Theorem 5.1 *The class of $M_{k,i}^{\perp}(L_{k,i})$ is independent of $L_{k,i}$, for general $L_{k,i}$, and is given by*

$$[M_{k,i}^{\perp}] = \nu_*^k\big(\{c(\mathcal{P}^k)s(\nu^{k*}\mathcal{O}_X(1))\}_i \cap [\widetilde{X}^k]\big).$$

Proof Let $\Sigma_{k,i}(L^{k,i}) \subset \mathrm{Grass}_{n-m_k+1}(V)$ denote the special Schubert variety consisting of the set of $(n - m_k)$-planes that meet the $(m_k - i)$-plane $L^{k,i}$. Then $M_{k,i}^{\perp}(L^{k,i}) = \psi_k^{-1}(\Sigma_{k,i}(L^{k,i}))$. The first statement follows, with the same reasoning as in [23, Prop. (1.2), p. 253].

Let $L^{k,i} = \mathbb{P}(W)$ and set $W' = \mathrm{Ker}(V \to W)$. The condition $N_{X,P}^k \cap L^{k,i} \neq \emptyset$ means that the rank of the composed map $W'_{\widetilde{X}^k} \to \mathcal{E}^k$ is $\leq n - m_k$. By Porteous' formula (see [8, Thm. 14.4, p. 254]), $M_{k,i}^{\perp}(L^{k,i})$ has class

$$[M_{k,i}^{\perp}] = \nu_*^k(s_i(\mathcal{E}^k) \cap [\widetilde{X}^k]) = \nu_*^k\big(\{s((\mathcal{K}^k)^\vee)s(\nu^{k*}\mathcal{O}_X(1))\}_i \cap [\widetilde{X}^k]\big).$$

Since the Segre class $s((\mathcal{K}^k)^\vee)$ is equal to the Chern class $c(\mathcal{P}^k)$, the formula follows. □

Corollary 5.2 *Let $h := c_1(\mathcal{O}_X(1))$ denote the hyperplane class. We have*

$$[M_{k,i}^{\perp}] = \sum_{j=0}^{i} h^{i-j} \cap [M_{k,j}],$$

and hence

$$\deg[M_{k,i}^{\perp}] = \sum_{j=0}^{i} \deg[M_{k,j}].$$

Proof This follows, using Theorem 3.1, since

$$s(\mathcal{O}_{\widetilde{X}^k}(1)) = 1 + c_1(\mathcal{O}_{\widetilde{X}^k}(1)) + c_1(\mathcal{O}_{\widetilde{X}^k}(1))^2 + \cdots$$

and $\mathcal{O}_{\widetilde{X}^k}(1) = \nu^{k*}\mathcal{O}_X(1)$. □

Remark 5.3 The (generic) *Euclidean distance degree* of $X \subset \mathbb{P}(V)$, introduced in [7], can be interpreted as the degree of $[M^{\perp}_{1,m}]$. The above formula says that this degree is equal to the sum of the degrees of the polar classes of X, as stated in [7, Thm. 5.4, p. 126] (see also [26]).

6 Curves

Let $X \subset \mathbb{P}(V)$ be a non-degenerate curve. At a general point $P \in X$ we have a complete flag

$$\{P\} \subseteq T_{X,P} = \mathrm{Osc}^1_{X,P} \subset \mathrm{Osc}^2_{X,P} \subset \cdots \subset \mathrm{Osc}^{n-1}_{X,P} \subset \mathbb{P}(V).$$

In this case $\nu^k : \widetilde{X}^k = \widetilde{X} \to X$ is the normalization of X, $m_k = k$, and $\dim L_{k,1} = n - k - 1$.

Note that the locus $M_{k,1}(L_{k,1})$ maps to kth order hyperosculating points on the image of X under the linear projection $f\colon \mathbb{P}(V) \dashrightarrow \mathbb{P}(V')$ with center $L_{k,1}$, Namely, let $P \in M_{k,1}(L_{k,1})$. Then $\mathrm{Osc}_{X,P} \cap L_{k,1} \neq \emptyset$, and hence $f(\mathrm{Osc}_{X,P})$ has dimension $< k$. This means that the kth jet map of $f(X)$ has rank $< k+1$ at the point $f(P)$. For example, for $k = 1$, $f(P)$ is a cusp on $f(X)$, and for $k = 2$, $f(P)$ is an inflection point.

Recall that the *kth rank* r_k of the curve X is defined as the degree of the kth osculating developable of X, i.e., as the number of kth order osculating spaces intersecting a given, general linear space of dimension $n - k - 1$ [2, pp. 199–200]. Hence $r_k = \deg \nu^k_*(c_1(\mathcal{P}^k) \cap [\widetilde{X}])$, and r_k is also equal to the degree of the *kth associated curve* of X (see [22, Prop. (3.1), p. 480] and [28], [21]).

Let $d := r_0$ denote the degree of X. Since $[M_{k,1}] = \nu^k_*(c_1(\mathcal{P}^k) \cap [\widetilde{X}])$, we get $\deg[M_{k,1}] = r_k$. We have $r_k = (k+1)(d+k(g-1)) - \sum_{j=0}^{k-1}(k-j)\kappa_j$, where g is the genus of $\widetilde{X}$ and κ_j is the jth stationary index of X (see [22, Thm. (3.2), p. 481]). It follows that $\deg[M^{\perp}_{k,1}] = \deg[M_{k,0}] + \deg[M_{k,1}] = d + r_k$. For more on ranks, duality, projections, and sections, see [22]. For example, the ranks of the *strict dual curve* $X^{(n-1)} \subset \mathbb{P}(V)^{\vee}$ satisfy $r_k(X^{(n-1)}) = r_{n-k-1}(X)$.

Example 6.1 If $X \subset \mathbb{P}(V)$ is a *rational normal curve* of degree n, then $\deg[M_{k,1}] = r_k = (k+1)(n-k)$ and $\deg[M^{\perp}_{k,1}] = n + r_k = n + (k+1)(n-k)$.

Note that X is *toric* and $(n-1)$*-self dual*: $X^{(n-1)} \subset \mathbb{P}(V)^{\vee}$ is a rational normal curve of degree n.

7 Scrolls

Ruled varieties – scrolls – are examples of varieties that are not generically k-regular, for $k \geq 2$. Hence we cannot hope to use the bundles of principal parts to compute the degrees of the higher polar classes for such varieties. However, in several cases we have results. In particular, since the degree of the "top" kth order polar class is the same as the degree of the kth dual variety, we can consider the following situations.

7.1 Rational normal scrolls

Let $X = \mathbb{P}(\oplus_{i=1}^m \mathcal{O}_{\mathbb{P}^1}(d_i)) \subset \mathbb{P}(V)$ be a rational normal scroll of type $(d_1,\ldots,d_m)$, with $m \geq 2$, $0 < d_1 \leq \cdots \leq d_m$, and $n = \sum_{i=1}^m (d_i + 1) - 1$. Let $d := d_1 + \cdots + d_m$ denote the degree of X. The higher order dual varieties of rational normal scrolls were studied in [27]. For example, for k such that $k \leq d_1$, we have

$$\deg[M_{k,m}] = \deg X^{(k)} = kd - k(k-1)m,$$

where the first equality follows from Theorem 3.5 and the second was computed in [5, Prop. 4.1, p. 389]. In the case $k = d_1 = \cdots = d_m$, this gives

$$\deg[M_{k,m}] = \deg X^{(k)} = kd - k(k-1)m = md_1^2 - d_1(d_1-1)m = md_1 = d.$$

Indeed, in this case X is k-selfdual: $X^{(k)} \subset \mathbb{P}(V)^\vee$ is a rational normal scroll of the same type as X.

We refer to [27] and [5] for other cases.

7.2 Elliptic normal surface scrolls

Higher order dual varieties of elliptic normal scrolls were studied in [16]. Let C be a smooth elliptic curve and $\mathcal{E}$ a rank two bundle on C. Assume $H^0(C,\mathcal{E}) \neq 0$, but that $H^0(C,\mathcal{E} \otimes \mathcal{L}) = 0$ for all invertible sheaves $\mathcal{L}$ with $\deg \mathcal{L} < 0$. Then let $e := \deg \mathcal{E}$ denote the Atiyah invariant. There are two cases. Either $\mathcal{E}$ is decomposable – then $\mathcal{E} = \mathcal{O}_C \oplus \mathcal{L}$ and $e = -\deg \mathcal{L} \geq 0$, or $\mathcal{E}$ is indecomposable, in which case $e = 0$ or $e = -1$. Now let $\mathcal{M}$ be an invertible sheaf on C of degree d. If $d \geq e + 3$, then $\mathcal{O}_{\mathbb{P}(\mathcal{E}\otimes\mathcal{M})}(1)$ yields an embedding of $\mathbb{P}(\mathcal{E} \otimes \mathcal{M})$ as a linearly normal scroll $X \subset \mathbb{P}(V) \cong \mathbb{P}^{2d-e-1}$ of degree $2d - e$. The following holds [16, Thm. 1, Thm. 2, p. 150].

If $\mathcal{E}$ is decomposable:

(i) if $e = 0$, then $\deg[M_{d-1,2}] = \deg X^{(d-1)} = 2d(d-1)$;
(ii) if $e = 1$, then $\deg[M_{d-2,2}] = \deg X^{(d-2)} = 2d^2 - 5d + 2$;
(iii) if $e \geq 2$, then $\deg[M_{d-2,2}] = \deg X^{(d-2)} = d(d-1)$.

If $\mathcal{E}$ is indecomposable:

(i) if $e = -1$, then $\deg[M_{d-1,2}] = \deg X^{(d-1)} = 2d^2 - 3$;
(ii) if $e = 0$, then $\deg[M_{d-1,2}] = \deg X^{(d-1)} = 2d^2 - d - 2$.

7.3 Scrolls over smooth curves

Consider now the more general situation where $X \subset \mathbb{P}(V)$ is a smooth scroll of dimension m and degree d over a smooth curve C of genus g. We showed in [14] that in this case, for k such that $m_k = km$, the kth jet map j_k factors via a bundle $\mathcal{P}^k$ of rank $km + 1$. Moreover, the Chern classes of $\mathcal{P}^k$ can be expressed in terms of d, m, k, g, the class of a fiber of $X \to C$, and the class of a hyperplane section of X. Thus we can get a formula for the "top" kth order polar class in terms of these numbers and the Segre classes of the inflection loci of X, see [14] for details.

7.4 Scrolls over smooth varieties

When we replace the curve C by a higher-dimensional smooth projective variety of dimension r, the situation gets more complicated, but it is again possible to find a kth osculating bundle $\mathcal{P}^k$ of rank $(m-r)\binom{r+k-1}{r} + \binom{r+k}{r}$, whose Chern classes can be computed, see [15].

8 Toric varieties

The Schwartz–MacPherson Chern class of a toric variety X with torus orbits $\{X_\alpha\}_\alpha$ is equal to, by Ehlers' formula (see [9, Lemma, p. 109], [3, Thm., p. 188], [18, Thm. 1.1; Cor. 1.2 (a)], [1, Thm. 4.2, p. 410]),

$$c^{\mathrm{SM}}(X) = \sum_\alpha [\overline{X}_\alpha].$$

This implies (proof by the definition of $c^{\mathrm{SM}}(X)$ and induction on $\dim X$) that the Mather Chern class of a toric variety X is equal to

$$c^{\mathrm{M}}(X) = \sum_\alpha \mathrm{Eu}_X(X_\alpha)[\overline{X}_\alpha],$$

where $\mathrm{Eu}_X(X_\alpha)$ denotes the value of the local Euler obstruction of X at a point in the orbit X_α.

Therefore the polar classes of a toric variety X of dimension m are

$$[M_i] = \sum_{j=0}^{i} (-1)^j \binom{m-j+1}{m-i+1} h^{i-j} \cap \sum_\alpha \mathrm{Eu}_X(X_\alpha)[\overline{X}_\alpha],$$

and the reciprocal polar classes

$$[M_i^{\perp}] = \sum_{\ell=0}^{i} h^{i-\ell} \sum_{j=0}^{\ell} (-1)^j \binom{m-j+1}{m-\ell+1} h^{\ell-j} \cap \sum_{\alpha} \mathrm{Eu}_X(X_\alpha)[\overline{X}_\alpha]$$

$$= \sum_{j=0}^{i} (-1)^j \sum_{\ell=0}^{i-j} \binom{m-j+1}{\ell} h^{i-j} \cap \sum_{\alpha} \mathrm{Eu}_X(X_\alpha)[\overline{X}_\alpha],$$

where the second sum in each expression is over α such that $\operatorname{codim} X_\alpha = j$.

It follows that if $X = X_\Pi$ is a projective toric variety corresponding to a convex lattice polytope Π, then (cf. [17, Thm. 1.4, p. 2042])

$$\deg[M_i] = \sum_{j=0}^{i} (-1)^j \binom{m-j+1}{m-i+1} \mathrm{EVol}^j(\Pi),$$

and

$$\deg[M_i^{\perp}] = \sum_{j=0}^{i} (-1)^j \sum_{\ell=0}^{i-j} \binom{m-j+1}{\ell} \mathrm{EVol}^j(\Pi),$$

where $\mathrm{EVol}^j(\Pi) := \sum_\alpha \mathrm{Eu}_X(X_\alpha)\,\mathrm{Vol}(F_\alpha)$ denotes the sum of the lattice volumes of the faces F_α of Π of codimension j weighted by the local Euler obstruction $\mathrm{Eu}_X(X_\alpha)$ of X at a point of X_α, where X_α is the torus orbit of X corresponding to the face F_α. In particular, we get

$$\deg[M_m^{\perp}] = \sum_{j=0}^{m} (-1)^j (2^{m-j+1} - 1)\,\mathrm{EVol}^j(\Pi),$$

(cf. [11, Thm. 1.1, p. 215]).

Example 8.1 Nødland [20, 4.1] studied weighted projective threefolds. In particular he showed the following. Assume a, b, c are positive, pairwise relatively prime integers. The weighted projective threefold $\mathbb{P}(1, a, b, c)$ is the toric variety corresponding to the lattice polyhedron $\Pi := \mathrm{Conv}\{(0,0,0), (bc,0,0), (0,ac,0), (0,0,bc)\}$. It has isolated singularities at the three points corresponding to the three vertices other than $(0,0,0)$. Let $\mathrm{Vol}^j(\Pi)$ denote the sum of the lattice volumes of the faces of Π of codimension j. We have $\mathrm{Vol}^0(\Pi) = a^2b^2c^2$, $\mathrm{Vol}^1(\Pi) = abc(1+a+b+c)$, $\mathrm{Vol}^2(\Pi) = a+b+c+bc+ac+ab$, and $\mathrm{Vol}^3(\Pi) = 4$. The algorithm given in [20, A.2] can be used to compute the local Euler obstruction at the singular points. Nødland gave several examples of integers a, b, c such that the local Euler obstruction at each singular point is 1, thus providing counterexamples to a conjecture of Matsui and Takeuchi [17, p. 2063]. For example this holds for $a = 2, b = 3, c = 5$, so in this case $\mathrm{EVol}^j(\Pi) = \mathrm{Vol}^j(\Pi)$ and we can compute

$$\deg[M_0] = 900,\ \deg[M_1] = 3270,\ \deg[M_2] = 4451,\ \deg[M_3] = 2688,$$

and hence

$$\deg[M_3^\perp] = 11309.$$

In general, formulas for the degrees of higher order polar classes and reciprocal polar classes of toric varieties are not known. However, in some cases, they can be found. As we have seen, for smooth varieties (not necessarily toric), if the kth jet map is surjective (i.e., the embedded variety is k-regular), then the classes $[M_{k,i}]$ can be expressed in terms of Chern classes of the sheaf of principal parts $\mathcal{P}_X^k(1)$. Hence, in the case of a k-regular toric variety, they can be expressed in terms of lattice volumes of the faces of the corresponding polytope. The following two examples were worked out in [6, Rmk. 3.5, p. 385; Thm. 3.7, p. 387].

Example 8.2 Let $\Pi \subset \mathbb{R}^2$ be a smooth lattice polygon with edge lengths $\geq k$. Then X_Π is k-regular and

$$\deg[M_{k,1}] = \binom{k+2}{2} \operatorname{Vol}^0(\Pi) - \binom{k+2}{3} \operatorname{Vol}^1(\Pi),$$

$$\deg[M_{k,2}] = \binom{k+3}{4}\left(3 \operatorname{Vol}^0(\Pi) - 2k \operatorname{Vol}^1(\Pi) - \tfrac{1}{3}(k^2-4) \operatorname{Vol}^2(\Pi) + 4(k^2-1)\right).$$

Example 8.3 Let $\Pi \subset \mathbb{R}^3$ be a smooth lattice polyhedron with edge lengths ≥ 2. Then X_Π is 2-regular and

$$\deg[M_{2,1}] = 4 \operatorname{Vol}^0(\Pi) - \operatorname{Vol}^1(\Pi),$$

$$\deg[M_{2,2}] = 36 \operatorname{Vol}^0(\Pi) - 27 \operatorname{Vol}^1(\Pi) + 6 \operatorname{Vol}^2(\Pi) + 18 \operatorname{Vol}^0(\Pi_0) + 9 \operatorname{Vol}^1(\Pi_0),$$

$$\begin{aligned} \deg[M_{2,3}] = {} & 62 \operatorname{Vol}^0(\Pi) - 57 \operatorname{Vol}^1(\Pi) + 28 \operatorname{Vol}^2(\Pi) - 8 \operatorname{Vol}^3(\Pi) \\ & + 58 \operatorname{Vol}^0(\Pi_0) + 51 \operatorname{Vol}^1(\Pi_0) + 20 \operatorname{Vol}^2(\Pi_0), \end{aligned}$$

where $\Pi_0 := \operatorname{Conv}(\operatorname{int}(\Pi) \cap \mathbb{Z}^3)$ is the convex hull of the interior lattice points of Π.

Recall [5, Def. 1.1, p. 1760] that a variety $X \subset \mathbb{P}(V)$ is said to be *k-selfdual* if there exists a linear isomorphism $\phi : \mathbb{P}(V) \xrightarrow{\sim} \mathbb{P}(V)^\vee$ such that $\phi(X) = X^{(k)}$. If $m = \dim X$, then $\deg[M_{k,m}]$ is the degree of the k-dual variety $X^{(k)}$. So if X is k-selfdual, then $\deg[M_{k,m}] = \deg X$. We refer to [5] for examples of toric k-selfdual varieties.

References

[1] Aluffi, Paolo. 2006. Classes de Chern des variétés singulières, revisitées. *C. R. Math. Acad. Sci. Paris*, **342**(6), 405–410.

[2] Baker, H. F. 2010. *Principles of geometry. Volume 5. Analytical principles of the theory of curves*. Cambridge Library Collection. Cambridge University Press, Cambridge. Reprint of the 1933 original.

[3] Barthel, Gottfried, Brasselet, Jean-Paul, and Fieseler, Karl-Heinz. 1992. Classes de Chern des variétés toriques singulières. *C. R. Acad. Sci. Paris Sér. I Math.*, **315**(2), 187–192.

[4] Catanese, Fabrizio, and Trifogli, Cecilia. 2000. Focal loci of algebraic varieties. I. Special issue in honor of Robin Hartshorne. *Comm. Algebra*, **28**(12), 6017–6057.

[5] Dickenstein, Alicia, and Piene, Ragni. 2017. Higher order selfdual toric varieties. *Ann. Mat. Pura Appl. (4)*, **196**(5), 1759–1777.

[6] Dickenstein, Alicia, Di Rocco, Sandra, and Piene, Ragni. 2014. Higher order duality and toric embeddings. *Ann. Inst. Fourier (Grenoble)*, **64**(1), 375–400.

[7] Draisma, Jan, Horobeţ, Emil, Ottaviani, Giorgio, Sturmfels, Bernd, and Thomas, Rekha R. 2016. The Euclidean distance degree of an algebraic variety. *Found. Comput. Math.*, **16**(1), 99–149.

[8] Fulton, William. 1984. *Intersection theory*. Ergebnisse der Mathematik und ihrer Grenzgebiete (3) [Results in Mathematics and Related Areas (3)], vol. 2. Springer-Verlag, Berlin.

[9] Fulton, William. 1993. *Introduction to toric varieties*. Annals of Mathematics Studies, vol. 131. Princeton University Press, Princeton, NJ. The William H. Roever Lectures in Geometry.

[10] Haubrichs dos Santos, Cleber. 2015. *Étienne Bobillier (1798–1840): parcours mathématique, enseignant et professionnel*. Université de Lorraine. Thesis (Ph.D.).

[11] Helmer, Martin, and Sturmfels, Bernd. 2018. Nearest points on toric varieties. *Math. Scand.*, **122**(2), 213–238.

[12] Kleiman, Steven L. 1974. The transversality of a general translate. *Compositio Math.*, **28**, 287–297.

[13] Kleiman, Steven L. 1986. Tangency and duality. Pages 163–225 of: *Proceedings of the 1984 Vancouver conference in algebraic geometry*. CMS Conf. Proc., vol. 6. Amer. Math. Soc., Providence, RI.

[14] Lanteri, Antonio, Mallavibarrena, Raquel, and Piene, Ragni. 2008. Inflectional loci of scrolls. *Math. Z.*, **258**(3), 557–564.

[15] Lanteri, Antonio, Mallavibarrena, Raquel, and Piene, Ragni. 2012. Inflectional loci of scrolls over smooth, projective varieties. *Indiana Univ. Math. J.*, **61**(2), 717–750.

[16] Mallavibarrena, Raquel, and Piene, Ragni. 1991. Duality for elliptic normal surface scrolls. Pages 149–160 of: *Enumerative algebraic geometry (Copenhagen, 1989)*. Contemp. Math., vol. 123. Amer. Math. Soc., Providence, RI.

[17] Matsui, Yutaka, and Takeuchi, Kiyoshi. 2011. A geometric degree formula for A-discriminants and Euler obstructions of toric varieties. *Adv. Math.*, **226**(2), 2040–2064.

[18] Maxim, Laurenţiu G., and Schürmann, Jörg. 2015. Characteristic classes of singular toric varieties. *Comm. Pure Appl. Math.*, **68**(12), 2177–2236.

[19] Mork, Heidi Camilla, and Piene, Ragni. 2008. Polars of real singular plane curves. Pages 99–115 of: *Algorithms in algebraic geometry*. IMA Vol. Math. Appl., vol. 146. Springer, New York.

[20] Nødland, Bernt Ivar Utstøl. 2018. Local Euler obstructions of toric varieties. *J. Pure Appl. Algebra*, **222**(3), 508–533.

[21] Piene, Ragni. 1976. *Plücker formulas*. ProQuest LLC, Ann Arbor, MI. Thesis (Ph.D.)–Massachusetts Institute of Technology.

[22] Piene, Ragni. 1977. Numerical characters of a curve in projective n-space. Pages 475–495 of: *Real and complex singularities (Proc. Ninth Nordic Summer School/NAVF Sympos. Math., Oslo, 1976)*.

[23] Piene, Ragni. 1978. Polar classes of singular varieties. *Ann. Sci. École Norm. Sup. (4)*, **11**(2), 247–276.

[24] Piene, Ragni. 1983. A note on higher order dual varieties, with an application to scrolls. Pages 335–342 of: *Singularities, Part 2 (Arcata, Calif., 1981)*. Proc. Sympos. Pure Math., vol. 40. Amer. Math. Soc., Providence, RI.

[25] Piene, Ragni. 1988. Cycles polaires et classes de Chern pour les variétés projectives singulières. Pages 7–34 of: *Introduction à la théorie des singularités, II*. Travaux en Cours, vol. 37. Hermann, Paris.

[26] Piene, Ragni. 2015. Polar varieties revisited. Pages 139–150 of: *Computer algebra and polynomials*. Lecture Notes in Comput. Sci., vol. 8942. Springer, Cham.

[27] Piene, Ragni, and Sacchiero, Gianni. 1984. Duality for rational normal scrolls. *Comm. Algebra*, **12**(9-10), 1041–1066.

[28] Pohl, William Francis. 1962. Differential geometry of higher order. *Topology*, **1**, 169–211.

[29] Raynaud, Michel, and Gruson, Laurent. 1971. Critères de platitude et de projectivité. Techniques de "platification" d'un module. *Invent. Math.*, **13**, 1–89.

22

Characteristic Classes of Symmetric and Skew-symmetric Degeneracy Loci

Sutipoj Promtapan[a] and Richárd Rimányi[b]

Abstract. We give two formulas for the Chern–Schwartz–MacPherson class of symmetric and skew-symmetric degeneracy loci. We apply them in enumerative geometry, explore their algebraic combinatorics, and discuss K theory generalizations.

1 Introduction

Degeneracy loci formulas are universal expressions for the characteristic classes of certain degeneracy loci. The two most widely used such formulas are

- the Giambelli–Thom–Porteous formula [26]

$$[\overline{\Sigma}_r] = s_{(r+l)^r},$$

  and
- formulas of Józefiak–Lascoux–Pragacz and Harris–Tu [19, 18, 13, 8]

$$[\overline{\Sigma}_r^{\wedge}] = s_{r-1,r-2,\ldots,2,1}, \qquad [\overline{\Sigma}_r^{S}] = 2^{r-1} s_{r,r-1,\ldots,2,1}.$$

Some explanations are in order.

1.1 Degeneracy loci interpretation

First we explain the two formulas above in the language of "degeneracy loci". Let $\psi : A^n \to B^{n+l}$ ($l \geqslant 0$) be a vector bundle map over the base space M, and let Σ_r be the set of points x in M over which ψ_x has rank $n-r$ (that is, corank r). Then under a suitable assumption on M and a transversality assumption on ψ, the above Giambelli–Thom–Porteous formula

[a] University of North Carolina, Chapel Hill
[b] University of North Carolina, Chapel Hill

holds for the *fundamental cohomology class* $[\overline{\Sigma}_r] \in H^*(M)$ of $\overline{\Sigma}_r$, where $s_{\lambda_1,\ldots,\lambda_k} = \det(c_{\lambda_i+j-i})_{i,j=1,\ldots,k}$ and c_i is defined by

$$1 + c_1 t + c_2 t^2 + \cdots = \frac{1 + c_1(B)t + c_2(B)t^2 + \ldots}{1 + c_1(A)t + c_2(A)t^2 + \ldots}. \tag{1}$$

Now let A be a rank n vector bundle, and $\psi : A^* \to A$ be a *skew-symmetric* or *symmetric* vector bundle map over the base space M, and let Σ^r be the set of points x in M over which ψ_x has corank r (in the skew-symmetric case $n-r$ is necessarily even). Then, under a suitable assumption on M and a transversality assumption on ψ, the Józefiak–Lascoux–Pragacz–Harris–Tu formulas hold, where s_λ is the same as above, with c_i the ith Chern class of A.

1.2 Equivariant cohomology interpretation

Consider the $G = \mathrm{GL}_n(\mathbb{C}) \times \mathrm{GL}_{n+l}(\mathbb{C})$ action on $\mathrm{Hom}(\mathbb{C}^n, \mathbb{C}^{n+l})$ by $(A, B) \cdot X = BXA^{-1}$, and let Σ_r be the subset in $\mathrm{Hom}(\mathbb{C}^n, \mathbb{C}^{n+l})$ of matrices of corank r. Then the Giambelli–Thom–Porteous formula holds for the equivariant fundamental class of $[\overline{\Sigma}_r]$ in $H^*(BG)$. The classes c_i are as in (1) where a_i and b_i are the Chern classes of the tautological rank n and rank $n+l$ vector bundles over BG.

Similarly, consider the $G = \mathrm{GL}_n(\mathbb{C})$ action on the set of skew-symmetric or symmetric $n \times n$ matrices by $A \cdot X = A^T X A$ and let $\Sigma_r^\wedge, \Sigma_r^S$ be the set of those of corank r. Then for the *G-equivariant fundamental classes* $[\overline{\Sigma}_r^\wedge], [\overline{\Sigma}_r^S]$ the Józefiak–Lascoux–Pragacz–Harris–Tu formulas hold in $H^*(BG)$, where c_i is the i'th Chern class of the tautological bundle over BG.

1.3 MacPherson deformation of the fundamental class

The notion of fundamental class has an inhomogeneous deformation, called Chern–Schwartz–MacPherson class (CSM), denoted

$$\mathrm{c}^{\mathrm{sm}}(\Sigma) = \mathrm{c}^{\mathrm{sm}}(\Sigma \subset M) = [\overline{\Sigma}] + \text{higher order terms}.$$

The CSM class encodes more geometric and enumerative properties of the singular variety Σ than its lowest degree term, the fundamental class. It is also related with symplectic topology and representation theory (through Maulik-Okounkov's notion of "stable envelope class" [21]), at least in Schubert calculus settings, see [29, 14, 4].

The CSM version of the Giambelli–Thom–Porteous formula is calculated in [25], see also [14, 37]. To give a sample of that result we introduce the Segre–Schwartz–MacPherson class (SSM): $\mathrm{s}^{\mathrm{sm}}(\Sigma \subset M) = \mathrm{c}^{\mathrm{sm}}(\Sigma \subset$

$M)/c(TM)$. This carries the same information as the CSM class, but certain theorems are phrased more elegantly for SSM classes. We have

$$\mathrm{s}^{\mathrm{sm}}(\Sigma_0 \subset \mathrm{Hom}(\mathbb{C}^n,\mathbb{C}^{n+1})) = s_0 - s_2 + (2s_3 + s_{21}) \\ + (-3s_4 - 3s_{31} - s_{211}) + \cdots$$

(Those looking for positivity properties of such expansions may find this formula disappointing, but luckily positivity can be saved, see [14, Section 1.5].)

The goal of this paper is to calculate the CSM (or equivalently, the SSM) deformations of the Józefiak–Lascoux–Pragacz–Harris–Tu formulas. The reader is invited to jump ahead and see sample results in Section 6.

1.4 Plan of the paper

After introducing our geometric settings (the $\Lambda^2\mathbb{C}^n$ and $S^2\mathbb{C}^n$ representations) in Section 2, we recall the notion of Chern–Schwartz–MacPherson class in Section 3. In particular, first we recall the traditional approach to CSM classes via resolutions and push-forward, and then we recall the recent development, triggered by Maulik-Okounkov's notion of stable envelopes, claiming that CSM classes are the unique solutions to some interpolation problems.

We follow the traditional approach in Section 4, and we follow the interpolation approach in Section 5. Both yield formulas for CSM classes of the orbits of $\Lambda^2\mathbb{C}^n$ and $S^2\mathbb{C}^n$. The fact that the formulas obtained in the two approaches are equal is not obvious algebraically from their form. The one obtained from interpolation seems better: it is "one summation shorter", also, the other one is an exclusion-inclusion formula (sum of terms with alternating signs), hence it is not obviously suitable for further combinatorial study.

In Section 6 we make the first steps towards the algebraic combinatorics of the obtained formulas. We discuss stability, normalization, and most importantly positivity properties. The positivity properties can be studied for Schur expansions, or for the more conceptual expansions in terms of $\tilde{s}_\lambda$ functions.

In Section 7 we show sample applications in geometry of the calculated CSM classes. Namely, we focus on the most direct consequences, the Euler characteristics of general linear sections of symmetric and skew-symmetric degeneracy loci.

Finally, in Section 8 we discuss two natural directions for future study. First we give sample results about the closely related Chern–Mather classes of symmetric and skew-symmetric degeneracy loci. Then we explore the natural K theory analogue of CSM class, the so-called *motivic Chern class*. The traditional approach to motivic Chern classes is similar to the traditional approach to CSM classes (roughly speaking, replace the notion of Euler characteristic with that of chi-y-genus). Hence the K theory analogs of the results in Section 4 are promising. However, the interpolation approach to motivic Chern classes is more sophisticated [16], hence finding analogs of the results in Section 5 remains a challenge.

Acknowledgment. The first author was supported by the he Development and Promotion of Science and Technology Talents Project (Royal Government of Thailand scholarship) during his doctoral studies at UNC Chapel Hill. The second author is supported by a Simons Foundation grant.

Notation. Denote $[n] = \{1, \dots, n\}$. The set of r-element subsets of $[n]$ will be denoted by $\binom{[n]}{r}$. For $I \in \binom{[n]}{r}$ let $\bar{I} = [n] - I$. Varieties are considered over the complex numbers, and cohomology is meant with rational coefficients.

2 The representations $\Lambda^2\mathbb{C}^n$, $S^2\mathbb{C}^n$

Consider the action of $GL_n(\mathbb{C})$ on the vector space of skew-symmetric $n \times n$ matrices, and on the vector space of symmetric $n \times n$ matrices, by $A \cdot X = A^T X A$. These representations will be denoted by $\Lambda^2\mathbb{C}^n$ and $S^2\mathbb{C}^n$ respectively. The orbits of both of these representations are determined by rank (see Linear Algebra textbooks, e.g. [32, Sect. 9]).

- For $0 \leqslant r \leqslant n$, $n - r$ even, the orbit of rank $n - r$ ("corank r") matrices in $\Lambda^2\mathbb{C}^n$ will be denoted by $\Sigma^{\wedge}_{n,r}$. For example

$$X^{\wedge}_{n,r} = \underbrace{H \oplus \cdots \oplus H}_{(n-r)/2} \oplus \underbrace{0 \oplus \cdots 0}_{r} \in \Sigma^{\wedge}_{n,r},$$

  where $H = \begin{pmatrix} 0 & 1 \\ -1 & 0 \end{pmatrix}$. We have $\operatorname{codim}(\Sigma^{\wedge}_{n,r} \subset \Lambda^2\mathbb{C}^n) = \binom{r}{2}$.
- For $0 \leqslant r \leqslant n$, the orbit of rank $n - r$ ("corank r") matrices in $S^2\mathbb{C}^n$ will be denoted by $\Sigma^S_{n,r}$. For example $X^S_{n,r} = \underbrace{1 \oplus \cdots \oplus 1}_{n-r} \oplus \underbrace{0 \oplus \cdots 0}_{r} \in \Sigma^S_{n,r}$.

  We have $\operatorname{codim}(\Sigma^S_{n,r} \subset S^2\mathbb{C}^n) = \binom{r+1}{2}$.

In later sections we will approach the geometric study of these orbits by constructing resolutions of their closures, and by studying their stabilizer groups.

3 Chern–Schwartz–MacPherson classes

Deligne and Grothendieck conjectured [34] and MacPherson proved [20] the existence of a unique natural transformation $C_* : \mathcal{F}(-) \to H_*(-)$ from the covariant functor of constructible functions to the covariant functor of Borel-Moore homology, satisfying certain properties. Independently, Schwartz [33] introduced the notion of 'obstruction class' (for the extension of stratified radial vector frames over a complex algebraic variety), and later Brasselet and Schwartz [10] proved that C_* and the obstruction class essentially coincide, via Alexander duality. We will study the *equivariant co*homology version of the resulting Chern–Schwartz–MacPherson (CSM) class, due to Ohmoto [22, 23, 24].

3.1 Equivariant CSM class, after MacPherson, Ohmoto

Let G be an algebraic group acting on the smooth algebraic variety M, and let f be a G-invariant constructible function f on M (say, to $\mathbb{Q}$). The associated *G-equivariant Chern–Schwartz–MacPherson class* $\mathrm{c}^{\mathrm{sm}}(f)$ is an element of $H_G^*(M) = H_G^*(M, \mathbb{Q})$.

Before further discussing this notion let us consider a version of it, the G-equivariant Segre-Schwartz-MacPherson (SSM) class $\mathrm{s}^{\mathrm{sm}}(f) = \mathrm{c}^{\mathrm{sm}}(f)/c(TM) \in H_G^{**}(M)$.[1] Also, for an invariant (not necessarily closed) subvariety $\Sigma \subset M$ denote $\mathrm{c}^{\mathrm{sm}}(\Sigma) = \mathrm{c}^{\mathrm{sm}}(\Sigma \subset M) = \mathrm{c}^{\mathrm{sm}}(\mathbf{1}_\Sigma) \in H_G^*(M)$, and $\mathrm{s}^{\mathrm{sm}}(\Sigma) = \mathrm{s}^{\mathrm{sm}}(\Sigma \subset M) = \mathrm{s}^{\mathrm{sm}}(\mathbf{1}_\Sigma) \in H_G^*(M)$, where $\mathbf{1}_\Sigma$ is the indicator function of Σ.

We will sketch Ohmoto's definition in a Remark at the end of Section 3.1. For our purposes the following (defining) properties will be sufficient.

(i) (additivity) For equivariant constructible functions f and g on M we have

$$\mathrm{c}^{\mathrm{sm}}(f+g) = \mathrm{c}^{\mathrm{sm}}(f) + \mathrm{c}^{\mathrm{sm}}(g), \text{ and } \mathrm{c}^{\mathrm{sm}}(\lambda \cdot f) = \lambda \cdot \mathrm{c}^{\mathrm{sm}}(f) \text{ for } \lambda \in \mathbb{Q}.$$

(ii) (normalization) For an equivariant proper embedding of a *smooth* subvariety $i : \Sigma \subset M$ we have

$$\mathrm{c}^{\mathrm{sm}}(\Sigma \subset M) = i_* c(T\Sigma) \in H_G^*(M).$$

[1] Since we divided by the equivariant total Chern class "$c(TM) = 1+$higher order terms", the SSM class may be non-zero in arbitrarily high degrees: it lives in the completion, as indicated. In the rest of the paper we will not indicate this completion, and write $\mathrm{c}^{\mathrm{sm}}(f), \mathrm{s}^{\mathrm{sm}}(f) \in H_G^*(M)$.

(iii) (functoriality) For a G-equivariant proper map between smooth G-varieties $\eta : Y \to M$,

$$\eta_*(c(TY)) = \sum_j j \cdot \mathrm{c}^{\mathrm{sm}}(M_j) \tag{2}$$

where $M_j = \{x \in M : \chi(\eta^{-1}(x)) = j\}$.

The named three properties uniquely define the CSM class (the uniqueness is obvious, the existence is the content of the arguments of MacPherson, Brasselet-Schwartz, Ohmoto). The CSM class, however, satisfies another key property [22, Theorem 4.2], [24, Proposition 3.8]:

(iv) Let $\Sigma \subset M$ be a closed invariant subvariety with an invariant Whitney stratification. For an equivariant map between smooth manifolds $\eta : Y \to M$ that is transversal to the strata of Σ, we have

$$\mathrm{s}^{\mathrm{sm}}(\eta^{-1}(\Sigma)) = \eta^*(\mathrm{s}^{\mathrm{sm}}(\Sigma)).$$

Remark 3.1 The most natural characteristic class, the fundamental class

$$[\Sigma \subset M] \in H^{2\,\mathrm{codim}(\Sigma\subset M)}(M)$$

of a closed subvariety $\Sigma \subset M$ behaves nicely with respect to both push-forward and pull-back. The CSM "deformation" of the notion of fundamental class is forced to behave nicely with respect to push-forward (see axiom (iii)). Yet, it is rather remarkable that it remains well-behaving with respect to pull-back as well (see property (iv)).

We called the CSM class a 'deformation' of the fundamental class because we also have [22, Section 4.1]:

(v) For a subvariety $\Sigma \subset M$ the lowest degree term of $\mathrm{c}^{\mathrm{sm}}(\Sigma \subset M)$ is $[\overline{\Sigma} \subset M]$. Terms of degree higher than $\dim M$ are 0. For projective M the integral of (the term of degree $\dim M$ of) $\mathrm{c}^{\mathrm{sm}}(\Sigma \subset M)$ is the topological Euler characteristic of Σ.

Remark 3.2 We have set up the CSM classes in cohomology, but their natural habitat is homology. Following MacPherson, Ohmoto defines equivariant CSM classes by first proving the existence and uniqueness of a natural transformation $C_*^G : \mathcal{F}^G(-) \to H_*^G(-)$ from the abelian group of G-equivariant constructible functions to the G-equivariant homology (some non-trivial definitions are needed to make this work!), satisfying axioms analogous to (i)–(iii) above. Then the cohomology version considered in this paper is obtained by composing C_*^G with homology push-forward to the ambient space, and Poincaré-duality in the smooth ambient space.

3.2 Interpolation characterization of CSM classes

Under certain circumstances, equivariant CSM classes are also determined by a set of interpolation properties [29, 14].

Consider a complex, linear algebraic group G and its linear representation on the complex vector space V. For an orbit Σ, and $x \in \Sigma$ let $G_\Sigma \leqslant G$ be the stabilizer subgroup of x. Let $T_\Sigma = T_x\Sigma$ be the tangent space of Σ at x, and $N_\Sigma = T_xV/T_\Sigma$, both G_Σ-representations (these definitions are independent of the choice of x in Σ up to natural isomorphisms). We will use the G_Σ equivariant Euler and total Chern classes of these representations. The inclusion $G_\Sigma \leqslant G$ induces a map $\phi_\Sigma : H^*(BG) \to H^*(BG_\Sigma)$—which is also independent of the choice of $x \in \Sigma$.

Assumption 3.3 *We assume that the representation $G \circlearrowright V$ has finitely many orbits, the orbits are cones (i.e. invariant under the dilation action of $\mathbb{C}^*$), and that the Euler class $e(N_\Sigma) \neq 0$ for all Σ.*

Theorem 3.4 ([14]) *Under Assumption 3.3 the CSM class of the orbit Σ is uniquely determined by the conditions*

(1) $\phi_\Sigma(\mathrm{c}^{\mathrm{sm}}(\Sigma)) = c(T_\Sigma)e(N_\Sigma)$ in $H^*(BG_\Sigma)$;
(2) for any orbit Ω, $c(T_\Omega)$ divides $\phi_\Omega(\mathrm{c}^{\mathrm{sm}}(\Sigma))$ in $H^*(BG_\Omega)$;
(3) for any orbit $\Omega \neq \Sigma$, $\deg(\phi_\Omega(\mathrm{c}^{\mathrm{sm}}(\Sigma))) < \deg(c(T_\Omega)e(N_\Omega))$.

In condition (3) by “deg” of a possibly inhomogeneous cohomology class we mean the degree of it highest degree non-zero degree component (and $\deg 0 = -\infty$).

It also follows from these conditions that for an orbit $\Omega \not\subset \overline{\Sigma}$ we have $\phi_\Omega(\Sigma) = 0$.

4 Sieve formula for the CSM classes

Our goal is to calculate the $\mathrm{GL}_n(\mathbb{C})$-equivariant CSM (or SSM) class of the orbits of $\Lambda^2\mathbb{C}^n$ and $S^2\mathbb{C}^n$. They are classes in

$$H^*_{\mathrm{GL}_n(\mathbb{C})}(\Lambda^2\mathbb{C}^n) = H^*_{\mathrm{GL}_n(\mathbb{C})}(S^2\mathbb{C}^n) = H^*(B\mathrm{GL}_n(\mathbb{C})) = \mathbb{Q}[\alpha_1, \dots, \alpha_n]^{S_n},$$

where α_i's are the Chern roots of the tautological n-bundle over $B\mathrm{GL}_n(\mathbb{C})$. The tautological n-bundle over $B\mathrm{GL}_n(\mathbb{C}) = \mathrm{Gr}_n(\mathbb{C}^\infty)$ is the $N \to \infty$ limit of the tautological subbundle $\{(W, w) \in \mathrm{Gr}_n(\mathbb{C}^N) \times \mathbb{C}^N : w \in W\} \to \mathrm{Gr}_n(\mathbb{C}^N)$.

In the whole paper c_k will denote the kth Chern class of that bundle, ie. the kth elementary symmetric polynomial of the α_i's.

In this section we make calculations using "traditional methods", and achieve an exclusion-inclusion type formula (Theorems 4.5, 4.7), then in the next section we solve the relevant interpolation problem and find improved formulas.

4.1 Fibered resolution

Consider a G-representation V, and an invariant closed subvariety $\Sigma \subset V$. The G-equivariant map $\eta : \widetilde{\Sigma} \to V$ is called a fibered resolution of Σ if there exists a G-equivariant commutative diagram

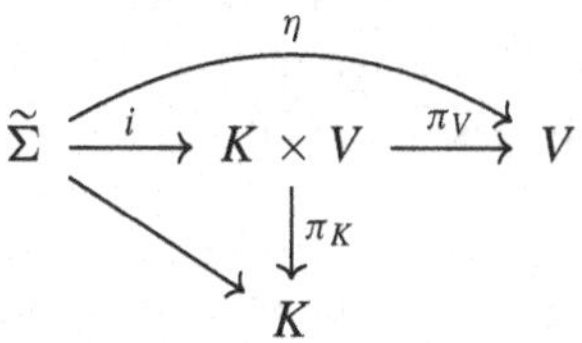

where η is a resolution of singularities of Σ, π_V is the projection to V, π_K is the projection to K, K is a smooth projective G-variety, the map $\widetilde{\Sigma} \to K$ is a G-vector bundle, i is a G-equivariant embedding of vector bundles, and $\eta = \pi_V \circ i$. Let $\nu = (K \times V \to K)/(\widetilde{\Sigma} \to K)$ be the G-equivariant quotient bundle over K. A pullback of the bundle ν to $\widetilde{\Sigma}$ is the normal bundle of the embedding $i : \widetilde{\Sigma} \to K \times V$. Define

$$\Phi_\Sigma := \frac{\eta_*(c(T\widetilde{\Sigma}))}{c(V)} = \eta_* \left(\frac{c(T\widetilde{\Sigma})}{c(V)} \right)$$

$$= \eta_*(c(-\nu)c(TK)) = \int_K e(\nu)c(-\nu)c(TK). \quad (3)$$

The equality of the displayed expressions is detailed in [14, Section 10.1].

The significance of the class Φ_Σ is that, on the one hand, one can write formulas for it (due to its last displayed expression), and, on the other hand, the CSM class of Σ is a linear combination of Φ-classes of some varieties contained in Σ.

4.2 Sieve formula for SSM classes of orbits of $\Lambda^2\mathbb{C}^n$

Elements $X \in \Lambda^2\mathbb{C}^n$ will be identified with skew-symmetric bilinear forms on $\mathbb{C}^{n*}$, and in turn, with skew-symmetric linear maps $\mathbb{C}^{n*} \to \mathbb{C}^n$, without further notation. For $0 \leqslant r \leqslant n$, $n - r$ even, define

$$\widetilde{\Sigma}^{\wedge}_{n,r} := \{(W,X) \in \mathrm{Gr}_r(\mathbb{C}^{n*}) \times \Lambda^2\mathbb{C}^n, X|_W = 0\},$$

and consider the diagram

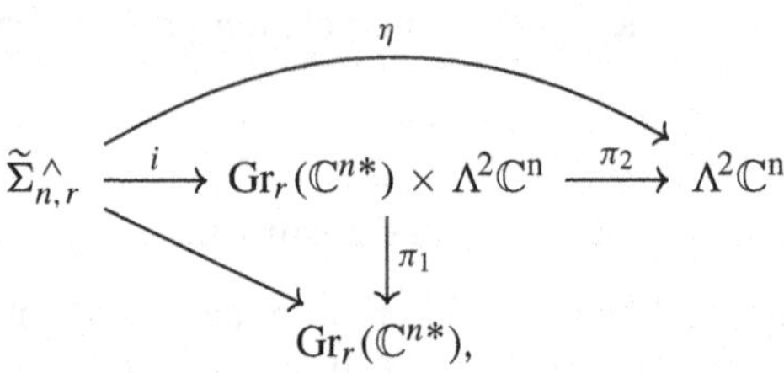

where i is the inclusion, π_1, π_2 are projection maps on the first and second coordinates, respectively. This diagram is a fibered resolution of $\overline{\Sigma}^{\wedge}_{n,r}$. Consider the corresponding Φ-class (see (3))

$$\Phi^{\wedge}_{n,r} = \int_{\mathrm{Gr}_r(\mathbb{C}^{n*})} e(\nu)c(-\nu)c(T\,\mathrm{Gr}_r(\mathbb{C}^{n*})),$$

where ν is the quotient bundle

$$(\mathrm{Gr}_r(\mathbb{C}^{n*}) \times \Lambda^2\mathbb{C}^n \to \mathrm{Gr}_r(\mathbb{C}^{n*}))/(\widetilde{\Sigma}^{\wedge}_{n,r} \to \mathrm{Gr}_r(\mathbb{C}^{n*})).$$

Proposition 4.1 *For $0 \leqslant r \leqslant n$, $n-r$ even, we have*

$$\Phi^{\wedge}_{n,r} = \sum_{\substack{I \subset [n] \\ |I|=r}} \left(\prod_{i<j \in I} \frac{\alpha_i + \alpha_j}{1+\alpha_i+\alpha_j} \prod_{i \in I} \prod_{j \in \bar{I}} \frac{(\alpha_i+\alpha_j)(1-\alpha_j+\alpha_i)}{(1+\alpha_i+\alpha_j)(-\alpha_j+\alpha_i)} \right). \tag{4}$$

Proof The fiber of the bundle $\widetilde{\Sigma}^{\wedge}_{n,r} \to \mathrm{Gr}_r(\mathbb{C}^{n*})$ over $W \in \mathrm{Gr}_r(\mathbb{C}^{n*})$ is $\{X \in \Lambda^2\mathbb{C}^n : X|_W = 0\} = \Lambda^2(\mathrm{Ann}(W))$ where $\mathrm{Ann}(W) = \{v \in \mathbb{C}^n : \phi(v) = 0 \text{ for all } \phi \in W\} \in \mathrm{Gr}_{n-r}(\mathbb{C}^n)$. Hence the bundle $\widetilde{\Sigma}^{\wedge}_{n,r} \to \mathrm{Gr}_r(\mathbb{C}^{n*})$ is $\Lambda^2(Q^*)$, where $0 \to S \to \mathbb{C}^{n*} \to Q \to 0$ is the tautological exact sequence of bundles over $\mathrm{Gr}_r(\mathbb{C}^{n*})$.

Hence we have

$$\begin{aligned}\nu = \Lambda^2\mathbb{C}^n - \widetilde{\Sigma}^{\wedge}_{n,r} &= \left(\Lambda^2(S^*) \oplus \Lambda^2(Q^*) \oplus (S^* \otimes Q^*)\right) - \Lambda^2(Q^*) \\ &= \Lambda^2(S^*) \oplus (S^* \otimes Q^*).\end{aligned}$$

Let $\delta_1, \ldots, \delta_r$ be the Chern roots of the bundle S, and let $\omega_1, \ldots, \omega_{n-r}$ be the Chern roots of Q. Then

$$e(\nu) = \prod_{1 \leqslant i < j \leqslant r} (-\delta_i - \delta_j) \prod_{i=1}^{r} \prod_{j=1}^{n-r} (-\omega_j - \delta_i),$$

$$c(\nu) = \prod_{1 \leqslant i < j \leqslant r} (1 - \delta_i - \delta_j) \prod_{i=1}^{r} \prod_{j=1}^{n-r} (1 - \omega_j - \delta_i),$$

$$c(T\,\mathrm{Gr}_r(\mathbb{C}^{n*})) = \prod_{i=1}^{r} \prod_{j=1}^{n-r} (1 + \omega_j - \delta_i),$$

and from the definition of $\Phi^{\wedge}_{n,r}$ we obtain

$$\Phi^{\wedge}_{n,r} = \int_{\mathrm{Gr}_r(\mathbb{C}^{n*})} \prod_{1 \leqslant i < j \leqslant r} \frac{-\delta_i - \delta_j}{1 - \delta_i - \delta_j} \prod_{i=1}^{r} \prod_{j=1}^{n-r} \frac{(-\omega_j - \delta_i)(1 + \omega_j - \delta_i)}{1 - \omega_j - \delta_i}.$$

The equivairant localization formula for this integral in exactly (4). □

Proposition 4.2 *For $0 \leqslant r \leqslant n$, $n - r$ even, we have*

$$\Phi^{\wedge}_{n,r} = \sum_{i=0}^{\frac{n-r}{2}} \binom{r+2i}{r} \mathrm{s}^{\mathrm{sm}}(\Sigma^{\wedge}_{n,r+2i})$$

$$= \sum_{i=0}^{\frac{n-r}{2}} \left(\binom{r+2i}{r} - \binom{r+2i-2}{r} \right) \mathrm{s}^{\mathrm{sm}}(\overline{\Sigma}^{\wedge}_{n,r+2i}).$$

Proof The closure $\overline{\Sigma}^{\wedge}_{n,r} = \Sigma^{\wedge}_{n,r} \cup \Sigma^{\wedge}_{n,r+2} \cup \cdots \cup \Sigma^{\wedge}_{n,n}$ is the image of η. For each $r \leqslant k \leqslant n$ and $n-k$ even, a preimage of each point in $\Sigma^{\wedge}_{n,k}$ is isomorphic to the space $\mathrm{Gr}_r(\mathbb{C}^{k*})$. Note that the Euler characteristic of the Grassmannian of all r-dimensional subspaces of a k-dimensional vector space over $\mathbb{C}$ is $\binom{k}{r}$. Using property (iii) of CSM classes from Section 3.1 we have

$$\eta_*(c(T\widetilde{\Sigma}^{\wedge}_{n,r})) = \sum_{i=0}^{\frac{n-r}{2}} \binom{r+2i}{r} \mathrm{c}^{\mathrm{sm}}(\Sigma^{\wedge}_{n,r+2i}).$$

Dividing both sides by $c(\Lambda^2\mathbb{C}^n)$ proves the first equality of the proposition.

By the additivity property of SSM classes (see Section 3.1 (i)), we have

$$\sum_{i=0}^{\frac{n-r}{2}} \binom{r+2i}{r} s^{\mathrm{sm}}(\Sigma^{\wedge}_{n,r+2i})$$

$$= \sum_{i=0}^{\frac{n-r}{2}} \binom{r+2i}{r} \left(s^{\mathrm{sm}}(\overline{\Sigma}^{\wedge}_{n,r+2i}) - s^{\mathrm{sm}}(\overline{\Sigma}^{\wedge}_{n,r+2i+2}) \right)$$

$$= \sum_{i=0}^{\frac{n-r}{2}} \left(\binom{r+2i}{r} - \binom{r+2i-2}{r} \right) s^{\mathrm{sm}}(\overline{\Sigma}^{\wedge}_{n,r+2i}),$$

which completes the proof. □

Proposition 4.2 expresses the Φ-classes as linear combinations of the SSM-classes. Inverting the matrix of these linear combinations will therefore express the SSM classes in terms of the Φ-classes.

Definition 4.3 Define the Euler numbers E_n by

$$\frac{1}{\cosh(x)} = \sum_{n=0}^{\infty} \frac{E_n}{n!} x^n.$$

For odd n the number E_n is zero. For even n Euler numbers form an alternating sequence: $E_0 = 1, E_2 = -1, E_4 = 5, E_6 = -61, E_8 = 1385,\ E_{10} = -50512, \ldots$. For explicit formulas for the Euler numbers see e.g. [35] and references therein.

Proposition 4.4 *We have* $\left(\binom{2j}{2i}\right)^{-1}_{0\leqslant i,j\leqslant m} = \left(\binom{2j}{2i} E_{2j-2i}\right)_{0\leqslant i,j\leqslant m}$ *and* $\left(\binom{2j+1}{2i+1}\right)^{-1}_{0\leqslant i,j\leqslant m} = \left(\binom{2j+1}{2i+1} E_{2j-2i}\right)_{0\leqslant i,j\leqslant m}$.

Proof The product of $\left(\binom{2j}{2i}\right)_{0\leqslant i,j\leqslant m}$ and $\left(\binom{2j}{2i} E_{2j-2i}\right)_{0\leqslant i,j\leqslant m}$ is upper triangular. For $i \leqslant j$ its (i,j)'th entry is

$$\sum_{k=i}^{j} \binom{2k}{2i}\binom{2j}{2k} E_{2j-2k} = \binom{2j}{2i} \sum_{k=i}^{j} \binom{2j-2i}{2j-2k} E_{2j-2k} = \begin{cases} 1 & i=j \\ 0 & i<j, \end{cases}$$

where the last equality follows from the defining equation

$$\left(1 + \frac{t^2}{2!} + \frac{t^4}{4!} + \ldots\right)\left(1 + \frac{E_2 t^2}{2!} + \frac{E_4 t^4}{4!} + \ldots\right) = 1.$$

The proof of the other statement is similar. □

For example

$$\begin{pmatrix} 1 & 1 & 1 & 1 \\ 0 & 1 & 6 & 15 \\ 0 & 0 & 1 & 15 \\ 0 & 0 & 0 & 1 \end{pmatrix}^{-1} = \begin{pmatrix} \binom{0}{0} & \binom{2}{0} & \binom{4}{0} & \binom{6}{0} \\ 0 & \binom{2}{2} & \binom{4}{2} & \binom{6}{2} \\ 0 & 0 & \binom{4}{4} & \binom{6}{4} \\ 0 & 0 & 0 & \binom{6}{6} \end{pmatrix}^{-1}$$

$$\begin{pmatrix} \binom{0}{0}E_0 & \binom{2}{0}E_2 & \binom{4}{0}E_4 & \binom{6}{0}E_6 \\ 0 & \binom{2}{2}E_0 & \binom{4}{2}E_2 & \binom{6}{2}E_4 \\ 0 & 0 & \binom{4}{4}E_0 & \binom{6}{4}E_2 \\ 0 & 0 & 0 & \binom{6}{6}E_0 \end{pmatrix} = \begin{pmatrix} 1 & -1 & 5 & -61 \\ 0 & 1 & -6 & 75 \\ 0 & 0 & 1 & -15 \\ 0 & 0 & 0 & 1 \end{pmatrix}.$$

Theorem 4.5 *For $0 \leqslant r \leqslant n$, $n - r$ even, we have*

$$s^{\mathrm{sm}}(\Sigma^{\wedge}_{n,r}) = \sum_{i=0}^{\frac{n-r}{2}} \binom{r+2i}{r} E_{2i} \cdot \Phi^{\wedge}_{n,r+2i}. \tag{5}$$

Proof This is a consequence of Proposition 4.2 and Proposition 4.4. □

This theorem, together with the expression (4) for the Φ-classes is our first formula for the SSM classes $\Sigma^{\wedge}_{n,r}$. We call it the sieve formula, because the coefficients in the summation (5) have alternating signs.

Example 4.6 For $n = 2$ we have

$$\Phi^{\wedge}_{2,0} = 1, \qquad \Phi^{\wedge}_{2,2} = \frac{\alpha_1 + \alpha_2}{1 + \alpha_1 + \alpha_2} = c_1 - c_1^2 + c_1^3 - c_1^4 + \cdots,$$

and therefore

$$s^{\mathrm{sm}}(\Sigma^{\wedge}_{2,0}) = \Phi^{\wedge}_{2,0} - \Phi^{\wedge}_{2,2} = 1 - c_1 + c_1^2 - c_1^3 + c_1^4 - \cdots,$$
$$s^{\mathrm{sm}}(\Sigma^{\wedge}_{2,2}) = \Phi^{\wedge}_{2,2} = c_1 - c_1^2 + c_1^3 - c_1^4 + \cdots.$$

For $n = 3$ we have

$$\begin{aligned} \Phi^{\wedge}_{3,1} &= 1 + (2c_1c_2 - 2c_3) + (-4c_1^2c_2 + 4c_1c_3) \\ &\quad + (4c_1^3c_2 + 2c_1c_2^2 - 4c_1^2c_3 - 2c_2c_3) + (-10c_1^2c_2^2 + 12c_1c_2c_3 - 2c_3^2) \\ &\quad + (-8c_1^5c_2 + 24c_1^3c_2^2 + 2c_1c_2^3 + 8c_1^4c_3 - 32c_1^2c_2c_3 + 8c_1c_3^2) + \cdots, \\ \Phi^{\wedge}_{3,3} &= (c_1c_2 - c_3) + (-2c_1^2c_2 + 2c_1c_3) \\ &\quad + (2c_1^3c_2 + c_1c_2^2 - 2c_1^2c_3 - c_2c_3) + (-5c_1^2c_2^2 + 6c_1c_2c_3 - c_3^2) \\ &\quad + (-4c_1^5c_2 + 12c_1^3c_2 + c_1c_2^3 + 4c_1^4c_3 - 16c_1^2c_2c_3 - c_2^2c_3 + 4c_1c_3^2) + \cdots, \end{aligned}$$

and

$$s^{\mathrm{sm}}(\Sigma^{\wedge}_{3,1}) = \Phi^{\wedge}_{3,1} - 3\Phi^{\wedge}_{3,3}, \qquad s^{\mathrm{sm}}(\Sigma^{\wedge}_{3,3}) = \Phi^{\wedge}_{3,3}.$$

4.3 Sieve formula for SSM classes of orbits of $S^2\mathbb{C}^n$

Arguments analogous to those in Section 4.2 give the following theorem, we leave the details to the reader.

Theorem 4.7 *For $0 \leqslant r \leqslant n$, we have*

$$s^{\mathrm{sm}}(\Sigma^S_{n,r}) = \sum_{i=0}^{n-r} (-1)^i \binom{r+i}{r} \Phi^S_{n,r+i},$$

$$s^{\mathrm{sm}}(\overline{\Sigma}^S_{n,r}) = \sum_{i=0}^{n-r} (-1)^i \binom{r+i-1}{r-1} \Phi^S_{n,r+i}.$$

where

$$\Phi^S_{n,r} = \sum_{\substack{I\subset[n]\\ |I|=r}} \left(\prod_{i\leqslant j\in I} \frac{\alpha_i+\alpha_j}{1+\alpha_i+\alpha_j} \prod_{i\in I}\prod_{j\in \bar{I}} \frac{(\alpha_i+\alpha_j)(1-\alpha_j+\alpha_i)}{(1+\alpha_i+\alpha_j)(-\alpha_j+\alpha_i)} \right). \quad \square$$

The fact that the sieve coefficients for symmetric loci are simple binomial coefficients, not Euler numbers are due to the fact that $\Sigma^S_{n,r}$ orbits exist for all r independent of parity, so at a certain point in the argument we need to invert the matrix of *all* binomial coefficients, not only the even ones.

Example 4.8 For $n = 2$, we have

$$\begin{aligned}
\Phi^S_{2,0} &= 1,\\
\Phi^S_{2,1} &= \frac{2(\alpha_1+\alpha_2)(1+\alpha_1+\alpha_2+4\alpha_1\alpha_2)}{(1+2\alpha_1)(1+2\alpha_2)(1+\alpha_1+\alpha_2)}\\
&= 2c_1 - 4c_1^2 + 8c_1^3 + (-16c_1^4 + 8c_1^2c_2) + (32c_1^5 - 40c_1^3c_2) + \cdots,\\
\Phi^S_{2,2} &= \frac{4\alpha_1\alpha_2(\alpha_1+\alpha_2)}{(1+2\alpha_1)(1+2\alpha_2)(1+\alpha_1+\alpha_2)}\\
&= 4c_1c_2 - 12c_1^2c_2 + (28c_1^3c_2 - 16c_1c_2^2) + \cdots,
\end{aligned}$$

and

$$\begin{aligned}
s^{\mathrm{sm}}(\Sigma^S_{2,0}) &= \Phi^S_{2,0} - \Phi^S_{2,1} + \Phi^S_{2,2} = 1 - 2c_1 + 4c_1^2 + (-8c_1^3 + 4c_1c_2)\\
&\quad + (16c_1^4 - 20c_1^2c_2) + \cdots,\\
s^{\mathrm{sm}}(\Sigma^S_{2,1}) &= \Phi^S_{2,1} - 2\Phi^S_{2,2} = 2c_1 - 4c_1^2 + (8c_1^3 - 8c_1c_2)\\
&\quad + (-16c_1^4 + 32c_1^2c_2) + \cdots,\\
s^{\mathrm{sm}}(\Sigma^S_{2,2}) &= \Phi^S_{2,2} = 4c_1c_2 - 12c_1^2c_2 + (28c_1^3c_2 - 16c_1c_2^2) + \cdots.
\end{aligned}$$

5 Interpolation formula for CSM classes

In Section 5.1 we define some functions, that we call W-functions because of their vague similarity to *weight functions* in [30]. Then in Section 5.2 we show that they represent CSM classes.

5.1 The W-functions

For $I \subset [n]$ let $\boldsymbol{\alpha}_I = \{\alpha_i : i \in I\}$. A permutation $\tau \in S_k$ acts on a rational function $f(\alpha_1, \ldots, \alpha_k)$ by $(\tau \cdot f)(\alpha_1, \alpha_2, \ldots, \alpha_k) = f(\alpha_{\tau(1)}, \alpha_{\tau(2)}, \ldots, \alpha_{\tau(k)})$.

Definition 5.1 For $0 \leqslant r \leqslant n$, $n - r$ even, define the skew-symmetric W-function

$$W^{\wedge}_{n,r}(\boldsymbol{\alpha}_{[n]}) = \sum_{\substack{|I|=r \\ I \subset [n]}} \left(W^{\wedge}_{n-r}(\boldsymbol{\alpha}_{\bar{I}}) \prod_{i<j \in I} (\alpha_i + \alpha_j) \times \prod_{i \in I} \prod_{j \in \bar{I}} \frac{(\alpha_i + \alpha_j)(1 + \alpha_i + \alpha_j)}{\alpha_i - \alpha_j} \right)$$

where

$$W^{\wedge}_k(\boldsymbol{\alpha}_{[k]}) = \frac{1}{2^{\frac{k}{2}} \cdot \left(\frac{k}{2}\right)!} \sum_{\tau \in S_k} \tau \left(\prod_{1 \leqslant i < j \leqslant k} \frac{(1 + \alpha_i + \alpha_j)(\alpha_i + \alpha_j)}{\alpha_i - \alpha_j} \times \prod_{i=1}^{\frac{k}{2}} \frac{\alpha_{2i-1} - \alpha_{2i}}{(\alpha_{2i-1} + \alpha_{2i})(1 + \alpha_{2i-1} + \alpha_{2i})} \right).$$

Despite its appearance, $W^{\wedge}_{n,r}$ is a polynomial (denominators cancel), it is in fact an integer coefficient symmetric polynomial in $\boldsymbol{\alpha}_{[n]}$, of highest degree term of degree $\frac{1}{2}(n^2 - 2n + r)$. The function $W^{\wedge}_k(\boldsymbol{\alpha}_{[k]})$ can be rewritten as

$$\frac{1}{\left(\frac{k}{2}\right)!} \sum_{\substack{|I_1| = \cdots = |I_{\frac{k}{2}}| = 2 \\ I_1 \cup \cdots \cup I_{\frac{k}{2}} = [k]}} \left(\prod_{1 \leqslant i < j \leqslant \frac{k}{2}} \prod_{\substack{i' \in I_i \\ j' \in I_j}} \frac{(\alpha_{i'} + \alpha_{j'})(1 + \alpha_{i'} + \alpha_{j'})}{\alpha_{i'} - \alpha_{j'}} \right).$$

Example 5.2 We have $W^{\wedge}_{2,0} = 1, W^{\wedge}_{2,2} = \alpha_1 + \alpha_2 = c_1$, and

$$\begin{aligned}
W^{\wedge}_{3,1} &= 1 + 2\alpha_1 + 2\alpha_2 + 2\alpha_3 + \alpha_1^2 + \alpha_2^2 + \alpha_3^2 + 3\alpha_1\alpha_2 + 3\alpha_1\alpha_3 + 3\alpha_2\alpha_3 \\
&= 1 + 2c_1 + c_1^2 + c_2, \\
W^{\wedge}_{3,3} &= \alpha_1^2\alpha_2 + \alpha_1^2\alpha_3 + \alpha_2^2\alpha_3 + \alpha_1\alpha_2^2 + \alpha_1\alpha_3^2 + \alpha_2\alpha_3^2 + 2\alpha_1\alpha_2\alpha_3 \\
&= c_1c_2 - c_3, \\
W^{\wedge}_{4,0} &= 1 + 2c_1 + c_1^2 + 2c_2 + 2c_1c_2 + c_2^2 + c_1c_3 - 4c_4, \\
W^{\wedge}_{4,2} &= c_1 + 2c_1^2 + c_1^3 + 2c_1c_2 + 2c_1^2c_2 + c_1c_2^2 + c_1^2c_3 - 4c_1c_4, \\
W^{\wedge}_{4,4} &= c_1c_2c_3 - c_1^2c_4 - c_3^2.
\end{aligned}$$

Definition 5.3 For $0 \leqslant r \leqslant n$, define the symmetric W-function

$$W^S_{n,r}(\boldsymbol{\alpha}_{[n]}) = \sum_{\substack{|I|=r \\ I\subset[n]}} \left(W^S_{n-r}(\boldsymbol{\alpha}_{\bar{I}}) \prod_{i\leqslant j\in I} (\alpha_i + \alpha_j) \times \prod_{i\in I} \prod_{j\in\bar{I}} \frac{(\alpha_i + \alpha_j)(1 + \alpha_i + \alpha_j)}{\alpha_i - \alpha_j} \right)$$

where

$$W^S_k(\boldsymbol{\alpha}_{[k]}) = \frac{1}{\lfloor \frac{k}{2} \rfloor !} \sum_{\tau\in S_k} \tau \left(\prod_{1\leqslant i<j\leqslant k} \frac{(1 + \alpha_i + \alpha_j)(\alpha_i + \alpha_j)}{\alpha_i - \alpha_j} \times \prod_{i=1}^{\lfloor \frac{k}{2} \rfloor} \frac{-\alpha_{2i}(1 + 2\alpha_{2i-1})(1 - \alpha_{2i-1} + \alpha_{2i})}{(\alpha_{2i-1} + \alpha_{2i})(1 + \alpha_{2i-1} + \alpha_{2i})} \right).$$

The function $W^S_{n,r}$ is also a symmetric polynomial in $\boldsymbol{\alpha}_{[n]}$, of highest degree component of degree $\frac{n(n+1)}{2} - \lfloor \frac{n-r+1}{2} \rfloor$.

Example 5.4 We have $W^S_{2,0} = 1 + \alpha_1 + \alpha_2 + 4\alpha_1\alpha_2 = 1 + c_1 + 4c_2$, $W^S_{2,1} = 2\alpha_1 + 2\alpha_2 + 2\alpha_1^2 + 4\alpha_1\alpha_2 + 2\alpha_2^2 = 2c_1 + 2c_1^2$, $W^S_{2,2} = 4\alpha_1^2\alpha_2 + 4\alpha_1\alpha_2^2 = 4c_1c_2$.

5.2 CSM classes in $\Lambda^2\mathbb{C}^n$ and $S^2\mathbb{C}^n$ as W-functions

Theorem 5.5 *We have* $c^{sm}(\Sigma^{\wedge}_{n,r}) = W^{\wedge}_{n,r}$ *and* $c^{sm}(\Sigma^S_{n,r}) = W^S_{n,r}$.

Proof We will show that $W^{\wedge}_{n,r}$ satisfies the three properties in Theorem 3.4 for $\Lambda^2\mathbb{C}^n$; along the way we will see that the representation $\Lambda^2\mathbb{C}^n$ satisfies

Assumption 3.3, hence the verification of the three properties proves $c^{\mathrm{sm}}(\Sigma^{\wedge}_{n,r}) = W^{\wedge}_{n,r}$.

Choosing the matrix $X^{\wedge}_{n,r}$ (see Section 2) as the representative of the orbit $\Sigma^{\wedge}_{n,r}$ we can read the following data:

- The stabilizer group $\mathrm{GL}_n(\mathbb{C})_{\Sigma^{\wedge}_{n,r}}$ deformation retracts to $\mathrm{Sp}(n-r,\mathbb{C}) \times \mathrm{GL}_r(\mathbb{C})$. The maximal torus of $\mathrm{Sp}(n-r,\mathbb{C}) \times \mathrm{GL}_r(\mathbb{C})$ embeds into the maximal torus of $\mathrm{GL}_n(\mathbb{C})$ by $(s_1, s_2, \ldots, s_{(n-r)/2}, a_{n-r+1}, \ldots, a_n) \mapsto (s_1, -s_1, s_2, -s_2, \ldots, s_{(n-r)/2}, -s_{(n-r)/2}, a_{n-r+1}, \ldots, a_n)$ and hence the map $\phi_{\Sigma^{\wedge}_{n,r}}$ on Chern roots is

$$(\alpha_1, \ldots, \alpha_{n-r}, \alpha_{n-r+1}, \ldots, \alpha_n) \mapsto (\sigma_1, -\sigma_1, \ldots, \sigma_{(n-r)/2}, -\sigma_{(n-r)/2}, \alpha_{n-r+1}, \ldots, \alpha_n).$$

  Equivalently, we have

$$\phi_{\Sigma^{\wedge}_{n,r}} : \prod_{i=1}^{n}(1+\alpha_i) \mapsto \prod_{i=1}^{\frac{n-r}{2}}(1-\sigma_i^2) \prod_{i=n-r+1}^{n} (1+\alpha_i).$$

- $T_{\Sigma^{\wedge}_{n,r}} = \mathrm{span}(e_i \otimes e_j - e_j \otimes e_i : 1 \leqslant i < j \leqslant n, i \leqslant n-r)$, $N_{\Sigma^{\wedge}_{n,r}} = \mathrm{span}(e_i \otimes e_j - e_j \otimes e_i : n-r+1 \leqslant i < j \leqslant n)$ and hence

$$\begin{aligned} c(T_{\Sigma^{\wedge}_{n,r}}) &= \prod_{1 \leqslant i < j \leqslant \frac{n-r}{2}} (1 \pm \sigma_i \pm \sigma_j) \prod_{i=1}^{\frac{n-r}{2}} \prod_{j=n-r+1}^{n} (1 \pm \sigma_i + \alpha_j), \\ e(N_{\Sigma^{\wedge}_{n,r}}) &= \prod_{n-r+1 \leqslant i < j \leqslant n} (\alpha_i + \alpha_j), \end{aligned}$$

  where we used the short-hand notations $(x \pm \sigma) := (x+\sigma)(x-\sigma)$ and $(1 \pm \sigma_i \pm \sigma_j) := (1+\sigma_i+\sigma_j)(1+\sigma_i-\sigma_j)(1-\sigma_i+\sigma_j)(1-\sigma_i-\sigma_j)$.

We see that $e(N_{\Sigma^{\wedge}_{n,r}}) \neq 0$ which proves that the representation $\Lambda^2\mathbb{C}^n$ satisfies Assumption 3.3.

Now we need to prove that the function $W^{\wedge}_{n,r}$ satisfies properties (1)–(3) of Theorem 3.4. Towards this goal, first we verify this claim for the special case of $r = 0$ (and necessarily n even). Recall that

$$W^{\wedge}_{n,0}(\boldsymbol{\alpha}_{[n]}) = \frac{1}{2^{\frac{n}{2}}\left(\frac{n}{2}\right)!} \sum_{\sigma \in S_n} \sigma \left(\prod_{1 \leqslant i < j \leqslant \frac{n}{2}} f_{2i-1,2j-1} f_{2i-1,2j} f_{2i,2j-1} f_{2i,2j} \right)$$

where $f_{i,j} = \frac{(1+\alpha_i+\alpha_j)(\alpha_i+\alpha_j)}{(\alpha_i-\alpha_j)}$. Applying $\phi_{\Sigma^{\wedge}_{k,0}}$ (as described above) to this expression term-by-term we obtian many 0 terms. The only non-zero terms correspond to $\sigma \in S_n$ such that $\sigma = \tau_1 \dots \tau_\ell$ for some $1 \leqslant \ell \leqslant \frac{n}{2}$ where $\tau_i = (2i'-1, 2i')$ for some $1 \leqslant i' \leqslant \frac{n}{2}$. There are $2^{\frac{n}{2}}(\frac{n}{2})!$ such terms, all of the same value, hence

$$\begin{aligned}\phi_{\Sigma^{\wedge}_{n,0}}(W^{\wedge}_{n,0}) &= \frac{1}{2^{\frac{n}{2}}\left(\frac{n}{2}\right)!}\cdot 2^{\frac{n}{2}}\left(\frac{n}{2}\right)! \prod_{1\leqslant i<j\leqslant \frac{n}{2}} \frac{(1\pm\sigma_i\pm\sigma_j)(\pm\sigma_i\pm\sigma_j)}{(\pm\sigma_i\pm\sigma_j)} \\ &= \prod_{1\leqslant i<j\leqslant \frac{n}{2}} (1\pm\sigma_i\pm\sigma_j) \\ &= c(T_{\Sigma^{\wedge}_{n,0}}) = c(T_{\Sigma^{\wedge}_{n,0}})e(N_{\Sigma^{\wedge}_{n,0}}),\end{aligned}$$

which verifies property (1).

To show that $W^{\wedge}_{k,0}$ satisfies properties (2) and (3), consider the restriction map $\phi_{\Sigma^{\wedge}_{n,m}}$ where $0 \leqslant m \leqslant n$, $n-m$ even. The non-zero terms in the image $\phi_{\Sigma^{\wedge}_{n,m}}(W^{\wedge}_{n,0})$ are those with $\sigma = \tau_1 \dots \tau_\ell \in S_n$ for some $1 \leqslant \ell \leqslant \frac{n-m}{2}$ such that $\tau_i = (2i'-1, 2i')$ for some $1 \leqslant i' \leqslant \frac{n-m}{2}$. The $\phi_{\Sigma^{\wedge}_{n,m}}$-image of $(1 + \alpha_{2i-1} + \alpha_{2j-1})(1+\alpha_{2i-1}+\alpha_{2j})(1+\alpha_{2i}+\alpha_{2j-1})(1+\alpha_{2i}+\alpha_{2j})$ is

$$\begin{cases}(1\pm\sigma_i\pm\sigma_j) & \text{if } 1\leqslant i<j\leqslant \frac{k-m}{2},\\ (1\pm\sigma_i+\alpha_{2j-1})(1\pm\sigma_i+\alpha_{2j}) & \text{if } 1\leqslant i\leqslant \frac{k-m}{2} < j \leqslant \frac{k}{2},\end{cases}$$

and we see that $\prod_{1\leqslant i<j\leqslant\frac{n-m}{2}}(1\pm\sigma_i\pm\sigma_j)\prod_{i=1}^{\frac{n-m}{2}}\prod_{j=n-m+1}^{n}(1\pm\sigma_i+\alpha_j) = c(T_{\Sigma^{\wedge}_{n,m}})$ is a common factor in every term of $\phi_{\Sigma^{\wedge}_{n,m}}(W^{\wedge}_{n,0})$, hence $W^{\wedge}_{n,0}$ satisfies property (2).

Assume that $0 < m \leqslant k$. Each term in the image $\phi_{\Sigma^{\wedge}_{n,m}}(W^{\wedge}_{n,0})$ has degree at most $4\cdot\binom{n/2}{2}$. Therefore

$$\deg(\phi_{\Sigma^{\wedge}_{n,m}}(W^{\wedge}_{n,0})) \leqslant 4\cdot\binom{n/2}{2} < \binom{n}{2} - \frac{n-m}{2} = \deg(c(T_{\Sigma^{\wedge}_{n,m}})e(N_{\Sigma^{\wedge}_{n,m}})),$$

and therefore $W^{\wedge}_{n,0}$ satisfies property (3).

Next we show that the general $W^{\wedge}_{n,r}$ satisfies the properties (1)–(3) of Theorem 3.4. First consider $\phi_{\Sigma^{\wedge}_{n,r}}(W^{\wedge}_{n,r})$. Due to the factor $\prod_{i\in I}\prod_{j\in\bar{I}}(\alpha_i + \alpha_j)$ in the numerator of $W^{\wedge}_{n,r}$ it follows that only one term has non-zero $\phi_{\Sigma^{\wedge}_{n,r}}$-image, and we obtain

$$\begin{aligned}\phi_{\Sigma^{\wedge}_{n,r}}(W^{\wedge}_{n,r}) &= \phi_{\Sigma^{\wedge}_{n,r}}(W^{\wedge}_{n-r,0}(\boldsymbol{\alpha}_{\bar I}))\\ &\quad\times \prod_{n-r+1\leqslant i<j\leqslant n}(\alpha_i+\alpha_j)\prod_{i=n-r+1}^{n}\prod_{j=1}^{\frac{n-r}{2}}\frac{(\alpha_i\pm\sigma_j)(1+\alpha_i\pm\sigma_j)}{(\alpha_i\pm\sigma_j)}\\ &= \prod_{1\leqslant i<j\leqslant\frac{n-r}{2}}(1\pm\sigma_i\pm\sigma_j)\prod_{n-r+1\leqslant i<j\leqslant n}(\alpha_i+\alpha_j)\\ &\quad\times\prod_{i=n-r+1}^{n}\prod_{j=1}^{\frac{n-r}{2}}(1+\alpha_i\pm\sigma_j) = c(T_{\Sigma^{\wedge}_{n,r}})e(N_{\Sigma^{\wedge}_{n,r}}).\end{aligned}$$

This proves property (1).

Now let $0\leqslant m\leqslant n$, $n-m$ even, and $m\neq r$, and we study $\phi_{\Sigma^{\wedge}_{n,m}}(W^{\wedge}_{n,r})$. If $m<r$, then $\phi_{\Sigma^{\wedge}_{n,m}}(W^{\wedge}_{n,r})=0$ either because of the factor $\prod_{i<j\in I}(\alpha_i+\alpha_j)$ or because of the factor $\prod_{i\in I}\prod_{j\in\bar I}(\alpha_i+\alpha_j)$ in $W^{\wedge}_{n,r}$. If $m>r$ then all the non-zero terms in the image come from the terms in $W^{\wedge}_{n,r}$ with I such that $I\cap\{1,\ldots,n-m\}=\varnothing$. Let $I\subset\{n-m+1,\ldots,n\}$, and $|I|=r$. Then

$$\phi_{\Sigma^{\wedge}_{n,m}}(W^{\wedge}_{n-r,0}(\boldsymbol{\alpha}_{\bar I})) = \prod_{1\leqslant i<j\leqslant\frac{n-m}{2}}(1\pm\sigma_i\pm\sigma_j)\prod_{j\in\bar I-[n-m]}\prod_{k=1}^{\frac{n-m}{2}}(1+\alpha_j\pm\sigma_k),$$

and

$$\begin{aligned}\phi_{\Sigma^{\wedge}_{n,m}}\left(\prod_{i<j\in I}(\alpha_i+\alpha_j)\prod_{i\in I}\prod_{j\in\bar I}\frac{(\alpha_i+\alpha_j)(1+\alpha_i+\alpha_j)}{\alpha_j-\alpha_i}\right) &= \prod_{i<j\in I}(\alpha_i+\alpha_j)\\ \times\prod_{i\in I}\prod_{j\in\bar I-[n-m]}\prod_{k=1}^{\frac{n-m}{2}}\frac{(\alpha_i+\alpha_j)(\alpha_i\pm\sigma_k)(1+\alpha_i+\alpha_j)(1+\alpha_i\pm\sigma_k)}{(\alpha_j-\alpha_i)(\pm\sigma_k-\alpha_i)}&.\end{aligned}$$

The factor $\prod_{1\leqslant i<j\leqslant\frac{n-m}{2}}(1\pm\sigma_i\pm\sigma_j)\prod_{j=n-m+1}^{n}\prod_{k=1}^{\frac{n-m}{2}}(1+\alpha_j\pm\sigma_k) = c(T_{\Sigma^{\wedge}_{n,m}})$ is a common factor in all non-zero terms of $\phi_{\Sigma^{\wedge}_{n,m}}(W^{\wedge}_{n,r})$, which proves property (2).

Now we consider the degree of $\phi_{\Sigma^{\wedge}_{n,m}}(W^{\wedge}_{n,r})$ for $m\neq r$. We have $\deg(\phi_{\Sigma^{\wedge}_{n,m}}(W^{\wedge}_{n-r,0}))<\binom{n-r}{2}-\frac{n-m}{2}$, and hence

$$\begin{aligned}\deg(\phi_{\Sigma^{\wedge}_{n,m}}(W^{\wedge}_{n,r})) &< \binom{n-r}{2}-\frac{n-m}{2}+\binom{r}{2}+r(n-r)\\ &= \binom{n}{2}-\frac{n-m}{2}\\ &= \deg(c(T_{\Sigma^{\wedge}_{n,m}})e(N_{\Sigma^{\wedge}_{n,m}})),\end{aligned}$$

proving property (3). This completes the proof of the first statement, $c^{sm}(\Sigma^{\wedge}_{n,r}) = W^{\wedge}_{n,r}$, of the theorem.

The proof of the second statement, the case of $S^2\mathbb{C}^n$, is analogous, we leave it to the reader (or see [27]). □

6 Towards the algebraic combinatorics of CSM classes of $\Lambda^2\mathbb{C}^n$, $S^2\mathbb{C}^n$

Characteristic classes of geometrically relevant varieties usually display stabilization and positivity properties. We can expect stabilization properties from the SSM versions, not from the CSM versions, because the SSM version is the one consistent with pull-back (and hence transversal intersection). Also, traditionally the combinatorics of characteristic classes show their true nature when they are expanded in Schur basis.

Our formulas for the SSM classes obtained in Sections 4 and 5 can be expanded in Schur basis (to fix our conventions, note that $s_{11} = \sum_{i<j} \alpha_i\alpha_j$, $s_2 = \sum_{i\leqslant j} \alpha_i\alpha_j$), and we obtain

$$
\begin{aligned}
s^{sm}(\Sigma^{\wedge}_{2,0}) &= s_0 - s_1 + (s_2 + s_{11}) - (s_3 + 2s_{21}) + (s_4 + 2s_{22} + 3s_{31}) - \cdots \\
s^{sm}(\Sigma^{\wedge}_{4,0}) &= s_0 - s_1 + (s_2 + s_{11}) - (s_3 + 2s_{21} + s_{111}) \\
&\quad + (s_4 + 2s_{22} + 3s_{31} + 3s_{211} + s_{1111}) \\
&\quad - (s_5 + 5s_{32} + 4s_{41} + 5s_{221} + 6s_{311} + 4s_{2111}) + \cdots \\
s^{sm}(\Sigma^{\wedge}_{2,2}) &= s_1 - (s_2 + s_{11}) + (s_3 + 2s_{21}) - (s_4 + 2s_{22} + 3s_{31}) + \cdots \\
s^{sm}(\Sigma^{\wedge}_{4,2}) &= s_1 - (s_2 + s_{11}) + (s_3 + 2s_{21} + s_{111}) \\
&\quad - (s_4 + 2s_{22} + 3s_{31} + 3s_{211} + s_{1111}) \\
&\quad + (s_5 + 5s_{32} + 4s_{41} + 5s_{221} + 6s_{311} + 4s_{2111}) - \cdots \\
s^{sm}(\Sigma^{\wedge}_{4,4}) &= s_{321} - (3s_{322} + 3s_{331} + 3s_{421} + 3s_{3211}) + (10s_{332} + 10s_{422} \\
&\quad + 10s_{431} + 6s_{521} + 10s_{3221} + 10s_{3311} + 10s_{4211}) - \cdots
\end{aligned}
$$

$$
\begin{aligned}
s^{sm}(\Sigma^{S}_{1,0}) &= s_0 - 2s_1 + 4s_2 - 8s_3 + 16s_4 - \cdots \\
s^{sm}(\Sigma^{S}_{2,0}) &= s_0 - 2s_1 + (4s_2 + 4s_{11}) - (8s_3 + 12s_{21}) \\
&\quad + (16s_4 + 12s_{22} + 28s_{31}) - \cdots \\
s^{sm}(\Sigma^{S}_{3,0}) &= s_0 - 2s_1 + (4s_2 + 4s_{11}) - (8s_3 + 12s_{21} + 8s_{111}) \\
&\quad + (16s_4 + 12s_{22} + 28s_{31} + 28s_{211}) - \cdots \\
s^{sm}(\Sigma^{S}_{1,1}) &= 2s_1 - 4s_2 + 8s_3 - 16s_4 + \cdots
\end{aligned}
$$

$$\begin{aligned}
\mathrm{s}^{\mathrm{sm}}(\Sigma^S_{2,1}) &= 2s_1 - (4s_2 + 4s_{11}) + (8s_3 + 8s_{21}) - (16s_4 + 16s_{31}) + \cdots \\
\mathrm{s}^{\mathrm{sm}}(\Sigma^S_{3,1}) &= 2s_1 - (4s_2 + 4s_{11}) + (8s_3 + 8s_{21} + 8s_{111}) \\
&\quad - (16s_4 + 16s_{31} + 16s_{211}) + \cdots \\
\mathrm{s}^{\mathrm{sm}}(\Sigma^S_{2,2}) &= 4s_{21} - (12s_{22} + 12s_{31}) + (40s_{32} + 28s_{41}) - \cdots \\
\mathrm{s}^{\mathrm{sm}}(\Sigma^S_{3,2}) &= 4s_{21} - (12s_{22} + 12s_{31} + 12s_{211}) \\
&\quad + (40s_{32} + 28s_{41} + 40s_{221} + 40s_{311}) - \cdots .
\end{aligned}$$

Here are some observations on these expressions:

- **(stabilization)** For $n < m$ the formula for $\mathrm{s}^{\mathrm{sm}}(\Sigma_{m,r})$ also works for $\mathrm{s}^{\mathrm{sm}}(\Sigma_{n,r})$ (either $\Lambda^2\mathbb{C}^n$ or $S^2\mathbb{C}^n$). Of course, some s_λ functions may be non-0 for m variables, but 0 for n variables, so some terms of $\mathrm{s}^{\mathrm{sm}}(\Sigma_{m,r})$ are not necessary to name $\mathrm{s}^{\mathrm{sm}}(\Sigma_{n,r})$. The reason for this stabilization, on the one hand, is that $\Lambda^2\mathbb{C}^n$ can be viewed as a linear section of $\Lambda^2\mathbb{C}^m$ (consistent with the group actions) such that orbits of $\Lambda^2\mathbb{C}^n$ are transversal to this linear space, and on the other hand, SSM classes are consistent with transversal intersection, see (iv) in Section 3.1. Due to this stabilization one may consider the limit objects (formal power series) $\mathrm{s}^{\mathrm{sm}}(\Sigma_{\infty,r})$. In [27] generating functions (in the "iterated residue" sense) are presented for these limit power series.
- **(positivity)** We expect that the coefficients of SSM classes in Schur basis have predictable signs. All the above examples support

  Conjecture 6.1 *The Schur expansions of* $\mathrm{s}^{\mathrm{sm}}(\Sigma^{\wedge}_{n,r})$ *and* $\mathrm{s}^{\mathrm{sm}}(\Sigma^S_{n,r})$ *have alternating signs.*

  The sign behavior of SSM classes is determined under very general circumstances in [5]. It would be interesting to check whether results in that paper imply our conjecture.

- **(lowest degree terms)** The lowest degree term of either a CSM or an SSM class is the fundamental class of the closure of the orbit. Hence the expressions above are all of the form $\mathrm{s}^{\mathrm{sm}}(\Sigma^{\wedge}_{n,r}) = s_{r-1,r-2,\dots,1} + h.o.t.$, $\mathrm{s}^{\mathrm{sm}}(\Sigma^S_{n,r}) = 2^{r-1}s_{r,r-1,\dots,1} + h.o.t.$.
- **(normalization)** The additivity property of CSM classes imply that the sum of the SSM classes of all orbits of a representation (with finitely many orbits) is 1—we encourage the reader to verify this property in the above examples. Normalization and positivity properties together indicate a formal similarity between SSM theory and probability theory, cf. [14, Remark 8.8].

Our choice above for expanding in terms of *Schur functions* is essentially due to tradition. Schur functions are the fundamental classes of so-called matrix Schubert varieties. Those varieties are indeed very basic ones, but the choice of choosing their *fundamental class* as our basic polynomials might be improved. It might be more natural that for SSM classes of geometrically relevant varieties the "right" choice of expansion is in terms of the *SSM classes* of matrix Schubert varieties. These functions, named $\tilde{s}_\lambda$, are defined and calculated in [14, Definition 8.2]. Moreover, the $\tilde{s}_\lambda = s_\lambda + h.o.t.$ functions are themselves Schur alternating (conjectured in [14], proved in [5]). Remarkably, SSM classes of some quiver loci are proved to be $\tilde{s}_\lambda$-positive—indicating a two-step positivity structure of SSM classes (see the Introduction of [14]). The following are $\tilde{s}_\lambda$-expansions:

$$\begin{aligned}
\mathrm{s}^{\mathrm{sm}}(\Sigma^{\wedge}_{\infty,0}) &= \tilde{s}_0 + \tilde{s}_{22} + (\tilde{s}_{44} + \tilde{s}_{2222}) + (\tilde{s}_{66} + \tilde{s}_{4422} + \tilde{s}_{222222}) + \cdots,\\
\mathrm{s}^{\mathrm{sm}}(\Sigma^{\wedge}_{\infty,1}) &= \tilde{s}_0 + \tilde{s}_1 + (\tilde{s}_2 + \tilde{s}_{11}) + (\tilde{s}_3 + \tilde{s}_{111}) + (\tilde{s}_4 + \tilde{s}_{1111}) + (\tilde{s}_5 + \tilde{s}_{11111})\\
&\quad + (\tilde{s}_6 + \tilde{s}_{33} + \tilde{s}_{222} + \tilde{s}_{111111}) + \cdots\\
\mathrm{s}^{\mathrm{sm}}(\Sigma^{\wedge}_{\infty,2}) &= \tilde{s}_1 + (\tilde{s}_2 + \tilde{s}_{11}) + (\tilde{s}_3 + \tilde{s}_{21} + \tilde{s}_{111}) + (\tilde{s}_4 + \tilde{s}_{31} + \tilde{s}_{211} + \tilde{s}_{1111})\\
&\quad + (\tilde{s}_5 + \tilde{s}_{41} + \tilde{s}_{32} + \tilde{s}_{311} + \tilde{s}_{221} + \tilde{s}_{2111} + \tilde{s}_{11111}) + \cdots,
\end{aligned}$$

$$\begin{aligned}
\mathrm{s}^{\mathrm{sm}}(\Sigma^{S}_{\infty,0}) &= \tilde{s}_0 - \tilde{s}_1 + (\tilde{s}_2 + \tilde{s}_{11}) - (\tilde{s}_3 + \tilde{s}_{21} + \tilde{s}_{111})\\
&\quad + (\tilde{s}_4 + \tilde{s}_{31} + \tilde{s}_{22} + \tilde{s}_{211} + \tilde{s}_{1111})\\
&\quad - (\tilde{s}_5 + \tilde{s}_{41} + \tilde{s}_{32} + \tilde{s}_{311} + \tilde{s}_{221} + \tilde{s}_{2111} + \tilde{s}_{11111}) + \cdots,\\
\mathrm{s}^{\mathrm{sm}}(\Sigma^{S}_{\infty,1}) &= 2\tilde{s}_1 + (2\tilde{s}_3 - 2\tilde{s}_{21} + 2\tilde{s}_{111})\\
&\quad + (2\tilde{s}_5 - 2\tilde{s}_{41} + 2\tilde{s}_{32} + 2\tilde{s}_{311} + 2\tilde{s}_{221} - 2\tilde{s}_{2111} + 2\tilde{s}_{11111})\\
&\quad - 4\tilde{s}_{321} + \cdots,\\
\mathrm{s}^{\mathrm{sm}}(\Sigma^{S}_{\infty,2}) &= 4\tilde{s}_{21} + (4\tilde{s}_{41} + 4\tilde{s}_{2111}) - 4\tilde{s}_{321}\\
&\quad + (4\tilde{s}_{61} + 4\tilde{s}_{43} + 4\tilde{s}_{4111} + 4\tilde{s}_{2221} + 4\tilde{s}_{211111})\\
&\quad - (4\tilde{s}_{521} + 4\tilde{s}_{32111}) + \cdots,\\
\mathrm{s}^{\mathrm{sm}}(\Sigma^{S}_{\infty,3}) &= 8\tilde{s}_{321} + (8\tilde{s}_{521} + 8\tilde{s}_{32111})\\
&\quad + (8\tilde{s}_{721} + 8\tilde{s}_{541} - 8\tilde{s}_{4321} + 8\tilde{s}_{32221} + 8\tilde{s}_{3211111})\\
&\quad + (8\tilde{s}_{921} + 8\tilde{s}_{741} - 8\tilde{s}_{6321} + 8\tilde{s}_{543} + 8\tilde{s}_{33321}\\
&\quad - 8\tilde{s}_{432111} + 8\tilde{s}_{3222111} + 8\tilde{s}_{321111111}) + \cdots,\\
\mathrm{s}^{\mathrm{sm}}(\Sigma^{S}_{\infty,4}) &= 16\tilde{s}_{4321} + (16\tilde{s}_{6321} + 16\tilde{s}_{432111}) + \cdots.
\end{aligned}$$

It is worth verifying in these examples the normalization properties

$$\sum_i \mathrm{s}^{\mathrm{sm}}(\Sigma^{S}_{\infty,i}) = \sum_i \mathrm{s}^{\mathrm{sm}}(\Sigma^{\wedge}_{\infty,2i}) = \sum_i \mathrm{s}^{\mathrm{sm}}(\Sigma^{\wedge}_{\infty,2i+1}) = \sum_\lambda \tilde{s}_\lambda = 1.$$

The calculated $\tilde{s}_\lambda$-expansions (the ones above and many more) display several patterns; let us phrase two of them as conjectures.

Conjecture 6.2 • *The $\tilde{s}_\lambda$-expansions of* $\mathrm{s}^{\mathrm{sm}}(\Sigma^{\wedge}_{\infty,i})$ *and* $\mathrm{s}^{\mathrm{sm}}(\Sigma^{S}_{\infty,i})$ *are invariant under* $\lambda \mapsto \lambda^T$ *(=the transpose partition, eg.* $(6321)^T = 432111$*). That is, in both expansions the coefficient of* $\tilde{s}_\lambda$ *and the coefficient of* $\tilde{s}_{\lambda^T}$ *are the same.*

- *The $\tilde{s}_\lambda$-expansion of* $\mathrm{s}^{\mathrm{sm}}(\Sigma^{\wedge}_{\infty,i})$ *has non-negative coefficients. The* $\tilde{s}_\lambda$*-expansion of* $\mathrm{s}^{\mathrm{sm}}(\Sigma^{S}_{\infty,2i})$ *has alternating coefficients.*

7 Applications

We will apply the calculated CSM classes to find the Euler characteristics of general linear sections of the projectivizations of $\Sigma^{\wedge}_{n,r}$, $\Sigma^{S}_{n,r}$. First we study the relation between characteristic classes in vector spaces and those in projective spaces.

7.1 Characteristic classes before vs after projectivization

Consider the algebraic representation $G \circlearrowright V = \mathbb{C}^N$, with $T = (\mathbb{C}^*)^m \leqslant G$ the maximal torus, and weights σ_j. Suppose the representation contains the scalars, that is, there is a map $\phi : \mathbb{C}^* \to T$, $\phi(s) = (s^{w_1}, s^{w_2}, \ldots, s^{w_m})$ such that $\phi(s)$ acts on V with multiplication by s^w ($w \neq 0$). Then the G-invariant subsets $\Sigma \subset V$ are necessarily cones.

We want to compare the G-equivariant characteristic classes of $\Sigma \subset V$ with those of $\mathbb{P}\Sigma \subset \mathbb{P}\mathrm{V}$. The first one lives in the ring $H^*_G(V) = H^*(BG)$, while the second one lives in

$$H^*_G(\mathbb{P}\mathrm{V}) = H^*_G(\mathbb{P}\mathrm{V}) = H^*(BG)[\xi] \,/\, \textstyle\prod_j(\xi - \sigma_j), \tag{6}$$

where ξ is the first Chern class of the G-equivariant tautological line bundle over $\mathbb{P}\mathrm{V}$. Here $H^*(BG)$ is a subring of $H^*(BT) = \mathbb{Q}[\alpha_1, \ldots, \alpha_m]$, and the weights σ_j are linear combinations of the α_i's.

Theorem 7.1 *The substitutions*

$$[\Sigma]|_{\alpha_i \mapsto \alpha_i + \frac{w_i}{w}\xi}, \qquad \mathrm{s}^{\mathrm{sm}}(\Sigma)|_{\alpha_i \mapsto \alpha_i + \frac{w_i}{w}\xi}$$

represent $[\mathbb{P}\Sigma \subset \mathbb{P}\mathrm{V}]$ *and* $\mathrm{s}^{\mathrm{sm}}(\mathbb{P}\Sigma \subset \mathbb{P}\mathrm{V})$*, respectively, in* (6).

The first statement is [12, Theorem 6.1], and the proof there holds for SSM classes as well, because the proof given there only uses the pull-back property which [ ] shares with SSM classes.

The non-equivariant ("ordinary") characteristic classes are always obtained from the equivariant ones by substituting 0 in the equivariant variables. Therefore, the non-equivariant SSM class $\mathrm{s}_0^{\mathrm{sm}}(\mathbb{P}\Sigma \subset \mathbb{P}V)$ of $\mathbb{P}\Sigma$ living in $H^*(\mathbb{P}V) = \mathbb{Q}[\xi]/\xi^N$ is obtained as

$$\mathrm{s}_0^{\mathrm{sm}}(\mathbb{P}\Sigma \subset \mathbb{P}V) = \mathrm{s}^{\mathrm{sm}}(\Sigma \subset V)|_{\alpha_i \mapsto \frac{w_i}{w}\xi}.$$

Remarkably, the same holds for CSM classes too: $\mathrm{c}_0^{\mathrm{sm}}(\mathbb{P}\Sigma \subset \mathbb{P}V) = \mathrm{c}^{\mathrm{sm}}(\Sigma \subset V)|_{\alpha_i \mapsto \frac{w_i}{w}\xi}$, which follows from the calculation

$$\begin{aligned}\mathrm{c}_0^{\mathrm{sm}}(\mathbb{P}\Sigma) &= \mathrm{s}_0^{\mathrm{sm}}(\mathbb{P}\Sigma)c(\mathbb{P}V) = \mathrm{s}^{\mathrm{sm}}(\Sigma)|_{\alpha_i \mapsto \frac{w_i}{w}\xi} \cdot (1+\xi)^N \\ &= \left.\frac{\mathrm{c}^{\mathrm{sm}}(\Sigma)}{\prod_j (1+\sigma_j)}\right|_{\alpha_i \mapsto \frac{w_i}{w}\xi} \cdot (1+\xi)^N = \mathrm{c}^{\mathrm{sm}}(\Sigma)|_{\alpha_i \mapsto \frac{w_i}{w}\xi},\end{aligned}$$

where the last equality used the defining property of w_i, w, namely: $1+\sigma_j|_{\alpha_i \mapsto \frac{w_i}{w}\xi} = 1+\xi$ (for all j).

7.2 Non-equivariant CSM classes of symmetric and skew-symmetric determinantal varieties

Let us study the projectivizations of $\Sigma^{\wedge}_{n,r}$ and $\Sigma^{S}_{n,r}$. Some of these projective varieties are well known: $\mathbb{P}\,\Sigma^{\wedge}_{n,n-2}$ is the Plücker embedding of $\mathrm{Gr}_2\,\mathbb{C}^n$, and $\mathbb{P}\,\Sigma^{S}_{n,n-1}$ is the Veronese embedding of $\mathbb{P}\,\mathbb{C}^n$. Applying the result of the preceding section to $\Lambda^2\mathbb{C}^n$ and $S^2\mathbb{C}^n$ we find that the ordinary CSM classes of the orbits are obtained by

$$\begin{aligned}\mathrm{c}_0^{\mathrm{sm}}(\mathbb{P}\,\Sigma^{\wedge}_{n,r} \subset \mathbb{P}\,\Lambda^2\mathbb{C}^n) &= \mathrm{c}^{\mathrm{sm}}(\Sigma^{\wedge}_{n,r} \subset \Lambda^2\mathbb{C}^n)|_{\alpha_i \mapsto \xi/2},\\ \mathrm{c}_0^{\mathrm{sm}}(\mathbb{P}\,\Sigma^{S}_{n,r} \subset \mathbb{P}\,S^2\mathbb{C}^n) &= \mathrm{c}^{\mathrm{sm}}(\Sigma^{S}_{n,r} \subset S^2\mathbb{C}^n)|_{\alpha_i \mapsto \xi/2}.\end{aligned}$$

For example from the explicit formula for $\mathrm{c}^{\mathrm{sm}}(\Sigma^{S}_{3,r} \subset S^2\mathbb{C}^3)$ given in Definition 5.3, after substituting $\alpha_i = \xi_i/2$ for all i we obtain

$$\begin{aligned}\mathrm{c}_0^{\mathrm{sm}}(\mathbb{P}\,\Sigma^{S}_{3,0}) &= 1 + 3\xi + 6\xi^2 + 6\xi^3 + 3\xi^4,\\ \mathrm{c}_0^{\mathrm{sm}}(\mathbb{P}\,\Sigma^{S}_{3,1}) &= 3\xi + 9\xi^2 + 10\xi^3 + 6\xi^4 + 3\xi^5,\\ \mathrm{c}_0^{\mathrm{sm}}(\mathbb{P}\,\Sigma^{S}_{3,2}) &= 4\xi^3 + 6\xi^4 + 3\xi^5.\end{aligned} \tag{7}$$

As we know, the degrees of the lowest degree terms above, namely $0, 1, 3$, are the codimensions of the given orbits. The coefficients of the lowest degree terms, namely $1,3,4$, are the degrees of the closures of those orbits. The integral, ie. the coefficients of ξ^5, namely $0,3,3$ are the Euler characteristics of the orbits. Observe also, that the sum of the three classes above are $1+6\xi+15\xi^2+20\xi^3+15\xi^4+6\xi^5 = c(T\,\mathbb{P}^5) = (1+\xi)^6 \in \mathbb{Q}[\xi]/(\xi^6)$.

The codimensions, the degrees, and the Euler characteristics of the orbits of $\mathbb{P}\,\Lambda^2\mathbb{C}^n$ and $\mathbb{P}\,S^2\mathbb{C}^n$ are

$$\begin{aligned}
\operatorname{codim}(\mathbb{P}\,\Sigma^{\wedge}_{n,r} \subset \mathbb{P}\,\Lambda^2\mathbb{C}^n) &= \tbinom{r}{2},\\
\operatorname{codim}(\mathbb{P}\,\Sigma^{S}_{n,r} \subset \mathbb{P}\,S^2\mathbb{C}^n) &= \tbinom{r+1}{2},\\
\deg(\mathbb{P}\,\Sigma^{\wedge}_{n,r} \subset \mathbb{P}\,\Lambda^2\mathbb{C}^n) &= \tfrac{1}{2^{r-1}} \prod_{i=0}^{r-2} \frac{\binom{n+i}{r-1-i}}{\binom{2i+1}{i}},\\
\deg(\mathbb{P}\,\Sigma^{S}_{n,r} \subset \mathbb{P}\,S^2\mathbb{C}^n) &= \prod_{i=0}^{r-1} \frac{\binom{n+i}{r-i}}{\binom{2i+1}{i}},
\end{aligned}$$

$$\chi(\mathbb{P}\,\Sigma^{\wedge}_{n,r}) = \begin{cases} \binom{n}{2} & \text{for } r = n-2 \\ 0 & \text{for } r < n-2, \end{cases}$$

$$\chi(\mathbb{P}\,\Sigma^{S}_{n,r}) = \begin{cases} n & \text{for } r = n-1 \\ \binom{n}{2} & \text{for } r = n-2 \\ 0 & \text{for } r < n-2. \end{cases}$$

These formulas are well known, and can be found either by classical methods or using our formula, see details in [27]. The question is: what geometric information is carried by the 'middle' terms of the c_0^{sm} classes like (7). This will be answered in the next section.

7.3 Euler characteristics of general linear sections

Let $X \subset \mathbb{P}^N$ be a locally closed set, and let $X_r = X \cap H_1 \cap \cdots \cap H_r$ be the intersection with r general hyperplanes. Following [1] define the *Euler characteristic polynomial of* X to be

$$\chi_X(t) = \sum_{i=0}^{N} \chi(X_i)(-t)^i.$$

From the non-equivariant CSM class of X, $c_0^{\mathrm{sm}}(X \subset \mathbb{P}^N) = \sum_{i=0}^{N} a_i \xi^i$ define $\gamma_X(t) = \sum_{i=0}^{N} a_i t^{N-i}$. Aluffi showed that the two polynomials χ_X and γ_X are related as follows. For a polynomial $p(t)$ define

$$\mathcal{J}(p)(t) = \frac{tp(-t-1)+p(0)}{t+1}.$$

The operation $\mathcal{J}$ is a degree-preserving linear involution on polynomials in t.

Theorem 7.2 [1, Theorem 1.1] *For every locally closed subset* X *of* $\mathbb{CP}^N$, *we have*

$$\mathcal{J}(\chi_X(t)) = \gamma_X(t) \quad \textit{and} \quad \mathcal{J}(\gamma_X(t)) = \chi_X(t).$$

Putting our results together with Theorem 7.2 we have an algorithm to find the Euler characteristics of general linear sections of the orbits of $\mathbb{P}\,\Sigma^{\wedge}_{n,r}$, $\mathbb{P}\,\Sigma^{S}_{n,r}$. Namely: Formulas for $\mathrm{GL}_n(\mathbb{C})$-equivariant CSM classes of $\Sigma^{\wedge}_{n,r}$, $\Sigma^{S}_{n,r}$ are given in Sections 4 and 5. Those formulas turn to formulas for non-equivariant CSM classes for $\mathbb{P}\,\Sigma^{\wedge}_{n,r}$, $\mathbb{P}\,\Sigma^{S}_{n,r}$ in Section 7.2. According to Theorem 7.2, the coefficients of the $\mathcal{J}$-operation of those non-equivariant CSM classes are the Euler characteristics of the general linear sections.

Example 7.3 From the calculations in (7) we get

$$\gamma_{\mathbb{P}\Sigma^S_{3,0}}(t) = 3t + 6t^2 + 6t^3 + 3t^4 + t^5,$$
$$\gamma_{\mathbb{P}\Sigma^S_{3,1}}(t) = 3 + 6t + 10t^2 + 9t^3 + 3t^4,$$
$$\gamma_{\mathbb{P}\Sigma^S_{3,2}}(t) = 3 + 6t + 4t^2.$$

After applying the involution $\mathcal{J}$ we get the Euler characteristic polynomials

$$\chi_{\mathbb{P}\Sigma^S_{3,0}}(t) = (-t) - (-t)^2 + 3(-t)^3 - (-t)^4 + (-t)^5,$$
$$\chi_{\mathbb{P}\Sigma^S_{3,1}}(t) = 3 + 2(-t) + (-t)^2 + 3(-t)^4,$$
$$\chi_{\mathbb{P}\Sigma^S_{3,2}}(t) = 3 + 2(-t) + 4(-t)^2,$$

and the Euler characteristics of Table 1. Similar calculation yields e.g. the Euler characteristics presented in Table 2.

It is worth verifying that the sum of columns (in both tables) is the Euler characteristic of the appropriate projective linear space.

Table 1. *Euler characteristics of general linear sections of the orbits in* $\mathbb{P}\,S^2\mathbb{C}^3$.

X	$\chi(X)$	$\chi(X_1)$	$\chi(X_2)$	$\chi(X_3)$	$\chi(X_4)$	$\chi(X_5)$
$\mathbb{P}\Sigma^S_{3,0}$	0	1	−1	3	−1	1
$\mathbb{P}\Sigma^S_{3,1}$	3	2	1	0	3	0
$\mathbb{P}\Sigma^S_{3,2}$	3	2	4	0	0	0

Table 2. *Euler characteristics of general linear sections of the orbits in* $\mathbb{P}\Lambda^2\mathbb{C}^6$. *Column sums are* $\chi(\mathbb{P}^{14-i})$.

X	$\chi(X)$	$\chi(X_1)$	$\chi(X_2)$	$\chi(X_3)$	$\chi(X_4)$
$\mathbb{P}\Sigma^{\wedge}_{6,0}$	0	–1	1	–3	5
$\mathbb{P}\Sigma^{\wedge}_{6,2}$	0	3	0	9	–6
$\mathbb{P}\Sigma^{\wedge}_{6,4}$	15	12	12	6	12

X	$\chi(X_5)$	$\chi(X_6)$	$\chi(X_7)$	$\chi(X_8)$	$\chi(X_9)$
$\mathbb{P}\Sigma^{\wedge}_{6,0}$	–11	21	–29	29	–21
$\mathbb{P}\Sigma^{\wedge}_{6,2}$	27	–36	51	–36	27
$\mathbb{P}\Sigma^{\wedge}_{6,4}$	–6	24	–14	14	0

X	$\chi(X_{10})$	$\chi(X_{11})$	$\chi(X_{12})$	$\chi(X_{13})$	$\chi(X_{14})$
$\mathbb{P}\Sigma^{\wedge}_{6,0}$	11	–5	3	–1	1
$\mathbb{P}\Sigma^{\wedge}_{6,2}$	–6	9	0	3	0
$\mathbb{P}\Sigma^{\wedge}_{6,4}$	0	0	0	0	0

8 Future directions

8.1 Chern–Mather classes

Another reason for studying CSM classes of singular varieties is the relation with their Chern–Mather classes. For the role of Chern–Mather classes in geometry see the recent paper [2] and references therein. One approach to Chern–Mather classes is the construction called Nash blow-up, another one is the natural transformation $C_* : \mathcal{F}^G(-) \to H_*^G(-)$ mentioned in the Remark at the end of Section 3.1. As we know, the (homology) CSM class of a closed subvariety W is $C_*(\mathbf{1}_W)$. The Chern–Mather class $\mathrm{c}^{\mathrm{M}}(W)$ of W is the C_*-image of another remarkable constructible function, the so-called local Euler obstruction function, Eu_W. Hence if Eu_W can be calculated, i.e. expressed as a linear combination of $\mathbf{1}_{V_i}$'s (for locally closed set V_i), then the same linear relation holds among $\mathrm{c}^{\mathrm{M}}(W)$ and the CSM classes of V_i's. Arguments along these lines are carried out in [27] resulting the following theorem.

Theorem 8.1 [27, Thms 9.11, 9.18] *For* $0 \leqslant r \leqslant n$ *we have*

$$\mathrm{Eu}_{\overline{\Sigma}^{\wedge}_{n,r}} = \sum_{k=0}^{\frac{n-r}{2}} \binom{\lfloor \frac{r}{2} \rfloor + k}{\lfloor \frac{r}{2} \rfloor} \mathbf{1}_{\Sigma^{\wedge}_{n,r+2k}},$$

and hence

$$\mathrm{c^M}(\overline{\Sigma}^{\wedge}_{n,r}) = \sum_{k=0}^{\frac{n-r}{2}} \binom{\lfloor \frac{r}{2} \rfloor + k}{\lfloor \frac{r}{2} \rfloor} \mathrm{c^{sm}}(\Sigma^{\wedge}_{n,r+2k}).$$

The authors do not know the local Euler obstructions for the orbit closures in $\mathrm{S^2\mathbb{C}^n}$.

8.2 K theory generalization: motivic Chern classes

There is a natural generalization of the cohomological notion of CSM class to K theory, called *motivic Chern class*. It was defined in [9] and the equivariant version is set up in [16, 6].

The equivariant motivic Chern class $\mathrm{mC}(\Sigma)$ of an invariant subvariety $\Sigma \subset M$ of the smooth ambient variety M lives in $K_G(M)[y]$. It is convenient to consider its Segre version, the motivic Segre class $\mathrm{mS}(\Sigma) = \mathrm{mC}(\Sigma)/c(TM)$ where $c(TM)$ is the K theoretic total Chern class. Hence, mC and mS of the orbits of $\Lambda^2\mathbb{C}^n$ and $\mathrm{S^2\mathbb{C}^n}$ are elements of (a completion of)

$$K_{\mathrm{GL}_n(\mathbb{C})}(\mathrm{pt}) = \mathbb{Z}[\alpha_1^{\pm 1}, \dots, \alpha_n^{\pm 1}]^{S_n}[y],$$

where α_i are the K theory Chern roots $\mathrm{GL}_n(\mathbb{C})$, ie. their sum is the tautological n-bundle over $B\mathrm{GL}_n(\mathbb{C})$.

The traditional approach to studying motivic Chern classes is through resolutions and a property similar to (2) (with the notion of Euler characteristic replaced with the notion of chi-y-genus). Our construction in Section 4.1 fits that approach, and hence arguments analogous to those in Section 4.2 can be carried out to obtain a sieve formula for the motivic Segre class mS of the orbits of $\Lambda^2\mathbb{C}^n$, $\mathrm{S^2\mathbb{C}^n}$. Here we present the result for $\Lambda^2\mathbb{C}^n$.

Theorem 8.2 [27, Cor. 10.15] *Let $0 \leqslant r \leqslant n$, $n - r$ even, and $q = -y$. We have*

$$\mathrm{mS}(\Sigma^{\wedge}_{n,r}) = \sum_{k=0}^{\frac{n-r}{2}} \binom{r+2k}{r}_q E_{2k}(q)\Phi^{\wedge}_{n,r+2k}, \tag{8}$$

where

$$\Phi^{\wedge}_{n,r} = \sum_{\substack{I \subset [n] \\ |I|=r}} \left(\prod_{i<j \in I} \frac{1-\frac{1}{\alpha_i\alpha_j}}{1+\frac{y}{\alpha_i\alpha_j}} \prod_{i\in I} \prod_{j \in \bar{I}} \frac{\left(1-\frac{1}{\alpha_i\alpha_j}\right)\left(1+\frac{y\alpha_j}{\alpha_i}\right)}{\left(1+\frac{y}{\alpha_i\alpha_j}\right)\left(1-\frac{\alpha_j}{\alpha_i}\right)} \right),$$

$$\binom{n}{m}_q = \frac{[n]_q!}{[m]_q!\,[n-m]_q!}, \qquad [n]_q! = [1]_q[2]_q \ldots [n]_q, \qquad [0]_q! = 1,$$

and the q-Euler numbers $E_n(q)$ are defined by

$$\frac{1}{\cosh_q(t)} = \sum_{n=0}^{\infty} \frac{E_n(q)}{[n]_q!} \cdot t^n, \qquad \cosh_q(t) = \sum_{i=0}^{\infty} \frac{t^{2n}}{[2n]_q!}.$$

Theorem 8.2 is just a (rather complicated) sieve formula. The desired formula would be analogous to the interpolation formula Theorem 5.5 for CSM classes. Although interpolation characterization of motivic Chern classes exist [16, Section 5.2], the solution of those interpolation constraints (involving Newton polytopes of specializations) is highly non-trivial, and hence is subject to future study. Initial results and conjectures are in [27].

Remark 8.3 As mentioned above, the proof of Theorem 8.2 is the K theory version of our cohomology arguments in Section 4.2—in particular, it uses push-forward morphisms. The notion of motivic Chern class is consistent with push-forward morphisms (in a sense generalizing (2), see [9, Section 2], [16, Section 2.3]). This is why the proof of Section 4.2 has a K theory counterpart, leading to Theorem 8.2. There is, however, a traditional notion of *K theory fundamental class* (namely, the class of the structure sheaf) which only behaves nicely under push-forward morphism if the varieties involved have rational singularities, c.f. [11], [28, §5]. In fact, the orbit closures in $S^2\mathbb{C}^n$ and $\Lambda^2\mathbb{C}^n$ *do* have rational singularities, see Sections 6.3 and 6.4 (especially the discussions following the proofs of Propositions 6.3.2 and 6.4.2) of [36]. Therefore the $y = 0$ specialization of (8) recovers the K theory fundamental class of the orbit closures. For a detailed study of the various notions of K theory fundamental classes see [11], for more work on K theory fundamental classes of symmetric and skew-symmetric degeneracy loci see [7].

References

[1] P. Aluffi. Euler characteristics of general linear sections and polynomial Chern classes. Rend. Circ. Mat. Palermo. (special issue) 62 (2013) 3–26.

[2] P. Aluffi. Projective duality and a Chern-Mather involution. Trans. AMS 370 (2018), no. 3, 1803–1822.

[3] P. Aluffi, L. C. Mihalcea. Chern–Schwartz–MacPherson classes for Schubert cells in flag manifolds. Compositio Math. 152 (2016), 2603–2625.

[4] P. Aluffi, L. C. Mihalcea, J. Schürmann, Ch. Su. Shadows of characteristic cycles, Verma modules, and positivity of Chern–Schwartz–MacPherson classes of Schubert cells. Preprint, arXiv:1708.08697.

[5] P. Aluffi, L. C. Mihalcea, J. Schürmann, Ch. Su. Positivity of Segre-MacPherson classes. In *Facets of Algebraic Geometry. A Collection in Honor of William Fulton's 80th Birthday*, Volume 1, pp. 1–28, Cambridge University Press, 2022.

[6] P. Aluffi, L. C. Mihalcea, J. Schürmann, Ch. Su. Motivic Chern classes of Schubert cells, Hecke algebras, and applications to Casselman's problem. Preprint, arXiv:1902.10101.

[7] D. Anderson. K-theoretic Chern class formulas for vexillary degeneracy loci. Adv. Math. 350 (2019), 440–485.

[8] D. Anderson, W. Fulton. Degeneracy Loci, Pfaffians, and Vexillary Signed Permutations in Types B, C, and D. Preprint, arXiv:1210.2066.

[9] J.-P. Brasselet, J. Schürmann, and Sh.Yokura. Hirzebruch classes and motivic Chern classes for singular spaces. *Journal of Topology and Analysis*, 2(1):1–55, March 2010.

[10] J.-P. Brasselet and M.-H. Schwartz. Sur les classes de Chern d'un ensemble analytique complexe. In *Caractéristique d'Euler-Poincaré - Séminaire E.N.S. 1978–1979*, number 82-83 in Astérisque, pages 93–147. Société Mathématique de France, 1981.

[11] L. M. Fehér. Motivic Chern classes of cones. In *Singularities and Their Interaction with Geometry and Low Dimensional Topology. In Honor of A. Némethi.* Eds. J. Fernández de Bobadilla, T. László, A. Stipsicz. Birkhäuser, pp. 181–205, 2021.

[12] L. Fehér, A. Némethi, R. Rimányi: Degeneracy of two and three forms. Canad. Math. Bull. Vol. 48 (4), 2005, 547–560.

[13] L. Fehér, R. Rimányi. Calculation of Thom polynomials and other cohomological obstructions for group actions, in Real and Complex Singularities (Sao Carlos, 2002) Ed. T.Gaffney and M.Ruas, Contemp. Math.,#354, Amer. Math. Soc., Providence, RI, June 2004, pp. 69–93.

[14] L. Fehér, R. Rimányi. Chern–Schwartz–MacPherson classes of degeneracy loci. Geometry and Topology 22 (2018) 3575–3622.

[15] L. M. Fehér, R. Rimányi, A. Weber. Characteristic classes of orbit stratifications, the axiomatic approach. In *Schubert Calculus and Its Applications in Combinatorics and Representation Theory, Guangzhou, China, November 2017.* Eds. J. Hu, C. Li, L. C. Mihalcea. Springer Proc. in Math. & Stat., Vol. 332, Springer, pp. 223–249, 2020.

[16] L. M. Fehér, R. Rimányi, A. Weber. Motivic Chern classes and K-theoretic stable envelopes. Proc. London Math. Soc, 122(1), January 2021, 153–189.

[17] W. Fulton, P. Pragacz. Schubert varieties and degeneracy loci. Springer LNM 1689 (1998).

[18] J. Harris, L. W. Tu. On symmetric and skew-symmetric determinantal varieties. Topology, Vol. 23, Issue 1, 1984, Pages 71–84.

[19] T. Józefiak, A. Lascoux, P. Pragacz. Classes of Determinantal Varieties Associated with Symmetric and Skew-Symmetric Matrices. Izvestiya: Mathematics, Vol. 18, Issue 3, 575–586, 1982.

[20] R. MacPherson. Chern classes for singular algebraic varieties. Annals of Mathematics, 100(2):423–432, 1974.

[21] D. Maulik, A. Okounkov. Quantum groups and quantum cohomology, Astérisque, Vol. 408, Société Mathématique de France, 2019.

[22] T. Ohmoto. Equivariant Chern classes of singular algebraic varieties with group actions. Math. Proc. of the Cambridge Phil. Soc., 140:115–134, 2006.

[23] T. Ohmoto. A note on Chern-Schwartz-MacPherson class. IRMA Lecture in Mathematics and Theoretical Physics, 20:117–132, 2012.

[24] T. Ohmoto. Singularities of maps and characteristic classes. In *School on Real and Complex Singularities in São Carlos, 2012*, pages 191–265, 2016. Mathematical Soc. Japan.

[25] A. Parusiński, P. Pragacz. Chern–Schwartz–MacPherson classes and the Euler characteristic of degeneracy loci and special divisors, JAMS 8 (1995), no. 4, 793–817.

[26] I. R. Porteous. Simple singularities of maps, in Proceedings of Liverpool Singularities Symposium, I (1969/70), Springer LNM 192, 286–307.

[27] S. Promtapan. Equivariant Chern–Schwartz–MacPherson classes of symmetric and skew-symmetric determinantal varieties, Ph.D. thesis, UNC Chapel Hill, 2019.

[28] R. Rimányi, A. Szenes. Residues, Grothendieck polynomials, and K-theoretic Thom polynomials. Preprint, arXiv:1811.02055.

[29] R. Rimányi, A. Varchenko. Equivariant Chern-Schwartz-MacPherson classes in partial flag varieties: interpolation and formulae, in Schubert Varieties, Equivariant Cohomology and Characteristic Classes, IMPANGA2015 (eds. J. Buczynski, M. Michalek, E. Postingel), EMS 2018, pp. 225–235.

[30] R. Rimányi, V. Tarasov, A. Varchenko. Partial flag varieties, stable envelopes and weight functions, Quantum Topol. 6 (2015), no. 2, 333–364.

[31] R. Rimányi, A. Weber. Elliptic classes of Schubert cells via Bott-Samelson resolution, J. of Topology, Vol. 13, Issue 3, September 2020, 1139–1182.

[32] D. J. S. Robinson. A course in linear algebra with applications. Word Scientific, 2nd ed. 2006.

[33] M.-H. Schwartz. Classes caractéristiques définies par une stratification d'une variété analytique complexe. *C. R. Math. Acad. Sci. Paris*, 260:3262–3264 and 3535–3537, 1965.

[34] D. Sullivan. Combinatorial invariants of analytic spaces. In C.T.C. Wall, editor, *Proceedings of Liverpool Singularities — Symposium I*, pages 165–177, Springer 1971.

[35] Ch.-F. Wei, F. Qi. Several closed expressions for the Euler numbers. *Journal of Inequalities and Applications*, 2015(1):219, 2015.

[36] J. Weyman. Cohomology of Vector Bundles and Syzygies. Cambridge University Press, 2009.

[37] X. Zhang. Chern classes and characteristic cycles of determinantal varieties Journal of Algebra, Vol. 497, 2018, Pages 55–91.

23

Equivariant Cohomology, Schubert Calculus, and Edge Labeled Tableaux

Colleen Robichaux[a], Harshit Yadav[b] and Alexander Yong[c]

To William Fulton on his eightieth birthday, for inspiring generations.

Abstract. This chapter concerns the concept of *edge labeled Young tableaux*, introduced by H. Thomas and the third author. It is used to model equivariant Schubert calculus of Grassmannians. We survey results, problems, conjectures, together with their influences from combinatorics, algebraic and symplectic geometry, linear algebra, and computational complexity. We report on a new shifted analogue of edge labeled tableaux. Conjecturally, this gives a Littlewood–Richardson rule for the structure constants of the Anderson–Fulton ring, which is related to the equivariant cohomology of isotropic Grassmannians.

1 Introduction

1.1 Purpose

Singular cohomology is a functor between the categories

$$\left\{\begin{array}{c}\text{topological spaces}\\ \text{and}\\ \text{continuous maps}\end{array}\right\} \rightarrow \left\{\begin{array}{c}\text{graded algebras}\\ \text{and}\\ \text{homomorphisms}\end{array}\right\}.$$

The cohomology functor links the geometry of Grassmannians to symmetric functions and Young tableaux. However, this does not take into account the large torus action on the Grassmannian. A similar functor for topological spaces with continuous group actions is equivariant cohomology.

[a] University of Illinois at Urbana-Champaign
[b] Rice University, Houston
[c] University of Illinois at Urbana-Champaign

What are the equivariant analogues for these centerpieces of algebraic combinatorics?

We posit a comprehensive answer, with applications, and future perspectives.

1.2 Schubert calculus

Let $X = \mathsf{Gr}_k(\mathbb{C}^n)$ be the Grassmannian of k-dimensional planes in $\mathbb{C}^n$. The group GL_n of invertible $n \times n$ matrices acts transitively on X by change of basis. Let $\mathsf{B}_- \subset \mathsf{GL}_n$ be an opposite Borel subgroup of lower triangular matrices. B_- acts on X with finitely many orbits X_λ° where λ is a partition (identified with its Young diagram, in English notation) that is contained in the $k \times (n-k)$ rectangle Λ. These *Schubert cells* satisfy

$$X_\lambda^\circ \cong \mathbb{C}^{k(n-k)-|\lambda|}$$

where $|\lambda| = \sum_i \lambda_i$. Their closures, the *Schubert varieties*, satisfy

$$X_\lambda := \overline{X_\lambda^\circ} = \coprod_{\mu \supseteq \lambda} X_\mu^\circ.$$

Let $\nu^\vee$ be the 180°-rotation of $k \times (n-k) \setminus \nu$. Suppose

$$|\lambda| + |\mu| + |\nu^\vee| = k(n-k) = \dim X.$$

By Kleiman transversality [26], there is a dense open

$$\mathcal{O} \subset \mathsf{GL}_n \times \mathsf{GL}_n \times \mathsf{GL}_n$$

such that

$$c_{\lambda,\mu}^{\nu} := \#\{g_1 \cdot X_\lambda \cap g_2 \cdot X_\mu \cap g_3 \cdot X_{\nu^\vee}\} \in \mathbb{Z}_{\geq 0}$$

is independent of $(g_1, g_2, g_3) \in \mathcal{O}$. Each $c_{\lambda,\mu}^{\nu}$ is called a *Littlewood–Richardson coefficient*. Modern Schubert calculus is concerned with these coefficients, as well as their generalizations/analogues (from varying the space X or cohomology theory).

Let $\sigma_\lambda \in H^{2|\lambda|}(X)$ be the Poincaré dual to X_λ. These *Schubert classes* form a $\mathbb{Z}$-linear basis of $H^*(X)$ and

$$\sigma_\lambda \smile \sigma_\mu = \sum_{\nu \subseteq \Lambda} c_{\lambda,\mu}^{\nu} \sigma_\nu.$$

The *Schur function* s_λ is the generating series

$$s_\lambda = \sum_T x^T$$

for semistandard Young tableaux of shape λ, *i.e.*, row weakly increasing and column strictly increasing fillings of λ with elements of $\mathbb{N}$. The weight of T is

$$x^T := \prod_i x_i^{\# i \in T}.$$

For example, if $\lambda = (2, 1)$, the semistandard tableaux are

$$\begin{array}{|c|c|}\hline 1 & 1 \\ \hline 2 \\ \cline{1-1}\end{array}\quad \begin{array}{|c|c|}\hline 1 & 2 \\ \hline 2 \\ \cline{1-1}\end{array}\quad \begin{array}{|c|c|}\hline 1 & 3 \\ \hline 2 \\ \cline{1-1}\end{array}\quad \begin{array}{|c|c|}\hline 1 & 1 \\ \hline 3 \\ \cline{1-1}\end{array}\quad \begin{array}{|c|c|}\hline 1 & 2 \\ \hline 3 \\ \cline{1-1}\end{array}\quad \begin{array}{|c|c|}\hline 1 & 3 \\ \hline 3 \\ \cline{1-1}\end{array}\quad \begin{array}{|c|c|}\hline 2 & 2 \\ \hline 3 \\ \cline{1-1}\end{array}\quad \begin{array}{|c|c|}\hline 2 & 3 \\ \hline 3 \\ \cline{1-1}\end{array}\ldots.$$

Hence

$$s_{(2,1)} = x_1^2x_2 + x_1x_2^2 + x_1x_2x_3 + x_1^2x_3 + x_1x_2x_3 + x_1x_3^2 + x_2^2x_3 + x_2x_3^2 + \cdots.$$

Schur functions form a $\mathbb{Z}$-linear basis of Sym, the ring of symmetric functions in infinitely many variables.

The map $\sigma_\lambda \mapsto s_\lambda$ induces a ring isomorphism

$$H^*(X) \cong \mathsf{Sym}/I$$

where I is the ideal $\langle s_\lambda : \lambda \not\subseteq \Lambda \rangle$. Therefore in Sym,

$$s_\lambda \cdot s_\mu = \sum_\nu c_{\lambda,\mu}^{\nu} s_\nu. \tag{1}$$

To compute Schubert calculus of X, it suffices to determine (1) by working with Schur *polynomials* in only finitely many variables $x_1, \ldots, x_k$. Better yet, the *jeu de taquin* theory of Young tableaux, introduced by M.-P. Schützenberger [54] gives a combinatorial rule for computing $c_{\lambda,\mu}^{\nu}$; this is summarized in Section 3. What we discussed thus far constitutes textbook material on Schubert calculus; see, *e.g.*, W. Fulton's [17].

1.3 Overview

This chapter describes an equivariant analogue of M.-P. Schützenberger's theory, due to H. Thomas and the third author; in short, one replaces Young tableaux with *edge labeled tableaux*. Now, we hasten to offer an *apologia*: such tableaux are not the only combinatorial model to compute equivariant Schubert calculus. For example, one has work of A. Molev-B. Sagan [39] and the *puzzles* of A. Knutson-T. Tao [30]. The latter has had important recent followup, see, *e.g.*, A. Knutson-P. Zinn-Justin's [33] and the references therein. One also has the tableaux of V. Kreiman [36] or A. Molev [38].

That said, we wish to argue how the edge labeled model is a handy and flexible viewpoint. It has been applied to obtain equivariant analogues of a number of theorems (delineated in Section 3 and 5). Another application, due to O. Pechenik and the third author [44], is to Schubert calculus for the equivariant K-theory of X. Translating the combinatorics into puzzle language allowed for a proof (of a correction) of a conjecture of A. Knutson-R. Vakil about the same structure constants [45]. However, as we wish to restrict to equivariant cohomology proper, this direction is not part of our discussion.

There is an important frontier to cross, that is, the still unsolved problem of finding a combinatorial rule for equivariant Schubert calculus of maximal orthogonal and Lagrangian Grassmannians. The non-equivariant story is explained in Section 7. We explain the problem in Section 8 together with some recent developments of C. Monical [40] and of the authors [51]. The latter work shows that the combinatorial problems concerning the two spaces are essentially equivalent.

This brings us to the principal new announcement of this work (Section 9): the notion of *shifted edge labeled tableaux*. We define an analogue of *jeu de taquin* and use this to conjecturally define an associative ring (which we prove to also be commutative). The introduction of this ring is stimulated by recent work of D. Anderson and W. Fulton (see Section 10) who define a ring connected to the equivariant cohomology of Lagrangian Grassmannians. Conjecturally, the two rings are isomorphic. This provides our strongest evidence to date of the applicability of the edge labeled approach to the aforementioned open problem; we know of no similar results using other combinatorial models.

As this work is partially expository and partly an announcement, we limited the number of complete proofs in order to keep the focus on the high-level research objectives. Where possible, we have sketched arguments (with references) and indicated those results which may be taken as an exercise for the interested reader. These exercises are warmups for the conjectures and open problems contained herein.

2 Equivariant cohomology of Grassmannians

2.1 Generalities

We recall some general notions about equivariant cohomology. References that we consulted are L. Tu's synopsis [64], J. Tymoczko's exposition [65] and A. Knutson-T. Tao's [30, Section 2].

Let $\mathcal{M}$ be a topological space with the continuous action of a topological group G. If G acts freely on $\mathcal{M}$, then in fact the equivariant cohomology ring $H^*_{\mathsf{G}}(\mathcal{M})$ is $H^*(\mathcal{M}/\mathsf{G})$, see [65, Proposition 2.1]. However, in general the action is not free, and $\mathcal{M}/\mathsf{G}$ might be, *e.g.*, non-Hausdorff. Borel's *mixing space construction* introduces a contractible space EG on which G acts freely. Thus G's diagonal action on $\mathsf{EG} \times \mathcal{M}$ is free and

$$H^*_{\mathsf{G}}(\mathcal{M}) := H^*(\mathsf{EG} \times \mathcal{M}/\mathsf{G}).$$

The space EG is the total space of the *universal principal* G*-bundle*

$$\pi : \mathsf{EG} \to \mathsf{BG}$$

where $\mathsf{BG} = \mathsf{EG}/\mathsf{G}$ is the classifying space of G. Here, universality means that if

$$\rho : P \to \mathcal{M}$$

is any G-bundle, there exists a unique map

$$f : \mathcal{M} \to \mathsf{BG}$$

(up to homotopy) such that $P \cong f^*(\mathsf{EG})$. By functoriality, the constant map $c : \mathcal{M} \to \{pt\}$ induces a homomorphism

$$c^* : H^*_{\mathsf{G}}(pt) \to H^*_{\mathsf{G}}(\mathcal{M}).$$

Hence $H^*_{\mathsf{G}}(\mathcal{M})$ is a module over $H^*_{\mathsf{G}}(pt)$ by $\beta \cdot \kappa := c^*(\beta)\kappa$ for $\beta \in H^*_{\mathsf{G}}(pt)$ and $\kappa \in H^*_{\mathsf{G}}(\mathcal{M})$. While ordinary (singular) cohomology of a point is $\mathbb{Z}$, $H^*_{\mathsf{G}}(pt)$ is *big*. For instance, if $\mathsf{G} = \mathsf{T}$ is an n torus $(S^1)^n$, then

$$H^*_{\mathsf{T}}(pt) = \mathbb{Z}[t_1, \ldots, t_n].$$

If we presume G is a algebraic group acting on a smooth algebraic variety M, these notions have versions in the algebraic category; see, *e.g.*, D. Anderson's [2].

2.2 The Grassmannian

Concretely, if $\lambda = (\lambda_1 \geq \lambda_2 \geq \ldots \geq \lambda_k \geq 0)$ then

$$X_\lambda = \{V \in X \,|\, \dim(V \cap F^{n-k+i-\lambda_i}) \geq i, 1 \leq i \leq k\}, \tag{2}$$

where $F^d = \text{span}(e_n, e_{n-1}, \ldots, e_{n-d+1})$ and e_i is the i-th standard basis vector; see [19, Section 9.4] for details.

Let $\mathsf{T} \subset \mathsf{GL}_n$ be the torus of invertible diagonal matrices. Then from (2), X_λ is T-stable. Therefore, X_λ admits an equivariant Schubert class ξ_λ in the T-equivariant cohomology ring $H^*_{\mathsf{T}}(X)$. By what we have recounted in Section 2.1, $H^*_{\mathsf{T}}(X)$ is a module over

$$H^*_{\mathsf{T}}(pt) := \mathbb{Z}[t_1, t_2, \ldots, t_n]. \tag{3}$$

The equivariant Schubert classes form a $H_{\mathsf{T}}^*(pt)$-module basis of $H_{\mathsf{T}}^*(X)$. Therefore,

$$\xi_\lambda \cdot \xi_\mu = \sum_{\nu \subseteq \Lambda} C_{\lambda,\mu}^{\nu} \xi_\nu, \tag{4}$$

where $C_{\lambda,\mu}^{\nu} \in H_{\mathsf{T}}^*(pt)$. For more details about equivariant cohomology specific to flag varieties we point the reader to [2] and S. Kumar's textbook [35, Chapter XI].

Let $\beta_i := t_i - t_{i+1}$. D. Peterson conjectured, and W. Graham [21] proved

Theorem 2.1 (Equivariant positivity [21])

$$C_{\lambda,\mu}^{\nu} \in \mathbb{Z}_{\geq 0}[\beta_1, \ldots, \beta_{n-1}].$$ [1]

In fact, $\deg C_{\lambda,\mu}^{\nu} = |\lambda| + |\mu| - |\nu|$ and $C_{\lambda,\mu}^{\nu} = 0$ unless $|\lambda| + |\mu| \geq |\nu|$. In the case $|\lambda| + |\mu| = |\nu|$, $C_{\lambda,\mu}^{\nu} = c_{\lambda,\mu}^{\nu}$ is the Littlewood–Richardson coefficient of Section 1.

Fix a grid with n rows and $m \geq n + \lambda_1 - 1$ columns. The *initial diagram* places λ in the northwest corner of this grid. For example, if $\lambda = (3,2,0,0)$, the initial diagram for λ is the first of the three below.

$$\begin{bmatrix} + & + & + & \cdot & \cdot & \cdot \\ + & + & \cdot & \cdot & \cdot & \cdot \\ \cdot & \cdot & \cdot & \cdot & \cdot & \cdot \\ \cdot & \cdot & \cdot & \cdot & \cdot & \cdot \end{bmatrix} \begin{bmatrix} + & + & \cdot & \cdot & \cdot & \cdot \\ + & \cdot & \cdot & + & \cdot & \cdot \\ \cdot & \cdot & \cdot & \cdot & \cdot & \cdot \\ \cdot & \cdot & \cdot & + & \cdot & \cdot \end{bmatrix} \begin{bmatrix} + & \cdot & \cdot & \cdot & \cdot & \cdot \\ \cdot & \cdot & + & \cdot & \cdot & \cdot \\ \cdot & + & \cdot & \cdot & + & \cdot \\ \cdot & \cdot & \cdot & + & \cdot & \cdot \end{bmatrix}$$

A *local move* is a change of any 2×2 subsquare of the form

$$\begin{matrix} + & \cdot \\ \cdot & \cdot \end{matrix} \quad \mapsto \quad \begin{matrix} \cdot & \cdot \\ \cdot & + \end{matrix}$$

A *plus diagram* is any configuration of $+$'s in the grid resulting from some number of local moves starting from the initial diagram for λ. We have given two more examples of plus diagrams for $\lambda = (3,2,0,0)$.

Let $\mathsf{Plus}(\lambda)$ denote the set of plus diagrams for λ. If $P \in \mathsf{Plus}(\lambda)$, let

$$\mathsf{wt}_x(P) = x_1^{\alpha_1} x_2^{\alpha_2} \cdots x_n^{\alpha_n}.$$ [i]

Here, α_i is the number of $+$'s in the ith row of P. For instance, if P is the rightmost diagram shown above, $\mathsf{wt}_x(P) = x_1 x_2 x_3^2 x_4$. A more refined statistic is

[1] Actually, W. Graham proved that for any generalized flag variety H/B, the equivariant Schubert structure constant is expressible as a nonnegative integer polynomial in the simple roots of the (complex, semisimple) Lie group H.

$$\mathsf{wt}_{x,y}(P) = \prod_{(i,j)} x_i - y_j.$$

The product is over those (i, j) with a $+$ in row i and column j of P. For the same P,

$$\mathsf{wt}_{x,y}(P) = (x_1 - y_1)(x_2 - y_3)(x_3 - y_2)(x_3 - y_5)(x_4 - y_4).$$

Let $X = \{x_1, x_2, \ldots, x_n\}$ and $Y = \{y_1, y_2, \ldots, y_{n+\lambda_1-1}\}$ be two collections of indeterminates. The *factorial Schur function* is

$$s_\lambda(X;Y) = \sum_{P \in \mathsf{Plus}(\lambda)} \mathsf{wt}_{x,y}(P).$$

This description arises in, *e.g.*, [28]. Moreover, it is an exercise to show

$$s_\lambda(X) = \sum_{P \in \mathsf{Plus}(\lambda)} \mathsf{wt}_x(P) = s_\lambda(X;0,0,\ldots).$$

The factorial Schur polynomials form a $\mathbb{Z}[Y]$-linear basis of $\mathsf{Sym} \otimes_{\mathbb{Q}} \mathbb{Z}[Y]$. In addition,

$$s_\lambda(X;Y)s_\mu(X;Y) = \sum_\nu C_{\lambda,\mu}^{\nu}\, s_\nu(X;Y). \tag{5}$$

For example, one checks that

$$s_{(1,0)}(x_1,x_2;Y)^2 = s_{(2,0)}(x_1,x_2;Y) + s_{(1,1)}(x_1,x_2;Y) + (y_3 - y_2)s_{(1,0)}(x_1,x_2;Y).$$

In view of (3), the definition of $C_{\lambda,\mu}^{\nu}$ in terms of (5) gives a definition of $H_{\mathsf{T}}^*(\mathsf{X})$ that suffices for our combinatorial ends.

2.3 Equivariant restriction

We may further assume G is an algebraic n-torus T which acts on $\mathcal{M}$ with finitely many isolated fixed points $\mathcal{M}^{\mathsf{T}}$. A feature of equivariant cohomology is that the inclusion $\mathcal{M}^{\mathsf{T}}$ into $\mathcal{M}$ induces an *injection*

$$H_{\mathsf{T}}^*(\mathcal{M}) \hookrightarrow H_{\mathsf{T}}^*(\mathcal{M}^{\mathsf{T}}) \cong \bigoplus_{\mathcal{M}^{\mathsf{T}}} H_{\mathsf{T}}^*(pt) \cong \bigoplus_{\mathcal{M}^{\mathsf{T}}} \mathbb{Z}[t_1, \ldots, t_n].$$

For each T-invariant cycle Y in $\mathcal{M}$, one has an equivariant cohomology class $[Y]_{\mathsf{T}} \in H_{\mathsf{T}}^*(\mathcal{M})$. This injection says that this class is a $\#\{\mathcal{M}^{\mathsf{T}}\}$-tuple of polynomials $[Y]|_x$ where $x \in \mathcal{M}^{\mathsf{T}}$. Each polynomial $[Y]|_x$ is an *equivariant restriction*. Under certain assumptions on $\mathcal{M}$, which cover all generalized flag manifolds (such as Grassmannians), one has a divisibility condition on the restrictions. This alludes to the general and influential package of ideas contained in Goresky–Kottwitz–MacPherson ("GKM") theory; we refer to [20] as well as J. Tymoczko's survey [65].

One has more precise results (predating [20]) for any generalized flag variety. Work of B. Kostant-S. Kumar [34] combined with a formula of H. Anderson-J. Jantzen-W. Soergel [4] describes $[Y]|_x$ where Y is a Schubert variety and x is one of the T-fixed points. For restriction formulas specific to the Grassmannian $M = \mathsf{Gr}_k(\mathbb{C}^n)$, see the formula of T. Ikeda-H. Naruse [25, Section 3] in terms of *excited Young diagrams*.[2] From either formula, one sees immediately that

$$\xi_\lambda|_\mu = 0 \text{ unless } \lambda \subseteq \mu, \tag{6}$$

and

$$\xi_\mu|_\mu \neq 0. \tag{7}$$

The central difference between this picture of equivariant cohomology of Grassmannians and the Borel-type presentation (due to A. Arabia [5]) is that multiplication can be done *without relations* and computed by pointwise multiplication of the restriction polynomials. In particular, each instance of (4) gives rise to $\binom{n}{k}$ many polynomial identities. For example, combining (4) and (6) gives

$$\xi_\lambda|_\mu \cdot \xi_\mu|_\mu = C_{\lambda,\mu}^{\mu} \xi_\mu|_\mu.$$

By (7), this implies

$$\xi_\lambda|_\mu = C_{\lambda,\mu}^{\mu}, \tag{8}$$

which is a fact first noted (for generalized flag varieties) by A. Arabia [5]. It follows that:

$$\xi_\lambda \cdot \xi_{(1)} = \xi_\lambda|_{(1)}\xi_\lambda + \sum_{\lambda^+} \xi_{\lambda^+}, \tag{9}$$

where λ^+ is obtained by adding a box to λ; see, *e.g.*, [30, Proposition 2]. Alternatively, it is an exercise to derive it from a vast generalization due to C. Lenart-A. Postnikov's [37, Corollary 1.2].

Since $H_\mathsf{T}^*(X)$ is an associative ring, one has

$$(\xi_\lambda \cdot \xi_\mu) \cdot \xi_{(1)} = \xi_\lambda \cdot (\xi_\mu \cdot \xi_{(1)}),$$

[2] One can give another formula in terms of certain specializations of the factorial Schur polynomial; see, *e.g.*, [25, Theorem 5.4] and the associated references.

which when expanded using (4) and (9) gives a recurrence that uniquely determines the structure coefficients; we call this the *associativity recurrence*. Since we will not explicitly need it in this chapter we leave it as an exercise (see [62, Lemma 3.3] and the references therein).[3]

3 Young tableaux and jeu de taquin

There are several combinatorial rules for the Littlewood–Richardson coefficient $c_{\lambda,\mu}^{\nu}$. The one whose theme pervades this chapter is the *jeu de taquin* rule. It is the first *proved* rule for the coefficients [54].

Let ν/λ be a skew shape. A *standard tableau* T of shape ν/λ is a bijective filling of ν/λ with $1, 2, \ldots, |\nu/\lambda|$ such that the rows and columns are increasing. Let $\mathsf{SYT}(\nu/\lambda)$ be the set of all such tableaux. An *inner corner* c of λ/μ is a maximally southeast box of μ. For $T \in \mathsf{SYT}(\lambda/\mu)$, a *jeu de taquin slide* $\mathsf{jdt}_{\mathsf{c}}(T)$ is obtained as follows. Initially place $\bullet$ in c, and apply one of the following *slides*, according to how T looks near c:

$$\text{(J1)} \quad \begin{array}{|c|c|}\hline \bullet & a \\ \hline b & \\ \cline{1-1}\end{array} \mapsto \begin{array}{|c|c|}\hline b & a \\ \hline \bullet & \\ \cline{1-1}\end{array} \text{(if } b < a \text{, or } a \text{ does not exist)}$$

$$\text{(J2)} \quad \begin{array}{|c|c|}\hline \bullet & a \\ \hline b & \\ \cline{1-1}\end{array} \mapsto \begin{array}{|c|c|}\hline a & \bullet \\ \hline b & \\ \cline{1-1}\end{array} \text{(if } a < b \text{, or } b \text{ does not exist)}$$

Repeat application of (J1) or (J2) on the new box c' where $\bullet$ arrives. End when $\bullet$ arrives at a box d of λ that has no labels south or east of it. Then $\mathsf{jdt}_{\mathsf{c}}(T)$ is obtained by erasing $\bullet$.

A *rectification* of $T \in \mathsf{SYT}(\lambda/\mu)$ is defined iteratively. Pick an inner corner c_0 of λ/μ and compute $T_1 := \mathsf{jdt}_{\mathsf{c}_0}(T) \in \mathsf{SYT}(\lambda^{(1)}/\mu^{(1)})$. Let c_1 be an inner corner of $\lambda^{(1)}/\mu^{(1)}$ and compute $T_2 := \mathsf{jdt}_{\mathsf{c}_1}(T_1) \in \mathsf{SYT}(\lambda^{(2)}/\mu^{(2)})$. Repeat $|\mu|$ times, arriving at a standard tableau of straight (*i.e.*, partition) shape. Let $\mathsf{Rect}_{\{\mathsf{c}_i\}}(T)$ be the result.

Theorem 3.1 (First fundamental theorem of jeu de taquin) $\mathsf{Rect}_{\{\mathsf{c}_i\}}(T)$ *is independent of the choice of sequence of successive inner corners* $\{\mathsf{c}_i\}$.

Theorem 3.1 permits one to speak of *the* rectification $\mathsf{Rect}(T)$.

[3] We give an analogue (16) of the associativity recurrence in our proof of Theorem 8.2.

Example 3.2 For instance, here are two different rectification orders for a tableau T.

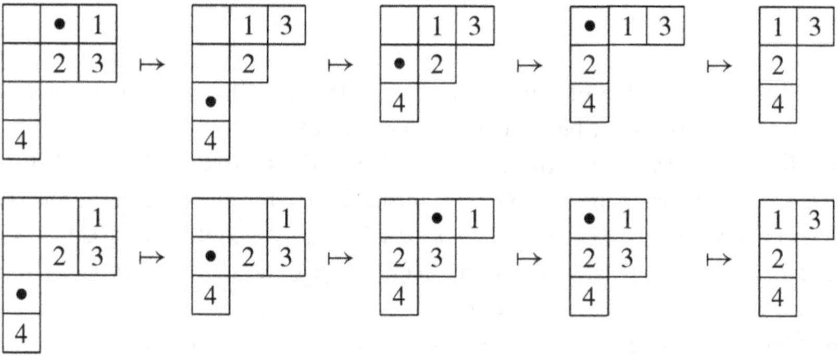

Theorem 3.3 (Second fundamental theorem of jeu de taquin) *The cardinality*

$$\#\{T \in \mathsf{SYT}(\nu/\lambda) : \mathsf{Rect}(T) = U\} \tag{10}$$

is independent of the choice of $U \in \mathsf{SYT}(\mu)$.

Example 3.4 Below are the tableaux $T \in \mathsf{SYT}((3,2,1)/(2,1))$ such that $\mathsf{Rect}(T) = U \in \mathsf{SYT}((2,1))$. Of the tableaux below, T_1, T_2 rectify to U_1 and T_3, T_4 rectify to U_2.

$$T_1 = \begin{array}{|c|c|c|}\hline \ & \ & 2 \\ \hline \ & 1 \\ \cline{1-2} 3 \\ \cline{1-1}\end{array} \quad T_2 = \begin{array}{|c|c|c|}\hline \ & \ & 2 \\ \hline \ & 3 \\ \cline{1-2} 1 \\ \cline{1-1}\end{array} \quad T_3 = \begin{array}{|c|c|c|}\hline \ & \ & 1 \\ \hline \ & 3 \\ \cline{1-2} 2 \\ \cline{1-1}\end{array} \quad T_4 = \begin{array}{|c|c|c|}\hline \ & \ & 3 \\ \hline \ & 1 \\ \cline{1-2} 2 \\ \cline{1-1}\end{array}$$

$$U_1 = \begin{array}{|c|c|}\hline 1 & 2 \\ \hline 3 \\ \cline{1-1}\end{array} \qquad U_2 = \begin{array}{|c|c|}\hline 1 & 3 \\ \hline 2 \\ \cline{1-1}\end{array}$$

For proofs of Theorems 3.1 and 3.3 we recommend the self-contained argument found in M. Haiman's [23], which is based on his theory of *dual equivalence*.

Theorem 3.5 ([54], Jeu de taquin computes the Littlewood–Richardson coefficient) *Fix $U \in \mathsf{SYT}(\mu)$. Then $c_{\lambda,\mu}^{\nu}$ equals the number* (10).

Example 3.6 Continuing Example 3.4, fix $U := U_1$. Then $c_{(2,1),(2,1)}^{(3,2,1)} = 2 = \#\{T_1, T_2\}$. $\square$

It is convenient to fix a choice of tableau U in Theorem 3.5. Namely, let $U = S_\mu$ be the superstandard tableau of shape μ, which is obtained by filling the boxes of μ in English reading order with $1, 2, 3, \ldots$. This is the choice made in Example 3.6. A larger instance is

$$S_{(5,3,1)} = \begin{array}{|c|c|c|c|c|}\hline 1 & 2 & 3 & 4 & 5 \\ \hline 6 & 7 & 8 \\ \cline{1-3} 9 \\ \cline{1-1}\end{array}.$$

There are a number of ways to prove a Littlewood–Richardson rule such as Theorem 3.5. We describe the two that we will refer to in this chapter:

• *"Bijective argument":* In terms of (1), the most direct is to establish a bijection between pairs (A, B) of semistandard tableau of shape λ and μ respectively and pairs (C, D) where $C \in \mathsf{SYT}(\nu/\lambda)$ such that $\mathsf{Rect}(C) = S_\mu$ and D is a semistandard tableau of shape ν. This can be achieved using the *Robinson–Schensted correspondence.*

• *"Associativity argument":* This was used by A. Knutson and T. Tao-C. Woodward [32] and A. Buch and A. Kresch-H. Tamvakis [12]. Define a putative ring $(R, +, \star)$ with additive basis $\{[\lambda] : \lambda \subseteq \Lambda\}$ and product

$$[\lambda] \star [\mu] := \sum_{\nu \subseteq \Lambda} \overline{c}_{\lambda,\mu}^{\nu}[\nu],$$

where $\overline{c}_{\lambda,\mu}^{\nu}$ is a collection of nonnegative integers. Assume one can prove $\star$ is commutative and associative and moreover agrees with Pieri's rule, *i.e.*, $\overline{c}_{\lambda,(p)}^{\nu} = 0$ unless ν/λ is a horizontal strip of size p, and equals 1 otherwise. Then it follows that $c_{\lambda,\mu}^{\nu} = \overline{c}_{\lambda,\mu}^{\nu}$. One can apply this to prove Theorem 3.5. The proofs of commutativity and associativity use Theorem 3.1. Typically, the hard step is the proof of associativity, which explains the nomenclature for this proof technique. □

Another formulation of the Littlewood–Richardson rule is in terms of semistandard Young tableaux of shape ν/λ and content μ. These are fillings T of ν/λ with μ_i many i's, and such that the rows are weakly increasing and columns are strictly increasing. The *row reading word* is obtained by reading the entries of T along rows, from right to left and from top to bottom. Such a word $(w_1, w_2, \ldots, w_{|\nu/\lambda|})$ is *ballot* if for every fixed $i, k \geq 1$,

$$\#\{j \leq k : w_j = i\} \geq \#\{j \leq k : w_j = i+1\}.$$

A tableau is ballot if its reading word is ballot.

Theorem 3.7 (Ballot version of the Littlewood–Richardson rule) *$c_{\lambda,\mu}^{\nu}$ equals the number of semistandard tableaux of shape ν/λ and content μ that are ballot.*

Example 3.8 Suppose $\lambda = (3,2,1), \mu = (3,2,1), \nu = (4,4,3,1)$. Below are the 3 semistandard tableaux of shape ν/λ and content μ that are ballot.

			1
		1	2
	1	2	
3			

			1
		1	2
	1	3	
2			

			1
		1	2
	2	3	
1			

Proof sketch for Theorem 3.7: Given a semistandard tableau T, one creates a standard tableau T' of the same shape by replacing all μ_1 many 1's by $1, 2, 3, \ldots, \mu_1$ from left to right and then replacing the (original) μ_2 many 2's by $\mu_1 + 1, \mu_1 + 2, \ldots, \mu_1 + \mu_2$ etc. This process is called *standardization*. Standardizing the tableaux in Example 3.8 respectively gives:

			3
		2	5
	1	4	
6			

			3
		2	5
	1	6	
4			

			3
		2	5
	4	6	
1			

We claim standardization induces a bijection between the rules of Theorem 3.7 and Theorem 3.5. More precisely, if T is furthermore ballot, then T' satisfies $\mathsf{rect}(T') = S_\mu$. We leave it as an exercise to establish this, *e.g.*, by induction on $|\lambda|$. □

There is a polytopal description of the Littlewood–Richardson rule derivable from Theorem 3.7. We first learned this from a preprint version of [42]. Suppose T is a semistandard tableau of shape ν/λ. Set

$$r_k^i = r_k^i(T) = \#\{k\text{'s in the } i\text{th row of } T\}.$$

Let μ denote the content of T and $\ell(\mu)$ be the number of nonzero parts of μ. By convention, let $r^i_{\ell(\mu)+1} = 0, r_k^{\ell(\nu)+1} = 0$.

Now consider the following linear inequalities, constructed to describe the tableaux from Theorem 3.7 that are counted by $c^\nu_{\lambda,\mu}$.

(A) Non-negativity: $r_k^i \geq 0, \ \forall i, k$.
(B) Shape constraints: $\lambda_i + \sum_k r_k^i = \nu_i, \ \forall i$.
(C) Content constraints: $\sum_i r_k^i = \mu_k, \ \forall k$.
(D) Tableau constraints: $\lambda_{i+1} + \sum_{j \leq k} r_j^{i+1} \leq \lambda_i + \sum_{j' < k} r_{j'}^i, \ \forall i$.
(E) Ballot constraints: $\sum_{i' < i} r_k^{i'} \geq r_{k+1}^i + \sum_{i' < i} r_{k+1}^{i'}, \ \forall i, k$.

Define a polytope

$$\mathcal{P}^\nu_{\lambda,\mu} = \{(r_k^i) : \text{(A)–(E)}\} \subseteq \mathbb{R}^{\ell(\nu)\cdot\ell(\mu)}.$$

The following is a straightforward exercise, once one assumes Theorem 3.7:

Theorem 3.9 (Polytopal Littlewood–Richardson rule) *$c^{\nu}_{\lambda,\mu}$ counts the number of integer lattice points in $\mathcal{P}^{\nu}_{\lambda,\mu}$.*

Example 3.10 Using λ,μ,ν as in Example 3.8, $\mathcal{P}^{\nu}_{\lambda,\mu}$ has 3 integer lattice points $R(T_j)=(r(T_j)_{ik})=(r^i_k(T_j))$ below. Each $r^i_k(T_j) = \#\{k'\text{s in the } i\text{th row of } T_j\}$, where the T_j are as in Example 3.8.

$$R(T_1)=\begin{pmatrix}1&0&0\\1&1&0\\1&1&0\\0&0&1\end{pmatrix}\quad R(T_2)=\begin{pmatrix}1&0&0\\1&1&0\\1&0&1\\0&1&0\end{pmatrix}\quad R(T_3)=\begin{pmatrix}1&0&0\\1&1&0\\0&1&1\\1&0&0\end{pmatrix}.$$

Another conversation concerns $\mathsf{LR}_r = \{(\lambda,\mu,\nu) : \ell(\lambda),\ell(\mu),\ell(\nu)\le r : c^{\nu}_{\lambda,\mu}\neq 0\}$.

Corollary 3.11 LR_r *is a semigroup,* i.e., *if* $(\lambda,\mu,\nu),(\alpha,\beta,\gamma)\in\mathsf{LR}_r$, *then* $(\lambda+\alpha,\mu+\beta,\nu+\gamma)\in\mathsf{LR}_r$.

Proof Since $(\lambda,\mu,\nu),(\alpha,\beta,\gamma)\in\mathsf{LR}_r$, by Theorem 3.9 there exists lattice points $(r^i_k)\in\mathcal{P}^{\nu}_{\lambda,\mu}$ and $(\overline{r}^i_k)\in\mathcal{P}^{\gamma}_{\alpha,\beta}$. By examination of the inequalities (A)-(E), clearly $(r^k_i+\overline{r}^k_i)$ is a lattice point in $\mathcal{P}^{\nu+\gamma}_{\lambda+\alpha,\mu+\beta}$, and we are done by another application of Theorem 3.9. □

This *Littlewood–Richardson semigroup* LR_r is discussed in A. Zelevinsky's article [69]. That work concerns the Horn and saturation conjectures (we will discuss these in Section 5). The point that a polytopal rule for $c^{\nu}_{\lambda,\mu}$ implies the semigroup property already appears in *ibid.* It is also true that LR_r is *finitely generated.* This is proved in A. Elashvili's [15], who credits the argument to M. Brion and F. Knop from "August-September, 1989". This argument (which applies more generally to tensor product multiplicities of any reductive Lie group) is not combinatorial. For another demonstration, see the proof of Proposition 6.8. Despite subsequent advances in understanding Littlewood–Richardson coefficients, the following remains open:

Problem 3.12 (*cf.* Problems A and C of [69]) *Explicitly give a finite (minimal) list of generators of* LR_r.

In connection to the work of Section 5, a closely related problem has been solved by P. Belkale [9]. His paper determines the extremal rays of the rational polyhedral cone defined by the points of LR_r.

4 Edge labeled tableaux and jeu de taquin

The history of the combinatorics of $C^{\nu}_{\lambda,\mu}$ is interesting in its own right. The first combinatorial rule for $C^{\nu}_{\lambda,\mu}$ is due to Molev–Sagan [39], who solved an even more general problem. Under the obvious specialization, this rule is not positive in the sense of Theorem 2.1. The first such rule, in terms of *puzzles*, was found by A. Knutson and T. Tao [30]. Subsequently, visibly equivalent tableaux rules were independently found by V. Kreiman [36] and A. Molev [38]. Later, P. Zinn-Justin [70] studied the puzzle rule of [30] based on the quantum integrability of the tiling model that underlies puzzles. Coming full circle, a point made in [70, Section 6.5] is that the rule of [39] *does* provide a positive rule after all, under the "curious identity" of [70, Section 6.4].

This work takes a different view. It is about the *edge labeled tableaux* introduced by H. Thomas and the third author [62]. A *horizontal edge* of ν/λ is an east-west line segment which either lies along the lower boundary of λ or ν, or which separates two boxes of ν/λ. An *equivariant filling* of ν/λ is an assignment of elements of $[N] := \{1, 2, \ldots, N\}$ to the boxes of ν/λ or to a horizontal edge of ν/λ. While every box contains a single label, each horizontal edge holds an element of $2^{[N]}$. An equivariant filling is *standard* if every label in $[N]$ appears exactly once and moreover any box label is:

- strictly smaller than any label in its southern edge and the label in the box immediately below it;
- strictly larger than any label in its northern edge and the label in the box immediately above it; and
- weakly smaller than the label in the box immediately to its right.

(No condition is placed on the labels of adjacent edges.) Let $\mathsf{EqSYT}(\nu/\lambda, \ell)$ be the set of equivariant standard tableaux with entries from $[N]$.

Example 4.1 Let $\nu/\lambda = (4,3,2)/(3,2,1) \subseteq \Lambda = 3 \times 4$ and

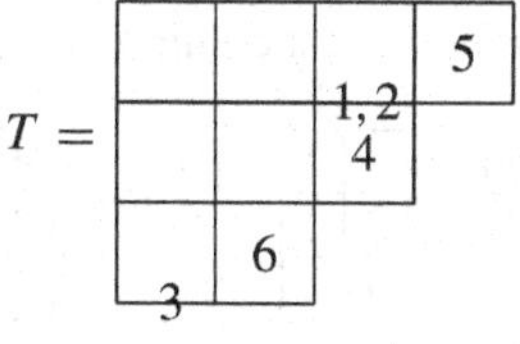

Then $T \in \mathsf{EqSYT}(\nu/\lambda, 6)$. □

As with ordinary tableaux, a rectification of $T \in \mathsf{EqSYT}(\nu/\lambda, N)$ is defined iteratively. Given an inner corner c and $T \in \mathsf{EqSYT}(\nu/\lambda, N)$, if none of the following possibilities applies, terminate the slide. Otherwise use the unique applicable case (in the left tableaux listed to the below, c contains the $\bullet$):

(J1) $\begin{array}{|c|c|}\hline \bullet & a \\ \hline b & \multicolumn{1}{c}{} \\ \cline{1-1} \end{array} \mapsto \begin{array}{|c|c|}\hline b & a \\ \hline \bullet & \multicolumn{1}{c}{} \\ \cline{1-1} \end{array}$ (if $b < a$, or a does not exist)

(J2) $\begin{array}{|c|c|}\hline \bullet & a \\ \hline b & \multicolumn{1}{c}{} \\ \cline{1-1} \end{array} \mapsto \begin{array}{|c|c|}\hline a & \bullet \\ \hline b & \multicolumn{1}{c}{} \\ \cline{1-1} \end{array}$ (if $a < b$, or b does not exist)

(J3) $\begin{array}{|c|c|}\hline \bullet & a \\ \hline \end{array}_{S} \mapsto \begin{array}{|c|c|}\hline a & \bullet \\ \hline \end{array}_{S}$ (if $a < \min(S)$)

(J4) $\begin{array}{|c|c|}\hline \bullet & a \\ \hline \end{array}_{S} \mapsto \begin{array}{|c|c|}\hline s & a \\ \hline \end{array}_{S'}$ (if $s := \min(S) < a$ and $S' := S \setminus \{s\}$)

This *equivariant jeu de taquin slide* into c is denoted by $\mathsf{Ejdt}_{\mathsf{c}}(T)$. Clearly, $\mathsf{Ejdt}_{\mathsf{c}}(T)$ is also a standard equivariant filling.

The rectification of T, denoted $\mathsf{Erect}(T)$, is the result of successively using $\mathsf{Ejdt}_{\mathsf{c}}$ by choosing c that is eastmost among all choices of inner corners at each stage.

Example 4.2 Continuing Example 4.1, we use "•" to indicate the boxes that are moved into during $\mathsf{Erect}(T)$. The rectification of the third column is as follows:

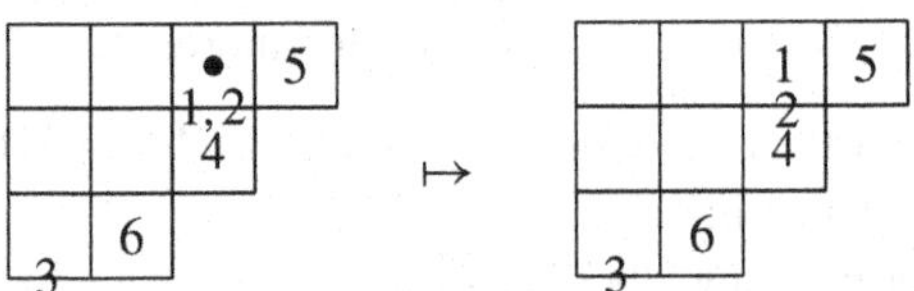

The rectification of the second column given by:

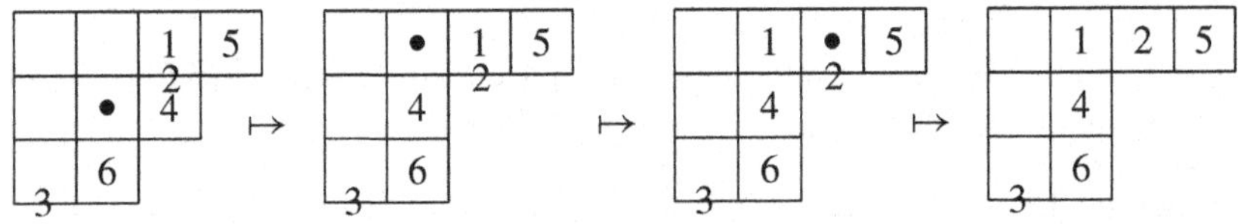

and finally the rectification of the first column given by:

$\begin{array}{|c|c|c|c|}\hline & 1 & 2 & 5 \\ \hline & 4 \\ \cline{1-2} \bullet_{3} & 6 \\ \cline{1-2} \end{array} \mapsto \begin{array}{|c|c|c|c|}\hline & 1 & 2 & 5 \\ \hline \bullet & 4 \\ \cline{1-2} 3 & 6 \\ \cline{1-2} \end{array} \mapsto \begin{array}{|c|c|c|c|}\hline & 1 & 2 & 5 \\ \hline 3 & 4 \\ \cline{1-2} \bullet & 6 \\ \cline{1-2} \end{array} \mapsto \begin{array}{|c|c|c|c|}\hline \bullet & 1 & 2 & 5 \\ \hline 3 & 4 \\ \cline{1-2} 6 \\ \cline{1-1} \end{array} \mapsto \cdots \mapsto \begin{array}{|c|c|c|}\hline 1 & 2 & 5 \\ \hline 3 & 4 \\ \cline{1-2} 6 \\ \cline{1-1} \end{array}$

Here "$\mapsto \cdots \mapsto$" refers to slides moving the • right in the first row. □

We now define $\mathsf{Ejdtwt}(T) \in \mathbb{Z}[t_1, \ldots, t_n]$ for a standard tableau T. Each box x in $\Lambda = k \times (n-k)$ has a (Manhattan) *distance* from the lower-left box: suppose x has matrix coordinates (i, j), then

$$\mathsf{dist}(x) := k + j - i.$$

Next, assign $x \in \Lambda$ the weight $\beta(x) = t_{\mathsf{dist}(x)} - t_{\mathsf{dist}(x)+1}$.

If after rectification of a column, the label $\mathfrak{l}$ still remains an edge label, $\mathsf{Ejdtfactor}(\mathfrak{l})$ is declared to be zero. Otherwise, say an edge label $\mathfrak{l}$ *passes* through a box x if it occupies x during the equivariant rectification of the column of T in which $\mathfrak{l}$ begins. Let the boxes passed be $x_1, x_2, \ldots, x_s$. Also, when the rectification of a column is complete, suppose the filled boxes strictly to the right of x_s are $y_1, \ldots, y_t$. Set

$$\mathsf{Ejdtfactor}(\mathfrak{l}) = (\beta(x_1) + \beta(x_2) + \cdots + \beta(x_s)) + (\beta(y_1) + \beta(y_2) + \cdots + \beta(y_t)).$$

Notice that since the boxes $x_1, \ldots, x_s, y_1, \ldots, y_t$ form a hook inside ν, $\mathsf{Ejdtfactor}(i) = t_e - t_f$ with $e < f$. Now define

$$\mathsf{Ejdtwt}(T) := \prod_{\mathfrak{l}} \mathsf{Ejdtfactor}(\mathfrak{l}),$$

where the product is over all edge labels $\mathfrak{l}$ of T.

Theorem 4.3 (Edge labeled jeu de taquin rule [62])

$$C_{\lambda,\mu}^{\nu} = \sum_{T} \mathsf{Ejdtwt}(T),$$

where the sum is over all $T \in \mathsf{EqSYT}(\nu/\lambda, |\mu|)$ *such that* $\mathsf{Erect}(T) = S_\mu$.

Since each $\mathsf{Ejdtfactor}(\mathfrak{l})$ is a positive sum of the indeterminates $\beta_i = t_i - t_{i+1}$, Theorem 4.3 expresses $C_{\lambda,\mu}^{\nu}$ as a polynomial with positive coefficients in the β_i's, in agreement with Theorem 2.1.

Example 4.4 Below each $x \in \Lambda = 2 \times 3$ is filled with $\beta(x)$:

t_2-t_3	t_3-t_4	t_4-t_5
t_1-t_2	t_2-t_3	t_3-t_4

T_1, T_2 below are those $T \in \mathsf{EqSYT}((3,3)/(2,2), 3)$ such that $\mathsf{Erect}(T) = S_{(2,1)}$ with nonzero weight:

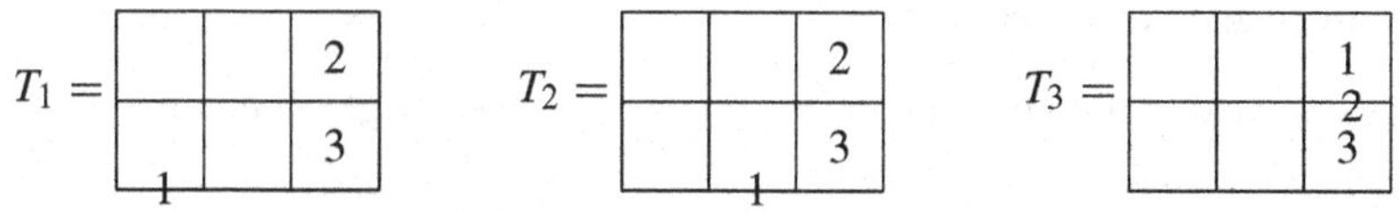

$T_1 =$

		2
		3
1 (edge)		

$T_2 =$

		2
		3
	1 (edge)	

$T_3 =$

		1
		2 (edge)
		3

While T_3 also satisfies $\mathsf{Erect}(T) = S_{(2,1)}$, it has weight zero since there is an edge label when one has completed rectifying the third column.

Theorem 4.3 asserts

$$\begin{aligned} C^{(3,3)}_{(2,2),(2,1)} &= \mathsf{Ejdtwt}(T_1) + \mathsf{Ejdtwt}(T_2) \\ &= [((t_1 - t_2) + (t_2 - t_3)) + (t_3 - t_4)] \\ &\quad + [((t_2 - t_3) + (t_3 - t_4)) + (t_4 - t_5)] \\ &= (t_2 - t_5) + (t_1 - t_4). \end{aligned}$$

In order to prove Theorem 4.3, one wishes to adapt the general strategy indicated at the end of Section 2.3. If we define $\overline{C}^{\nu}_{\lambda,\mu} := \sum_T \mathsf{Ejdtwt}(T)$ (as in Theorem 4.3) and show this collection of numbers satisfies the associativity recurrence, one shows $C^{\nu}_{\lambda,\mu} = \overline{C}^{\nu}_{\lambda,\mu}$, as desired.[4]

The rule of Theorem 4.3 appears to be too "rigid" to carry out this strategy. Instead, in [62], a more "flexible version" in terms of semistandard edge labeled tableaux is introduced, together with a corresponding collection of jeu de taquin slides. While we do not wish to revisit the rather technical list of slide rules here, one of the consequences is a ballot rule for $C^{\nu}_{\lambda,\mu}$, generalizing Theorem 3.7, which we explain now.

An equivariant tableau is *semistandard* if the box labels weakly increase along rows (left to right), and all labels strictly increase down columns. A single edge may be labeled by a *set* of integers, without repeats; the smallest of them must be strictly greater than the label of the box above, and the largest must be strictly less than the label of the box below.

Example 4.5 Below is an equivariant semistandard Young tableau on $(4,2,2)/(2,1)$.

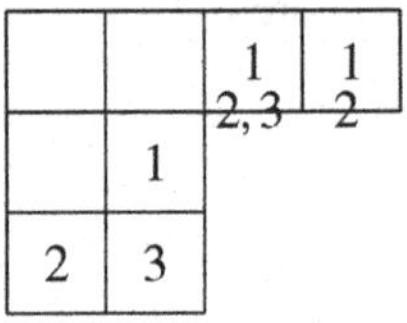

<table>
<tr><td></td><td></td><td>1
2,3</td><td>1
2</td></tr>
<tr><td></td><td>1</td><td></td><td></td></tr>
<tr><td>2</td><td>3</td><td></td><td></td></tr>
</table>

The content of this tableau is $(3,3,2)$. □

Let $\mathsf{EqSSYT}(\nu/\lambda)$ be the set of all equivariant semistandard Young tableaux of shape ν/λ. A tableau $T \in \mathsf{EqSSYT}(\nu/\lambda)$ is *ballot* if, for every column c and every label ℓ,

$$(\#\ell\text{'s weakly right of column } c) \geq (\#(\ell+1)\text{'s weakly right of column } c).$$

[4] It would be quite interesting to give a "bijective argument" (in the sense of Section 3) using the combinatorial description of the factorial Schur polynomials.

Given a tableau $T \in \mathsf{EqSSYT}(\nu/\lambda)$, a (box or edge) label ℓ is *too high* if it appears weakly above the upper edge of a box in row ℓ. In the above example, all edge labels are too high. (When there are no edge labels, the semistandard and ballot conditions imply no box label is too high, but in general the three conditions are independent.)

Suppose an edge label ℓ lies on the bottom edge of a box x in row i. Let $\rho_\ell(x)$ be the number of times ℓ appears as a (box or edge) label strictly to the right of x. We define

$$\mathsf{Eballotfactor}(\ell, x) = t_{\mathsf{dist}(x)} - t_{\mathsf{dist}(x)+i-\ell+1+\rho_\ell(x)}.$$

When the edge label is not too high, this is always of the form $t_p - t_q$, for $p < q$. (In particular, it is nonzero.) Finally, define

$$\mathsf{Eballotwt}(T) = \prod \mathsf{Eballotfactor}(\ell, x),$$

the product being over all edge labels ℓ.

Theorem 4.6 (Edge labeled ballot rule [62, Theorem 3.1])

$$C^{\nu}_{\lambda,\mu} = \sum_T \mathsf{Eballotwt}(T),$$

where the sum is over all $T \in \mathsf{EqSSYT}(\nu/\lambda)$ of content μ that are ballot and have no label which is too high.

Example 4.7 In Example 4.4, we saw

$$C^{(3,3)}_{(2,2),(2,1)} = (t_2 - t_5) + (t_1 - t_4).$$

Below has $x \in \Lambda$ is filled with $\mathsf{dist}(x)$.

2	3	4
1	2	3

We see T_1, T_2 below are those $T \in \mathsf{EqSSYT}((3,3)/(2,2))$ of content $(2,1)$ that are ballot and have no label which is too high.

$T_1 =$

		1
		2
1 (edge)		

$T_2 =$

		1
		2
	1 (edge)	

Thus we see as Theorem 4.3 states, since for both cases $p_\ell(x) = p_1(x) = 1$,

$$\begin{aligned} C^{(3,3)}_{(2,2),(2,1)} &= \mathsf{Eballotwt}(T_1) + \mathsf{Eballotwt}(T_2) \\ &= \mathsf{Eballotfactor}(1,(2,1)) + \mathsf{Eballotfactor}(1,(2,2)) \\ &= (t_1 - t_{1+2-1+1+1}) + (t_2 - t_{2+2-1+1+1}) \\ &= (t_1 - t_4) + (t_2 - t_5), \end{aligned}$$

so this rule agrees.

5 Nonvanishing of Littlewood–Richardson coefficients, Saturation and Horn inequalities

For which triples of partitions (λ, μ, ν) *does* $c^{\nu}_{\lambda,\mu} \neq 0$*?*

The importance of this question comes from a striking equivalence to a 19th century question about linear algebra. This equivalence was first suggested as a question by R. C. Thompson in the 1970s; see R. Bhatia's survey [10, pg. 308] (another survey on this topic is W. Fulton's [18]). Suppose A, B, C are three $r \times r$ Hermitian matrices and $\lambda, \mu, \nu \in \mathbb{R}^r$ are the respective lists of eigenvalues (written in decreasing order). The *eigenvalue problem for Hermitian matrices* asks

Which eigenvalues (λ, μ, ν) *can occur if* $A + B = C$*?*

After work of H. Weyl, K. Fan, V. B. Lidskii-H. Weilandt and others, A. Horn recursively defined a list of inequalities on triples $(\lambda, \mu, \nu) \in \mathbb{R}^{3r}$ (see Theorem 5.2(2) below). He conjectured that these give a complete solution to the eigenvalue problem [24]. That these inequalities (or equivalent ones) are necessary has been proved by several authors, including B. Totaro [63] and A. Klyachko [27]. A. Klyachko also established that his list of inequalities is sufficient, giving the first solution to the eigenvalue problem.

Also, A. Klyachko showed that his inequalities give an asymptotic solution to the problem of which Littlewood–Richardson coefficients $c^{\nu}_{\lambda,\mu}$ are nonzero. That is, suppose λ, μ, ν are partitions with at most r parts. A. Klyachko proved that if $c^{\nu}_{\lambda,\mu} \neq 0$, then $(\lambda, \mu, \nu) \in \mathbb{Z}^{3r}_{\geq 0}$ satisfies his inequalities. Conversely, he showed that if $(\lambda, \mu, \nu) \in \mathbb{Z}^{3r}_{\geq 0}$ satisfy his inequalities then $c^{N\nu}_{N\lambda, N\mu} \neq 0$ for some $N \in \mathbb{N}$. (Here, $N\lambda$ is the partition with each part of λ stretched by a factor of N.) Subsequently, A. Knutson and T. Tao [29] sharpened the last statement, and established:

Theorem 5.1 (Saturation theorem) $c_{\lambda,\mu}^{\nu} \neq 0$ *if and only if* $c_{N\lambda,N\mu}^{N\nu} \neq 0$ *for any* $N \in \mathbb{N}$.

Combined with [27], it follows that A. Klyachko's solution agrees with A. Horn's conjectured solution.[5] Let $[r] := \{1, 2, \ldots r\}$. For any

$$I = \{i_1 < i_2 < \cdots < i_d\} \subseteq [r]$$

define the partition

$$\tau(I) := (i_d - d \geq \cdots \geq i_2 - 2 \geq i_1 - 1).$$

This bijects subsets of $[r]$ of cardinality d with partitions whose Young diagrams are contained in a $d \times (r - d)$ rectangle. The following combines the main results of [27, 29]:

Theorem 5.2 ([27], [29]) *Let* λ, μ, ν *be partitions with at most* r *parts such that*

$$|\lambda| + |\mu| = |\nu|. \tag{11}$$

The following are equivalent:

1. $c_{\lambda,\mu}^{\nu} \neq 0$.
2. *For every* $d < r$*, and every triple of subsets* $I, J, K \subseteq [r]$ *of cardinality* d *such that* $c_{\tau(I),\tau(J)}^{\tau(K)} \neq 0$*, we have*

$$\sum_{i \in I} \lambda_i + \sum_{j \in J} \mu_j \geq \sum_{k \in K} \nu_k. \tag{12}$$

3. *There exist* $r \times r$ *Hermitian matrices* A, B, C *with eigenvalues* λ, μ, ν *such that* $A + B = C$.

Remark 5.3 The logic of the proof of Theorem 5.2 in [29] is to show Theorem 5.1. In fact, Theorem 5.1 also follows from the equivalence (1) $\iff$ (2) of Theorem 5.2. This is since the Horn inequalities from (2) are homogeneous. This point seems to have been first noted in P. Belkale's [8] which moreover gives a geometric proof of Theorem 5.2. □

Remark 5.4 The equivalence (1) $\iff$ (2) immediately implies the semigroup property of LR_r (Corollary 3.11). □

[5] The proof in a preprint version of [29] used the polytopal Littlewood–Richardson rule of Bernstein–Zelevinsky; we refer to the survey of A. Buch [11]. The published proof is formulated in terms of the *Honeycomb model*. It is an easy exercise to prove the "⇒" of the equivalence using either of the Littlewood–Richardson rules found in Section 3. A proof of the converse using such rules would be surprising. Another solution, due to H. Derksen-J. Weyman [14] was given in the setting of *semi-invariants of quivers*.

Remark 5.5 P. Belkale's doctoral thesis [6] (published in [7]) shows that a much smaller list of inequalities than those in Theorem 5.2(2) suffice. Namely, replace the condition "$c^{\tau(\nu)}_{\tau(\lambda),\tau(\mu)} \neq 0$" with "$c^{\tau(\nu)}_{\tau(\lambda),\tau(\mu)} \neq 1$". A. Knutson-T. Tao-C. Woodward [31] showed that the inequalities in this shorter list are minimal, *i.e.*, none can be dispensed with. □

Remark 5.6 Theorem 5.2 gives a different proof that LR_r is finitely generated as a semigroup. We will give the argument in the proof of Proposition 6.8, below.

While Theorem 5.2 characterizes nonvanishing of $c^{\nu}_{\lambda,\mu}$, the inequalities are recursive and non-transparent to work with. The Littlewood–Richardson rules of Section 3 require one to search for a valid tableau in a possibly large search space. K. Purbhoo [48] (see also [47]) developed a general and intriguing *root game*, which in the case of Grassmannians can be "won" if and only if $c^{\nu}_{\lambda,\mu} \neq 0$.

Theorem 5.1 permits a determination of the *formal* computational complexity of the nonvanishing decision problem "$c^{\nu}_{\lambda,\mu} \neq 0$" (in the bit length of the input (λ,μ,ν), where one assumes arithmetic operations take constant time). This was resolved independently by T. McAllister-J. De Loera [13] and K. D. Mulmuley-H. Narayanan-M. Sohoni, [42], by a neat argument that combines Theorem 5.1 with celebrated developments in linear programming:

Theorem 5.7 *The decision problem of determining if $c^{\nu}_{\lambda,\mu} \neq 0$ is in the class* P *of polynomial problems.*

Proof By Theorem 3.9, $c^{\nu}_{\lambda,\mu} \neq 0$ if and only if the polytope $\mathcal{P}^{\nu}_{\lambda,\mu}$ has a lattice point. Clearly, if $\mathcal{P}^{\nu}_{\lambda,\mu} \neq \emptyset$, it has a rational vertex. In this case, a dilation $N\mathcal{P}^{\nu}_{\lambda,\mu}$ contains a lattice point. One checks from the definitions that $N\mathcal{P}^{\nu}_{\lambda,\mu} = \mathcal{P}^{N\nu}_{N\lambda,N\mu}$, which means $c^{N\nu}_{N\lambda,N\mu} \neq 0$. Thus, by Theorem 5.1,

$$c^{\nu}_{\lambda,\mu} \neq 0 \iff c^{N\nu}_{N\lambda,N\mu} \neq 0 \iff \mathcal{P}^{\nu}_{\lambda,\mu} \neq \emptyset.$$

To determine if $\mathcal{P}^{\nu}_{\lambda,\mu} \neq \emptyset$, one needs to decide feasibility of any linear programming problem involving $\mathcal{P}^{\nu}_{\lambda,\mu}$. One appeals to ellipsoid/interior point methods for polynomiality.[6] Actually, our inequalities are of the form $A\mathbf{x} \leq \mathbf{b}$ where the vector $\mathbf{b}$ is integral and the entries of A are from $\{-1,0,1\}$. Hence our polytope is *combinatorial* and so one achieves a strongly polynomial time complexity using É. Tardos' algorithm; see [60, 22]. □

[6] The Klee–Minty cube shows that the practically efficient simplex method has exponential worst-case complexity.

In contrast, H. Narayanan [43] proved that counting $c_{\lambda,\mu}^{\nu}$ is a #P-complete problem in L. Valiant's complexity theory of counting problems [66]. In particular, this means that no polynomial time algorithm for computing $c_{\lambda,\mu}^{\nu}$ can exist unless $\mathsf{P} = \mathsf{NP}$ (it is widely expected that $\mathsf{P} \neq \mathsf{NP}$). It is curious that the counting problem is (presumably) hard, whereas the nonzeroness version is polynomial time. This already occurs for the original #P-complete problem from [66], *i.e.*, to compute the permanent of an $n \times n$ matrix $M = (m_{ij})$ where $m_{ij} \in \{0, 1\}$. Now, determining if $\text{per}(M) > 0$ is equivalent to deciding the existence of matching in a bipartite graph that has incidence matrix M; the algorithm of J. Edmonds and R. Karp provides the polynomial-time algorithm.

6 Equivariant nonvanishing, saturation, and Friedland's inequalities

We now turn to the equivariant analogues of results from the previous section.

For which triples of partitions (λ, μ, ν) *does* $C_{\lambda,\mu}^{\nu} \neq 0$*?*

In [3] it was shown that this Schubert calculus question is also essentially equivalent to an eigenvalue problem. Recall that a Hermitian matrix M *majorizes* another Hermitian matrix M' if $M - M'$ is positive semidefinite (its eigenvalues are all nonnegative). In this case, we write $M \geq M'$. S. Friedland [16] studied the following question:

Which eigenvalues (λ, μ, ν) *can occur if* $A + B \geq C$*?*

His solution, given as linear inequalities, includes Klyachko's inequalities, a trace inequality and some extra inequalities. Later, W. Fulton [19] proved the extra inequalities are unnecessary (where his inequalities are those of Theorem 6.2(2) below), leading to a natural extension of the equivalence (2) $\Longleftrightarrow$ (3) of Theorem 5.2.

One would like an extension of the equivalence with (1) of Theorem 5.2 as well. Now, D. Anderson, E. Richmond and the third author [3] proved:

Theorem 6.1 (Equivariant saturation) $C_{\lambda,\mu}^{\nu} \neq 0$ *if and only if* $C_{N\lambda,N\mu}^{N\nu} \neq 0$ *for any* $N \in \mathbb{N}$.[7]

[7] In our notation, $C_{\lambda,\mu}^{\nu}$ depends on the indices k and n of the Grassmannian $\mathsf{Gr}_k(\mathbb{C}^n)$. However, the edge labeled rule of Theorem 4.3 has the property that $C_{\lambda,\mu}^{\nu}$ is that for fixed k, the polynomial is independent of the choice of n provided both are sufficiently large so that $\lambda, \mu, \nu \subseteq k \times (n-k)$. Indeed, the coefficients $C_{\lambda,\mu}^{\nu}$ for such values are the structure constants for the Schubert basis in the graded inverse limit of equivariant cohomology rings under the standard embedding $\iota : \mathsf{Gr}_k(\mathbb{C}^n) \hookrightarrow \mathsf{Gr}_k(\mathbb{C}^{n+1})$; see [3, Section 1.2].

Actually, Theorem 6.1 is proved by establishing the equivalence (1) $\iff$ (2) below.

Theorem 6.2 ([3], [16], [19]) *Let λ, μ, ν be partitions with at most r parts such that*

$$|\lambda| + |\mu| \geq |\nu| \text{ and } \max\{\lambda_i, \mu_i\} \leq \nu_i \text{ for all } i \leq r. \tag{13}$$

The following are equivalent:

1. $C^{\nu}_{\lambda,\mu} \neq 0$.
2. *For every $d < r$, and every triple of subsets $I, J, K \subseteq [r]$ of cardinality d such that $c^{\tau(K)}_{\tau(I),\tau(J)} \neq 0$, we have*

$$\sum_{i \in I} \lambda_i + \sum_{j \in J} \mu_j \geq \sum_{k \in K} \nu_k.$$

3. *There exist $r \times r$ Hermitian matrices A, B, C with eigenvalues λ, μ, ν such that $A + B \geq C$.*

Theorem 6.2 states that the main inequalities controlling nonvanishing of $C^{\nu}_{\lambda,\mu}$ are just Horn's inequalities (12).

Remark 6.3 The second condition in (13) is unnecessary in Theorem 5.2 since it is already implied by (11) combined with (12). The condition is not required for the equivalence (2) $\iff$ (3). However it is needed for the equivalence with (1). For example, the 1×1 matrices $A = [1], B = [1], C = [0]$ satisfy $A + B \geq C$, but $C^{(0)}_{(1),(1)} = 0$. This is not in contradiction with Theorem 6.2 since $\max\{1, 1\} \leq 0$ is violated. □

Just as in Remark 5.3, the equivalence (1) $\iff$ (2) and the homogeneity of the Friedland–Fulton inequalities implies Theorem 6.1. On the other hand the equivalence (1) $\iff$ (2) relies on the classical Horn theorem (Theorem 5.2) and the edge labeled ballot tableau rule Theorem 4.6. To give the reader a sense of the proof we now provide:

Sketch of proof that (1) $\Longrightarrow$ (2): Using Theorem 4.6, it is an exercise to show that

Claim 6.4 *If $C^{\nu}_{\lambda,\mu} \neq 0$ and $|\nu| < |\lambda| + |\mu|$, then for any s such that $|\nu| - |\lambda| \leq s < |\mu|$, there is a $\mu^{\downarrow} \subset \mu$ with $|\mu^{\downarrow}| = s$ and $C^{\nu}_{\lambda,\mu^{\downarrow}} \neq 0$.*

Since $C^{\nu}_{\lambda,\mu} \neq 0$, by the claim (and induction) there exists $\lambda^{\downarrow} \subseteq \lambda$ such that $|\lambda^{\downarrow}| + |\mu| = |\nu|$ and $C^{\nu}_{\lambda^{\downarrow},\mu} \neq 0$. This latter number is a classical

Littlewood–Richardson coefficient, we can apply Theorem 5.2 to conclude that for any triple (I, J, K) with $c^{\tau(K)}_{\tau(I),\tau(J)} \neq 0$ one has

$$\sum_{i\in I} \lambda_i^{\downarrow} + \sum_{i\in J} \mu_j \geq \sum_{k\in K} \nu_k.$$

Now we are done since $\sum_{i\in I} \lambda_i \geq \sum_{i\in I} \lambda_i^{\downarrow}$. □

The converse (2) $\Longrightarrow$ (1) uses another exercise that can be proved using Theorem 4.6:

Claim 6.5 *If $C^{\nu}_{\lambda,\mu} \neq 0$ then $C^{\nu}_{\lambda,\mu^{\uparrow}} \neq 0$ for any $\mu \subset \mu^{\uparrow} \subseteq \nu$.*

See [3, Section 2] for proofs of Claim 6.4 and Claim 6.5.

Remark 6.6 Just as with Theorem 5.2, the list of inequalities in Theorem 6.2(2) contain redundancies. Building from the results discussed in Remark 5.5, W. Fulton [19] shows that one can also replace the "$c^{\tau(\nu)}_{\tau(\lambda),\tau(\mu)} \neq 0$" with "$c^{\tau(\nu)}_{\tau(\lambda),\tau(\mu)} = 1$". W. Fulton's work shows that if the second condition in (13) is ignored, the inequalities are minimal for the equivalence (2) $\Longleftrightarrow$ (3). □

Let $\mathsf{EqLR}_r = \{(\lambda, \mu, \nu) : \ell(\lambda), \ell(\mu), \ell(\nu) \leq r, C^{\nu}_{\lambda,\mu} \neq 0\}$.

Corollary 6.7 EqLR_r *is a semigroup.*

Proof Just as Corollary 3.11 clearly follows from the first two equivalences of Theorem 5.2 (Remark 5.4), the present claim holds by the first two equivalences of Theorem 6.2. However, we can also prove this directly from Claims 6.4 and 6.5: By applying Claim 6.4 there exists $\lambda^{\circ} \subseteq \lambda$ and $\alpha^{\circ} \subseteq \alpha$ such that $C^{\nu}_{\lambda^{\circ},\mu} = c^{\nu}_{\lambda^{\circ},\mu} \neq 0$ and $C^{\gamma}_{\alpha^{\circ},\beta} = c^{\gamma}_{\alpha^{\circ},\beta} \neq 0$. By Corollary 3.11,

$$(\lambda^{\circ} + \alpha^{\circ}, \mu + \beta, \nu + \gamma) \in \mathsf{LR}_r.$$

Now, since $\lambda^{\circ} + \alpha^{\circ} \subseteq \lambda + \alpha \subseteq \nu + \gamma$, we can apply Claim 6.5 to conclude $(\lambda + \alpha, \mu + \beta, \nu + \gamma) \in \mathsf{EqLR}_r$, as desired. □

Proposition 6.8 EqLR_r *is finitely generated.*

The argument we give is based on discussion with S. Fomin and A. Knutson. It applies *mutatis mutandis* to prove that LR_r is finitely generated:

Proof Since the inequalities from Theorem 6.2 are finite in number, and each inequality has its bounding hyperplane containing the origin, the set $C \subseteq \mathbb{R}^{3r}$ they define is a polyhedral cone. Since the inequalities have rational coefficients, by definition, $\mathcal{C}$ is rational. Moreover, $\mathcal{C}$ is also clearly pointed, *i.e.*, $\mathcal{C} \cap -\mathcal{C} = \{\mathbf{0}\}$. Now apply [53, Theorem 16.4]. □

Naturally, one would like a solution for the generalization of Problem 3.12 to EqLR_r.

Using Theorem 6.1 and Theorem 4.6, A. Adve together with the first and third authors [3] prove a generalization of Theorem 5.7:

Theorem 6.9 ([3]) *The decision problem of determining $C_{\lambda,\mu}^{\nu} \neq 0$ is in the class* P *of polynomial problems.*

Sketch of proof: Using Theorem 5.7 one can construct a polytope $\mathcal{Q}_{\lambda,\mu}^{\nu}$ analogous to $\mathcal{P}_{\lambda,\mu}^{\nu}$. The main property is that $C_{\lambda,\mu}^{\nu} \neq 0$ if and only if $\mathcal{Q}_{\lambda,\mu}^{\nu}$ contains a lattice point (in particular, in contrast to Theorem 3.9, the *number* of lattice points of $\mathcal{Q}_{\lambda,\mu}^{\nu}$ is not well-understood). The remainder of the proof proceeds exactly as in the proof of Theorem 5.7, except that we use Theorem 6.1 in place of Theorem 5.1. □

For the remainder of this section, assume $C_{\lambda,\mu}^{\nu}$ is expressed (uniquely) as a polynomial in the variables $\beta_i := t_i - t_{i+1}$. We now state a few open problems/conjectures introduced in [51], for the special case of Grassmannians.

A refinement of the nonvanishing question is:

Question 6.10 *What is the computational complexity of determining if $[\beta_1^{i_1} \cdots \beta_{n-1}^{i_{n-1}}]C_{\lambda,\mu}^{\nu} \neq 0$?*

This question concerns the Newton polytope of $C_{\lambda,\mu}^{\nu}$. Recall, the *Newton polytope* of

$$f = \sum_{(n_1,\ldots,n_r)\in\mathbb{Z}_{\geq 0}^r} c_{n_1,\ldots,n_r} \prod_{j=1}^{r} \alpha_j^{n_j} \in \mathbb{R}[\alpha_1,\ldots,\alpha_r]$$

is $\mathsf{Newton}(f) := \mathsf{conv}\{(n_1,\ldots,n_r) : c_{n_1,\ldots,n_r} \neq 0\} \subseteq \mathbb{R}^r$. f has *saturated Newton polytope* (SNP) [41] if $c_{n_1,\ldots,n_r} \neq 0 \iff (n_1,\ldots,n_r) \in \mathsf{Newton}(f)$.

Conjecture 6.11 *$C_{\lambda,\mu}^{\nu}$ has SNP.*

This raises the question:

Problem 6.12 *Give a half space description of* $\mathsf{Newton}(C_{\lambda,\mu}^{\nu})$.

A proof of Conjecture 6.11 together with any reasonable solution to Problem 6.12 would imply that the decision problem in Question 6.10 is in the computational complexity class NP ∩ coNP. This would strongly suggest that the decision problem is not NP-complete, and in fact suggest the problem is in P. We refer the reader to [1, Section 1] for elaboration on these points.

7 Maximal orthogonal and Lagrangian Grassmannians

7.1 Goals in the sequel

Beyond Grassmannians $\mathsf{Gr}_k(\mathbb{C}^n)$, the *maximal orthogonal Grassmannians* and *Lagrangian Grassmannians* have been of significant interest. Their classical (non-equivariant) Schubert calculus shares many analogies with the Grassmannian case. They concern the Q-Schur polynomials of I. Schur [55], and the tableau combinatorics of D. Worley [67], B. Sagan [52] and J. Stembridge [59]. Although these combinatorial results were originally developed to study projective representations of the symmetric group, the connection to Schubert calculus of these spaces was established by P. Pragacz [46].

It is therefore natural to seek extensions of the results from Sections 1-6. Discussion of efforts toward this goal occupy the remainder of this work.

7.2 Definition of the spaces

Consider the two classical Lie groups of non-simply laced type: $\mathsf{G} = \mathsf{SO}_{2n+1}(\mathbb{C})$ and $\mathsf{G} = \mathsf{Sp}_{2n}(\mathbb{C})$. These are the automorphism groups of a non-degenerate bilinear form $\langle\cdot,\cdot\rangle$. In the former case, $\langle\cdot,\cdot\rangle$ is symmetric, and on $W = \mathbb{C}^{2n+1}$. In the latter case, $\langle\cdot,\cdot\rangle$ is skew-symmetric, and on $W = \mathbb{C}^{2n}$. A subspace $V \subseteq W$ is called *isotropic* if $\langle v_1, v_2\rangle = 0$ for all $v_1, v_2 \in V$. The maximum dimension of an isotropic subspace of W is n.

Let $Y = \mathsf{OG}(n, 2n+1)$ be the *maximal orthogonal Grassmannian* of n-dimensional isotropic subspaces of $\mathbb{C}^{2n+1}$; this space has an action of $\mathsf{G} = \mathsf{SO}_{2n+1}(\mathbb{C})$. Similarly, let $Z = \mathsf{LG}(n, 2n)$ be the *Lagrangian Grassmannian* of n-dimensional isotropic subspaces of $\mathbb{C}^{2n}$. In either case, the (opposite) Borel subgroup $\mathsf{B}_- \leq \mathsf{G}$ consists of the lower triangular matrices in G. The maximal torus T are the diagonal matrices in G. Just as in the case of the Grassmannian, the corresponding B_- acts on Y (resp. Z) with finitely many orbits Y_λ° (resp. Z_λ°); these are the *Schubert cells*.

In both cases, the Schubert cells and thus Schubert varieties $Y_\lambda = \overline{Y_\lambda^\circ}$ (resp. $Z_\lambda = \overline{Z_\lambda^\circ}$) are indexed by strict partitions fitting inside the shifted staircase

$$\rho_n = (n, n-1, n-2, \ldots, 3, 2, 1).$$

A *strict partition* is an integer partition $\lambda = (\lambda_1 > \lambda_2 > \cdots > \lambda_\ell > 0)$. Identify λ with its shifted shape, which is the usual Young diagram (in English notation) but where the i-th row from the top is indented $i-1$ many spaces. We refer to, *e.g.*, [25, Section 6] and the references therein for additional details.

Let $\sigma_\lambda(Y) \in H^{2|\lambda|}(Y)$ be the Poincaré dual to Y_λ. These *Schubert classes* form a $\mathbb{Z}$-linear basis of $H^*(Y)$, and we define the structure constants by

$$\sigma_\lambda(Y) \smile \sigma_\mu(Y) = \sum_{\nu \subseteq \rho_n} o_{\lambda,\mu}^{\nu} \sigma_\nu(Y).$$

Similarly, the Schubert classes $\sigma_\lambda(Z)$ form a $\mathbb{Z}$-linear basis of $H^*(Z)$, and

$$\sigma_\lambda(Z) \smile \sigma_\mu(Z) = \sum_{\nu \subseteq \rho_n} l_{\lambda,\mu}^{\nu} \sigma_\nu(Z).$$

7.3 Schur $P-$ and $Q-$ functions; P. Pragacz's theorem

Let

$$q_r(x_1, \ldots, x_n) = 2\sum_{i=1}^{n} x_i^r \prod_{i \neq j} \frac{x_i + x_j}{x_i - x_j}.$$

This is clearly symmetric and in fact a polynomial. Next, set

$$Q_{(r,s)} = q_r q_s + 2\sum_{i=1}^{s} (-1)^i q_{r+i} q_{s-i}.$$

Recall the *Pfaffian* of a $2t \times 2t$ skew-symmetric matrix $M = (m_{ij})$

$$\mathsf{Pf}(M) = \sum_{\sigma \in \mathfrak{S}_{2t}} \mathsf{sgn}(\sigma) \prod_{i=1}^{t} m_{\sigma(2i-1),\sigma(2i)},$$

where σ satisfies $\sigma(2i-1) < \sigma(2i)$ for $1 \leq i \leq m$ and $\sigma(1) < \sigma(3) < \cdots < \sigma(2i-3) < \sigma(2i-1)$. Then for $\lambda = (\lambda_1 > \lambda_2 > \cdots > \lambda_\ell > 0)$, the *Schur $Q-$function* [55] is

$$Q_\lambda = \mathsf{Pf}(Q_{(\lambda_i,\lambda_j)}).$$

If $\ell(\lambda)$ is odd, we add a 0 at the end of λ. The Schur $Q-$ functions linearly span the subalgebra $\Gamma \subset \mathsf{Sym}$ generated by the q_i's. In fact, $Q_{(i)} = q_i$. The *Schur $P-$function* is

$$P_\lambda := 2^{-\ell(\lambda)} Q_\lambda. \tag{14}$$

P. Pragacz [46] proved that the Schur $P-$ and Schur $Q-$ polynomials represent the Schubert classes of Y and Z respectively. That is,

$$H^*(Y) \cong \Gamma/J,$$

where J is the ideal $\langle P_\lambda : \lambda \not\subseteq \rho_n \rangle$, and $\sigma_\lambda(Y)$ maps to $P_\lambda + J$ under this isomorphism. Moreover,

$$P_\lambda P_\mu = \sum_\nu o^\nu_{\lambda,\mu} P_\nu.$$

Similarly,

$$Q_\lambda Q_\mu = \sum_\nu l^\nu_{\lambda,\mu} Q_\nu.$$

J. R. Stembridge [59] proved

$$P_\lambda = \sum_T x^T,$$

where the sum is over semistandard fillings of the shifted shape λ. That is, fill each box of λ with a label from the ordered set $1' < 1 < 2' < 2 < 3' < 3 < \cdots$ such that the rows and columns are weakly increasing, two i'-s cannot appear in the same row and two i-s cannot appear in the same column. Moreover, there are no primed entries on the main diagonal. For example, if $\lambda = (2,1)$, the shifted semistandard tableaux are

$$\begin{array}{|c|c|}\hline 1 & 1 \\ \hline \multicolumn{1}{c|}{} & 2 \\ \cline{2-2}\end{array}\quad \begin{array}{|c|c|}\hline 1 & 2' \\ \hline \multicolumn{1}{c|}{} & 2 \\ \cline{2-2}\end{array}\quad \begin{array}{|c|c|}\hline 1 & 1 \\ \hline \multicolumn{1}{c|}{} & 3 \\ \cline{2-2}\end{array}\quad \begin{array}{|c|c|}\hline 1 & 2' \\ \hline \multicolumn{1}{c|}{} & 3 \\ \cline{2-2}\end{array}\quad \begin{array}{|c|c|}\hline 1 & 2 \\ \hline \multicolumn{1}{c|}{} & 3 \\ \cline{2-2}\end{array}\quad \begin{array}{|c|c|}\hline 1 & 3' \\ \hline \multicolumn{1}{c|}{} & 3 \\ \cline{2-2}\end{array}\quad \begin{array}{|c|c|}\hline 2 & 2 \\ \hline \multicolumn{1}{c|}{} & 3 \\ \cline{2-2}\end{array}\quad \begin{array}{|c|c|}\hline 2 & 3' \\ \hline \multicolumn{1}{c|}{} & 3 \\ \cline{2-2}\end{array}\ \ldots.$$

Hence $P_{(2,1)}(x_1, x_2, x_3) = x_1^2 x_2 + x_1 x_2^2 + x_1^2 x_3 + x_1 x_2 x_3 + x_1 x_2 x_3 + x_1 x_3^2 + x_2^2 x_3 + x_2 x_3^2 + \cdots$. If we allow primed entries on the diagonal, we get the Schur $Q-$ function. It is easy to see that this definition satisfies (14).

7.4 Shifted Littlewood–Richardson rules

D. Worley [67] introduced a *jeu de taquin* theory for shifted shapes. A standard tableau T of shifted skew shape ν/λ is a filling of λ with the labels $1, 2, 3, \ldots, |\nu/\lambda|$ that is increasing along rows and columns. Let $\mathsf{shSYT}(\nu/\lambda)$ denote the set of these tableaux. The notions of slides and rectification are just as in the unshifted case, using (J1) and (J2). With this, the exact analogues of Theorems 3.1 and 3.3 hold and one can define ShEjdt and shRect etc, in the obvious manner. Indeed, one has the following combinatorial rule for $o^\nu_{\lambda,\mu}$:

Theorem 7.1 (Shifted jeu de taquin Littlewood–Richardson rule) *Fix* $U \in \mathsf{shSYT}(\mu)$. *Then*

$$o^{\nu}_{\lambda,\mu} = \#\{T \in \mathsf{shSYT}(\nu/\lambda) : \mathsf{shRect}(T) = U\}.$$

Example 7.2 Let $\lambda = (3,1), \mu = (3,1), \nu = (4,3,1)$. The following are the 2 shifted tableaux of shape ν/λ that rectify to $U = \begin{array}{|c|c|c|}\hline 1 & 2 & 3 \\ \hline \multicolumn{1}{c|}{} & 4 & \multicolumn{1}{c}{} \\ \cline{2-2}\end{array}$. Thus, $o^{\nu}_{\lambda,\mu} = 2$.

$$\begin{array}{|c|c|c|c|}\hline & & & 1 \\ \hline \multicolumn{1}{c|}{} & & 2 & 3 \\ \cline{2-4} \multicolumn{1}{c}{} & \multicolumn{1}{c|}{} & 4 & \multicolumn{1}{c}{} \\ \cline{3-3}\end{array} \qquad \begin{array}{|c|c|c|c|}\hline & & & 3 \\ \hline \multicolumn{1}{c|}{} & & 1 & 4 \\ \cline{2-4} \multicolumn{1}{c}{} & \multicolumn{1}{c|}{} & 2 & \multicolumn{1}{c}{} \\ \cline{3-3}\end{array}$$

Define the *reading word* of a (possibly skew) shifted tableau T to be the word obtained by reading the rows of T from left to right starting with the bottom row. For a word $w = w_1 w_2 \dots w_n$, define $\mathsf{JS}_i(j)$ for $1 \le j \le 2n, i \ge 1$, depending on w:

$$\mathsf{JS}_i(j) := \text{number of occurrences of } i \text{ in } w_{n-j+1}, \dots, w_n,\ 0 \le j \le n,$$

and

$$\mathsf{JS}_i(n+j) := \mathsf{JS}_i(n) + \text{number of occurrences of } i' \text{ in } w_1, \dots, w_j, \quad 0 < j \le n.$$

The word w is *proto-ballot* if, when $\mathsf{JS}_i(j) = \mathsf{JS}_{i-1}(j)$, both of these statements hold:

$$w_{n-j} \neq i, i', \ \text{if } 0 \le j < n;$$
$$w_{j-n+1} \neq i-1, i', \ \text{if } n \le j < 2n.$$

Let $|w|$ be the word obtained from w by removing all primes. Now, w is *ballot* if it is *proto-ballot* and the leftmost i of $|w|$ is unprimed in w for all i. In his work on projective representation theory of symmetric groups, J. Stembridge [59, Theorem 8.3] gave the following semistandard analogue of Theorem 7.1.

Theorem 7.3 (Shifted ballot Littlewood–Richardson rule) $o^{\nu}_{\lambda,\mu}$ *counts the number of shifted semistandard tableaux of shape* ν/λ *of content* μ *that are ballot.*

Example 7.4 Let $\lambda = (3,1), \mu = (3,1), \nu = (4,3,1)$, then the following are the only 2 shifted semistandard tableaux of shape ν/λ of content μ that are ballot.

			1′
		1	1
		2	

			1
		1′	2
		1	

It follows from (14) that

$$l^{\nu}_{\lambda,\mu} = 2^{\ell(\nu)-\ell(\lambda)-\ell(\mu)} o^{\nu}_{\lambda,\mu}. \tag{15}$$

Thus the above rules give a rule to compute $l^{\nu}_{\lambda,\mu}$ in a manifestly positive manner, as well.

7.5 Nonvanishing

K. Purbhoo and F. Sottile [49, 50] gave an extension of the Horn recursion to describe when $o^{\nu}_{\lambda,\mu}$ (or equivalently $l^{\nu}_{\lambda,\mu}$) is nonzero. Fix n and r. Suppose α is a (ordinary) partition whose (unshifted) Young diagram is contained in $r \times (n-r)$. Let

$$I_n(\alpha) := \{n-r+1-\alpha_1, n-r+2-\alpha_2, \ldots, n-\alpha_r\}.$$

Index the corners of ρ_n top to bottom from 1 to n. For $\lambda \subseteq \rho_n$, $0 < r < n$ and $\alpha \subseteq r \times (n-r)$ let $[\lambda]_\alpha$ be the number of boxes of λ that survive after crossing out the rows to the right and columns above the corners indexed by $I_n(\alpha)$. Define λ^c to be the complement of λ in ρ_n (after reflecting), whereas $\alpha^c = \alpha^\vee$ is the rotation of the complement of α in $r \times (n-r)$.

Example 7.5 Suppose that $n = 6, r = 3$ and $\alpha = (3,2,1)$. Then

$$I_n(\alpha) = \{6-3+1-3, 6-3+2-2, 6-3+3-1\} = \{1,3,5\}.$$

Suppose $\lambda = (6,4,3,1) \subseteq \rho_n$. In the figure below, shaded boxes are the ones that are crossed out. Thus, $[\lambda]_\alpha = 3$.

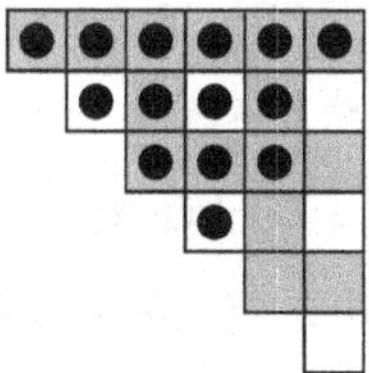

Theorem 7.6 (K. Purbhoo-F. Sottile's theorem) *For $\lambda, \mu, \nu \subseteq \rho_n$, $o^{\nu^c}_{\lambda,\mu} \neq 0$ if and only if*

- $|\lambda| + |\mu| + |\nu| = \dim Y = \binom{n+1}{2}$, *and*
- *for all $0 < r < n$ and all $\alpha, \beta, \gamma \subset r \times (n-r)$ such that $c^{\gamma^c}_{\alpha,\beta} \neq 0$, one has* $[\lambda]_\alpha + [\mu]_\beta + [\nu]_\gamma \leq \binom{n+1-r}{2}$.

Remark 7.7 The obvious analogue of saturation does *not* hold. For example, take $\lambda = (2,1), \mu = (2), \nu = (3,2)$. Then $o^{\nu}_{\lambda,\mu} \neq 0$, but $o^{2\nu}_{2\lambda,2\mu} = o^{(6,4)}_{(4,2),(4)} = 0$. □

Inspired by the complexity results concerning $c^{\nu}_{\lambda,\mu}$, we take this opportunity to pose:

Problem 7.8 *Is the decision problem of determining if $o^{\nu}_{\lambda,\mu} \neq 0$ in the class* P *of polynomial time problems?*

Remark 7.7 implies that the argument used in the proofs of Theorems 5.7 and 6.9 cannot work.

Problem 7.9 *Is counting $o^{\nu}_{\lambda,\mu}$ in the class of* #P*-complete problems?*

8 Equivariant Schubert calculus of Y and Z

One is interested in the equivariant cohomology of Y and Z. As with Grassmannians (Section 2), one has structure constants with respect to the Schubert basis,

$$\xi_\lambda(Y) \cdot \xi_\mu(Y) = \sum_{\nu \subseteq \rho_n} O^{\nu}_{\lambda,\mu} \xi_\nu(Y) \text{ and } \xi_\lambda(Z) \cdot \xi_\mu(Z) = \sum_{\nu \subseteq \rho_n} L^{\nu}_{\lambda,\mu} \xi_\nu(Z).$$

If $|\lambda| + |\mu| = |\nu|$ then $O^{\nu}_{\lambda,\mu} = o^{\nu}_{\lambda,\mu}$ and $L^{\nu}_{\lambda,\mu} = l^{\nu}_{\lambda,\mu}$. For sake of brevity, we refer to [25] and the references therein.

Suppose that $H^*_{\mathsf{T}}(pt) = \mathbb{Z}[t_1, \ldots, t_n]$. The general form of Theorem 2.1 (see the attached footnote to that result) states that if $\gamma_1 = t_1$ and for $i > 1$, $\gamma_i = t_i - t_{i-1}$, then

$$O^{\nu}_{\lambda,\mu} \in \mathbb{Z}_{\geq 0}[\gamma_1, \gamma_2, \ldots, \gamma_n].$$

Similarly, if $\alpha_1 = 2t_1, \alpha_2 = t_2 - t_1 \ldots, \alpha_n = t_n - t_{n-1}$ then

$$L^{\nu}_{\lambda,\mu} \in \mathbb{Z}_{\geq 0}[\alpha_1, \alpha_2, \ldots, \alpha_n].$$

Problem 8.1 *Give a combinatorial rule for $O^{\nu}_{\lambda,\mu}$ and/or $L^{\nu}_{\lambda,\mu}$.*

Naturally, we desire a rule in terms of shifted edge labeled tableaux. Such a rule (or any combinatorial rule) has eluded us. The reader wishing to give Problem 8.1 a try might find Table 1 useful.

Let $\tilde{L}^{\nu}_{\lambda,\mu}$ be the polynomial obtained from $L^{\nu}_{\lambda,\mu}$ after substitutions $\alpha_1 \mapsto 2\gamma_1$ and $\alpha_i \mapsto \gamma_i$ for $i > 1$. The following is a refinement of (15):

Theorem 8.2 (*cf.* Theorem 1.1 of [51]) $O^{\nu}_{\lambda,\mu} = 2^{\ell(\nu)-\ell(\lambda)-\ell(\mu)} \tilde{L}^{\nu}_{\lambda,\mu}$.

Table 1. *Table of products for $n = 2$.*

λ	μ	ν	$O^{\nu}_{\lambda,\mu}$	$L^{\nu}_{\lambda,\mu}$
[1]	[1]	[1]	γ_1	α_1
[1]	[1]	[1]	γ_1	α_1
[1]	[1]	[2]	1	2
[1]	[2]	[2]	$\gamma_1 + \gamma_2$	$\alpha_1 + 2\alpha_2$
[2]	[1]	[2]	$\gamma_1 + \gamma_2$	$\alpha_1 + 2\alpha_2$
[2]	[2]	[2]	$\gamma_1\gamma_2 + \alpha_2^2$	$\alpha_1\alpha_2 + 2\alpha_2^2$
[1]	[2]	[2, 1]	1	1
[1]	[2, 1]	[2, 1]	$2\gamma_1 + \gamma_2$	$2\alpha_1 + 2\alpha_2$
[2]	[1]	[2, 1]	1	1
[2]	[2]	[2, 1]	$2\gamma_1 + 2\gamma_2$	$\alpha_1 + 2\alpha_2$
[2]	[2, 1]	[2, 1]	$2\gamma_1^2 + 3\gamma_1\gamma_2 + \gamma_2^2$	$\alpha_1^2 + 3\alpha_1\alpha_2 + 2\alpha_2^2$
[2, 1]	[1]	[2, 1]	$2\gamma_1 + \gamma_2$	$2\alpha_1 + 2\alpha_2$
[2, 1]	[2]	[2, 1]	$2\gamma_1^2 + 3\gamma_1\gamma_2 + \gamma_2^2$	$\alpha_1^2 + 3\alpha_1\alpha_2 + 2\alpha_2^2$
[2, 1]	[2, 1]	[2, 1]	$2\gamma_1^3 + 3\gamma_1^2\gamma_2 + \gamma_1\gamma_2^2$	$\alpha_1^3 + 3\alpha_1^2\alpha_2 + 2\alpha_1\alpha_2^2$

Thus, the $O^{\nu}_{\lambda,\mu}$ and $L^{\nu}_{\lambda,\mu}$ versions of Problem 8.1 are essentially equivalent.

Theorem 8.2 was stated in a weaker form as a conjecture in C. Monical's doctoral thesis [40, Conjecture 5.1]. A proof of a generalization was given in [51, Theorem 1.1]. Below, we offer another proof that uses a variation of the associativity recurrence alluded to at the end of Section 2.3. This recurrence should be useful to prove any guessed rule for $O^{\nu}_{\lambda,\mu}$ or $L^{\nu}_{\lambda,\mu}$, so we wish to explicate it here.

Proof This will serve as the base case of the associativity recurrence below:

Lemma 8.3 $O^{\lambda}_{\lambda,\mu} = 2^{\ell(\nu)-\ell(\lambda)-\ell(\mu)}\tilde{L}^{\lambda}_{\lambda,\mu}$.

Proof By the same reasoning as the derivation of (8), we have $L^{\lambda}_{\lambda,\mu} = \xi_\mu(Z)|_\lambda$ and $O^{\lambda}_{\lambda,\mu} = \xi_\mu(Y)|_\lambda$. The lemma thus holds since, by [25, Theorem 3], $\xi_\mu(Y)|_\lambda = 2^{-\ell(\mu)}\xi_\mu(Z)|_\lambda$. □

Assign weights to each box of the staircase ρ_n as follows. For Y, the boxes on the main diagonal are assigned weight γ_1. The boxes on the next diagonal are assigned γ_2, etc. For Z, the boxes on the main diagonal are assigned α_1 whereas the boxes on the second diagonal are assigned $2\alpha_2$, and the third diagonal $2\alpha_3$, etc. Let $\beta_Y := \rho_n \to \{\gamma_i\}$ and $\beta_Z : \rho_n \to \{\alpha_1, 2\alpha_2, \ldots, 2\alpha_n\}$ be these two assignments.

Thus, when $n = 3$ the assignment is

$$\begin{array}{|c|c|c|}\hline \gamma_1 & \gamma_2 & \gamma_3 \\ \hline \multicolumn{1}{c|}{} & \gamma_1 & \gamma_2 \\ \cline{2-3} \multicolumn{1}{c}{} & \multicolumn{1}{c|}{} & \gamma_1 \\ \cline{3-3} \end{array} \quad \text{(for } Y\text{)} \quad \text{and} \quad \begin{array}{|c|c|c|}\hline \alpha_1 & 2\alpha_2 & 2\alpha_3 \\ \hline \multicolumn{1}{c|}{} & \alpha_1 & 2\alpha_2 \\ \cline{2-3} \multicolumn{1}{c}{} & \multicolumn{1}{c|}{} & \alpha_1 \\ \cline{3-3} \end{array} \quad \text{(for } Z\text{).}$$

For a straight shape $\lambda \subseteq \rho_n$, define

$$\mathtt{wt}_Y(\lambda) = \sum_{x\in\lambda} \beta_Y(x).$$

For a skew shape $\nu/\lambda \subseteq \rho_n$,

$$\mathtt{wt}_Y(\nu/\lambda) := \mathtt{wt}_Y(\nu) - \mathtt{wt}_Y(\lambda).$$

Similarly, one defines $\mathtt{wt}_Z(\nu/\lambda)$.

Let λ^+ be λ with a box added. Also let ν^- be ν with a box removed. We claim that

$$\sum_{\lambda^+} O^{\nu}_{\lambda^+,\mu} = O^{\nu}_{\lambda,\mu}\mathtt{wt}_Y(\nu/\lambda) + \sum_{\nu^-} O^{\nu^-}_{\lambda,\mu}. \tag{16}$$

This is proved by the considering the associativity relation

$$(\xi_\lambda(Y)\cdot\xi_{(1)}(Y))\cdot\xi_\mu(Y) = \xi_\lambda(Y)\cdot(\xi_{(1)}(Y)\cdot\xi_\mu(Y)),$$

and using the Pieri rule for Y:

$$\xi_{(1)}(Y)\cdot\xi_\lambda(Y) = \mathtt{wt}_Y(\lambda)\xi_\lambda(Y) + \sum_{\lambda^+}\xi_{\lambda^+}(Y). \tag{17}$$

The proof of (17) can be obtained starting with the same reasoning as the derivation of (9). Alternatively, it can be deduced by specializing more general formulas such as C. Lenart-A. Postnikov's [37, Corollary 1.2].

Similarly, the Pieri rule for Z reads

$$\xi_{(1)}(Z)\cdot\xi_\lambda(Z) = \mathtt{wt}_Z(\lambda)\xi_\lambda(Z) + \sum_{\lambda^+} 2^{\ell(\lambda)+1-\ell(\lambda^+)}\xi_{\lambda^+}(Z).$$

Consequently, by the same reasoning we obtain

$$\sum_{\lambda^+} L^{\nu}_{\lambda^+,\mu}2^{\ell(\lambda)+1-\ell(\lambda^+)} = L^{\nu}_{\lambda,\mu}\mathtt{wt}_Z(\nu/\lambda) + \sum_{\nu^-} L^{\nu^-}_{\lambda,\mu}2^{\ell(\nu^-)+1-\ell(\nu)}. \tag{18}$$

Now, to complete the proof by induction we start from (18). This is an identity of polynomials and remains so after the substitution $\alpha_1 \mapsto 2\gamma_1$ and $\alpha_i \mapsto \gamma_i$ for $i > 1$. That is,

$$\sum_{\lambda^+} \widetilde{L}^{\nu}_{\lambda^+,\mu} 2^{\ell(\lambda)+1-\ell(\lambda^+)} = \widetilde{L}^{\nu}_{\lambda,\mu} \widetilde{\mathtt{wt}}_Z(\nu/\lambda) + \sum_{\nu^-} \widetilde{L}^{\nu^-}_{\lambda,\mu} 2^{\ell(\nu^-)+1-\ell(\nu)}, \tag{19}$$

where $\widetilde{\mathtt{wt}}_Z(\nu/\lambda)$ is $\mathtt{wt}_Z(\nu/\lambda)$ with the same substitution. Note that

$$\frac{1}{2}\widetilde{\mathtt{wt}}_Z(\nu/\lambda) = \mathtt{wt}_Y(\nu/\lambda).$$

Now multiply both sides of (19) by $\frac{1}{2} \times 2^{\ell(\nu)-\ell(\lambda)-\ell(\mu)}$. This gives

$$\sum_{\lambda^+} \widetilde{L}^{\nu}_{\lambda^+,\mu} 2^{\ell(\nu)-\ell(\lambda^+)-\ell(\mu)} = 2^{\ell(\nu)-\ell(\lambda)-\ell(\mu)} \widetilde{L}^{\nu}_{\lambda,\mu} \mathtt{wt}_Y(\nu/\lambda) + \sum_{\nu^-} \widetilde{L}^{\nu^-}_{\lambda,\mu} 2^{\ell(\nu^-)-\ell(\lambda)-\ell(\mu)}.$$

By induction,

$$\sum_{\lambda^+} O^{\nu}_{\lambda^+,\mu} = 2^{\ell(\nu)-\ell(\lambda)-\ell(\mu)} \widetilde{L}^{\nu}_{\lambda,\mu} \mathtt{wt}_Y(\nu/\lambda) + \sum_{\nu^-} O^{\nu^-}_{\lambda,\mu}. \tag{20}$$

Comparing (20) and (16) we deduce that

$$O^{\nu}_{\lambda,\mu} = 2^{\ell(\nu)-\ell(\lambda)-\ell(\mu)} \widetilde{L}^{\nu}_{\lambda,\mu},$$

as needed. □

Turning to nonvanishing, clearly:

Corollary 8.4 $[\gamma_1^{i_1} \cdots \gamma_n^{i_n}] O^{\nu}_{\lambda,\mu} \neq 0 \iff [\alpha_1^{i_1} \cdots \alpha_n^{i_n}] L^{\nu}_{\lambda,\mu} \neq 0$; *in particular* $O^{\nu}_{\lambda,\mu} \neq 0 \iff L^{\nu}_{\lambda,\mu} \neq 0$.

Moreover, C. Monical [40] gave a conjectural equivariant extension of Theorem 7.6.

Conjecture 8.5 (C. Monical's Horn-type conjecture) *For* $\lambda, \mu, \nu \subseteq \rho_n$ *(and not a smaller staircase),* $O^{\nu^c}_{\lambda,\mu} \neq 0$ *if and only if for* $k = |\lambda| + |\mu| + |\nu| - \binom{n+1}{2}$,

- $k \geq 0$, *and*
- *for all* $0 < r < n$ *and all* $\alpha, \beta, \gamma \subseteq r \times (n-r)$ *with* $|\alpha| + |\beta| + |\gamma| = r(n-r)$ *and* $c^{\gamma^c}_{\alpha,\beta} \neq 0$ *we have* $[\lambda]_\alpha + [\mu]_\beta + [\nu]_\gamma - k \leq \binom{n+1-r}{2}$.

In *loc. cit.*, C. Monical reports checking this conjecture for all $\lambda, \mu, \nu \subseteq \rho_5$. Now, from Corollary 8.4 we obtain:

Corollary 8.6 (*cf.* Conjecture 5.3 of [40]) *C. Monical's inequalities characterize* $O^{\nu}_{\lambda,\mu} \neq 0$ *if and only if they characterize* $L^{\nu}_{\lambda,\mu} \neq 0$.

9 Shifted edge labeled tableaux

In this section, we define *shifted edge labeled tableaux*. At present, we do not know a good theory when edge labels are permitted on arbitrary horizontal edges. However, our central new idea is to *restrict edge labels to diagonal boxes*. This restriction gives rise to a combinatorial rule which defines a commutative and (conjecturally) associative ring.

9.1 Main definitions

If $\mu \subseteq \lambda$, then λ/μ is the skew-shape consisting of boxes of λ not in μ. The boxes in matrix position (i,i) are the *diagonal boxes*. A *diagonal edge* of λ/μ refers to the southern edge of a diagonal box of λ. If $\mu = \emptyset$, we call $\lambda = \lambda/\mu$ a *straight shape*.

For example if $\lambda = (6,3,1)$ and $\mu = (3,1)$, the shape λ/μ consists of the six unmarked boxes shown below

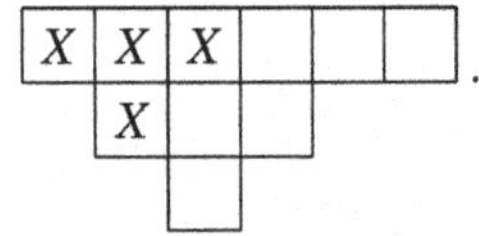

This has one diagonal box but three diagonal edges.

A *shifted edge labeled tableau* of shape λ/μ is a filling of the boxes of λ/μ and southern edges of the diagonal boxes with the labels $[N] = \{1,2,3,\ldots,N\}$ such that:

(S1) Every box of λ/μ is filled.
(S2) Each *diagonal* edge contains a (possibly empty) subset of $[N]$.
(S3) $1,2,\ldots,N$ appears exactly once.
(S4) The labels strictly increase left to right along rows and top to bottom along columns. In particular, each label of a diagonal edge is strictly larger than the box labels in the same column.

These conditions imply that $N \geq |\lambda/\mu|$. Let $\mathsf{eqShSYT}(\lambda/\mu, N)$ be the set of all such tableaux. If we restrict to tableaux satisfying only (S1), (S3) and (S4), then $N = |\lambda/\mu|$ and we obtain the notion of shifted standard Young tableaux from Section 7.

An *inner corner* c of λ/μ is a maximally southeast box of μ. For $T \in \mathsf{eqShSYT}(\lambda/\mu, N)$, we define a *(shifted, edge labeled) jeu de taquin slide* $\mathsf{shEjdt}_{\mathsf{c}}(T)$, obtained as follows. Initially place $\bullet$ in c, and apply one of the following *slides*, depending on what T looks like locally around c:

(J1) $\begin{array}{|c|c|}\hline \bullet & a \\ \hline b & \multicolumn{1}{c}{} \\ \cline{1-1}\end{array} \mapsto \begin{array}{|c|c|}\hline b & a \\ \hline \bullet & \multicolumn{1}{c}{} \\ \cline{1-1}\end{array}$ (if $b < a$, or a does not exist)

(J2) $\begin{array}{|c|c|}\hline \bullet & a \\ \hline b & \multicolumn{1}{c}{} \\ \cline{1-1}\end{array} \mapsto \begin{array}{|c|c|}\hline a & \bullet \\ \hline b & \multicolumn{1}{c}{} \\ \cline{1-1}\end{array}$ (if $a < b$, or b does not exist)

(J3′) $\begin{array}{|c|c|}\hline \bullet & a \\ \hline \multicolumn{1}{c}{S} \end{array} \mapsto \begin{array}{|c|c|}\hline a & \bullet \\ \hline \multicolumn{1}{c}{S} \end{array}$ (if c is a *diagonal* box and $a < \min(S)$)

(J4′) $\begin{array}{|c|c|}\hline \bullet & a \\ \hline \multicolumn{1}{c}{S} \end{array} \mapsto \begin{array}{|c|c|}\hline s & a \\ \hline \multicolumn{1}{c}{S'} \end{array}$ (if c is a *diagonal* box, $s := \min(S) < a$, and $S' := S \setminus \{s\}$)

Repeat the above sliding procedure on the new box c' containing the new position $\bullet$ until $\bullet$ arrives at a box or diagonal edge d of λ that has no labels immediately south or east of it. Then $\mathsf{shEjdt}_{\mathsf{c}}(T)$ is obtained by erasing $\bullet$.

A *rectification* of $T \in \mathsf{eqShSYT}(\lambda/\mu, N)$ is defined as usual: Choose an inner corner c_0 of λ/μ and compute $T_1 := \mathsf{shEjdt}_{\mathsf{c}_0}(T)$, which has shape $\lambda^{(1)}/\mu^{(1)}$. Now let c_1 be an inner corner of $\lambda^{(1)}/\mu^{(1)}$ and compute $T_2 := \mathsf{shEjdt}_{\mathsf{c}_1}(T_1)$. Repeat $|\mu|$ times, arriving at a standard tableau of straight shape. Let $\mathsf{shEqRect}_{\{\mathsf{c}_i\}}(T)$ be this tableau.

In general, $\mathsf{shEqRect}$ is not independent of rectification order, when $N > |\lambda/\mu|$:

Example 9.1 The reader can check that if one uses column rectification order (picking the rightmost inner corner at each step) then

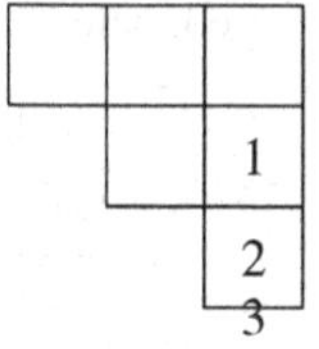

rectifies to $\begin{array}{|c|c|c|}\hline 1 & 2 & 3 \\ \hline\end{array}$ while row rectification (choosing the southmost inner corner at each step) gives $\begin{array}{|c|c|}\hline 1 & 2 \\ \hline \multicolumn{1}{c|}{} & 3 \\ \cline{2-2}\end{array}$. □

We will define $\mathsf{shEqRect}(T)$ to be the rectification under row rectification order.

9.2 A (putative) commutative ring structure

Let S_μ be the superstandard tableau of shifted shape μ, which is obtained by filling the boxes of μ in English reading order with $1, 2, 3, \ldots$. For example,

$$S_{(5,3,1)} = \begin{array}{ccccc} 1 & 2 & 3 & 4 & 5 \\ & 6 & 7 & 8 & \\ & & 9 & & \end{array}.$$

Define

$$d^{\nu}_{\lambda,\mu} := \#\{T \in \mathsf{eqShSYT}(\nu/\lambda, |\mu|) : \mathsf{shEqRect}(T) = S_{\mu}\}.$$

Let

$$\Delta(\nu;\lambda,\mu) := |\lambda| + |\mu| - |\nu| \text{ and } L(\nu;\lambda,\mu) := \ell(\lambda) + \ell(\mu) - \ell(\nu).$$

Introduce an indeterminate z and set

$$D^{\nu}_{\lambda,\mu} := 2^{L(\nu;\lambda,\mu)-\Delta(\nu;\lambda,\mu)} z^{\Delta(\nu;\lambda,\mu)} d^{\nu}_{\lambda,\mu}.$$

Next we define formal symbols $[\lambda]$ for each $\lambda \subseteq \rho_n$. Let R_n be the free $\mathbb{Z}[z]$-module generated by these. We declare a product structure on R_n

$$[\lambda] \star [\mu] = \sum_{\nu} D^{\nu}_{\lambda,\mu}[\nu].$$

While positivity of $D^{\nu}_{\lambda,\mu}$ is immediate from the definition, the following is not:

Conjecture 9.2 $D^{\nu}_{\lambda,\mu} \in \mathbb{Z}[z]$.

Example 9.3 Suppose that $\lambda = (2,1), \mu = (3,1), \nu = (3,1)$. Then $\Delta(\nu;\lambda,\mu) = 3+4-4 = 3$, $L(\nu;\lambda,\mu) = 2+2-2 = 2$. Also, $d^{\nu}_{\lambda,\mu} = 2$ because the following are the only 2 shifted edge labeled tableaux which rectify to S_{μ}.

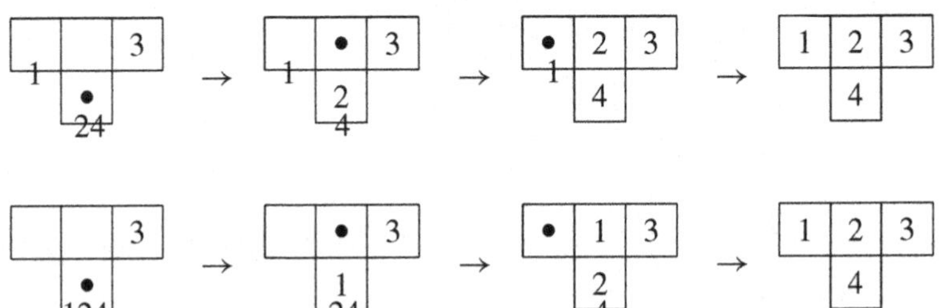

Thus $D^{\nu}_{\lambda,\mu} = 2^{2-3} \times z^3 \times 2 = z^3$. □

In the previous example, $2^{L(\nu;\lambda,\mu)-\Delta(\nu;\lambda,\mu)} = 2^{-1}$. Further, in the tableaux counting $d^{\nu}_{\lambda,\mu}$, only the edge labels differed. In the case that $L(\nu;\lambda,\mu) - \Delta(\nu;\lambda,\mu) = -k < 0$ one might wonder if the tableaux counting $d^{\nu}_{\lambda,\mu}$, namely

$$F(\lambda,\mu;\nu) = \{T \in \mathsf{shEqSYT}(\nu/\lambda, |\mu|) : \mathsf{shEqRect}(T) = S_{\mu}\}$$

can be sorted into equivalence classes of size 2^k by ignoring edge labels. The following example shows this is false in general:

Table 2. *Table of products for* $n = 3$.

λ	μ	$\lambda \star \mu$
[1]	[1]	$z[1] + 2[2]$
[1]	[2]	$z[2] + [2,1] + 2[3]$
[1]	[2, 1]	$2z[2,1] + 2[3,1]$
[1]	[3]	$z[3] + [3,1]$
[1]	[3, 1]	$2z[3,1] + 2[3,2]$
[1]	[3, 2]	$2z[3,2] + [3,2,1]$
[1]	[3, 2, 1]	$3z[3,2,1]$
[2]	[2]	$z[2,1] + z[3] + 2[3,1]$
[2]	[2, 1]	$z^2[2,1] + 3z[3,1] + 2[3,2]$
[2]	[3]	$z[3,1] + [3,2]$
[2]	[3, 1]	$z^2[3,1] + 3z[3,2] + [3,2,1]$
[2]	[3, 2]	$z^2[3,2] + 2z[3,2,1]$
[2]	[3, 2, 1]	$3z^2[3,2,1]$
[2, 1]	[2, 1]	$z^3[2,1] + 3z^2[3,1] + 6z[3,2]$
[2, 1]	[3]	$z^2[3,1] + z[3,2] + [3,2,1]$
[2, 1]	[3, 1]	$z^3[3,1] + 3z^2[3,2] + 3z[3,2,1]$
[2, 1]	[3, 2]	$z^3[3,2] + 3z^2[3,2,1]$
[2, 1]	[3, 2, 1]	$4z^3[3,2,1]$
[3]	[3]	$z[3,2]$
[3]	[3, 1]	$z^2[3,2] + z[3,2,1]$
[3]	[3, 2]	$z^2[3,2,1]$
[3]	[3, 2, 1]	$z^3[3,2,1]$
[3, 1]	[3, 1]	$z^3[3,2] + 3z^2[3,2,1]$
[3, 1]	[3, 2]	$2z^3[3,2,1]$
[3, 1]	[3, 2, 1]	$2z^4[3,2,1]$
[3, 2]	[3, 2]	$z^4[3,2,1]$
[3, 2]	[3, 2, 1]	$z^5[3,2,1]$
[3, 2, 1]	[3, 2, 1]	$z^6[3,2,1]$

Example 9.4 Suppose that $\lambda = (3), \mu = (3,2,1), \nu = (4,2,1)$. Then $\Delta(\nu;\lambda,\mu) = 3 + 6 - 7 = 2$ and $L(\nu;\lambda,\mu) = 1 + 3 - 3 = 1$, so $k = 1$. Below a $T \in F_{(3),(3,2,1);(4,2,1)}$. Any $T' \in \mathsf{eqShSYT}((4,2,1)/(3),6)$ formed by moving the edge labels of T is not in $F_{(3),(3,2,1);(4,2,1)}$.

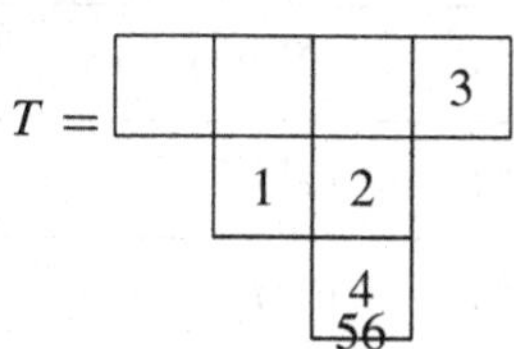

While Conjecture 9.2 is a purely combinatorial question, it would also follow from a conjectural connection to equivariant Schubert calculus, through work of D. Anderson-W. Fulton presented in Section 10.

The next result gives a further consistency check of our combinatorics. It was suggested by H. Thomas (private communication):

Theorem 9.5 *R_n is commutative,* i.e., $D^{\nu}_{\lambda,\mu} = D^{\nu}_{\mu,\lambda}$.

Proof The argument is based on a variation of S. Fomin's *growth diagram* formulation of *jeu de taquin*; see, *e.g.*, [58, Appendix 1].

Given a tableau $T \in \mathsf{eqShSYT}(\lambda/\theta, n)$, define the corresponding *e-partition* (e for "edge") to be $\mathsf{epart}(T) := (\lambda_1^{i_1}, \lambda_2^{i_2}, \ldots)$ where i_k = number of edge labels on the kth diagonal edge. Each such T can be encoded as a sequence of e-partitions starting with the (usual) partition $\theta = (\theta_1^0, \theta_2^0, \ldots)$: If 1 appears in a box in row i then the next e-partition has an extra box in this position, *i.e.*, we replace θ_i^0 with $(\theta_i + 1)^0$. Otherwise 1 appears on the edge of a diagonal box in row i, in which case, the one-larger e-partition has θ_i^0 replaced by θ_i^1. Repeat this process by looking at the position of 2 in T *etc.* Evidently, such an encoding of T is unique.

Example 9.6

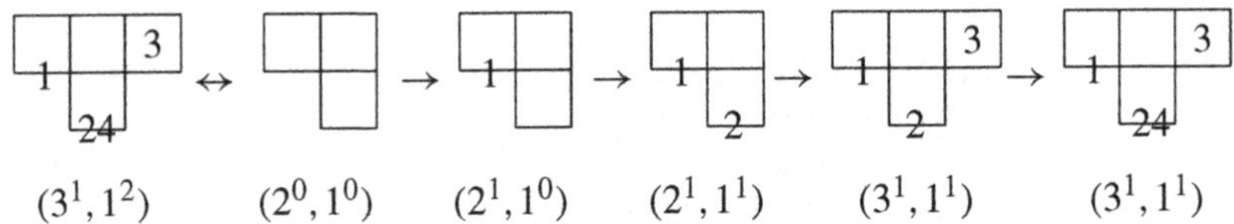

$(3^1,1^2)$ $\leftrightarrow$ $(2^0,1^0)$ $\rightarrow$ $(2^1,1^0)$ $\rightarrow$ $(2^1,1^1)$ $\rightarrow$ $(3^1,1^1)$ $\rightarrow$ $(3^1,1^1)$

Rectifying the left tableau above:

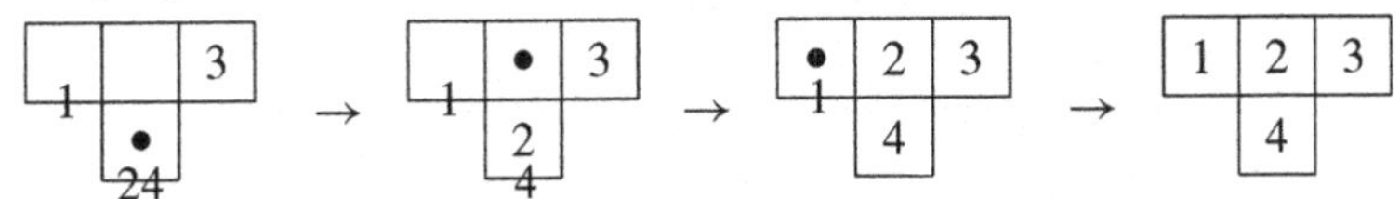

Each of these four tableaux also has an associated sequence of e-partitions. Place these atop of one another as below. The result is a *tableau rectification diagram*:

$(2^0,1^0)$	$(2^1,1^0)$	$(2^1,1^1)$	$(3^1,1^1)$	$(3^1,1^2)$
(2^0)	(2^1)	$(2^1,1^0)$	$(3^1,1^0)$	$(3^1,1^1)$
(1^0)	(1^1)	(2^1)	(3^1)	$(3^1,1^0)$
$\emptyset$	(1^0)	(2^0)	(3^0)	$(3^0,1^0)$

Given two e-partitions $\lambda = (\lambda_1^{i_1}, \lambda_2^{i_2}, \ldots)$ and $\mu = (\mu_1^{j_1}, \mu_2^{j_2}, \ldots)$, we will say μ *covers* λ if:

(i) there exists unique m such that $\lambda_m + 1 = \mu_m$ and $\lambda_k = \mu_k$ for $k \neq m$, and $i_k = j_k$ for all k; or
(ii) $\lambda_k = \mu_k$ for all k and there exists a unique m such that $i_m + 1 = j_m$ and $i_k = j_k$ for $k \neq m$.

In the case that μ covers λ we define μ/λ to be the extra box added in row m (if in cases (i) above), or the mth diagonal edge (in case (ii)). If x is a diagonal edge, *define* $\mathsf{shEjdt}_x(T) = T$. For two e-partitions, $\lambda = (\lambda_1^{i_1}, \lambda_2^{i_2}, \ldots)$ and $\mu = (\mu_1^{j_1}, \mu_2^{j_2}, \ldots)$, let

$$\lambda \vee \mu = (\mathsf{max}\{\lambda_1, \mu_1\}^{i_1+j_1}, \mathsf{max}\{\lambda_2, \mu_2\}^{i_2+j_2}, \ldots).$$

Consider the following local conditions on any 2×2 subsquare $\begin{array}{|c|c|}\hline \alpha & \beta \\ \hline \gamma & \delta \\ \hline \end{array}$ on a grid of e-partitions:

(G1) Each e-partition covers the e-partition immediately to its left or below.
(G2) $\delta = \gamma \vee \mathsf{epart}(\mathsf{shEjdt}_{\alpha/\gamma}(T))$, where T is the filling of β/α by 1. Similarly $\alpha = \gamma \vee \mathsf{epart}(\mathsf{shEjdt}_{\delta/\gamma}(T))$ where T is the filling of β/δ by 1.

Call any rectangular table of e-partitions satisfying (G1) and (G2) a *growth diagram.* By the symmetry in the definition of (G1) and (G2), if $\mathcal{G}$ is a growth diagram, then so is $\mathcal{G}$ reflected about its antidiagonal. The following is straightforward from the definitions:

Claim 9.7 *If* $\begin{array}{|c|c|}\hline \alpha & \beta \\ \hline \gamma & \delta \\ \hline \end{array}$ *is a* 2×2 *square in the tableau rectification diagram, then (G1) and (G2) hold.*

Proof Fix any two rows of the tableau rectification diagram; call this $2 \times (n+1)$ subdiagram $\mathcal{R}$. The higher of the two rows corresponds to some shifted edge labeled tableau U and the other row corresponds to $\mathsf{shEjdt}_{\mathsf{c}}(U)$ where c is a box (determined by the shapes in the leftmost column). Now U is filled by $1, 2, \ldots, n$. Notice that if we consider the submatrix $\mathcal{R}'$ of $\mathcal{R}$ consisting of the leftmost $k+1$ columns, then $\mathcal{R}'$ corresponds to the computation of $\mathsf{shEjdt}_{\mathsf{c}}(U')$ where U' is U with labels $k+1, k+2, \ldots, n$ removed. The upshot is that it suffices to prove the claim for the rightmost 2×2 square in $\mathcal{R}$, which we will label with shapes $\begin{array}{|c|c|}\hline \alpha & \beta \\ \hline \gamma & \delta \\ \hline \end{array}$. Let U_β be the tableau associated to the chain of e-partitions ending at β. Similarly define U_α. As well we have

$$U_\gamma = \mathsf{shEjdt}_{\mathsf{c}}(U_\alpha) \tag{21}$$

and

$$U_\delta = \mathsf{shEjdt}_{\mathsf{c}}(U_\beta). \tag{22}$$

Insofar as (G1) is concerned, it is obvious that β covers α and δ covers γ. That β covers δ follows from (22). Similarly, α covers γ because of (21).

Now we turn to the proof of the first sentence of (G2). Suppose that $S \in \mathsf{eqShSYT}(\lambda/\theta, n)$. Define $\bar{S}$ to be the tableau obtained by forgetting the entries $1, 2, \ldots n-1$ in S and replacing the n by 1. Also define $\tilde{S}$ to be the tableau obtained by forgetting the entry n in S. Then, it is clear that

$$\mathsf{epart}(S) = \mathsf{epart}(\tilde{S}) \vee \mathsf{epart}(\bar{S}) \tag{23}$$

By definition, $\tilde{U_\delta} = U_\gamma$. This combined with (23) applied to $S = U_\delta$ shows that to prove the claim it suffices to show that $\mathsf{shEjdt}_{\alpha/\gamma}(T) = \bar{U_\delta}$.

Case 1: (α covers γ by (i)) In the computation of $\mathsf{shEjdt}_{\mathsf{c}}(U_\alpha)$, the $\bullet$ (that starts at c) arrives at the outer corner *box* α/γ. By definition, U_β contains n at β/α. Therefore, if α/γ is not adjacent to β/α, clearly U_δ is U_γ with n adjoined at β/α. Thus, $\mathsf{shEjdt}_{\alpha/\gamma}(T) = \bar{U_\delta}$ as desired.

Otherwise α/γ is adjacent to β/α. Then by the definition of shEjdt, the position of 1 in $\mathsf{shEjdt}_{\alpha/\gamma}(T)$ is the same as the position of n in U_δ. Thus, $\mathsf{shEjdt}_{\alpha/\gamma}(T) = \bar{U_\delta}$.

Case 2: (α covers γ by (ii)) Then in the computation of $\mathsf{shEjdt}_{\mathsf{c}}(U_\alpha)$, the $\bullet$ must have arrived at a diagonal box, and $k(< n)$ is the smallest edge label of this same box, resulting in a (J4$'$) slide. Now regardless of where n is placed in U_β, it is clear that U_δ is U_γ with n adjoined in the same place as n's place in U_β, *i.e.*, β/α. In other words, $\mathsf{shEjdt}_{\alpha/\gamma}(T) = \bar{U_\delta}$.

Proof of the second sentence of (G2): For a pair of e-partitions λ, μ with μ covering λ, define $U_{\mu/\lambda}$ to be the tableau with 1 placed in the location μ/λ in μ. Clearly,

$$\alpha = \mathsf{epart}(U_\alpha) = \mathsf{epart}(U_\gamma) \vee \mathsf{epart}(U_{\alpha/\gamma}).$$

Thus, it suffices to show

$$\mathsf{shEjdt}_{\delta/\gamma}(T) = U_{\alpha/\gamma}. \tag{24}$$

Case 1:(δ covers γ by (i)) Then in U_δ, n occupies box δ/γ. If δ/γ is not adjacent to β/δ, it is clear that U_β and U_δ have n in the same place, *i.e.*, δ/γ. Thus β/δ and α/γ are the same box or edge position. Thus Equation (24) follows.

Otherwise δ/γ is adjacent to β/δ. Then,

$$\mathsf{shEjdt}_{\delta/\gamma}(T) = U_{\delta/\gamma}. \tag{25}$$

By the definition of shEjdt, it follows that in the computation of $\mathsf{shEjdt}_{\mathsf{c}}(U_\alpha)$, the • arrived at an outer corner α/γ. Using this, combined with the fact that δ/γ is adjacent to β/δ, we conclude that α/γ is a box in the same position as the box δ/γ. Now (25) is precisely (24).

Case 2: (δ covers γ by (ii)) Then δ/γ is an edge, so n occupies an edge in U_δ. Thus in $\mathsf{shEjdt}_{\mathsf{c}}(U_\beta)$, n is never moved. So, $\beta/\delta = \alpha/\gamma$. Since δ/γ is a diagonal edge, $\mathsf{shEjdt}_{\delta/\gamma}(T) = T := U_{\beta/\delta} = U_{\alpha/\gamma}$. □

Let $\mathsf{Growth}(\lambda, \mu; \nu)$ be the set of growth diagrams such that:

- the leftmost column encodes the superstandard tableau of shape λ;
- the bottom-most row encodes the superstandard tableau of shape μ;
- the shape of the e-partition in the top right corner is ν.

Claim 9.8 $\#\mathsf{Growth}(\lambda, \mu; \nu) = \#F(\lambda, \mu; \nu)$.

Proof Given $T \in F(\lambda, \mu; \nu)$, form the tableau rectification diagram $\mathcal{G}(T)$. Notice that since we are using row rectification order, the left side of the diagram will be the sequence for S_λ. Since T is assumed to rectify to S_μ, the bottom row of the diagram will be the sequence for S_μ. Hence by Claim 9.7, $\mathcal{G}(T) \in \mathsf{Growth}(\lambda, \mu; \nu)$, and thus $T \mapsto \mathcal{G}(T)$ is an injection implying $\#F(\lambda, \mu; \nu) \leq \#\mathsf{Growth}(\lambda, \mu; \nu)$.

For the reverse inequality, given any $\mathcal{G} \in \mathsf{Growth}(\lambda, \mu; \nu)$, by (G1), the top row defines $T(\mathcal{G}) \in \mathsf{eqShSYT}(\nu/\lambda, |\mu|)$. Then $T(\mathcal{G})$ has a tableau rectification diagram $\mathcal{G}'$. By Claim 9.7, $\mathcal{G}'$ is uniquely determined by its left and top borders together with (G2). Thus, since $\mathcal{G}$ and $\mathcal{G}'$ share the same left and top borders and both satisfy (G2), $\mathcal{G} = \mathcal{G}'$. In particular, $T(\mathcal{G}) \in F(\lambda, \mu; \nu)$. Thus, $\mathcal{G} \mapsto T(\mathcal{G})$ is an injection proving $\#\mathsf{Growth}(\lambda, \mu; \nu) \leq \#F(\lambda, \mu; \nu)$. □

To conclude, we must show that $d^{\nu}_{\lambda,\mu} = d^{\nu}_{\mu,\lambda}$. Since

$$d^{\nu}_{\lambda,\mu} = \#F(\lambda, \mu; \nu) = \#\mathsf{Growth}(\lambda, \mu; \nu),$$

it suffices to show that

$$\#\mathsf{Growth}(\lambda, \mu; \nu) = \#\mathsf{Growth}(\mu, \lambda; \nu).$$

Reflecting along the antidiagonal defines a bijection between $\mathsf{Growth}(\lambda, \mu\ ; \nu)$ and $\mathsf{Growth}(\mu, \lambda; \nu)$. □

Example 9.9 Under column rectification, Theorem 9.5 is false. Suppose $\lambda = (4,3)$, $\mu = (3,2,1)$ and $\nu = (4,3,2,1)$. The number of tableaux of shape ν/λ that column rectify to S_μ is 20 while the number of those with shape ν/μ column rectifying to S_λ is 16. □

Conjecture 9.10 *$(R_n, \star)$ is an associative ring.*

Additional support for Conjecture 9.10 comes from a conjectural connection to a commutative, associative ring studied by D. Anderson-W. Fulton, as described in the next section.

10 Conjectural connection to work of W. Fulton and D. Anderson and equivariant Schubert calculus

10.1 Results of W. Fulton and D. Anderson

For a strict shape $\lambda \subseteq \rho_n$, let $\sigma_\lambda = \mathsf{Pf}(c_{\lambda_i,\lambda_j})$ where

$$c_{p,q} = \sum_{0\leq a\leq b\leq q} (-1)^b \left(\binom{b}{a} + \binom{b-1}{a}\right) z^a c_{p+b-a} c_{q-b}.$$

If $\ell = \ell(\lambda)$ is odd, define $\lambda_{\ell+1} = 0$.

Recently, D. Anderson and W. Fulton have studied a $\mathbb{Z}[z]$-algebra

$$\mathfrak{P} = \mathbb{Z}[z, c_1, c_2, \ldots]/(c_{p,p} = 0, \forall p > 0)$$

and shown it has a basis over $\mathbb{Z}[z]$ of σ_λ. Define structure constants by

$$\sigma_\lambda \cdot \sigma_\mu = \sum_{\nu \subseteq \rho_n} \mathfrak{D}_{\lambda,\mu}^{\nu} \sigma_\nu.$$

Also let

$$\mathfrak{d}_{\lambda,\mu}^{\nu} := \frac{\mathfrak{D}_{\lambda,\mu}^{\lambda}}{2^{L(\lambda;\lambda,\mu)-\Delta(\lambda;\lambda,\mu)} z^{\Delta(\lambda;\lambda,\mu)}}.$$

Conjecture 10.1 *There is a ring isomorphism $\phi : R_n \to \mathfrak{P}$ that sends $[\lambda] \mapsto \sigma_\lambda$. Therefore, $\mathfrak{D}_{\lambda,\mu}^{\nu} = D_{\lambda,\mu}^{\nu}$ (equivalently $\mathfrak{d}_{\lambda,\mu}^{\nu} = d_{\lambda,\mu}^{\nu}$).*

We have exhaustively checked Conjecture 10.1 for all $n \leq 4$ and many $n = 5$ cases.

D. Anderson and W. Fulton (private communication) connected the above ring to equivariant Schubert calculus of Z. That is,

$$L_{\lambda,\mu}^{\nu}(\alpha_1 \mapsto z, \alpha_2 \mapsto 0, \ldots, \alpha_n \mapsto 0) = \mathfrak{D}_{\lambda,\mu}^{\nu}. \tag{26}$$

Proposition 10.2 *Conjecture 10.1* $\Longrightarrow$ *Conjecture 9.2.*

Proof By Theorem 2.1, $L^{\nu}_{\lambda,\mu}$ is a nonnegative integer polynomial in $\alpha_1,\ldots,\alpha_n$, the simples of the type C root system. Now apply (26). □

In turn, Conjecture 10.1 should follow from a proof that $D^{\nu}_{\lambda,(p)} = \mathfrak{D}^{\nu}_{\lambda,(p)}$, together with Conjecture 9.10, by a variation of the "associativity argument" of Section 3.

10.2 Two numerologically nice cases of Conjecture 10.1

Theorem 10.3 $d^{\lambda}_{\lambda,(p)} = \binom{\ell(\lambda)}{p}2^{p-1} = \mathfrak{d}^{\lambda}_{\lambda,(p)}$.

Proof We will use some results of Ikeda–Naruse [25] that we now recall. For a strict partition $\lambda = (\lambda_1 > \ldots > \lambda_r > 0)$, let D_λ denote the associated shifted shape. Explicitly,

$$D_\lambda = \{(i,j) \in \mathbb{Z}^2 | 1 \le i \le r, i \le j < \lambda_i + i\}.$$

For instance, $D_{(3,1)} =$. Given an arbitrary subset $C \subset D_\lambda$, if a box $x \in C$ satisfies either of the following conditions:

(I) $x = (i,i)$ and $(i,i+1), (i+1,i+1) \in D_\lambda \setminus C$
(II) $x = (i,j), j \neq i$ and $(i+1,i), (i,i+1), (i+1,i+1) \in D_\lambda \setminus C$

then set $C' = C \cup \{x + (1,1)\} \setminus \{x\}$. The procedure $C \to C'$ is called an *elementary excitation occurring* at $x \in C$. Any subset $C' \subset D_\lambda$ obtained from C by an application of successive elementary excitations is called an *excited Young diagram* (EYD) of C. Denote by $\mathcal{E}_\lambda(\mu)$ the set of all EYDs of D_μ contained in D_λ.

Example 10.4 Suppose $\lambda = (4,2,1), \mu = (2)$, then $\mathcal{E}_\lambda(\mu)$ consists of the following EYDs

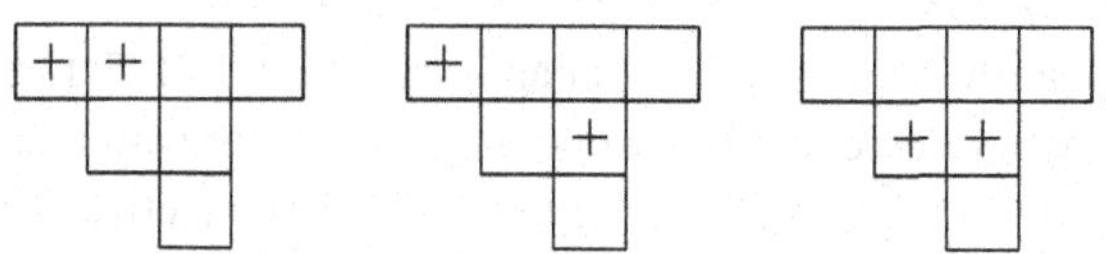

Lemma 10.5 $\mathfrak{D}^{\lambda}_{\lambda,\mu} = \#\mathcal{E}_{\rho_{\ell(\lambda)}}(\mu) \times z^{|\mu|}$.

Proof This follows from [25, Theorem 3] which gives a formula for $\xi_\lambda(Z)|_\mu = L^{\mu}_{\lambda,\mu}$ in terms of EYDs, combined with (26). We omit the details, which amount mostly to translating from the appropriate Weyl group elements to the associated partitions. □

By Lemma 10.5,

$$\begin{aligned}\mathfrak{d}^{\lambda}_{\lambda,\mu} &= \frac{\mathfrak{D}^{\lambda}_{\lambda,\mu}}{2^{L(\lambda;\lambda,\mu)-\Delta(\lambda;\lambda,\mu)}z^{\Delta(\lambda;\lambda,\mu)}}\\ &= \frac{\#\mathcal{E}_{\rho_{\ell(\lambda)}}(\mu)\times z^{|\mu|}}{2^{\ell(\mu)-|\mu|}z^{|\mu|}}\\ &= \#\mathcal{E}_{\rho_{\ell(\lambda)}}(\mu)\times 2^{|\mu|-\ell(\mu)}.\end{aligned}$$

As all boxes in $D_{(p)}$ are in distinct columns, they stay in distinct columns even after the application of the excitation moves (I) and (II). Therefore, $\mathcal{E}_{\rho_{\ell(\lambda)}}((p))$ contains at most $\binom{\ell(\lambda)}{p}$ elements, since, there are $\ell(\lambda)$ columns in $D_{\rho_{\ell(\lambda)}}$. It is not hard to see from (I) and (II) this upper bound is an equality. This proves the second equality of the theorem.

Let $N = \ell(\lambda)$; we now prove the first equality of the theorem statement by induction on $N+p$. When $N+p\leq 1$ the claim is obvious. When $N+p=2$, there is one case, namely, $\lambda=(1)$, $p=1$ and $d^{\lambda}_{\lambda,(p)}=1=\binom{1}{1}2^{1-1}$, as desired. Now suppose $N+p=k>2$ and the claim holds for smaller $N+p$.

Let $F(\lambda,(p);\lambda)$ be the tableaux enumerated by $d^{\lambda}_{\lambda,(p)}$. If $T\in F(\lambda,(p);\lambda)$ we say that a label q *appears in* row r if q is an edge label on the southern edge of the diagonal box in row r. Let $\overline{T}$ be T with the first row removed, this is of shape $\overline{\lambda}$. There are three disjoint cases that T can fall into:

1. (1 does not appear in row 1 of T and $\overline{T}\in F(\overline{\lambda},(p);\overline{\lambda})$): Then there are $d^{\overline{\lambda}}_{\overline{\lambda},(p)}$ many such choices; this equals $\binom{N-1}{p}2^{p-1}$, by induction.
2. (1 does not appear in row 1 of T and $\overline{T}\notin F(\overline{\lambda},(p);\overline{\lambda})$): Then it is straightforward to check (from the assumption that $T\in F(\lambda,(p);\lambda)$) that $\overline{T}$ row rectifies to $\overline{S}$ of shape $(p-1)$ where the first row consists of box labels $1,3,4,\ldots,p-1$ and has a 2 in the south edge of the first box. Notice that the choices for $\overline{T}$ are in bijection with $F(\overline{\lambda},(p-1);\overline{\lambda})$ where the map is to remove the edge label 2 and shift the labels $3,4,5\ldots,p$ down by one. This, combined with induction asserts that there are $d^{\overline{\lambda}}_{\overline{\lambda},(p-1)}=\binom{N-1}{p-1}2^{p-2}$ many choices.
3. (1 appears in row 1): No other label appears in row 1 of T. Let U be $\overline{T}$ with every entry decremented by one. It is straightforward that $U\in F(\overline{\lambda},(p-1);\overline{\lambda})$, and that the map $T\mapsto U$ is bijective. Thus, there are $d^{\overline{\lambda}}_{\overline{\lambda},(p-1)}=\binom{N-1}{p-1}2^{p-2}$ many tableaux in this case, by induction.

By Pascal's identity,

$$\binom{N}{p}2^{p-1}=\binom{N-1}{p}2^{p-1}+\binom{N-1}{p-1}2^{p-2}+\binom{N-1}{p-1}2^{p-2}.$$

This, combined with cases (1)-(3), completes the induction. □

Theorem 10.6 $d^{\rho_n}_{\rho_n,\rho_n} = 2^{\binom{n}{2}} = \eth^{\rho_n}_{\rho_n,\rho_n}$.

Proof sketch: In the special case when $\ell(\lambda) = \ell(\mu)$, it is easy to observe that $|\mathcal{E}^{\lambda}_{\mu}| = 1$. Thus in this case,

$$\eth^{\lambda}_{\lambda,\mu} = 2^{|\mu|-\ell(\mu)}.$$

Further when $\mu = \lambda = \rho_n$,

$$\eth^{\rho_n}_{\rho_n,\rho_n} = 2^{|\mu|-\ell(\mu)} = 2^{\binom{n}{2}}.$$

This proves the rightmost equality.

For the remaining equality, consider $T \in \mathsf{shEqSYT}(\rho_n/\rho_n, N)$ where $N = |\rho_n| = \binom{n+1}{2}$. For $1 \leq i \leq n$, let

$$E_i(T) = \{k \mid k \text{ lies on the } i\text{th diagonal edge of } T\}.$$

For $T \in \mathsf{shSYT}(\nu/\lambda)$, let $T(i,j)$ be the entry in box (i,j) (in matrix coordinates). Define $U_n \in \mathsf{shEqSYT}(\rho_n/\rho_n, N)$ by the requirement that $E_i(U_n) = \bigcup_{r=1}^{i} S_{\rho_n}(r,i)$. That is, the labels on the ith diagonal edge of U_n are precisely the labels appearing in column i of S_{ρ_n}.

For $T \in \mathsf{shEqSYT}(\rho_n/\rho_n, N)$ and $I \subseteq E_i(T)$ for some $i \in [n-1]$, define the *I-slide* of T, $\mathsf{Sl}_I(T) \in \mathsf{shEqSYT}(\rho_n/\rho_n, N)$, by

$$E_k(\mathsf{Sl}_I(T)) := \begin{cases} E_k(T) & \text{if } k \in [n] \setminus \{i, i+1\}, \\ E_k(T) \setminus I & \text{if } k = i, \\ E_k(T) \cup I & \text{if } k = i+1. \end{cases}$$

Example 10.7 Let $n = 4$. Taking $I = \{6\} \subseteq E_3(U_4) = \{3, 6, 8\}$, below we illustrate $\mathsf{Sl}_{\{6\}}(U_4)$.

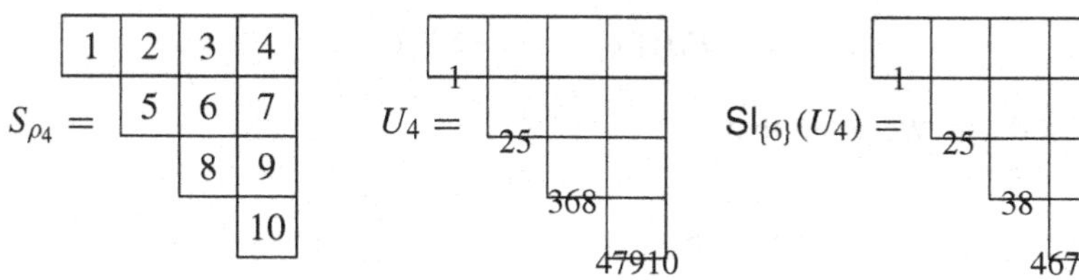

For $T \in \mathsf{shSYT}(\nu/\lambda)$, let

$$\mathsf{row}_k(T) = \{\text{entries in row } k \text{ of } T\}.$$

Say $I \subseteq E_i(T)$ is *n-slidable* if $1 \leq i < n$,

$$I \subseteq \bigcup_{k=1}^{i} \{\min(E_i(T) \cap \mathsf{row}_k(S_{\rho_n}))\}, \tag{27}$$

and for $i < k \leq n$, $E_k(T) = E_k(U_n)$.

Example 10.8 Consider T below and $i = 3$. Then to the right we have S_{ρ_4} with

$$\bigcup_{k=1}^{3} \{\min(E_3(T) \cap \mathsf{row}_k(S_{\rho_4}))\} = \bigcup_{k=1}^{3} \{\min(\{1,3,5,6,8\} \cap \mathsf{row}_k(S_{\rho_4}))\}$$
$$= \{1,5,8\}$$

lightly shaded and $\{3,6\}$, the remainder of entries of $E_3(T)$, shaded darker. Thus any $I \subseteq \{1,5,8\}$ is 4-slidable, so $\{1,8\}$ is 4-slidable but $\{1,3,8\}$ is not.

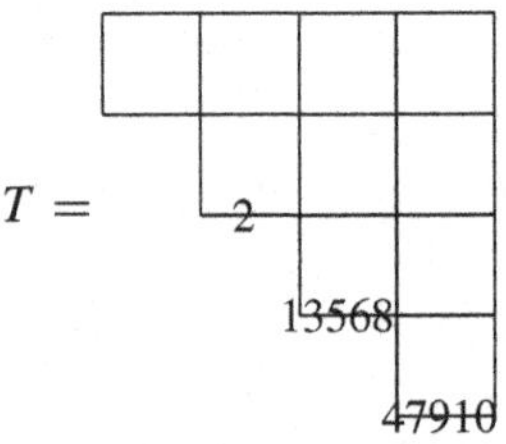

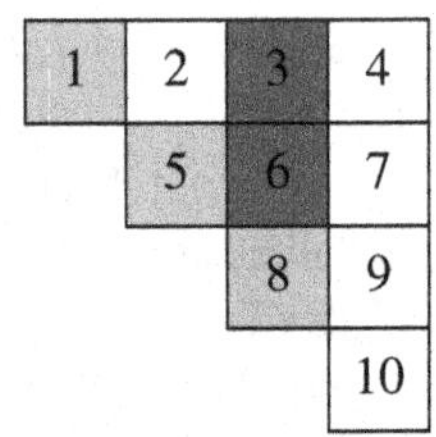

The proof of this claim is lengthy and will appear elsewhere:

Claim 10.9 *Fix* $T \in \mathsf{shEqSYT}(\rho_n/\rho_n, N)$. *Then* $\mathsf{shEqRect}(T) = S_{\rho_n}$ *if and only if* $T = \mathsf{Sl}_{I_{n-1}} \circ \mathsf{Sl}_{I_{n-2}} \circ \cdots \circ \mathsf{Sl}_{I_1}(U_n)$ *where each* $I_i \subseteq E_i(\mathsf{Sl}_{I_{i-1}} \circ \cdots \circ \mathsf{Sl}_{I_1}(U_n))$ *is n-slidable.*

By Claim 10.9, $d_{\rho_n,\rho_n}^{\rho_n}$ equals the number of sequences $\{I_i\}_{i=1}^{n-1}$ where

$$I_i \subseteq E_i(\mathsf{Sl}_{I_{i-1}} \circ \cdots \circ \mathsf{Sl}_{I_1}(U_n))$$

is n-slidable. We assert that

$$i = \#\bigcup_{k=1}^{i} \{\min(E_i(\mathsf{Sl}_{I_{i-1}} \circ \cdots \circ \mathsf{Sl}_{I_1}(U_n)) \cap \mathsf{row}_k(S_{\rho_n}))\}.$$

Indeed, to see this, note that $i = \#\bigcup_{k=1}^{i} \min(E_i(U_n) \cap \mathsf{row}_k(S_{\rho_n}))\}$ and, by definition of I-slidable, $E_i(\mathsf{Sl}_{I_{i-1}} \circ \cdots \circ \mathsf{Sl}_{I_1}(U_n)) \supseteq E_i(U_n)$. Hence, by (27), there are 2^i choices for each n-slidable I_i, so $d_{\rho_n,\rho_n}^{\rho_n} = 2^{\binom{n}{2}}$, as desired. □

We illustrate Claim 10.9 with the following example:

Example 10.10 Below is $T = \mathsf{Sl}_{I_3} \circ \mathsf{Sl}_{I_2} \circ \mathsf{Sl}_{I_1}(U_4)$ with the choices of I_i given above each arrow. Beneath each arrow, entries in

$$\bigcup_{k=1}^{i} \{\min(E_i(\mathsf{Sl}_{I_{i-1}} \circ \cdots \circ \mathsf{Sl}_{I_1}(U_4)) \cap \mathsf{row}_k(S_{\rho_n}))\}$$

are lightly shaded in S_{ρ_4} and the remaining entries of $E_i(T)$ are shaded darker. Thus I_i is 4-slidable if and only if all entries of I_i are lightly shaded. Thus in the example below, I_1, I_2, and I_3 are all 4-slidable. Therefore by Lemma 10.9, $\mathsf{shEqRect}(T) = S_{\rho_4}$.

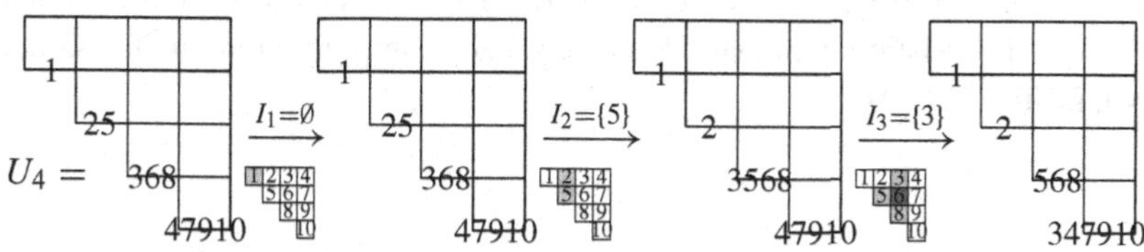

Now consider $T' = \mathsf{Sl}_{I'_3} \circ \mathsf{Sl}_{I'_2} \circ \mathsf{Sl}_{I'_1}(U_4)$ with the choices of I'_i given above each arrow. Here, I'_1, I'_2 are 4-slidable, but I'_3 is not. Thus by Claim 10.9, $\mathsf{shEqRect}(T') \neq S_{\rho_4}$.

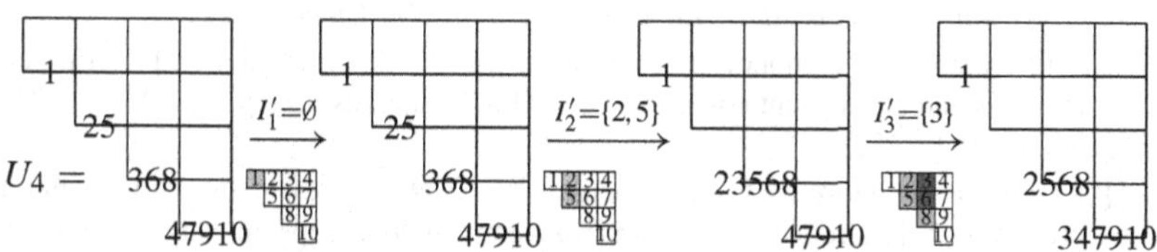

While $2^{\binom{n}{2}}$ is the number of labeled graphs on n vertices, consulting the Online Encyclopedia of Integer Sequences [56], one also finds that it counts the number of

- perfect matchings of order n Aztec diamond [57],
- Gelfand-Zeitlin patterns with bottom row $[1, 2, 3, \ldots, n]$ [68], and
- certain domino tilings [56, A006125]

among other things.

We end with a problem of enumerative combinatorics:

Problem 10.11 *Give bijections between the shifted edge labeled tableaux counted by $d_{\rho_n, \rho_n}^{\rho_n}$ and the equinumerous objects above.*

Acknowledgements

AY's thanks to Bill Fulton goes back to 1999. In May 2018, the authors attended a conference on Schubert calculus held at Ohio State University where Bill kindly shared the fine points of his ongoing work with David Anderson. This was the stimulus for this chapter. We also thank David Anderson, Soojin Cho, Sergey Fomin, Allen Knutson, William Linz, Alun Morris, Gidon Orelowitz, John Stembridge, Hugh Thomas, Brian Shin, and the anonymous referees for helpful remarks. We thank Anshul Adve, David Anderson, Cara Monical, Oliver Pechenik, Ed Richmond, and Hugh Thomas for their contributions reported here. AY was partially supported by an NSF grant, a UIUC Campus research board grant, and a Simons Collaboration Grant. This material is based upon work of CR supported by the National Science Foundation Graduate Research Fellowship Program under Grant No. DGE – 1746047.

References

[1] Adve, A., Robichaux, C., and Yong, A. 2019. Vanishing of Littlewood–Richardson polynomials is in P. *Comput. Complexity*, **28**, no. 2, 241–257.

[2] Anderson, D. 2012. Introduction to Equivariant Cohomology in Algebraic Geometry. *Contributions to Algebraic Geometry: Impanga Lecture Notes*.

[3] Anderson, D., Richmond, E., and Yong, A. 2013. Eigenvalues of Hermitian matrices and equivariant cohomology of Grassmannians. *Compositio Math.*, **149**, no. 9, 1569–1582.

[4] Andersen, H., Jantzen, J., and Soergel, W. 1994. Representations of quantum groups at a pth root of unity and of semisimple groups in characteristic p: independence of p. *Asterisque*, **220**, 321 pp.

[5] Arabia, A. 1989. Cohomologie T-équivariant de la variété de drapeaux d'un groupe de Kac–Moody. *Bull. Math. Soc. France*, **117**, 129–165.

[6] Belkale, P. 1999. Local systems on $\mathbb{P}^1 \setminus S$ for a finite set. Ph. D. thesis, University of Chicago.

[7] Belkale, P. 2001. Local systems on $\mathbb{P}^1 \setminus S$ for S a finite set. *Compositio Math.*, **129**, no. 1, 67–86.

[8] Belkale, P. 2006. Geometric proofs of Horn and saturation conjectures. *J. Algebraic Geom.*, **15**, no. 1, 133–173.

[9] Belkale, P. 2019. Extremal rays in the Hermitian eigenvalue problem. *Math. Ann.*, **373**, no. 3-4, 1103–1133.

[10] Bhatia, R. 2001. Linear algebra to quantum cohomology: the story of Alfred Horn's inequalities. *Amer. Math. Monthly*, **108**, no. 4, 289–318.

[11] Buch, A. S. 2000. The saturation conjecture (after Knutson, A. and Tao, T.). With an appendix by William Fulton. *Enseign. Math.*, **46**(2), no. 1-2, 43–60.

[12] Buch, A. S., Kresch, A., and Tamvakis, H. 2004. Littlewood–Richardson rules for Grassmannians. *Adv. Math.*, **185**, no. 1, 80–90.

[13] De Loera, J. A., and McAllister, T. B. 2006. On the computation of Clebsch–Gordan coefficients and the dilation effect. *Experiment. Math.*, **15**(1): 7–19.

[14] Derksen, H., and Weyman, J. 2000. Semi-invariants of quivers and saturation for Littlewood–Richardson coefficients. *J. Amer. Math. Soc.*, **13**, no. 3, 467–479.

[15] Elashvili, A. G. 1992. Invariant algebras. *Advances in Soviet Math.*, **8**, 57–64.

[16] Friedland, S. 2000. Finite and infinite dimensional generalizations of Klyachko's theorem. *Linear Algebra Appl.*, **319**, 3–22.

[17] Fulton, W. 1997. Young tableaux. With applications to representation theory and geometry. *London Mathematical Society Student Texts*, **35**. Cambridge University Press, Cambridge. ix+260pp.

[18] Fulton, W. 2000. Eigenvalues, invariant factors, highest weights, and Schubert calculus. *Bull. Amer. Math. Soc. (N.S.)*, **37**, no. 3, 209–249 (electronic).

[19] Fulton, W. 2000. Eigenvalues of majorized Hermitian matrices and Littlewood–Richardson coefficients. *Linear Algebra Appl.*, **319**, no. 1–3, 23–36.

[20] Goresky, M., Kottwitz, R., and MacPherson, R. 1998. Equivariant cohomology, Koszul duality, and the localization theorem. *Invent. Math.*, **131**, no. 1, 25–83.

[21] Graham, W. 2001. Positivity in equivariant Schubert calculus. *Duke Math. J.*, **109**, no. 3, 599–614.

[22] Grotschel, M., Lovasz, L., and Schrijver, A. 1993. Geometric algorithms and combinatorial optimization. *Algorithms and Combinatorics*, **2**. Springer Verlag.

[23] Haiman, M. D. 1992. Dual equivalence with applications, including a conjecture of Proctor. *Discrete Math.*, **99**, no. 1-3, 79–113.

[24] Horn, A. 1962. Eigenvalues of sums of Hermitian matrices. *Pacific J. Math.*, **12**, 225–241.

[25] Ikeda, T., and Naruse, H. 2009. Excited Young diagrams and equivariant Schubert calculus. *Trans. Amer. Math. Soc.*, **361**, no. 10, 5193–5221.

[26] Kleiman, S. L. 1974. The transversality of a general translate. *Compositio Math.*, **28**, 287–297.

[27] Klyachko, A. A. 1998. Stable vector bundles and Hermitian operators. *Selecta Math. (N.S.)*, **4**, 419–445.

[28] Knutson, A., Miller, E., and Yong, A. 2009. Gröbner geometry of vertex decompositions and of flagged tableaux. *J. Reine Angew. Math.*, **630**, 1–31.

[29] Knutson, A., and Tao, T. 1999. The honeycomb model of $GL_n(\mathbb{C})$ tensor products I: proof of the saturation conjecture. *J. Amer. Math. Soc.*, **12**, 1055–1090.

[30] Knutson, A., and Tao, T. 2003. Puzzles and (equivariant) cohomology of Grassmannians. *Duke Math. J.*, **119**, no. 2, 221–260.

[31] Knutson, A., Tao, T., and Woodward, C. 2004. The honeycomb model of $\mathrm{GL}_n(\mathbb{C})$ tensor products. II. Puzzles determine facets of the Littlewood–Richardson cone. *J. Amer. Math. Soc.*, **17**, no. 1, 19–48.

[32] Knutson, A., Tao, T., and Woodward, C. 2004. A positive proof of the Littlewood–Richardson rule using the octahedron recurrence. *Electron. J. Combin.*, **11**, no. 1, Research Paper 61, 18 pp.

[33] Knutson, A., and Zinn-Justin, P. 2017. Schubert puzzles and integrability I: invariant trilinear forms. preprint. arXiv:1706.10019.

[34] Kostant, B. and Kumar, S. 1990. T-equivariant K-theory of generalized flag varieties. *J. Differential Geom.*, **32**, no. 2, 549–603.

[35] S. Kumar, *Kac–Moody groups, their flag varieties and representation theory*, Progress in Mathematics, 204. Birkhäuser Boston, Inc., Boston, MA, 2002. xvi+606 pp.

[36] Kreiman, V. 2010. Equivariant Littlewood–Richardson skew tableaux. *Trans. Amer. Math. Soc.*, **362**, no. 5, 2589–2617.

[37] Lenart, C., and Postnikov, A. 2007. Affine Weyl groups in K-theory and representation theory. *Int. Math. Res. Not.*, no. 12, Art. ID rnm038, 65 pp.

[38] Molev, A. I. 2009. Littlewood–Richardson polynomials. *J. Algebra*, **321** no. 11, 3450–3468.

[39] Molev, A. I., and Sagan, B. E. 1999. A Littlewood–Richardson rule for factorial Schur functions. *Trans. Amer. Math. Soc.*, **351**, no. 11, 4429–4443.

[40] Monical, C. 2018. Polynomials in algebraic combinatorics. Ph.D thesis, University of Illinois at Urbana-Champaign.

[41] Monical, C., Tokcan, N., and Yong, A. 2019. Newton polytopes in algebraic combinatorics. *Sel. Math.*, **25**(5), no. 66, 37 pp.

[42] Mulmuley, K. D., Narayanan, H., and Sohoni, M. 2012. Geometric complexity theory III: on deciding nonvanishing of a Littlewood–Richardson coefficient. *J. Algebraic Combin.*, **36**, no. 1, 103–110.

[43] Narayanan, H. 2006. On the complexity of computing Kostka numbers and Littlewood–Richardson coefficients. *J. Algebraic Combin.*, **24**, no. 3, 347–354.

[44] Pechenik, O., and Yong, A. 2017. Equivariant K-theory of Grassmannians. *Forum Math. Pi*, **5**, e3, 128 pp.

[45] Pechenik, O., and Yong, A. 2017. Equivariant K-theory of Grassmannians II: the Knutson–Vakil conjecture. *Compos. Math.*, **153**, no. 4, 667–677.

[46] Pragacz, P. 1991. Algebro-geometric applications of Schur S- and Q-polynomials. Pages 130–191 of: *Topics in invariant theory* (Paris, 1989/1990), Lecture Notes in Math. **1478**, Springer, Berlin.

[47] Purbhoo, K. 2006. Vanishing and nonvanishing criteria in Schubert calculus. *Int. Math. Res. Not.*, Art. ID 24590, 38 pp.

[48] Purbhoo, K. 2007. Root games on Grassmannians. *J. Algebraic Combin.*, **25**, no. 3, 239–258.

[49] Purbhoo, K., and Sottile, F. 2006. The Horn recursion for Schur $P-$ and $Q-$ functions. Extended abstract. FPSAC San Diego, USA.

[50] Purbhoo, K., and Sottile, F. 2008. The recursive nature of cominuscule Schubert calculus. *Adv. Math.*, **217**, 1962–2004.

[51] Robichaux, C., Yadav, H., and Yong, A. 2019. The A·B·C·Ds of Schubert calculus. preprint. arXiv:1906.03646.

[52] Sagan, B. E. 1987. Shifted tableaux, Schur Q-functions, and a conjecture of R. Stanley. *J. Combin. Theory Ser. A*, **45**, no. 1, 62–103.

[53] Schrijver, A. 1986. Theory of linear and integer programming. *Wiley-Interscience Series in Discrete Mathematics. A Wiley-Interscience Publication*. John Wiley & Sons, Ltd., Chichester. xii+471 pp.

[54] Schützenberger, M.-P. 1977. La correspondance de Robinson. Pages 59–113 of: *Combinatoire et représentation du groupe symétrique Actes Table Ronde CNRS,*

Univ. Louis-Pasteur, Strasbourg, (French), Lecture Notes in Math., Vol. 579, (1976). Springer, Berlin.

[55] Schur, J. 1911. Über die Darstellung der symmetrischen und der alternierenden Gruppe durch gebrochene lineare Substitutionen. (German) *J. Reine Angew. Math.*, **139**, 155–250.

[56] Sloane, N. J. A. 2020. editor, The On-Line Encyclopedia of Integer Sequences, published electronically at https://oeis.org.

[57] Speyer, D. 2007. Perfect matchings and the octahedron recurrence. *J. Algebraic Combin.*, **25**, no. 3, 309–348.

[58] Stanley, R. P. 1999. Enumerative combinatorics. Vol. 2. With a foreword by Rota, G. C. and appendix 1 by Fomin, S. *Cambridge Studies in Advanced Mathematics*, **62**. Cambridge University Press, Cambridge. xii+581.

[59] Stembridge, J. R. 1989. Shifted Tableaux and the Projective Representations of Symmetric Groups. *Adv. Math.*, **74**, 87–134.

[60] Tardos, É. 1986. A Strongly Polynomial Algorithm to Solve Combinatorial Linear Programs. *Operations Research*, **34**(2), 250–256.

[61] Thomas, H., and Yong, A. 2009. A combinatorial rule for (co)minuscule Schubert calculus. *Adv. Math.*, **222**, no. 2, 596–620.

[62] Thomas, H., and Yong, A. 2018. Equivariant Schubert calculus and jeu de taquin. *Ann. Inst. Fourier (Grenoble)*, **68**, no. 1, 275–318.

[63] Totaro, B. 1994. Tensor products of semistables are semistable. Pages 242–250 of: Noguchi, T., Noguchi, J., and Ochiai, T. (eds) *Geometry and Analysis on Complex Manifolds, Festschrift for Professor S. Kobayashi's 60th Birthday*. World Scientific Publ. Co., Singapore.

[64] Tu, L. W. 2011. What is … equivariant cohomology?. *Notices Amer. Math. Soc.*, **58**(3), 423–426.

[65] Tymoczko, J. S. 2005. An introduction to equivariant cohomology and homology, following Goresky, Kottwitz, and Macpherson. preprint. arXiv:0503369.

[66] Valiant, L. G. 1979. The complexity of computing the permanent. *Theoret. Comput. Sci.*, **8**(2), 189–201.

[67] Worley, D. R. 1984. A theory of shifted Young tableaux. Thesis (Ph.D.), Massachusetts Institute of Technology. ProQuest LLC, Ann Arbor, MI.

[68] Zeilberger, D. 2005. Dave Robbins's Art of Guessing. *Adv. in Appl. Math.*, **34**, 939–954.

[69] Zelevinsky, A. 1999. Littlewood–Richardson semigroups. Pages 337–345 of : *New perspectives in algebraic combinatorics* (Berkeley, CA, 1996–97), *Math. Sci. Res. Inst. Publ.*, **38**, Cambridge Univ. Press, Cambridge.

[70] Zinn-Justin, P. 2009. Littlewood–Richardson coefficients and integrable tilings. *Electron. J. Combin.*, **16**(1), Research Paper 12, 33 pp.

24

Galois Groups of Composed Schubert Problems

Frank Sottile[a], Robert Williams[b] and Li Ying[c]

Abstract. Two Schubert problems on possibly different Grassmannians may be composed to obtain a Schubert problem on a larger Grassmannian whose number of solutions is the product of the numbers of the original problems. This generalizes a construction discovered while classifying Schubert problems on the Grassmannian of 4-planes in $\mathbb{C}^9$ with imprimitive Galois groups. We give an algebraic proof of the product formula. In a number of cases, we show that the Galois group of the composed Schubert problem is a subgroup of a wreath product of the Galois groups of the original problems, and is therefore imprimitive. We also present evidence for a conjecture that all composed Schubert problems have imprimitive Galois groups.

1 Introduction

In his 1870 book "Traité des Substitutions et des équations algébrique", C. Jordan [8] observed that problems in enumerative geometry have Galois groups (of associated field extensions) which encode internal structure of the enumerative problem. Interest in such enumerative Galois groups was revived by Joe Harris through his seminal paper [7] that revisited and extended some of Jordan's work and gave a modern proof that the Galois group equals a monodromy group. The Galois group of an enumerative problem is a subgroup of the symmetric group acting on its solutions. Progress in understanding enumerative Galois groups was slow until the 2000s when new methods were introduced by Vakil and others, some of which were particularly effective for Schubert problems on Grassmannians.

[a] Texas A & M University
[b] Sam Houston State University
[c] Vanderbilt University, Tennessee

Vakil's geometric Littlewood–Richardson rule [18] gave an algorithmic criterion that implies a Galois group of a Schubert problem (a *Schubert Galois group*) contains the alternating group (is *at least alternating*). This revealed some Schubert problems whose Galois groups might not be the full symmetric group, as well as many that were at least alternating [19]. A numerical computation [10] of monodromy groups of some Schubert problems found that all had full symmetric Galois groups, providing evidence that the typical Schubert Galois group is full symmetric. Vakil's experimentation suggested that Schubert Galois groups for Grassmannians of two- and three-dimensional linear subspaces should always be full symmetric, but that this no longer holds for four-planes in $\mathbb{C}^8$.

In fact (with help from Derksen), Vakil identified a Schubert problem with six solutions whose Galois group is the symmetric group S_4, and the action is that of S_4 on subsets of $\{1,2,3,4\}$ of cardinality two. It was later shown that all Schubert Galois groups for Grassmannians of 2-planes are at least alternating [3] and those for Grassmannians of 3-planes are doubly transitive [16], which supports Vakil's observations. All Schubert problems on the next two Grassmannians of 4-planes whose Galois groups are not at least alternating were classified, those in $\mathbb{C}^8$ in [12] and in $\mathbb{C}^9$ in [13]. Of over 35,000 nontrivial Schubert problems on these Grassmannians, only 163 had imprimitive Galois groups. Besides the example of Derksen/Vakil, this study found two distinct types of constructions giving Schubert problems with imprimitive Galois groups.

Our goal is to better understand one of these types, called Type I in [13]. We generalize the construction in [13] and call it *composition of Schubert problems* (Definition 3.7). We prove a product formula (Theorem 3.9) for the number of solutions to a composition of Schubert problems, which includes a combinatorial bijection involving sets of generalized Littlewood–Richardson tableaux (explained in Remark 3.13). In many cases, we show that the Galois group of a composition of Schubert problems is imprimitive by embedding it into a wreath product of Galois groups of Schubert problems on smaller Grassmannians.

In Section 2 we recall some standard facts about Schur functions and the Littlewood–Richardson formula, and then deduce some results when the indexing partitions involve large rectangles. In Section 3 we use these formulas to give some results in the cohomology ring of Grassmannians. We also define the main objects in this paper, a composable Schubert problem and a composition of Schubert problems, and prove our main theorem which is a product formula for the number of solutions to a composition of Schubert problems. In Section 4 we define a family (*block column Schubert problems*) of composable Schubert problems and prove a geometric result that implies the Galois group of a composition with a block column Schubert problem is imprimitive. We also present computational evidence for a conjecture that all nontrivial compositions have imprimitive Galois groups.

2 Schur functions and Littlewood–Richardson coefficients

We recall facts about Schur functions and Littlewood–Richardson coefficients which are found in [6, 11, 15, 17]. We use these to establish some formulas for products of Schur functions whose partitions contain large rectangles. This includes bijections involving generalized Littlewood–Richardson tableaux which count the coefficients in such products.

2.1 Symmetric functions

A *partition* λ is a weakly decreasing sequence $\lambda\colon \lambda_1 \geq \lambda_2 \geq \cdots \geq \lambda_k > 0$ of positive integers. We set $|\lambda| := \lambda_1 + \cdots + \lambda_k$. Here k is the *length* of λ, written $\ell(\lambda)$. Each integer λ_i is a *part* of λ. When $a > \ell(\lambda)$, we write $\lambda_a = 0$, and we identify partitions that differ only in trailing parts of size 0. The set of all partitions forms a partial order called *Young's lattice* in which $\lambda \subseteq \mu$ when $\lambda_i \leq \mu_i$ for all i. We often identify a partition with its *Young diagram*, which is a left-justified array of $|\lambda|$ boxes with λ_i boxes in the ith row. Here are three partitions and their Young diagrams:

$$(3) \;=\; \qquad (4,2) \;=\; \qquad (6,3,1) \;=\; .$$

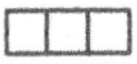

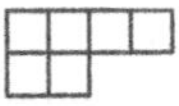

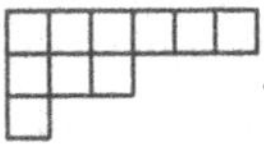

The partial order among partitions in Young's lattice becomes set-theoretic containment of their Young diagrams, which explains our notation $\lambda \subseteq \mu$. The *conjugate* λ' of a partition λ is the partition whose Young diagram is the matrix-transpose of the Young diagram of λ. For example:

$$(3)' \;=\; {}' \;=\; \qquad \text{and} \qquad (4,3,1)' \;=\; {}' \;=\; .$$

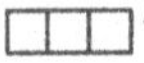

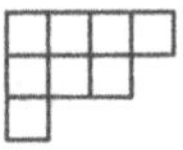

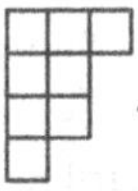

Conjugating a partition interchanges rows with columns.

A homogeneous formal power series $f \in \mathbb{Z}[\![x_1, x_2, \ldots]\!]$ is a *symmetric function* if it is invariant under permuting the variables. For any nonnegative integer n, the set Λ_n of symmetric functions of degree n is a finite rank free abelian group with a distinguished $\mathbb{Z}$-basis of *Schur functions* S_λ indexed by partitions λ with $|\lambda| = n$.

The graded algebra Λ of symmetric functions is the direct sum of all Λ_n for $n \geq 0$. Here, the empty partition $\emptyset$ is the unique partition of 0, and $S_\emptyset = 1$, the multiplicative identity for Λ. The set of all Schur functions forms a $\mathbb{Z}$-basis for Λ. Consequently, there are integer *Littlewood–Richardson coefficients* $c^\lambda_{\mu,\nu} \in \mathbb{Z}$ defined for triples λ, μ, ν of partitions by expanding the product $S_\mu \cdot S_\nu$ in the basis of Schur functions,

$$S_\mu \cdot S_\nu = \sum_\lambda c^\lambda_{\mu,\nu} S_\lambda.$$

Conjugation of partitions induces the *fundamental involution* ω on Λ defined by $\omega(S_\lambda) = S_{\lambda'}$. This is an algebra automorphism, and applying it to the product above shows that $c^\lambda_{\mu,\nu} = c^{\lambda'}_{\mu',\nu'}$. We will use this involution and identity throughout.

As Λ is graded, if $c^\lambda_{\mu,\nu} \neq 0$, then $|\lambda| = |\mu| + |\nu|$. More fundamentally, each $c^\lambda_{\mu,\nu}$ is nonnegative. We say that a Schur function S_λ *occurs* in a product $S_{\mu^1} \dots S_{\mu^r}$ of Schur functions if its coefficient in the expansion of that product in the Schur basis is positive.

The Littlewood–Richardson rule is a formula for Littlewood–Richardson coefficients $c^\lambda_{\mu,\nu}$. When $\nu \subset \lambda$, we write λ/ν for the set-theoretic difference of their diagrams, which is called a *skew shape*. We may fill the boxes in λ/ν with positive integers that weakly increase left-to-right across each row and strictly increase down each column, obtaining a *skew tableau*. The *content* of a skew tableau T is the sequence whose ith element is the number of occurrences of i in T. The word of a tableau T is the sequence of its entries read right-to-left, and from the top row of T to its bottom row. A *Littlewood–Richardson skew tableau T* is a skew tableau in which the content of any initial segment of its word is a partition. (We extend the notion of content to any sequence of positive integers.) That is, when reading T in the order of its word, at every position within the word, the number of i's encountered is always at least the number of $(i+1)$'s encountered, for every i. Below is the skew shape $(6,4,2,2)/(3,2)$ and two tableaux with that shape and content $(4,3,1,1)$ whose words are 111322142 and 111223142, respectively. Only the second is a Littlewood–Richardson tableau.

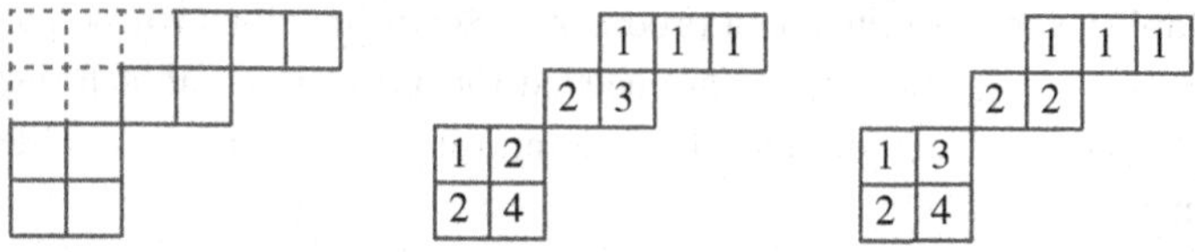

The Littlewood–Richardson coefficient $c^\lambda_{\mu,\nu}$ is the number of Littlewood–Richardson tableaux of shape λ/ν and content μ.

We record some consequences of this formula. First, if $c^{\lambda}_{\mu,\nu} \neq 0$, then $\mu, \nu \subseteq \lambda$ in Young's lattice. The set of partitions λ with $|\lambda|$ a fixed positive integer has a partial order, called *dominance*. For any positive integer k, write $|\lambda|_k$ for $\lambda_1 + \cdots + \lambda_k$. Given partitions λ and μ with $|\lambda| = |\mu|$, we say that μ *dominates* λ, written $\lambda \trianglelefteq \mu$ if, for all k, we have $|\lambda|_k \leq |\mu|_k$. Conjugation of partitions is an anti-involution of the dominance poset in that $\lambda \trianglelefteq \mu$ if and only if $\mu' \trianglelefteq \lambda'$.

Given partitions μ and ν, let (μ, ν) be the partition of $|\mu| + |\nu|$ obtained by sorting the sequence $(\mu_1, \ldots, \mu_{\ell(\mu)}, \nu_1, \ldots, \nu_{\ell(\nu)})$ in decreasing order. Let $\mu+\nu$ be the partition of $|\mu|+|\nu|$ whose kth part is $\mu_k+\nu_k$. Note that $(\mu, \nu)' = \mu' + \nu'$.

Proposition 2.1 *If λ, μ, and ν are partitions with $c^{\lambda}_{\mu,\nu} \neq 0$, then*

$$(\mu, \nu) \trianglelefteq \lambda \trianglelefteq \mu + \nu.$$

Furthermore, $c^{(\mu,\nu)}_{\mu,\nu} = c^{\mu+\nu}_{\mu,\nu} = 1$.

Proof In any Littlewood–Richardson skew tableau of shape λ/μ the first i rows can only contain entries $1, \ldots, i$, and thus if it has content ν, there are at most $\nu_1 + \cdots + \nu_i$ boxes in these first i rows. Since the ith row of $(\mu+\nu)/\mu$ has ν_i boxes, we see that there is a unique Littlewood–Richardson tableau of this shape and content ν, and that for any other Littlewood–Richardson skew tableau of shape λ/μ and content ν, we must have $\lambda \triangleleft \mu + \nu$. Applying conjugation gives the other inequality. □

Corollary 2.2 *If S_λ occurs in the product $S_\mu \cdot S_\nu$, then for any nonnegative integers a, b, we have $|\mu|_a + |\nu|_b \leq |\lambda|_{a+b}$.*

Proof As (μ, ν) is the decreasing rearrangement of the parts of μ and of ν, we have $|\mu|_a + |\nu|_b \leq |(\mu, \nu)|_{a+b}$. Since $(\mu, \nu) \triangleleft \lambda$ by Proposition 2.1, we have $|(\mu, \nu)|_{a+b} \leq |\lambda|_{a+b}$, which implies the inequality. □

2.2 Schur functions of partitions with large rectangles

We establish some results about products of Schur functions whose partitions contain large rectangles. These are needed for our main combinatorial result about the number of solutions to a composition of Schubert problems in Section 3.2.

Let a, d be nonnegative integers. Write $\square_{a,d}$ for the rectangular partition $(d, \ldots, d)$ with a parts, each of size d. Its Young diagram is an $a \times d$ rectangular array of ad boxes. Suppose that μ, α are partitions with $a \geq \ell(\mu)$

and $d \geq \alpha_1$. Then $\square_{a,d}+\mu$ is the partition whose kth part is $d + \mu_k$, for $k \leq a = \ell(\square_{a,d}+\mu)$. Also, $(\square_{a,d}+\mu, \alpha)$ is the partition whose parts are those of $\square_{a,d}+\mu$ followed by those of α, as $d \geq \alpha_1$. The Young diagram of $(\square_{a,d}+\mu, \alpha)$ is obtained from the rectangle $\square_{a,d}$ by placing the diagram of μ to the right of the rectangle and the diagram of α below the rectangle.

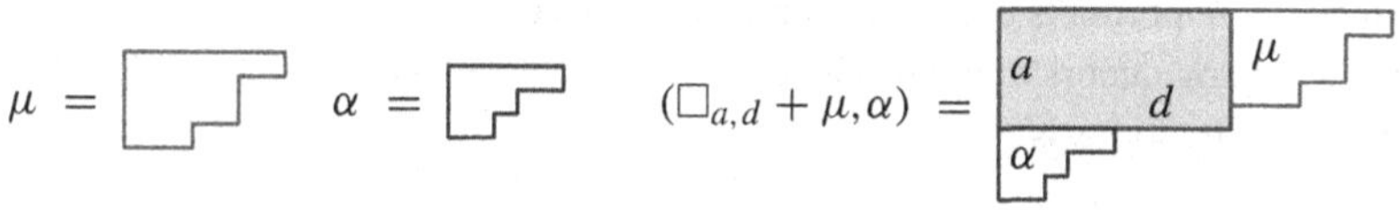

$\mu =$ $\quad$ $\alpha =$ $\quad$ $(\square_{a,d} + \mu, \alpha) =$

Observe that the conjugate of $(\square_{a,d} + \mu, \alpha)$ is $(\square_{d,a} + \alpha', \mu')$.

Lemma 2.3 *Let a, d be nonnegative integers and μ, α be partitions with $a \geq \ell(\mu)$ and $d \geq \alpha_1$. Then*

$$S_{\square_{a,d}+\mu} \cdot S_\alpha = S_{(\square_{a,d}+\mu,\alpha)} + \sum_{(\square_{a,d}+\mu,\alpha) \lhd \nu} c^{\nu}_{\square_{a,d}+\mu,\alpha} S_\nu,$$

and if ν indexes a nonzero term in the sum, then

$$|\nu|_a > |\square_{a,d} + \mu| = ad + |\mu|. \tag{1}$$

Proof By Proposition 2.1, the coefficient of $S_{(\square_{a,d}+\mu,\alpha)}$ in the product is 1, and the other nonzero terms are indexed by partitions ν that strictly dominate $(\square_{a,d} + \mu, \alpha)$. Thus,

$$|\nu|_a \geq |(\square_{a,d} + \mu, \alpha)|_a = |\square_{a,d} + \mu| = ad + |\mu|.$$

If the inequality is strict, then we have (1).

Suppose that $|\nu|_a = ad + |\mu|$. Since $\nu \geq \square_{a,d}+\mu$, and thus $\nu_i \geq d + \mu_i = (\square_{a,d}+\mu)_i$, this implies that the partition, $(\nu_1, \ldots, \nu_a)$, formed by the first a parts of ν equals $\square_{a,d}+\mu$. If we set $\beta = (\nu_{a+1}, \ldots)$, then $\nu = (\square_{a,d}+\mu, \beta)$. In particular $\beta_1 \leq d + \mu_a$ and $|\beta| = |\alpha|$. By the Littlewood–Richardson rule, $c^{\nu}_{\square_{a,d}+\mu,\alpha}$ counts the number of Littlewood–Richardson skew tableaux of shape $\nu/(\square_{a,d}+\mu) = \beta$ and content α. But a Littlewood–Richardson tableau of partition shape λ must have content λ, which shows that $\beta = \alpha$ and completes the proof of the lemma. $\square$

Lemma 2.4 *Let $d, a_1, \ldots, a_r$ be nonnegative integers and $\mu^1, \ldots, \mu^r$, $\alpha^1, \ldots, \alpha^r$ be partitions with $a_i \geq \ell(\mu^i)$ and $d \geq \alpha^i_1$, for $i = 1, \ldots, r$. If S_ν occurs in the product $\prod_{i=1}^r S_{(\square_{a_i,d}+\mu^i,\alpha^i)}$ of r terms, then*

$$|\nu|_{a_1+\cdots+a_r} \geq d(a_1+\cdots+a_r) + |\mu^1|+\cdots+|\mu^r|.$$

Proof We prove this by induction on the number r of factors of the form $S_{(\square_{a_i,d}+\mu^i,\alpha^i)}$. There is nothing to prove when $r = 1$ as in this case $\nu = (\square_{a,d}+\mu^1,\alpha^1)$. Suppose that the statement holds for $r-1$ factors, and that S_ν occurs in the product of r factors. By the positivity of the coefficients of Schur functions in products of Schur functions, S_ν occurs in a product $S_{(\square_{a_r,d}+\mu^r,\alpha^r)} \cdot S_\lambda$, where S_λ occurs in the product with $r-1$ factors. By Corollary 2.2 and our induction hypothesis,

$$\begin{aligned}|\nu|_{a_1+\cdots+a_r} &\geq |(\square_{a_r,d}+\mu^r,\alpha^r)|_{a_r} + |\lambda|_{a_1+\cdots+a_{r-1}}\\ &\geq da_r + |\mu^r| + d(a_1+\cdots+a_{r-1}) + |\mu^1|+\cdots+|\mu^{r-1}|\\ &= d(a_1+\cdots+a_r) + |\mu^1|+\cdots+|\mu^r|,\end{aligned}$$

which completes the proof. □

We give a combinatorial description of some of the coefficients which appear in a product of the form $\prod_{i=1}^r S_{\square_{a_i,d}+\mu^i}$. A *generalized Littlewood–Richardson tableau* [2, Ex. 3.7] $T_\bullet$ of shape λ and content $\underline{\mu} = (\mu^1,\ldots,\mu^r)$ is a sequence of partitions,

$$\emptyset = \lambda^0 \subseteq \lambda^1 \subseteq \cdots \subseteq \lambda^r = \lambda,$$

together with Littlewood–Richardson tableaux $T_1,\ldots,T_r$ where T_i has shape λ^i/λ^{i-1} and content μ^i. A repeated application of the Littlewood–Richardson rule implies the following.

Proposition 2.5 *The coefficient $c^\lambda_{\underline{\mu}}$ of S_λ in the product $\prod_{i=1}^r S_{\mu^i}$ is equal to the number of generalized Littlewood–Richardson tableaux of shape λ and content $\underline{\mu}$.*

Let λ,μ,ν be partitions with $|\lambda| = |\mu|+|\nu|$ and $\mu,\nu \subset \lambda$. Suppose that l,m,n are integers with $l = m+n$ and $l \geq \ell(\lambda)$, $m \geq \ell(\mu)$, and $n \geq \ell(\nu)$. For any $d \geq 0$ the skew shape $(\square_{l,d}+\lambda)/(\square_{n,d}+\nu)$ is obtained by placing the rectangle $\square_{m,d}$ to the left of the skew shape λ/ν, with the last row of $\square_{m,d}$ at row l. Write $\square_{m,d}+\lambda/\nu$ for this shape. We illustrate this, shading the skew shapes.

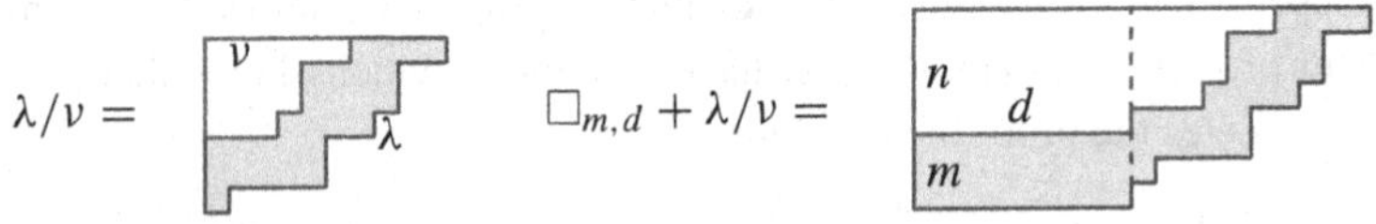

In any Littlewood–Richardson tableau T of shape $\square_{m,d}+\lambda/\nu$ with content $\square_{m,d}+\mu$, the rectangle $\square_{m,d}$ in T is *frozen* in that each of its d columns are filled with $1,\dots,m$. Removing the frozen rectangle gives a Littlewood–Richardson tableau of shape λ/ν and content μ (this may be proved by an induction on d, which we omit).

Lemma 2.6 *Adding or removing frozen rectangles from skew Littlewood–Richardson tableaux gives a bijection*

$$\left\{\begin{array}{c}\text{Littlewood–Richardson tableaux of}\\ \text{shape } \lambda/\nu \text{ and content } \mu.\end{array}\right\} \longleftrightarrow \left\{\begin{array}{c}\text{Littlewood–Richardson tableaux of shape}\\ \square_{m,d}{+}\lambda/\nu \text{ and content } \square_{m,d}{+}\mu.\end{array}\right\}.$$

Let $\underline{\mu} := (\mu^1,\dots,\mu^r)$ be a sequence of partitions and λ a partition with $|\mu^1|+\cdots+|\mu^r| = |\lambda|$. Let $\underline{a} := (a_1,\dots,a_r)$ be a sequence of integers with $a_i \ge \ell(\mu^i)$. Set $a := |a_1|+\cdots+|a_r|$.

Corollary 2.7 *For any $d \ge 0$, the coefficient $c^{\lambda}_{\underline{\mu}}$ of S_λ in the product $\prod_i S_{\mu^i}$ is equal to the coefficient of $S_{\square_{a,d}+\lambda}$ in the product $\prod_i S_{\square_{a_i,d}+\mu^i}$.*

Proof Let $\square_{\underline{a},d}{+}\underline{\mu}$ be the sequence of partitions whose ith element is $\square_{a_i,d}{+}\mu^i$. The bijection of Lemma 2.6 of adding or removing a frozen rectangle from Littlewood–Richardson tableaux extends to a bijection

$$\left\{\begin{array}{c}\text{Generalized Littlewood-}\\ \text{Richardson tableaux of shape } \lambda\\ \text{and content } \underline{\mu}.\end{array}\right\} \longleftrightarrow \left\{\begin{array}{c}\text{Generalized Littlewood–Richardson}\\ \text{tableaux of shape } \square_{a,d}+\lambda \text{ and content}\\ \square_{\underline{a},d}+\underline{\mu}.\end{array}\right\}. \tag{2}$$

By Proposition 2.5, these sets of tableaux have cardinality the two coefficients in the statement of the corollary, which completes the proof. □

Example 2.8 Let us consider Corollary 2.7 in the case when $|\lambda| = 3$, $\underline{\mu} = (\square,\square,\square)$, and $d = 2$. Using the Pieri formula, we have

$$S_{\square}\cdot S_{\square}\cdot S_{\square} \;=\; S_{(1,1,1)} + 2S_{(2,1)} + S_{(3)}, \tag{3}$$

and the generalized Littlewood–Richardson tableaux are the tableaux of content $(1,1,1)$,

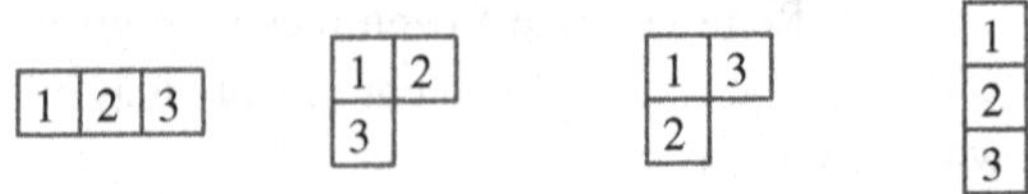

which shows (3). We let $\underline{a} = (1,1,1)$, so that $\square_{\underline{a},d} + \underline{\mu}$ consists of three identical tableaux $\square\square\square$. By the Pieri rule, a generalized Littlewood–Richardson tableau of shape κ and content $\square_{\underline{a},d} + \underline{\mu}$ is a tableau with content $(3,3,3)$. We show those of shape $\square_{3,2} + \lambda$.

1	1	1	2	3
2	2			
3	3			

1	1	1	2
2	2	3	
3	3		

1	1	1	3
2	2	2	
3	3		

1	1	1
2	2	2
3	3	3

The 'frozen rectangles' are shaded to better illustrate the bijection of Corollary 2.7.

3 Composition of Schubert problems

We define Schubert problems on Grassmannians and use results from Section 2 to establish some formulas in the cohomology ring of a Grassmannian. We next define the composition of two Schubert problems, illustrating this notion with several examples. We then formulate and prove our main theorem about the number of solutions to a composition of Schubert problems. We use facts about the Grassmannian and its Schubert varieties and cohomology ring, as may be found in [6].

3.1 Schubert problems on Grassmannians

For positive integers a,b, let $G(a,b)$ be the Grassmannian of a-dimensional linear subspaces of $\mathbb{C}^{a+b}$. This is a smooth complex algebraic variety of dimension ab. We will also need the alternative notation $Gr(a,V)$ of a-dimensional linear subspaces of a complex vector space V. Then $Gr(a,\mathbb{C}^{a+b}) = G(a,b)$.

Let $\lambda \subseteq \square_{a,b}$ be a partition so that $\ell(\lambda) \le a$ and $\lambda_1 \le b$. A complete flag $F_\bullet$ in $\mathbb{C}^{a+b}$ is a sequence of linear subspaces $F_\bullet\colon F_1 \subset F_2 \subset \cdots \subset F_{a+b} = \mathbb{C}^{a+b}$ with $\dim F_i = i$. Given a partition $\lambda \subseteq \square_{a,b}$ and a flag $F_\bullet$, we may define the *Schubert variety* to be

$$\Omega_\lambda F_\bullet := \{H \in G(a,b) \mid \dim H \cap F_{b+i-\lambda_i} \ge i \ \text{ for } i = 1,\dots,a\}. \tag{4}$$

This has codimension $|\lambda|$ in $G(a,b)$.

A *Schubert problem* on $G(a,b)$ is a list $\underline{\lambda} = (\lambda^1, \ldots, \lambda^r)$ of partitions with $\lambda^i \subseteq \Box_{a,b}$ such that $|\underline{\lambda}| := |\lambda^1| + \cdots + |\lambda^r| = ab$. We will call the partitions λ^i the *conditions* of the Schubert problem $\underline{\lambda}$. An *instance* of $\underline{\lambda}$ is determined by a list of flags $\mathcal{F}_\bullet := (F^1_\bullet, \ldots, F^r_\bullet)$. It consists of the points in the intersection,

$$\Omega_{\underline{\lambda}}\mathcal{F}_\bullet \ := \ \Omega_{\lambda^1}F^1_\bullet \cap \Omega_{\lambda^2}F^2_\bullet \cap \cdots \cap \Omega_{\lambda^r}F^r_\bullet. \tag{5}$$

When $\mathcal{F}_\bullet$ is general, this intersection is transverse [9]. As $\underline{\lambda}$ is a Schubert problem, $\Omega_{\underline{\lambda}}\mathcal{F}_\bullet$ is zero-dimensional and therefore consists of finitely many points. This number $\delta(\underline{\lambda})$ of points does not depend upon the choice of flags. It is the coefficient $c^{\Box_{a,b}}_{\underline{\lambda}}$ of the Schur function $S_{\Box_{a,b}}$ in the product $S_{\lambda^1}\cdots S_{\lambda^r}$. A Schubert problem $\underline{\lambda}$ is *trivial* if $\delta(\underline{\lambda}) \leq 1$.

We express this in the cohomology ring of the Grassmannian $G(a,b)$. Given a partition $\lambda \subseteq \Box_{a,b}$, the cohomology class associated to a Schubert variety $\Omega_\lambda F_\bullet$ is the *Schubert class* σ_λ. The class associated to an intersection of Schubert varieties given by general flags is the product of the associated Schubert classes, and $\sigma_{\Box_{a,b}}$ is the class of a point. The map defined on the Schur basis by

$$S_\lambda \longmapsto \begin{cases} \sigma_\lambda & \text{if } \lambda \subseteq \Box_{a,b} \\ 0 & \text{otherwise} \end{cases}$$

is a homomorphism from Λ onto this cohomology ring. Thus we may evaluate a product in cohomology by applying this map to the corresponding product in Λ, which removes all terms whose Schur function is indexed by a partition λ that is not contained in the $a \times b$ rectangle, $\Box_{a,b}$. Thus, for a Schubert problem $\underline{\lambda} = (\lambda^1, \ldots, \lambda^r)$ and general flags $\mathcal{F}_\bullet$, the class of the intersection $\Omega_{\underline{\lambda}}\mathcal{F}_\bullet$ (5) is

$$\sigma_{\lambda^1}\sigma_{\lambda^2}\cdots\sigma_{\lambda^r} \ = \ c^{\Box_{a,b}}_{\underline{\lambda}}\,\sigma_{\Box_{a,b}} \ = \ \delta(\underline{\lambda})\,\sigma_{\Box_{a,b}}.$$

The fundamental involution has a geometric counterpart. The map sending an a-dimensional subspace H of $V \simeq \mathbb{C}^{a+b}$ to its annihilator $H^\perp$ in the dual space V^* is an isomorphism $Gr(a,V) \xrightarrow{\sim} Gr(b,V^*)$. Given a flag $F_\bullet$ in V, its sequence of annihilators gives a flag $F^\perp_\bullet$ in V^* and for any partition $\lambda \subseteq \Box_{a,b}$ the map $H \mapsto H^\perp$ sends $\Omega_\lambda F_\bullet$ to $\Omega_{\lambda'}F^\perp_\bullet$. (Recall that λ' is conjugate of λ.) This induces an isomorphism on cohomology, sending the Schubert class σ_λ for $G(a,b)$ to the class $\sigma_{\lambda'}$ for $G(b,a)$. Because of this and the fundamental involution on Λ, results we prove for a Schubert problem $\underline{\lambda}$ on one Grassmannian $G(a,b)$ hold for its conjugate problem $\underline{\lambda}'$ on the dual Grassmannian $G(b,a)$.

For $\lambda \subseteq \Box_{a,b}$, let $\lambda^\vee \subseteq \Box_{a,b}$ be the partition whose ith part is $\lambda^\vee_i = b - \lambda_{a+1-i}$. The skew diagram $\Box_{a,b}/\lambda^\vee$ is the rotation λ° of the diagram of λ by $180°$. When $a = 5$, $b = 6$, and $\lambda = (4,3,1,1)$, here are λ and $\lambda^\vee$,

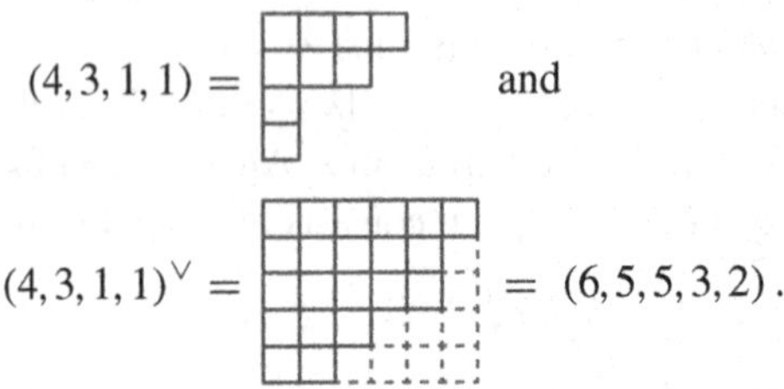

A key fact about the Schubert classes is Schubert's Duality Theorem [6, p. 149], a version of which is the following useful and well-known fact.

Lemma 3.1 *Let $\lambda, \mu \subseteq \Box_{a,b}$. Then $\sigma_\lambda \cdot \sigma_\mu \neq 0$ if and only if $\mu \subseteq \lambda^\vee$. Furthermore, $\sigma_\lambda \cdot \sigma_{\lambda^\vee} = \sigma_{\Box_{a,b}}$.*

We give a consequence of Lemma 3.1, which is also well-known.

Corollary 3.2 *The coefficient of σ_λ in a cohomology class σ equals the coefficient of $\sigma_{\Box_{a,b}}$ in the product $\sigma \cdot \sigma_{\lambda^\vee}$. In particular, $c^{\lambda^\vee}_{\mu,\nu}$ is the coefficient of $\sigma_{\Box_{a,b}}$ in the product $\sigma_\mu \cdot \sigma_\nu \cdot \sigma_\lambda$.*

We deduce a lemma which constrains the shape of partitions indexing certain nonzero products. This will be a key ingredient in our main theorem.

Lemma 3.3 *Suppose that $\kappa, \lambda \subseteq \Box_{a+c,b+d}$ are partitions with $\ell(\kappa) \leq a$ and $\lambda_1 \leq b$. If $|\kappa|+|\lambda| > ad+ab+cb$, then $\sigma_\kappa \cdot \sigma_\lambda = 0$ in the cohomology ring of $G(a+c,b+d)$. If $|\kappa|+|\lambda| = ad+ab+cb$ and $\sigma_\kappa \cdot \sigma_\lambda \neq 0$, then $\sigma_\kappa \cdot \sigma_\lambda = \sigma_{\Box^\vee_{c,d}}$ and there is partition $\alpha \subseteq \Box_{a,b}$ such that $\kappa = \Box_{a,d} + \alpha$ and $\lambda = (\Box_{c,b}, \alpha^\vee)$.*

Proof Let $\kappa, \lambda \leq \Box_{a+c,b+d}$ be partitions with $\ell(\kappa) \leq a$ and $\lambda_1 \leq b$. By Lemma 3.1, $\sigma_\kappa \cdot \sigma_\lambda \neq 0$ if and only if $\kappa \subseteq \lambda^\vee$. Recall that $\Box_{a+c,b+d}/\lambda^\vee = \lambda^\circ$, which is the Young diagram of λ rotated by 180° and placed in the southeast corner of $\Box_{a+c,b+d}$. Thus $\sigma_\kappa \cdot \sigma_\lambda \neq 0$ if and only if the diagrams of κ and λ° are disjoint.

As $\ell(\kappa) \leq a$, the Young diagram of κ lies in the first a rows of $\Box_{a+c,b+d}$. Similarly, λ° lies in the last b columns of $\Box_{a+c,b+d}$, as $\lambda_1 \leq b$. Since there are $ad+ab+cb$ boxes in these rows and columns, if

$$|\kappa| + |\lambda| > ad + ab + cb,$$

then the Young diagram of κ must contain a box of λ° and $\sigma_\kappa \cdot \sigma_\lambda = 0$.

Suppose that $|\kappa|+|\lambda| = ad+ab+cb$ and $\sigma_\kappa \cdot \sigma_\lambda \neq 0$. Then $\kappa \subseteq \lambda^\vee$, and the disjoint union $\kappa \cup \lambda^\circ$ covers the first a rows and last b columns of $\Box_{a+c,b+d}$. In particular, this implies that κ contains the northwest rectangle $\Box_{a,d}$ and λ° contains the southeast rectangle $\Box_{c,b}$. Thus there are partitions $\alpha, \beta \subseteq \Box_{a,b}$

such that $\kappa = \Box_{a,d} + \alpha$ and $\lambda = (\Box_{c,b}, \beta)$. As $\kappa \cup \lambda^\circ$ covers the northeast rectangle $\Box_{a,b}$, we see that $\beta = \alpha^\vee$.

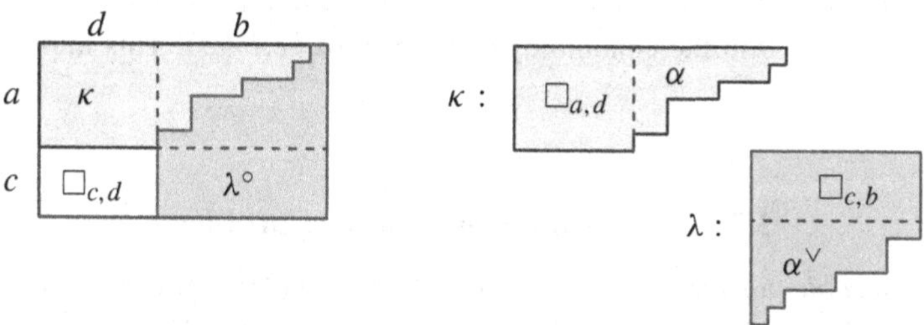

Let $\rho \subseteq \Box_{a+c,b+d}$ be a partition such that σ_ρ appears in the product $\sigma_\kappa \cdot \sigma_\lambda$. Let $\nu := \rho^\vee$. Then

$$|\nu| = (a+c)(b+d) - |\kappa| - |\lambda| = (a+c)(b+d) - ab - ad - cb = cd,$$

and κ, ν, λ form a Schubert problem with $\sigma_\kappa \cdot \sigma_\nu \cdot \sigma_\lambda \neq 0$. Thus $\sigma_\kappa \cdot \sigma_\nu \neq 0$. By Proposition 2.1, $\sigma_\kappa \cdot \sigma_\nu$ is a sum of terms $c^\gamma_{\kappa,\nu}\sigma_\gamma$ with $(\kappa,\nu) \trianglelefteq \gamma$. By Corollary 3.2, $c^{\lambda^\vee}_{\kappa,\nu}\sigma_{\Box_{a+c,b+d}} = \sigma_\kappa \cdot \sigma_\nu \cdot \sigma_\lambda$. As the product is nonzero, $\sigma_{\lambda^\vee}$ occurs in $\sigma_\kappa \cdot \sigma_\nu$ and so we have $(\kappa,\nu) \trianglelefteq \lambda^\vee$, by Proposition 2.1. Since $\kappa_i = \lambda^\vee_i$ for $i \leq a$, an induction on i shows that $(\kappa,\nu)_i = \kappa_i$ for $i \leq a$. As $\lambda^\vee_{a+j} = d$ for $j = 1, \ldots, c$, another induction shows that $\nu_j = d$ for $j = 1, \ldots, c$. Consequently, $\nu = \Box_{c,d}$ and $(\kappa,\nu) = \lambda^\vee$. By Proposition 2.1, $c^{\lambda^\vee}_{\kappa,\nu} = c^{(\kappa,\nu)}_{\kappa,\nu} = 1$ and thus $\sigma_\kappa \cdot \sigma_\nu \cdot \sigma_\lambda = \sigma_{\Box_{a+c,b+d}}$, by Corollary 3.2.

Thus we have shown that if $|\nu| = cd$, so that λ, κ, ν is a Schubert problem, then

$$\sigma_\kappa \cdot \sigma_\nu \cdot \sigma_\lambda = \begin{cases} 1 & \text{if } \nu = \Box_{c,d} \\ 0 & \text{otherwise} \end{cases}.$$

Thus by Corollary 3.2, $\sigma_\kappa \cdot \sigma_\lambda = \sigma_{\Box^\vee_{c,d}}$. □

For any partition λ and positive integer b, note that $|\lambda'|_b$ is the number of boxes in the first b columns of the Young diagram of λ.

Corollary 3.4 *Let $\rho, \tau \subseteq \Box_{a+c,b+d}$ be partitions with $|\rho|_a + |\tau'|_b > ad + cb + ab$. Then $\sigma_\rho \cdot \sigma_\tau = 0$ in the cohomology ring of $G(a+c, b+d)$.*

Proof Let $\kappa = (\rho_1, \ldots, \rho_a)$ be the partition formed by the first a parts of ρ and $\alpha = (\rho_{a+1}, \ldots)$ be the partition formed by the remaining parts of ρ. Similarly, let λ be the partition formed from the first b columns of the Young diagram of τ and β the partition formed by the remaining columns of τ. Then we have $\rho = (\kappa, \alpha)$ and $\tau = \lambda + \beta$. By Proposition 2.1, σ_ρ occurs in the product $\sigma_\kappa \cdot \sigma_\alpha$ and σ_τ occurs in the product $\sigma_\lambda \cdot \sigma_\beta$.

Thus the coefficient of any Schubert class σ_γ in the product $\sigma_\rho \cdot \sigma_\tau$ is at most the coefficient of σ_γ in the product $\sigma_\kappa \cdot \sigma_\alpha \cdot \sigma_\lambda \cdot \sigma_\beta = (\sigma_\kappa \cdot \sigma_\lambda) \cdot \sigma_\alpha \cdot \sigma_\beta$. As $\ell(\kappa) \leq a$ and $\lambda_1 \leq b$ and $|\kappa| + |\lambda| > ad + ab + cb$, Lemma 3.3 implies that $\sigma_\kappa \cdot \sigma_\lambda = 0$ in the cohomology ring of $G(a{+}c, b{+}d)$. This implies that $\sigma_\rho \cdot \sigma_\tau = 0$. □

3.2 Compositions of Schubert problems

A *composable partition* of a Schubert problem $\underline{\lambda}$ on $G(a,b)$ is a partition $\underline{\lambda} = (\underline{\mu}, \underline{\nu})$ of the conditions of $\underline{\lambda}$, where $\underline{\mu} = (\mu^1, \ldots, \mu^r)$ and $\underline{\nu} = (\nu^1, \ldots, \nu^s)$, together with two sequences $\underline{a} = (a_1, \ldots, a_r)$ and $\underline{b} = (b_1, \ldots, b_s)$ of nonnegative integers with $a = |\underline{a}| = a_1 + \cdots + a_r$ and $b = |\underline{b}| = b_1 + \cdots + b_s$. These data, $(\underline{\mu}, \underline{\nu})$, $\underline{a}$, and $\underline{b}$, further satisfy

$$a_i \;\geq\; \ell(\mu^i) \quad \text{for } i = 1, \ldots, r \qquad \text{and} \qquad b_j \;\geq\; \nu_1^j \quad \text{for } j = 1, \ldots, s. \tag{6}$$

We say that $\underline{\lambda}$ is *composable* if it admits a composable partition.

Example 3.5 For example, $\underline{\lambda} = (\underline{\mu}, \underline{\nu})$ with $\underline{\mu} = ((p),(q))$ and $\underline{\nu} = (\square, \ldots, \square)$ ($p{+}q$ occurrences of $\square$) is a composable Schubert problem on $G(2, p{+}q)$ with $\underline{a} = (1,1)$ and $\underline{b} = (1, \ldots, 1)$ ($p{+}q$ occurrences of 1). Also $\underline{\mu} = (\square\square, \square\square, \square)$ and $\nu = (\begin{smallmatrix}\square\\\square\end{smallmatrix}, \square, \square)$ with $\underline{a} = \underline{b} = (1,1,1)$ forms a composable partition of a Schubert problem on $G(3,3)$, as do $\underline{\mu} = (\square\square, \square\square, \square\square)$ and $\nu = (\square, \square, \square)$, with the same $\underline{a}$ and $\underline{b}$.

Remark 3.6 Not all Schubert problems are composable. For example, the Schubert problem $\underline{\lambda} = (\square, \square, \square, \square, \square, \square)$ on $G(2,3)$ is not composable as its number of conditions is 6, which exceeds the sum $2 + 3$. However, the Schubert problem $(\begin{smallmatrix}\square\square\square\\\square\square\end{smallmatrix}, \square, \square, \square, \square, \square)$ on $G(2,5)$ is equivalent to it, and this does have a composable partition, namely $\underline{\mu} = (\begin{smallmatrix}\square\square\square\\\square\square\end{smallmatrix})$ and $\underline{\nu} = (\square, \square, \square, \square, \square)$ with $\underline{a} = (2)$ and $\underline{b} = (1,1,1,1,1)$.

Geometrically equivalent means that every general instance of the Schubert problem $(\begin{smallmatrix}\square\square\square\\\square\square\end{smallmatrix}, \square, \square, \square, \square, \square)$ on $G(2,5)$ gives an instance of $(\square, \square, \square, \square, \square, \square)$ on $G(2,3)$ whose solutions correspond to solutions of the original problem, and every general instance of the second problem occurs in this way. This generalizes: Every Schubert problem that is not composable is geometrically equivalent to a composable Schubert problem on a larger Grassmannian. A proof of this claim may be modeled on the construction of Remark 3.6.

Definition 3.7 Suppose that $(\underline{\mu},\underline{\nu}),\underline{a},\underline{b}$ is a composable partition of a Schubert problem $\underline{\lambda}$ on $G(a,b)$, where $\underline{\mu} = (\mu^1,\ldots,\mu^r)$ and $\underline{\nu} = (\nu^1,\ldots,\nu^s)$. Let c,d be positive integers and let $\underline{\rho}$ be any Schubert problem on $G(c,d)$. Let us partition the conditions of $\underline{\rho}$ into three lists $\underline{\rho} = (\underline{\alpha},\underline{\beta},\underline{\gamma})$, where $\underline{\alpha} = (\alpha^1,\ldots,\alpha^r)$, $\underline{\beta} = (\beta^1,\ldots,\beta^s)$, and $\underline{\gamma} = (\gamma^1,\ldots,\gamma^t)$. The *composition* $\underline{\lambda}\circ\underline{\rho}$ of these two Schubert problems is the list of partitions

$$\big((\Box_{\underline{a},d} + \underline{\mu},\underline{\alpha})\,,\,(\Box_{c,\underline{b}},\underline{\nu}) + \underline{\beta}\,,\,\underline{\gamma}\big).$$

Here, we define $(\Box_{\underline{a},d}+\underline{\mu},\underline{\alpha})$ to be the sequence $((\Box_{a_1,d}+\mu^1,\alpha^1),\ldots,(\Box_{a_r,d}+\mu^r,\alpha^r))$, and similarly we define $(\Box_{c,\underline{b}},\underline{\nu}) + \underline{\beta}$ to be the sequence $\big((\Box_{c,b_1},\nu^1)+\beta^1,\ldots,(\Box_{c,b_s},\nu^s)+\beta^s\big)$. We always write a Schubert problem $\underline{\rho}$ on $G(c,d)$ to compose with $\underline{\lambda} = (\underline{\mu},\underline{\nu})$ as a triple $(\underline{\alpha},\underline{\beta},\underline{\gamma})$ with $\underline{\mu}$ and $\underline{\alpha}$ having the same number, r, of partitions, and both $\underline{\nu}$ and $\underline{\beta}$ have the same number, s, of partitions. We remark that any of the partitions μ^i, ν^j, α^i, β^j, or γ^k may be the zero partition (0), whose Young diagram is $\emptyset$.

Example 3.8 Let $\underline{\lambda} = (\underline{\mu},\underline{\nu})$ be the partition of a Schubert problem on $G(3,3)$ with $\underline{\mu} = (\ydiagram{2}, \ydiagram{2}, \ydiagram{1})$, $\nu = (\ydiagram{1,1}, \ydiagram{1}, \ydiagram{1})$, and $\underline{a} = \underline{b} = (1,1,1)$, which is from Example 3.5. Let $\underline{\rho}$ be the Schubert problem on $G(2,3)$ of Remark 3.6, expanded with some empty partitions, so that $\underline{\alpha} = (\Box,\emptyset,\Box)$, $\underline{\beta} = (\Box,\Box,\emptyset)$, and $\underline{\gamma} = (\Box,\Box)$. The composition $\underline{\lambda}\circ\underline{\rho}$ consists of the eight partitions

$$\ydiagram{5,1}\,,\ \ydiagram{5}\,,\ \ydiagram{4,1}\,,\ \ydiagram{2,1,1,1}\,,\ \ydiagram{2,1,1}\,,\ \ydiagram{1,1,1}\,,\ \Box\,,\ \Box\,.$$

We have shaded the rectangles $\Box_{a_i,3}$ and $\Box_{2,b_j}$ in the first six partitions.

Our main combinatorial result is a product formula for the number of solutions to a composition of Schubert problems.

Theorem 3.9 *Let $(\underline{\mu},\underline{\nu}),\underline{a},\underline{b}$ be a composable partition of a Schubert problem $\underline{\lambda}$ on $G(a,b)$ and $\underline{\rho} = (\underline{\alpha},\underline{\beta},\underline{\gamma})$ a Schubert problem on $G(c,d)$. The number of solutions to $\underline{\lambda}\circ\underline{\rho}$ is equal to the product of the number of solutions to $\underline{\lambda}$ with the number of solutions to $\underline{\rho}$,*

$$\delta(\underline{\lambda}\circ\underline{\rho}) \;=\; \delta(\underline{\lambda})\cdot\delta(\underline{\rho}).$$

Remark 3.10 Derksen and Weyman use representations of quivers to prove a result [5, Thm. 7.14] expressing certain Littlewood–Richardson coefficients as products of other Littlewood–Richardson coefficients. When the Schubert problems $\underline{\lambda}$ and $\underline{\rho}$ only involve three partitions each (e.g. $\underline{\lambda} = (\mu^1,\mu^2,\nu^1)$ and

$\underline{\rho} = (\alpha^1, \alpha^2, \beta^1)$), then Theorem 3.9 is a special case of this result of Derksen and Weyman.

More generally, their methods lead to an alternative proof of Theorem 3.9. First, their methods give an alternative proof of the enumerative consequence of Lemma 2.6 concerning the equality of Littlewood–Richardson numbers,

$$c^{\lambda}_{\mu,\nu} \;=\; c^{\square_{l,d}+\lambda}_{\square_{m,d}+\mu,\,\square_{n,d}+\nu},$$

where $l = m + n$, $l \geq \ell(\lambda)$, $m \geq \ell(\mu)$, and $n \geq \ell(\nu)$. Together with Lemmas 2.3 and 2.4 this implies Corollary 2.7, and eventually Theorem 3.9. We do not give the details of these arguments, as they require developing significant additional notation. We expect that representations of quivers may be used to give a direct proof of Theorem 3.9.

For an example of Theorem 3.9, observe that the Schubert problems $\underline{\lambda}$ and $\underline{\rho}$ of Example 3.8 both have five solutions, and thus $\underline{\lambda} \circ \underline{\rho}$ has 25 solutions. We first prove a preliminary result.

Lemma 3.11 *Let* $(\underline{\mu}, \underline{\nu}), \underline{a}, \underline{b}$ *be a composable partition of a Schubert problem* $\underline{\lambda}$ *on* $G(a,b)$ *and let* c,d *be nonnegative integers. In the cohomology ring of* $G(a+c, b+d)$ *we have*

$$\prod_{i=1}^{r} \sigma_{\square_{a_i,d}+\mu^i} \cdot \prod_{j=1}^{s} \sigma_{(\square_{c,b_j},\nu^j)} \;=\; \delta(\underline{\lambda})\sigma_{\square^\vee_{c,d}}, \tag{7}$$

where $\square^\vee_{c,d}$ *is the complement of* $\square_{c,d}$ *in* $\square_{a+c,b+d}$.

Let $\lambda = (\square, \square, \square, \square)$ be a composable Schubert problem on $G(2,2)$ with $\delta(\lambda) = 2$. Consider applying Lemma 3.11 to λ with $c = d = 2$. Then the product (7) becomes,

$$\begin{aligned}\sigma_{3} \cdot \sigma_{3} \cdot \sigma_{111} \cdot \sigma_{111} &= \left(\sigma_{42} + \sigma_{33}\right) \cdot \left(\sigma_{2211} + \sigma_{222}\right) \\ &= \sigma_{42} \cdot \sigma_{2211} + \sigma_{42} \cdot \sigma_{222} + \sigma_{33} \cdot \sigma_{2211} + \sigma_{33} \cdot \sigma_{222} \\ &= 2\sigma_{4422} \;=\; 2\sigma_{4422}\,,\end{aligned} \tag{8}$$

as the first and the last of the products on the second line are zero, by Corollary 3.4, while both middle ones are equal to the Schubert class σ_{4422}, by Lemma 3.3.

Proof of Lemma 3.11 Let $\square_{\underline{a},d} + \underline{\mu}$ be the list of partitions appearing in the first product and $(\square_{c,\underline{b}}, \underline{\nu})$ be the list of partitions appearing in the second product. Suppose that σ_κ appears in the first product. An induction based on Proposition 2.1 implies that

$$\ell(\kappa) \leq \sum_{i=1}^{r} \ell(\Box_{a_i,d} + \mu^i) = \sum_{i=1}^{r} a_i = a.$$

Applying the fundamental involution ω shows that if σ_λ occurs in the second product, then $\lambda_1 \leq \sum_j b_j = b$.

Observe that $|\kappa| = |\Box_{\underline{a},d} + \underline{\mu}|$, so that

$$|\kappa| = \sum_{i=1}^{r} |\Box_{a_i,d} + \mu^i| = \sum_{i=1}^{r} a_i d + |\underline{\mu}| = ad + |\underline{\mu}|, \tag{9}$$

and similarly $|\lambda| = bc + |\underline{\nu}|$. Thus $|\kappa| + |\lambda| = ad + cb + ab$ as $|\underline{\mu}| + |\underline{\nu}| = ab$. By Lemma 3.3, if $\sigma_\kappa \cdot \sigma_\lambda \neq 0$, then there exists a partition $\alpha \subseteq \Box_{a,b}$ for $G(a,b)$ such that $\kappa = \Box_{a,d} + \alpha$ and $\lambda = (\Box_{c,b}, \alpha^\vee)$, and $\sigma_\kappa \cdot \sigma_\lambda = \sigma_{\Box^\vee_{c,d}}$. Since (9) holds, we see that $|\alpha| = |\underline{\mu}|$ and $|\alpha^\vee| = |\underline{\nu}|$. By Corollary 2.7, the coefficient of σ_κ in $\prod_i \sigma_{\Box_{a_i,d}+\mu^i}$ is $c^\alpha_{\underline{\mu}}$. Applying the fundamental involution $\omega\colon \sigma_\rho \mapsto \sigma_{\rho'}$ shows that the coefficient of σ_λ in $\prod_j \sigma_{(\Box_{c,b_j},\nu^j)}$ is $c^{\alpha^\vee}_{\underline{\nu}}$.

Consequently, if we expand each product in the expression (7) as a sum of Schubert classes σ_ρ and then expand the product of those sums, the only terms which contribute are $c^\alpha_{\underline{\mu}} \cdot c^{\alpha^\vee}_{\underline{\nu}} \cdot \sigma_{\Box_{a,d}+\alpha} \cdot \sigma_{(\Box_{c,b},\alpha^\vee)}$, for $\alpha \subseteq \Box_{a,b}$ with $|\alpha| = |\underline{\mu}|$. That is,

$$\prod_{i=1}^{r} \sigma_{\Box_{a_i,d}+\mu^i} \cdot \prod_{j=1}^{s} \sigma_{(\Box_{c,b_j},\nu^j)} = \Big(\sum_\rho c^\rho_{\Box_{\underline{a},d}+\underline{\mu}} \sigma_\rho\Big) \cdot \Big(\sum_\tau c^\tau_{(\Box_{c,\underline{b}},\underline{\nu})} \sigma_\tau\Big)$$
$$= \sum_{\substack{\alpha\subseteq\Box_{a,b}\\ |\alpha|=|\underline{\mu}|}} c^\alpha_{\underline{\mu}} \cdot c^{\alpha^\vee}_{\underline{\nu}} \cdot \sigma_{\Box_{a,d}+\alpha} \cdot \sigma_{(\Box_{c,b},\alpha^\vee)} = \sum_{\substack{\alpha\subseteq\Box_{a,b}\\ |\alpha|=|\underline{\mu}|}} c^\alpha_{\underline{\mu}} \cdot c^{\alpha^\vee}_{\underline{\nu}} \cdot \sigma_{\Box^\vee_{c,d}},$$

the last equality by Lemma 3.3. A similar expansion of $\delta(\underline{\lambda})\sigma_{\Box_{a,b}} = \prod_i \sigma_{\mu^i} \cdot \prod_j \sigma_{\nu^j}$ gives

$$\prod_{i=1}^{r} \sigma_{\mu^i} \cdot \prod_{j=1}^{s} \sigma_{\nu^j} = \Big(\sum_\rho c^\rho_{\underline{\mu}} \sigma_\rho\Big) \cdot \Big(\sum_\tau c^\tau_{\underline{\nu}} \sigma_\tau\Big) = \sum_{\substack{\alpha\subseteq\Box_{a,b}\\ |\alpha|=|\underline{\mu}|}} c^\alpha_{\underline{\mu}} \cdot c^{\alpha^\vee}_{\underline{\nu}} \cdot \sigma_{\Box_{a,b}}.$$

As this is $\delta(\underline{\lambda})\sigma_{\Box_{a,b}}$, we conclude that the sum of the product of coefficients in both expressions equals $\delta(\underline{\lambda})$, which completes the proof of the lemma. □

Example 3.12 The proof of Theorem 3.9 involves expanding a product in cohomology in two different ways. Let us begin with an example, where

$\underline{\lambda} = \underline{\rho} = (\square, \square, \square, \square)$ and $a = b = c = d = 2$. Then $\underline{\mu} = \underline{\nu} = (\square, \square)$ and $\underline{a} = \underline{b} = (1,1)$, and we have $\square_{\underline{a},d} + \underline{\mu} = ((3),(3))$ and $(\square_{c,\underline{b}}, \underline{\nu}) = ((1,1,1),(1,1,1))$. The product that we expand is

$$\begin{aligned}
&\sigma_{(3)} \cdot \sigma_{(3)} \cdot \sigma_{(1,1,1)} \cdot \sigma_{(1,1,1)} \cdot \sigma_{\square} \cdot \sigma_{\square} \cdot \sigma_{\square} \cdot \sigma_{\square} \\
&\quad = 2\sigma_{(4,4,2,2)} \cdot \left(2\sigma_{(2,2)} + \sigma_{(1,1,1,1)} + 3\sigma_{(2,1,1)} + 3\sigma_{(3,1)} + \sigma_{(4)}\right) \\
&\quad = 2\sigma_{(4,4,2,2)} \cdot 2\sigma_{(2,2)} = 4\sigma_{\square_{4,4}}.
\end{aligned} \tag{10}$$

The first equality uses (8) to evaluate the product of the first four terms and expands the next four using the Pieri formula and (3). On the other hand,

$$\sigma_{(3)} \cdot \sigma_{\square} = \sigma_{(3,1)} + \sigma_{(4)} \qquad \text{and} \qquad \sigma_{(1,1,1)} \cdot \sigma_{\square} = \sigma_{(2,1,1)} + \sigma_{(1,1,1,1)}.$$

We may rearrange (10) to get

$$\left(\sigma_{(3)} \cdot \sigma_{\square}\right)^2 \cdot \left(\sigma_{(1,1,1)} \cdot \sigma_{\square}\right)^2 = \left(\sigma_{(3,1)} + \sigma_{(4)}\right)^2 \cdot \left(\sigma_{(2,1,1)} + \sigma_{(1,1,1,1)}\right)^2 = \sigma_{(3,1)}^2 \cdot \sigma_{(2,1,1)}^2,$$

as the other terms in the product vanish in the cohomology of $G(4,4)$. Since

$$\begin{aligned}
\sigma_{(3,1)}^2 &= \sigma_{(3,3,1,1)} + \sigma_{(4,2,2)} + \sigma_{(3,3,2)} + \sigma_{(4,2,1,1)} + \sigma_{(4,4)} + 2\sigma_{(4,3,1)}, \text{ and} \\
\sigma_{(2,1,1)}^2 &= \sigma_{(3,3,1,1)} + \sigma_{(4,2,2)} + \sigma_{(4,2,1,1)} + \sigma_{(3,3,2)} + \sigma_{(2,2,2,2)} + 2\sigma_{(3,2,2,1)},
\end{aligned}$$

if we use Lemma 3.1 to expand $\sigma_{(3,1)}^2 \cdot \sigma_{(2,1,1)}^2$, the only non-zero products in the expansion are of the form $\sigma_\lambda \cdot \sigma_{\lambda^\vee} = \sigma_{\square_{4,4}}$, where σ_λ occurs in $\sigma_{(3,1)}^2$ and $\sigma_{\lambda^\vee}$ occurs in $\sigma_{(2,1,1)}^2$. The only such pairs are each of the first four terms in the two expressions above. Thus we again see that (10) equals $4\sigma_{\square_{4,4}}$.

Proof of Theorem 3.9 We compute the product corresponding to the Schubert problem $(\square_{\underline{a},d} + \underline{\mu}, (\square_{c,\underline{b}}, \underline{\nu}), \underline{\rho})$ in two different ways. Using Lemma 3.11, in the cohomology of $G(a+c, b+d)$ we have

$$\prod_{i=1}^{r} \sigma_{\square_{a_i,d}+\mu^i} \cdot \prod_{j=1}^{s} \sigma_{(\square_{c,b_j}, \nu^j)} \cdot \prod_{\rho \in \underline{\rho}} \sigma_\rho = \delta(\underline{\lambda}) \sigma_{\square_{c,d}^\vee} \cdot \prod_{\rho \in \underline{\rho}} \sigma_\rho . \tag{11}$$

Since $\underline{\rho}$ is a Schubert problem on $G(c,d)$, the only partition $\kappa \subseteq \square_{c,d}$ with σ_κ appearing in the last product is $\square_{c,d}$ and the corresponding term is $\delta(\underline{\rho})\sigma_{\square_{c,d}}$. By Lemma 3.1, the product in (11) is $\delta(\underline{\lambda}) \cdot \delta(\underline{\rho}) \cdot \sigma_{\square_{a+c,b+d}}$.

As $\underline{\rho} = (\underline{\alpha}, \underline{\beta}, \underline{\gamma})$, we may rearrange the product in (11) to obtain

$$\left(\prod_{i=1}^{r} \sigma_{\square_{a_i,d}+\mu^i} \cdot \sigma_{\alpha^i}\right) \cdot \left(\prod_{j=1}^{s} \sigma_{(\square_{c,b_j},\nu^j)} \cdot \sigma_{\beta^j}\right) \cdot \prod_{k=1}^{t} \sigma_{\gamma^k} . \tag{12}$$

By Lemma 2.3, each term in the first product expands as

$$\sigma_{(\square_{a_i,d}+\mu^i,\alpha^i)} + \sum_{\kappa} c^{\kappa}_{\square_{a_i,d}+\mu^i,\alpha^i} \sigma_{\kappa} , \tag{13}$$

where if κ indexes a nonzero term in the sum, then $|\kappa|_{a_i} > a_i d + |\mu^i|$.

Expanding the product of these expressions gives a sum of products of the form

$$\left(\prod_{i=1}^{r} c^{\kappa^i}_{\square_{a_i,d}+\mu^i,\alpha^i}\right) \sigma_{\kappa^1}\sigma_{\kappa^2}\cdots\sigma_{\kappa^r}, \tag{14}$$

where for each i, $|\kappa^i|_{a_i} \geq a_i d + |\mu^i|$, and we have equality only when $\kappa^i = (\square_{a_i,d} + \mu^i, \alpha^i)$. Suppose that a Schubert class σ_ρ occurs in the expansion of such a product (14). Then by Corollary 2.2, $|\rho|_a \geq ad + |\underline{\mu}|$. If $|\rho|_a = ad + |\mu|$, then $\kappa^i = (\square_{a_i,d} + \mu^i, \alpha^i)$ for each i, and so σ_ρ occurs in the product $\prod_i \sigma_{(\square_{a_i,d}+\mu^i,\alpha^i)}$ of the first terms from (13) and the other terms do not contribute.

Applying the fundamental involution ω, each term in the second product may be expanded to give

$$\sigma_{(\square_{c,b_j},\nu^j)+\beta^j} + \sum_{\kappa} c^{\kappa}_{(\square_{c,b_j},\nu^j),\beta^j} \sigma_{\kappa} , \tag{15}$$

where if κ indexes a term in the sum, then $|\kappa'|_{b_j} > cb_j + |\nu^j|$. The same reasoning shows that if σ_τ appears in the expansion of the second product in (12), then $|\tau'|_b \geq cb + |\underline{\nu}|$. Furthermore, if $|\tau'|_b = cb + |\underline{\nu}|$, then σ_τ occurs in the product $\prod_j \sigma_{(\square_{c,b_j},\nu^j)+\beta^j}$ of the first terms in (15) and the other terms do not contribute.

Now suppose that σ_ρ occurs in the expansion of the first product and σ_τ in the second. By Corollary 3.4, if $|\rho|_a + |\tau'|_b > ad + cb + ab = ad + cb + |\underline{\mu}| + |\underline{\nu}|$, then the product $\sigma_\rho \cdot \sigma_\tau = 0$ in the cohomology ring of $G(a+c, b+d)$. Thus if $\sigma_\rho \cdot \sigma_\tau \neq 0$, then $|\rho|_a = ad + |\mu|$ and $|\tau'|_b = cb + |\underline{\nu}|$. These arguments and observations imply the identity in the cohomology ring of $G(a+c, b+d)$,

$$\left(\prod_{i=1}^{r} \sigma_{\square_{a_i,d}+\mu^i} \cdot \sigma_{\alpha^i}\right) \cdot \left(\prod_{j=1}^{s} \sigma_{(\square_{c,b_j},\nu^j)} \cdot \sigma_{\beta^j}\right)$$
$$= \prod_{i=1}^{r} \sigma_{(\square_{a_i,d}+\mu^i,\alpha^i)} \cdot \prod_{j=1}^{s} \sigma_{(\square_{c,b_j},\nu^j)+\beta^j} .$$

Thus in the cohomology ring of $G(a+c, b+d)$ the product (11) equals

$$\prod_{i=1}^{r} \sigma_{(\Box_{a_i,d}+\mu^i,\alpha^i)} \cdot \prod_{j=1}^{s} \sigma_{(\Box_{c,b_j},\nu^j)+\beta^j} \cdot \prod_{k=1}^{t} \sigma_{\gamma^k}\,, \tag{16}$$

which is the product of Schubert classes in the composition $\underline{\lambda} \circ \underline{\rho}$. Thus the two Schubert problems $\underline{\kappa} = (\Box_{\underline{a},d} + \underline{\mu}, (\Box_{c,\underline{b}}, \underline{\nu}), \underline{\rho})$ (11) and $\underline{\lambda} \circ \underline{\rho}$ (16) have the same number of solutions, so that $\delta(\underline{\lambda}) \cdot \delta(\underline{\rho}) = \delta(\underline{\kappa}) = \delta(\underline{\lambda} \circ \underline{\rho})$, which completes the proof. □

Remark 3.13 Under the bijection of (2) involving frozen rectangles, and its conjugate version, we get a bijection between generalized Littlewood–Richardson tableaux of shape $\Box_{a+c,b+d}$ and content $\underline{\lambda} \circ \underline{\rho}$ and pairs (S,T) of generalized Littlewood–Richardson tableaux, where S has shape $\Box_{a,b}$ and content $\underline{\lambda}$ and T has shape $\Box_{c,d}$ and content $\underline{\rho}$. One of us (Sottile) observed such a bijection in 2012, and this led eventually to the notion of composition of Schubert problem. Let us illustrate this in the product in Example 3.12 for the composed Schubert problem $\underline{\lambda} \circ \underline{\lambda}$, where $\underline{\lambda} = (\Box, \Box, \Box, \Box)$.

We first shade the 'frozen rectangles' in the partitions encoding terms of (8), filling in the remaining boxes with $1, 2, 3, 4$, one number for each partition (▭▭▭, ▭▭▭, ▯, ▯) indexing the product, and omitting terms that do not contribute. We write the product schematically as sums of indexing partitions.

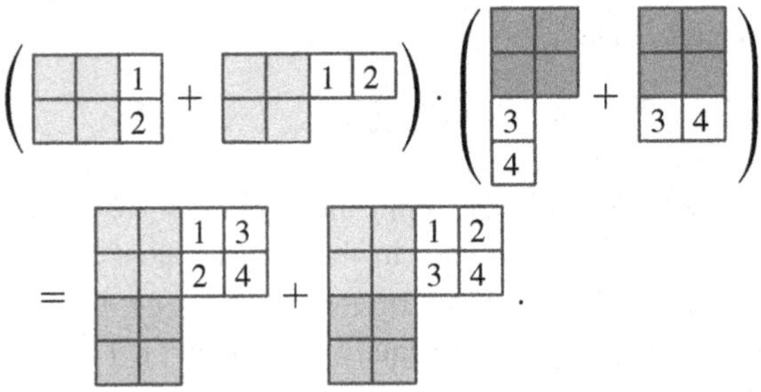

We similarly expand $\sigma_{\Box}^{4}$ using the letters $\alpha, \beta, \gamma, \delta$ instead of numbers to get

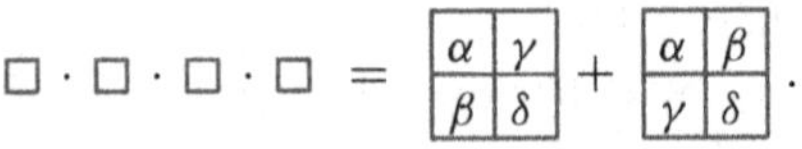

$$\Box \cdot \Box \cdot \Box \cdot \Box =$$

Similarly, if we expand the product of the eight terms coming from $\underline{\lambda} \circ \underline{\lambda}$, and express it schematically, we get

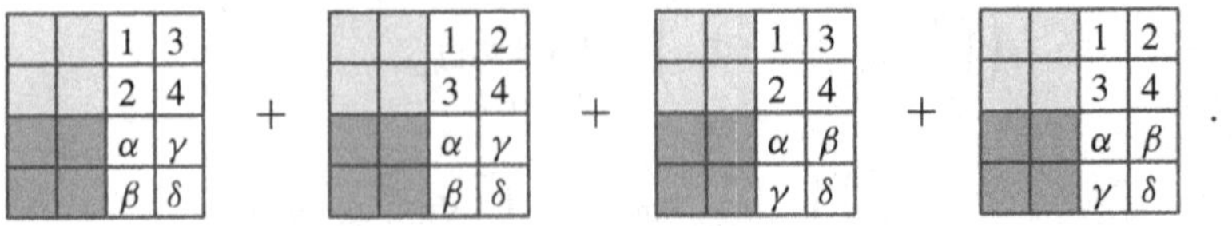

4 Galois Groups of Composed Schubert Problems

Given a composable Schubert problem $\underline{\lambda}$ and any other Schubert problem $\underline{\rho}$, Theorem 3.9 shows that $\delta(\underline{\lambda} \circ \underline{\rho}) = \delta(\underline{\lambda}) \cdot \delta(\underline{\rho})$. It was motivated by results in [13, Sec. 3.1], which we may now interpret as geometric proofs of this product identity, for the nontrivial composable Schubert problems $\underline{\lambda}$ on $G(2,2)$ and on $G(2,3)$. A consequence of those geometric results is that the Galois group of such a composition is imprimitive, specifically, it is a subgroup of a wreath product,

$$\mathrm{Gal}_{\underline{\lambda}\circ\underline{\rho}} \subset \mathrm{Gal}_{\underline{\rho}} \wr \mathrm{Gal}_{\underline{\lambda}} := (\mathrm{Gal}_{\underline{\rho}})^{\delta(\underline{\lambda})} \rtimes \mathrm{Gal}_{\underline{\lambda}} . \tag{17}$$

Further lemmas in [13, Sec. 3.1] established equality for $\underline{\lambda}$ a composable Schubert problem on $G(2,2)$ or $G(2,3)$.

We discuss Galois groups of Schubert problems, and following [13, Sec. 3] identify a structure (a fibration of Schubert problems) which implies the imprimitivity of a Galois group. We then describe a class of Schubert problems all of whose compositions are fibered, and close with computational evidence that more general compositions of Schubert problems have imprimitive Galois groups.

4.1 Galois groups and fibrations of Schubert problems

Let $\underline{\lambda} = (\lambda^1, \ldots, \lambda^r)$ be a Schubert problem on $G(a,b)$ with $\delta(\underline{\lambda})$ solutions. We write $\mathbb{F}\ell_{a+b}$ for the manifold of complete flags in $\mathbb{C}^{a+b}$, $\mathcal{F}_\bullet := (F^1_\bullet, \ldots, F^r_\bullet) \in (\mathbb{F}\ell_{a+b})^r$ for a list of flags, and

$$\Omega_{\underline{\lambda}}\mathcal{F}_\bullet := \Omega_{\lambda^1}F^1_\bullet \cap \Omega_{\lambda^2}F^2_\bullet \cap \cdots \cap \Omega_{\lambda^r}F^r_\bullet,$$

for the instance (5) of the Schubert problem $\underline{\lambda}$ given by $\mathcal{F}_\bullet$. Consider the total space of the Schubert problem $\underline{\lambda}$,

$$\Omega_{\underline{\lambda}} := \{(H, \mathcal{F}_\bullet) \in G(a,b) \times (\mathbb{F}\ell(a+b))^r \mid H \in \Omega_{\underline{\lambda}}\mathcal{F}_\bullet\}.$$

This is irreducible, as the fiber of the projection $\Omega_{\underline{\lambda}} \to G(a,b)$ over a point $H \in G(a,b)$ is a product $Z_{\lambda^1}H \times \cdots \times Z_{\lambda^r}H$, where

$$Z_\lambda H := \{F_\bullet \mid H \in \Omega_\lambda F_\bullet\}$$

is a Schubert variety of the flag variety $\mathbb{F}\ell(a+b)$. Since each Schubert variety is irreducible, $\Omega_{\underline{\lambda}}$ is irreducible as it is fibered over an irreducible variety $G(a,b)$ with irreducible fibers.

The fiber $\pi^{-1}(\mathcal{F}_\bullet)$ of the projection $\pi : \Omega_{\underline{\lambda}} \to (\mathbb{F}\ell(a+b))^r$ is the instance $\Omega_{\underline{\lambda}}\mathcal{F}_\bullet$ of the Schubert problem $\underline{\lambda}$ given by the flags $\mathcal{F}_\bullet$. When the flags are general, this is a transverse intersection and consists of $\delta(\underline{\lambda})$ points, and therefore π is a branched cover of degree $\delta(\underline{\lambda})$. The projection map π induces an inclusion of function fields $\pi^*\colon \mathbb{C}\big((\mathbb{F}\ell_{a+b})^r\big) \hookrightarrow \mathbb{C}(\Omega_{\underline{\lambda}})$, with the extension of degree $\delta(\underline{\lambda})$. The *Galois group* $\mathrm{Gal}_{\underline{\lambda}}$ of $\underline{\lambda}$ is the Galois group of the Galois closure of this extension.

Such *Schubert Galois groups* are beginning to be studied [3, 10, 12, 13, 16, 19], and while $\mathrm{Gal}_{\underline{\lambda}}$ is typically the full symmetric group $S_{\delta(\underline{\lambda})}$, it is common for $\mathrm{Gal}_{\underline{\lambda}}$ to be imprimitive. Let us recall some theory of permutation groups, for more, see [20]. A permutation group is a subgroup G of some symmetric group S_δ, so that G has a faithful action on $[\delta] := \{1,\ldots,\delta\}$. The group G is *transitive* if it has a single orbit on $[\delta]$. Galois groups are transitive permutation groups. A *block* of a permutation group G is a subset B of $[\delta]$ such that for every $g \in G$ either $gB \cap B = \emptyset$ or $gB = B$. The orbit of a block under G generates a partition of $[\delta]$ into blocks. The group G is primitive if its only blocks are singletons or $[\delta]$ itself, and *imprimitive* otherwise.

When G is imprimitive, there is a factorization $\delta = p \cdot q$ and an identification $[\delta] = [p] \times [q]$ with $p,q > 1$. If we let $\pi\colon [p] \times [q] \to [p]$ be the projection onto the first factor, then the fibers are blocks of G and the action of G on $[p] \times [q]$ preserves this fibration. Conversely, if G preserves such a nontrivial fibration, then it acts imprimitively.

A Schubert Galois group is imprimitive if its Schubert problem forms a fiber bundle whose base and fibers are nontrivial Schubert problems in smaller Grassmannians. Such a structure is called a decomposable projection in [1], which is equivalent to the Galois group being imprimitive [4, 14]. The existence of such a structure implies that $\mathrm{Gal}_{\underline{\lambda}}$ is a subgroup of the wreath product of the Galois groups of the two smaller Schubert problems. We give the definition of a fibration of Schubert problems from [13].

Definition 4.1 Let $\underline{\kappa}$, $\underline{\lambda}$, and $\underline{\rho}$ be nontrivial Schubert problems on $G(a{+}c,b{+}d)$, $G(a,b)$, and $G(c,d)$, respectively. We say that *$\underline{\kappa}$ is fibered over $\underline{\lambda}$ with fiber $\underline{\rho}$* if the following hold.

1. For every general instance $\mathcal{F}_\bullet \in (\mathbb{F}\ell_{a+c+b+d})^r$ of $\underline{\kappa}$, there is a subspace $V \subset \mathbb{C}^{a+c+b+d}$ of dimension $a{+}b$ and an instance $\mathcal{E}_\bullet$ of $\underline{\lambda}$ in $Gr(a,V)$ such that for every $H \in \Omega_{\underline{\kappa}}\mathcal{F}_\bullet$, we have $H \cap V \in \Omega_{\underline{\lambda}}\mathcal{E}_\bullet$.
2. If we set $W := \mathbb{C}^{a+c+b+d}/V$, then for any $h \in \Omega_{\underline{\lambda}}\mathcal{E}_\bullet$, there is an instance $\mathcal{F}_\bullet(h)$ of $\underline{\rho}$ in $Gr(c,W)$ such that if $h := H \cap V$, then $H/h \in Gr(c,W)$ is a solution to $\Omega_{\underline{\rho}}\mathcal{F}_\bullet(h)$.
3. The association $H \mapsto (h, H/h)$, where $h := H \cap V$, is a bijection between the sets of solutions $\Omega_{\underline{\kappa}}\mathcal{F}_\bullet$ and $\Omega_{\underline{\lambda}}\mathcal{E}_\bullet \times \Omega_{\underline{\rho}}\mathcal{F}_\bullet(h)$.

4. For a given $V \simeq \mathbb{C}^{a+b}$, all general instances $\mathcal{E}_\bullet$ of $\underline{\lambda}$ on $Gr(a,V)$ may be obtained in (1). For a given general instance $\mathcal{E}_\bullet$ of $\underline{\lambda}$ and $h \in \Omega_{\underline{\lambda}}\mathcal{E}_\bullet$, the instances $\mathcal{F}_\bullet(h)$ of $\underline{\rho}$ which arise are also general.

This is called a fibration as the second instance $\mathcal{F}_\bullet(h)$ depends upon h.

The following consequence of a fibration is Lemma 15 of [13].

Lemma 4.2 *If $\underline{\kappa}$ is a Schubert problem fibered over $\underline{\lambda}$ with fiber $\underline{\rho}$, then $\delta(\underline{\kappa}) = \delta(\underline{\lambda}) \cdot \delta(\underline{\rho})$ and $\mathrm{Gal}_{\underline{\kappa}}$ is a subgroup of the wreath product $\mathrm{Gal}_{\underline{\lambda}} \wr \mathrm{Gal}_{\underline{\rho}}$. If $\underline{\lambda}$ and $\underline{\rho}$ are nontrivial, then $\mathrm{Gal}_{\underline{\kappa}}$ is imprimitive. Also, the image of $\mathrm{Gal}_{\underline{\kappa}}$ under the map to $\mathrm{Gal}_{\underline{\lambda}}$ is surjective, and the kernel, which is a subgroup of $(\mathrm{Gal}_{\underline{\rho}})^{\delta(\underline{\lambda})}$, has image $\mathrm{Gal}_{\underline{\rho}}$ under projection to any factor.*

We make the following two conjectures.

Conjecture 4.3 *The Galois group $\mathrm{Gal}_{\underline{\lambda}\circ\underline{\rho}}$ of a composition of non-trivial Schubert problems is imprimitive.*

By Lemma 4.2, this is implied by a second, stronger conjecture.

Conjecture 4.4 *A composed Schubert problem $\underline{\lambda} \circ \underline{\rho}$ is fibered over $\underline{\lambda}$ with fiber $\underline{\rho}$.*

We prove Conjecture 4.4 for a class of composable Schubert problems in the next section, and provide computational evidence for Conjecture 4.3 in Section 4.4.

4.2 Block column Schubert problems

We identify a family of composable Schubert problems that we call block column Schubert problems. This includes all composable Schubert problems on $G(2,2)$ and $G(2,3)$ that were studied in [13, Sect. 3.1]. We show that for any block column Schubert problem $\underline{\lambda}$ and any Schubert problem $\underline{\rho}$, the composition $\underline{\lambda} \circ \underline{\rho}$ is a Schubert problem that is fibered over $\underline{\lambda}$ with fiber ρ. This proves that the Galois group of the composition $\underline{\lambda} \circ \underline{\rho}$ is imprimitive when both $\underline{\lambda}$ and $\underline{\rho}$ are nontrivial.

Definition 4.5 A *block column Schubert problem* on $G(a,b)$ is a Schubert problem $\underline{\lambda} = (\underline{\mu}, \underline{\nu})$ of the following form. The first part $\underline{\mu}$ consists of two rectangular partitions, $\underline{\mu} = (\square_{a_1,b_1}, \square_{a_2,b_2})$, where $a = a_1 + a_2$ and $b = b_1 + b_2$. The second part $\underline{\nu}$ consists of two families of rectangular partitions, $\underline{\nu} = (\square_{a_2,m_1}, \ldots, \square_{a_2,m_p}, \square_{a_1,n_1}, \ldots, \square_{a_1,n_q})$, where $m_1 + \cdots + m_p = b_1$ and $n_1 + \cdots + n_q = b_2$.

Lemma 4.6 *A block column Schubert problem is composable.*

Proof Let $\underline{\lambda} = (\underline{\mu}, \underline{\nu})$ be a block column Schubert problem. Let $\underline{a} = (a_1, a_2)$ and $\underline{b} = (m_1, \dots, m_p, n_1, \dots, n_q)$. Since these are the numbers of rows in the partitions of $\underline{\mu}$ and columns in the partitions of $\underline{\nu}$, respectively, we need only check their sums. Noting that $a_1 + a_2 = a$ and

$$m_1 + \cdots + m_p \;+\; n_1 + \cdots + n_q \;=\; b_1 + b_2 \;=\; b,$$

completes the proof. $\square$

Let us consider some examples of block column Schubert problems. One straightforward class is when $a_1 = a_2 = a$ and $b_1 = b_2 = b$ and $\underline{\lambda} = (\underline{\mu}, \underline{\nu})$ with $\underline{\mu} = \underline{\nu} = (\square_{a,b}, \square_{a,b})$, which is a Schubert problem on $G(2a, 2b)$. This class includes the Schubert problem $(\square, \square\,,\, \square, \square)$ on $G(2,2)$. Two more are

$$(\ydiagram{2}, \ydiagram{2}\,,\, \ydiagram{2}, \ydiagram{2}) \text{ on } G(2,4) \quad \text{and} \quad (\ydiagram{2,2}, \ydiagram{2,2}\,,\, \ydiagram{2,2}, \ydiagram{2,2}) \text{ on } G(4,4).$$

Two others are $(\ydiagram{2}, \square\,,\, \square, \square, \square)$ and $(\ydiagram{2}, \square\,,\, \ydiagram{2}, \square)$ on $G(2,3)$; these are two of the three composable Schubert problems on $G(2,3)$. The third, $(\ydiagram{2,1}, \square, \square, \square)$, is geometrically equivalent to $(\square, \square\,,\, \square, \square)$ on $G(2,2)$, as is $(\ydiagram{2}, \square\,,\, \ydiagram{2}, \square)$. A more interesting example is on $G(5,7)$ where $a_1 = 2$, $a_2 = 3$, $b_1 = 4$, and $b_2 = 3$, with $\underline{\mu} = (\square_{2,4}, \square_{3,3})$ and $\underline{\nu} = (\square_{3,2}, \square_{3,1}, \square_{3,1}\,,\, \square_{2,2}, \square_{2,1})$, which are the following partitions

$$\ydiagram{4,4}\,,\ \ydiagram{3,3,3}\ ,\ \ydiagram{2,2,2}\,,\ \ydiagram{1,1,1}\,,\ \ydiagram{1,1,1}\ ,\ \ydiagram{2,2}\,,\ \ydiagram{1,1}\,.$$

Not all composable Schubert problems are block column. The first Schubert problem of Example 3.5, $\underline{\mu} = ((p), (q))$ and $\underline{\nu} = (\square, \dots, \square)$ ($p+q$ occurrences of $\square$), is block column, but the remaining two are not.

To understand the structure of these Schubert problems, let $\underline{\lambda} = (\underline{\mu}, \underline{\nu})$ be a block column Schubert problem as in Definition 4.5. Observe that if we place the two rectangles in $\underline{\mu}$ in opposite corners of the rectangle $\square_{a,b}$ as $\square_{a_1,b_1}$ and $(\square_{a_2,b_2})^\circ$, then they lie along the main diagonal, meeting at their corners. There are two remaining rectangles $\square_{a_2,b_1}$ and $\square_{a_1,b_2}$ along the antidiagonal. As $m_1 + \cdots + m_p = b_1$, the first block $(\square_{a_2,m_1}, \dots, \square_{a_2,m_p})$ of $\underline{\nu}$ fills the first rectangle, with each partition spanning all a_2 rows, and the second block $(\square_{a_1,n_1}, \dots, \square_{a_1,n_q})$ of $\underline{\nu}$ similarly fills the second rectangle. We show this for the block column Schubert problems given above. In each, the two partitions in $\underline{\mu}$ are shaded.

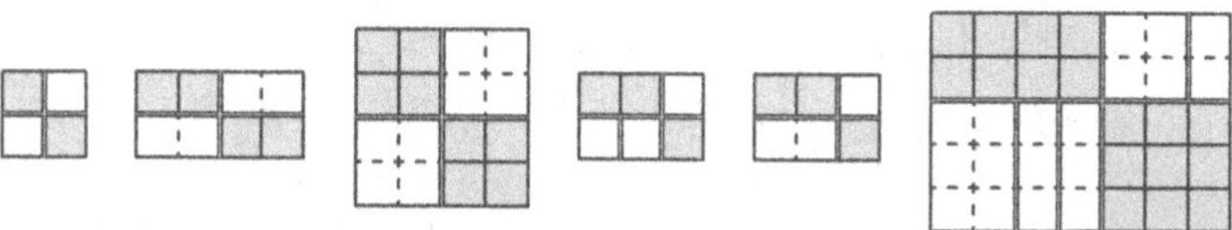

Lemma 4.7 *Let $\underline{\lambda} = (\underline{\mu}, \underline{\nu})$ be a block column Schubert problem on $G(a,b)$ with parameters $a_1, a_2, b_1, b_2, m_1, \ldots, m_p, n_1, \ldots, n_q$ as in Definition 4.5. Let $\mathcal{F}_\bullet = (E^1_\bullet, E^2_\bullet, F^1_\bullet, \ldots, F^p_\bullet, G^1_\bullet, \ldots, G^q_\bullet)$ be $2 + p + q$ general flags. The corresponding instance $\Omega_{\underline{\lambda}}\mathcal{F}_\bullet$ of $\underline{\lambda}$ consists of the set*

$$\begin{aligned}\{H \in G(a,b) \mid\ & \dim H \cap E^1_{a_1+b_2} \geq a_1\,,\ \dim H \cap E^2_{a_2+b_1} \geq a_2\,,\\ & \dim H \cap F^i_{b+a_2-m_i} \geq a_2\,,\ \dim H \cap G^j_{b+a_1-n_j} \geq a_1\\ & \textit{for } i \in [p] \textit{ and } j \in [q]\,\}.\end{aligned} \tag{18}$$

Proof The partition $\square_{c,d}$ has c parts, each of size d. Thus if $F_\bullet$ is a flag, then by (4),

$$\Omega_{\square_{c,d}} F_\bullet \;=\; \{H \in G(a,b) \mid \dim H \cap F_{b+i-d} \geq i \ \text{ for } i = 1, \ldots, c\}.$$

The conditions on H are implied by $\dim H \cap F_{b+c-d} \geq c$. The description (18) follows. □

Theorem 4.8 *Let $\underline{\lambda}$ be a block column Schubert problem on $G(a,b)$ and $\underline{\rho}$ a Schubert problem on $G(c,d)$. Then the composed Schubert problem $\underline{\lambda} \circ \underline{\rho}$ on $G(a{+}c, b{+}d)$ is fibered over $\underline{\lambda}$ with fiber $\underline{\rho}$.*

Corollary 4.9 *Suppose that $\underline{\lambda}$ is a nontrivial block column Schubert problem on $G(a,b)$ and $\underline{\rho}$ is a nontrivial Schubert problem on $G(c,d)$. Then the Galois group $\mathrm{Gal}_{\underline{\lambda}\circ\underline{\rho}}$ is imprimitive.*

Remark 4.10 Conjugating every partition in a block column Schubert problem gives a *block row Schubert problem.* A block row Schubert problem is composable, and the conclusions of Theorem 4.8 and Corollary 4.9 hold for block row Schubert problems. The reason for this is that the isomorphism between $G(a,b)$ and $G(b,a)$ induced by duality corresponds to conjugation $\lambda \mapsto \lambda'$ of Young diagrams.

4.3 Proof of Theorem 4.8

Let $\underline{\lambda} = (\underline{\mu}, \underline{\nu})$ be a block column Schubert problem on $G(a,b)$ with the parameters $a_1, a_2, b_1, b_2, m_1, \ldots, m_p, n_1, \ldots, n_q$ as in Definition 4.5. Let $\underline{\rho} = (\underline{\alpha}, \underline{\beta}, \underline{\gamma})$ be a Schubert problem on $G(c,d)$ with parameters $r = 2$, $s = p{+}q$, and t that we may compose with $\underline{\lambda}$. Then the composition $\underline{\lambda} \circ \underline{\rho}$ consists of the partitions

$$(\square_{a_1,(d+b_1)},\alpha^1),\ (\square_{a_2,(d+b_2)},\alpha^2),\ \square_{(c+a_2),m_1}+\beta^1,\ \ldots,$$
$$\square_{(c+a_2),m_p}+\beta^p,\ \square_{(c+a_1),n_1}+\beta^{p+1},\ \ldots,\ \square_{(c+a_1),n_q}+\beta^{p+q},\ \gamma^1,\ldots,\gamma^t.$$

Here is a schematic illustrating the partitions in the first three groups

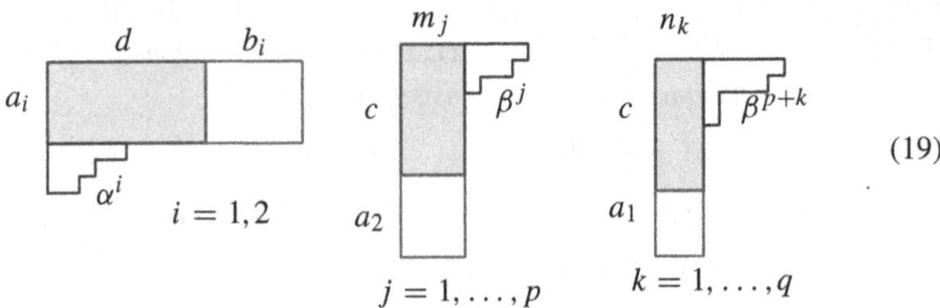

(19)

Let $\mathcal{F}_\bullet = (F^1_\bullet, F^2_\bullet, K^1_\bullet, \ldots, K^p_\bullet, K^{p+1}_\bullet, \ldots, K^{p+q}_\bullet, L^1_\bullet, \ldots, L^t_\bullet)$ be general flags in $\mathbb{C}^{a+b+c+d}$. We will use that they are general without comment in making assertions about the dimensions of intersections and spans. Generality also implies that the dimension inequalities in the definition of Schubert variety (4) all hold with equality, which we also use.

By Definition 4.1, to show that $\underline{\lambda} \circ \underline{\rho}$ is fibered over $\underline{\lambda}$, we must do the following.

1. Construct a linear subspace $V \simeq \mathbb{C}^{a+b}$ and an instance $\mathcal{E}_\bullet$ in V of $\underline{\lambda}$ such that if $H \in \Omega_{\underline{\lambda}\circ\underline{\rho}}\mathcal{F}_\bullet$, then $h := H \cap V$ is an element of $\Omega_{\underline{\lambda}}\mathcal{E}_\bullet$.
2. Let $W := \mathbb{C}^{a+c+b+d}/V$. Then, for any $h \in \Omega_{\underline{\lambda}}\mathcal{E}_\bullet$, construct an instance $\mathcal{F}_\bullet(h)$ of $\underline{\rho}$ in $Gr(c, W)$ such that if $h := H \cap V$, then $H/h \in \Omega_{\underline{\rho}}\mathcal{F}_\bullet(h)$.
3. Prove that the map $H \mapsto (h, H/h)$, where $h := H \cap V$, is a bijection between $\Omega_{\underline{\kappa}}\mathcal{F}_\bullet$ and $\Omega_{\underline{\lambda}}\mathcal{E}_\bullet \times \Omega_{\underline{\rho}}\mathcal{F}_\bullet(h)$ (actually this is a fiber bundle over $\Omega_{\underline{\lambda}}$.)
4. Show that the instances $\mathcal{E}_\bullet$ and $\mathcal{F}_\bullet(h)$ are sufficiently general.

We prove Steps (1)–(4) in separate headings below.

Remark 4.11 A (partial) flag $F_\bullet$ in $\mathbb{C}^n$ is a nested sequence of subspaces, where not all dimensions need occur. For any subspace $h \subset \mathbb{C}^n$, the sequence of subspaces $h + F_i$ for $F_i \in F_\bullet$ forms another flag $h+F_\bullet$ in $\mathbb{C}^n$ with smallest subspace h. If $V \subset \mathbb{C}^n$ is a subspace, then the subspaces $V \cap F_i$ form a flag $V \cap F_\bullet$ in V. If $\mathbb{C}^n \twoheadrightarrow V$ is surjective, then the image of $F_\bullet$ is a flag in V.

Step 1. Set $V := F^1_{b_2+a_1} + F^2_{b_1+a_2}$. As the flags are general, this is a direct sum and V has dimension $a_1+a_2+b_1+b_2 = a+b$. Define $\mathcal{E}_\bullet$ by $E^i_\bullet := F^i_\bullet \cap V$ for $i = 1,2$ and $E^{2+j}_\bullet := K^j_\bullet \cap V$ for $j = 1, \ldots, p+q$. Fixing $F^1_{b_2+a_1}$ and $F^2_{b_1+a_2}$

and thus V, every possible collection of flags $\mathcal{E}_\bullet$ in V can occur, which proves part of Assertion (4) from Definition 4.1.

Let $H \in \Omega_{\underline{\lambda}\circ\underline{\rho}}\mathcal{F}_\bullet$. Since $H \in \Omega_{(\Box_{a_i,(d+b_i)},\alpha^1)}F^i_\bullet$, for $i = 1,2$, it satisfies

$$\dim H \cap F^1_{d+b-(d+b_1)+a_1} = a_1 \text{ and } \dim H \cap F^2_{d+b-(d+b_2)+a_2} = a_2.$$

The subspaces here are just $F_{b_2+a_1}$ and $F_{b_1+a_2}$. Thus the dimension of $h := H \cap V$ is $a = a_1+a_2$. This implies that $h \in Gr(a,V)$, and that $h \in \Omega_{\Box_{a_i,b_i}}E^i_\bullet$ for $i = 1,2$.

We show that $h \in \Omega_{\underline{\lambda}}\mathcal{E}_\bullet$. First let $1 \le j \le p$. Since H lies in the Schubert variety $\Omega_{\Box_{(c+a_2),m_j}+\beta^j}K^j_\bullet$, we have that

$$\dim H \cap K^j_{b+d-m_j+c+a_2} = c + a_2.$$

Since V has codimension $c+d$, for any i we have $\dim V \cap K^j_{c+d+i} = i$. Thus $E^j_{b-m_j+a_2} = V \cap K^j_{b+d-m_j+c+a_2}$. Since h has codimension c in H, $\dim h \cap K^j_{b+d-m_j+c+a_2} = a_2$. Putting these dimension calculations together shows that

$$\dim h \cap E^j_{b-m_j+a_2} = a_2,$$

and thus $h \in \Omega_{\Box_{a_2,m_j}}E^j_\bullet$. Similar arguments for $k = 1,\dots,q$ show that $h \in \Omega_{\Box_{a_1,n_k}}E^{p+k}_\bullet$, and thus $h \in \Omega_{\underline{\lambda}}\mathcal{E}_\bullet$. This completes the proof of Assertion (1) in Definition 4.1.

Step 2. Let $h \in \Omega_{\underline{\lambda}}\mathcal{E}_\bullet$. Since $\dim h \cap K^j_{b+d+c+a_2-m_j} = a_2$ and $\dim h = a$, we have that

$$\dim\bigl(h + K^j_{b+d+c+a_2-m_j}\bigr) = a+b+c+d-m_j,$$

That is, it has codimension m_j. Similarly $h+K^{p+k}_{b+d+c+a_1-n_k}$ has codimension n_k. Since the sum of the m_j and of the n_k is b, the intersection $W(h)$ of these spaces,

$$\left(\bigcap_{j=1}^{p}\left(h + K^j_{b+d+c+a_2-m_j}\right)\right) \cap \left(\bigcap_{k=1}^{q}\left(h + K^{p+k}_{b+d+c+a_1-n_k}\right)\right), \tag{20}$$

has codimension b. Note that $W(h)$ contains h and $W(h)/h \simeq W = \mathbb{C}^{a+b+c+d}/V$. For any flag $F_\bullet$ in $\mathbb{C}^{a+b+c+d}$, let $F_\bullet(h)$ be the image of the flag $h + F_\bullet$ in W. We claim that $H \in \Omega_{\underline{\lambda}\circ\underline{\rho}}\mathcal{F}_\bullet$, and $h = H \cap V$, then $H/h \in \Omega_{\underline{\rho}}\mathcal{F}_\bullet(h)$ in $Gr(c, W(h)/h)$.

We show this for the first two conditions $\underline{\alpha} = (\alpha^1, \alpha^2)$ in $\underline{\rho}$. The first two conditions $(\Box_{a_i, d+b_i}, \alpha^i)$ of $\underline{\lambda} \circ \underline{\rho}$ are depicted on the left of (19). As $H \in \Omega_{(\Box_{a_1,d+b_1},\alpha^1)} F^1_\bullet$, we have,

$$\dim H \cap F^1_{b_2+a_1} \;=\; a_1 \quad \text{and}$$
$$\dim H \cap F^1_{d+b+a_1+j-\alpha^1_j} \;=\; a_1 + j \;\text{ for } j = 1, \ldots, c.$$

(The first is from Step 1.) Since $H \cap F^1_{b_2+a_1} = h \cap F^1_{b_2+a_1}$, we have

$$\dim H \cap (h + F^1_{d+b+a_1+j-\alpha^1_j}) \;=\; a + j \;\text{ for } j = 1, \ldots, c.$$

Then $\dim(h + F^1_{d+b+a_1+j-\alpha^1_j}) = d + b + a + j - \alpha^1_j$, so that its image in $W(h)/h$ has dimension $d + j - \alpha^1_j$, and thus

$$\dim\bigl(H/h \cap F^1(h)_{d+j-\alpha^1_j}\bigr) \;=\; j \;\text{ for } j = 1, \ldots, c,$$

which shows that $H/h \in \Omega_{\alpha^1} F^1_\bullet(h)$. The same arguments show that $H/h \in \Omega_{\alpha^2} F^2_\bullet(h)$.

We now consider the next $p+q$ conditions $\underline{\beta}$ in $\underline{\rho}$. Suppose that $1 \le j \le p$. As $H \in \Omega_{\Box_{c+a_2,m_j}+\beta^j} K^j_\bullet$, for each $i = 1, \ldots, c$, we have

$$\dim H \cap K^j_{b+d+i-m_j-\beta^j_i} \;=\; i.$$

Since V is in general position with respect to $K^j_\bullet$, $h \cap K^j_{b+d+i-m_j-\beta^j_i} = \{0\}$, so that

$$\dim\bigl(h + K^j_{b+d+i-m_j-\beta^j_i}\bigr) \;=\; a + b + d + i - m_j - \beta^j_i.$$

Since this is a subspace of $h + K^j_{b+d+c+a_2-m_j}$, which has codimension m_j and is one of the subspaces in the intersection (20) defining $W(h)$, we see that

$$\dim\Bigl(W(h) \;\cap\; \bigl(h + K^j_{b+d+i-m_j-\beta^j_i}\bigr)\Bigr) \;=\; a + d + i - \beta^j_i.$$

Thus the image of $h + K^j_{b+d+i-m_j-\beta^j_i}$ in $W(h)/h$ has dimension $d + i - \beta^j_i$, so that it is the subspace $K^j(h)_{d+i-\beta^j_i}$. We conclude that for $i = 1, \ldots, c$,

$$\dim\bigl(H/h \cap K^j(h)_{d+i-\beta^j_i}\bigr) \;=\; i.$$

Thus for $j = 1, \ldots, p$, we have $H/h \in \Omega_{\beta^j} K^j_\bullet(h)$. Similar arguments prove that $H/h \in \Omega_{\beta^j} K^j_\bullet(h)$ for $j = p + 1, \ldots, p + q$.

To complete this step, we show that $H/h \in \Omega_{\gamma^j} L^j_\bullet(h)$ for each $j = 1,\ldots,t$. Let $1 \le j \le t$. Then $H \in \Omega_{\gamma^j} L^j_\bullet$ and for each $i = 1,\ldots,c$, we have $\dim H \cap L^j_{b+d+i-\gamma^j_i} = i$. Again by the general position of V, we have that $\dim(h + L^j_{b+d+i-\gamma^j_i}) = a+b+d+i-\gamma^j_i$. Then the intersection of this subspace with $W(h)$ has dimension $a+d+i-\gamma^j_i$, so that its image in $W(h)/h$ has dimension $d+i-\gamma^j_i$, and is the subspace $L^j(h)_{d+i-\gamma^j_i}$ in the flag $L^j_\bullet(h)$. Almost as before, this implies that

$$\dim\bigl(H/h \cap L^j(h)_{d+i-\gamma^j_i}\bigr) = i \quad \text{for } i = 1,\ldots,c.$$

Thus we conclude that for $j = 1,\ldots,t$, we have that $H/h \in \Omega_{\gamma^j} L^j_\bullet(h)$. To complete the proof of Assertion (2) of Definition 4.1, we only need to identify $W(h)/h$ with W.

Step 3. The constructions in Steps 1 and 2 give us, for every $H \in \Omega_{\underline{\lambda}\circ\underline{\rho}}\mathcal{F}_\bullet$, a pair of subspaces $h := H \cap V$ and $H/h \in W(h)/h$, as well as instances $\mathcal{E}_\bullet$ of $\underline{\lambda}$ in $Gr(a,V)$ and $\mathcal{F}_\bullet(h)$ of $\underline{\rho}$ in $Gr(c, W(h)/h)$ such that $h \in \Omega_{\underline{\lambda}}\mathcal{E}_\bullet$ and $H/h \in \Omega_{\underline{\rho}}\mathcal{F}_\bullet(h)$.

The mapping $H \mapsto (h, H/h)$ is clearly injective. It is surjective, as by Theorem 3.9, there are $\delta(\underline{\lambda}\circ\underline{\rho}) = \delta(\underline{\lambda})\cdot\delta(\underline{\rho})$ elements H in $\Omega_{\underline{\lambda}\circ\underline{\rho}}\mathcal{F}_\bullet$, which is the number of pairs $(h, H/h)$ such that $h \in \Omega_{\underline{\lambda}}\mathcal{E}_\bullet$ and $H/h \in \Omega_{\underline{\rho}}\mathcal{F}_\bullet(h)$. This completes Assertion (3) of Definition 4.1.

Step 4. We already observed that $\mathcal{E}_\bullet$ is sufficiently general. For the same assertions regarding $\mathcal{F}_\bullet(h)$, we observe that the flags $L^j_\bullet$ are in general position with respect to V, as are the subspaces K^j_r for $r \le b+d+c-m_j$, and finally the subspaces $F^1_{b_2+a_1+r}/F^1_{b_2+a_1}$ for $r \ge 0$ are general in W, and the same for $F^2_{b_1+a_2+r}/F^2_{b_1+a_2}$. This completes the proof of Theorem 4.8, as well as Corollary 4.9. □

4.4 Computational evidence for Conjecture 4.3

We present evidence that nontrivial compositions $\underline{\lambda}\circ\underline{\rho}$ of Schubert problems have imprimitive Galois groups, even when $\underline{\lambda}$ is not a block column Schubert problem.

Let $\underline{\lambda} = (\underline{\mu}, \underline{\nu})$ with $\underline{\mu} = (\square\square, \square\square, \square)$, $\nu = (\boxminus, \square, \square)$, and $\underline{a} = \underline{b} = (1,1,1)$ be the composable partition of Example 3.5. Let $\underline{\rho}$ be the Schubert problem $(\square,\square,\square,\square)$ on $G(2,2)$, expanded with some empty partitions, so that $\underline{\alpha} = \underline{\beta} = (\square, \emptyset, \emptyset)$ and $\underline{\gamma} = (\square,\square)$. The composition $\underline{\lambda}\circ\underline{\rho}$ consists of the eight partitions

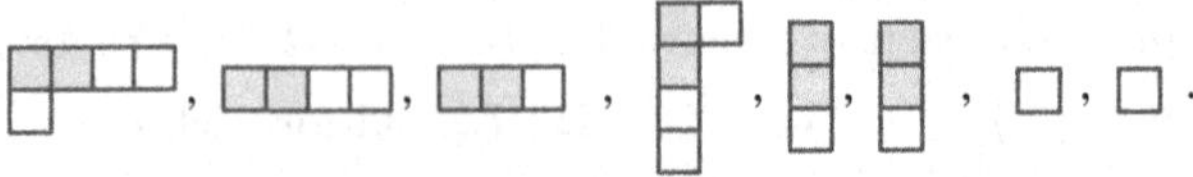

Algorithms in Schubert calculus imply that $\delta(\underline{\lambda}) = 5$ and $\delta(\underline{\rho}) = 2$, and we know from previous computations that both are full symmetric, $\mathrm{Gal}_{\underline{\lambda}} = S_5$ and $\mathrm{Gal}_{\underline{\rho}} = S_2$.

As explained in [13, § 2.2], we may use symbolic computation to study the Galois group of a Schubert problem $\underline{\kappa}$ of moderate size by computing cycle types of Frobenius lifts. This does not quite study $\mathrm{Gal}_{\underline{\kappa}}$, but rather $\mathrm{Gal}_{\underline{\kappa}}(\mathbb{Q})$, the Galois group of the branched cover $\Omega_{\underline{\kappa}} \to (\mathbb{F}\ell(a+b))^r$, where we have restricted scalars to $\mathbb{Q}$, instead of $\mathbb{C}$.

We computed 1 million Frobenius lifts in $\mathrm{Gal}_{\underline{\lambda}\circ\underline{\rho}}(\mathbb{Q})$, finding 24 cycle types. This computation supports the conjecture that $\mathrm{Gal}_{\underline{\lambda}\circ\underline{\rho}}(\mathbb{Q})$ is the expected wreath product $(S_2)^5 \rtimes S_5$, which is the hyperoctahedral group B_5. This group has $2^5 \cdot 5! = 3840$ elements. Table 1 shows the frequency of cycle types found. Each row is labeled by the cycle type, with the second column recording

Table 1. *Frequency of cycle types of* $\underline{\lambda} \circ \underline{\rho}$.

Cycles found in 1000000 instances of the Schubert problem $\underline{\lambda} \circ \underline{\rho}$							
Type	Freq.	Fraction	actual	Type	Freq.	Fraction	actual
(10)	99816	383.293	384	$(1^2,2^2,4)$	46399	178.172	180
$(2,8)$	62583	240.319	240	$(1^4,2,4)$	15367	59.009	60
$(1^2,8)$	62347	239.412	240	$(1^6,4)$	5157	19.803	20
$(4,6)$	41883	160.831	160	$(2^2,3^2)$	62596	240.369	240
$(2^2,6)$	62690	240.730	240	$(1^2,2,3^2)$	41447	159.156	160
$(1^2,2,6)$	41362	158.830	160	$(1^4,3^2)$	20932	80.379	80
$(1^4,6)$	20899	80.252	80	(2^5)	20960	80.486	81
(5^2)	100844	387.241	384	$(1^2,2^4)$	32217	123.713	125
$(2,4^2)$	78690	302.170	300	$(1^4,2^3)$	17806	68.375	70
$(1^2,4^2)$	78429	301.167	300	$(1^6,2^2)$	7706	29.591	30
$(3^2,4)$	42056	161.495	160	$(1^8,2)$	1261	4.842	5
$(2^3,4)$	36331	139.511	140	(1^{10})	222	0.852	1

the frequency. The empirical fraction of times the identity was obtained, $10^6/222 \simeq 4504.5$, suggests that the order of the group $|\mathrm{Gal}_{\underline{\lambda}\circ\underline{\rho}}(\mathbb{Q})|$ is a divisor of 10! near this number. Taking into account the empirical evidence from the computations in [13] that the identity is under sampled, we suppose that it is one of

$$3600, 3780, 3840, 4032, 4050, 4200, 4320, 4480.$$

Assuming that $|\mathrm{Gal}_{\underline{\lambda}\circ\underline{\rho}}(\mathbb{Q})| = 3840$, the third column gives the (normalized to 3840) fraction of times that cycle type was observed. The last column gives the number of elements in B_5 with the given cycle type in S_{10}.

References

[1] Améndola, C., Lindberg, J., and Rodriguez, J. I. 2016. *Solving parameterized polynomial systems with decomposable projections*. arXiv:1612.08807.

[2] Benkart, G., Sottile, F., and Stroomer, J. 1996. Tableau switching: Algorithms and applications. *J. Comb. Theory Ser. A*, **76**, 11–43.

[3] Brooks, C. J., Martín del Campo, A., and Sottile, F. 2015. Galois groups of Schubert problems of lines are at least alternating. *Trans. Amer. Math. Soc.*, **367**, 4183–4206.

[4] Brysiewicz, T., Rodriguez, J. I., Sottile, F., and Yahl, T. 2021. Solving decomposable sparse systems. *Numerical Algorithms*, **88**, 453–474.

[5] Derksen, Harm, and Weyman, Jerzy. 2011. The combinatorics of quiver representations. *Ann. Inst. Fourier (Grenoble)*, **61**(3), 1061–1131.

[6] Fulton, William. 1997. *Young tableaux*. London Mathematical Society Student Texts, vol. 35. Cambridge University Press, Cambridge.

[7] Harris, J. 1979. Galois Groups of Enumerative Problems. *Duke Math. J.*, **46**, 685–724.

[8] Jordan, C. 1870. *Traité des Substitutions et des équations algébrique*. Gauthier-Villars, Paris.

[9] Kleiman, S. L. 1974. The transversality of a general translate. *Compositio Math.*, **28**, 287–297.

[10] Leykin, A., and Sottile, F. 2009. Galois groups of Schubert problems via homotopy computation. *Math. Comp.*, **78**(267), 1749–1765.

[11] Macdonald, I. G. 2015. *Symmetric functions and Hall polynomials*. Second edn. Oxford Classic Texts in the Physical Sciences. The Clarendon Press, Oxford University Press, New York.

[12] Martín del Campo, A., and Sottile, F. 2016. Experimentation in the Schubert Calculus. Pages 295–336 of: Naruse, H., Ikeda, T., Masuda, M., and Tanisaki, T. (eds), *Schubert Calculus, Osaka 2012*. Advanced Studies in Pure Mathematics, vol. 71. Mathematical Society of Japan.

[13] Martín del Campo, A., Sottile, F., and Williams, R. 2019. *Classification of Schubert Galois groups in Gr(4,9)*. arXiv.org/1902.06809.

[14] Pirola, Gian Pietro, and Schlesinger, Enrico. 2005. Monodromy of projective curves. *J. Algebraic Geom.*, **14**(4), 623–642.

[15] Sagan, Bruce E. 2001. *The symmetric group*. Second edn. Graduate Texts in Mathematics, vol. 203. Springer-Verlag, New York.

[16] Sottile, F., and White, J. 2015. Double transitivity of Galois groups in Schubert calculus of Grassmannians. *Algebr. Geom.*, **2**(4), 422–445.

[17] Stanley, Richard P. 1999. *Enumerative combinatorics. Vol. 2*. Cambridge Studies in Advanced Mathematics, vol. 62. Cambridge University Press, Cambridge.

[18] Vakil, R. 2006a. A geometric Littlewood-Richardson rule. *Ann. of Math. (2)*, **164**(2), 371–421. Appendix A written with A. Knutson.

[19] Vakil, R. 2006b. Schubert induction. *Ann. of Math. (2)*, **164**(2), 489–512.

[20] Wielandt, Helmut. 1964. *Finite permutation groups*. Translated from the German by R. Bercov. Academic Press, New York-London.

25

A K-theoretic Fulton Class

Richard P. Thomas[a]

Abstract. Fulton defined classes in the Chow group of a quasi-projective scheme M which reduce to its Chern classes when M is smooth. When M has a perfect obstruction theory, Siebert gave a formula for its virtual cycle in terms of its total Fulton class.

We describe K-theory classes on M which reduce to the exterior algebra of differential forms when M is smooth. When M has a perfect obstruction theory, we give a formula for its K-theoretic virtual structure sheaf in terms of these classes.

1 Summary

Fix a quasi-projective scheme M over the complex numbers, and pick a global embedding in a smooth ambient variety A. Let $I \subset \mathcal{O}_A$ denote the ideal sheaf of M. We get the cone on the embedding $M \hookrightarrow A$,

$$C_M A := \operatorname{Spec} \bigoplus_{i \geq 0} I^i / I^{i+1}. \tag{1}$$

Then Fulton's total Chern class of M [7, Example 4.2.6] is

$$c_F(M) := c\big(T_A|_M\big) \cap s(C_M A) \in A_\bullet(M), \tag{2}$$

where c is the total Chern class and s denotes the Segre class. The result is independent of the choice of embedding. When M is smooth, $c_F(M)$ is just the total Chern class $c(T_M) \cap [M] = \sum_{i \geq 0} c_i(M) \cap [M]$ of M.

We define a K-theoretic analogue. For notation see Section 2; in particular t denotes the class of the weight one irreducible representation of $\mathbb{C}^*$.

[a] Imperial College London

Theorem 1.1 *Let $\mathbb{C}^*$ act trivially on M, with weight* 1 *on $\Omega_A|_M$, and with weight i on I^i/I^{i+1}. The K-theoretic Fulton class*

$$\Lambda_M := \Lambda^\bullet\Omega_A\big|_M \otimes \left(\bigoplus_{i\geq 0} I^i/I^{i+1}\right) \in K_0(M)[\![t]\!],$$

is independent of the ambient variety $A \supset M$ and is polynomial in t,

$$\Lambda_M = \sum_{i=0}^{d} (-1)^i \Lambda^i_M t^i \in K_0(M)[t],$$

where d is the embedding dimension of M. If M is smooth, $\Lambda^i_M = \Omega^i_M$.

In Proposition 3.1 we describe Λ_M in de Rham terms for M lci, but for more general M we have not seen these classes in the literature.

Suppose M has a perfect obstruction theory $E^\bullet \to \mathbb{L}_M$ [3] of virtual dimension vd := rank$(E^\bullet)$. Then we get a virtual cycle [3] for which Siebert [16] gave the following formula in terms of the Fulton class

$$[M]^{\mathrm{vir}} = \left[s\big((E^\bullet)^\vee\big) \cap c_F(M)\right]_{\mathrm{vd}} \in A_{\mathrm{vd}}(M). \tag{3}$$

A K-theoretic analogue is the following.

Theorem 1.2 *Given a perfect obstruction theory $E^\bullet \to \mathbb{L}_M$ the virtual structure sheaf can be calculated in terms of the K-theoretic Fulton class Λ_M as*

$$\mathcal{O}_M^{\mathrm{vir}} = \left[\frac{\Lambda_M}{\Lambda^\bullet(E^\bullet)}\right]_{t=1} = \left[\frac{\Lambda_M}{\Lambda^\bullet \mathbb{L}_M^{\mathrm{vir}}}\right]_{t=1}. \tag{4}$$

Siebert's formula (3) shows that $[M]^{\mathrm{vir}}$ depends only on the scheme structure of M and the K-theory class $[E^\bullet]$ of the virtual cotangent bundle $\mathbb{L}_M^{\mathrm{vir}} := E^\bullet$ (and not the map $E^\bullet \to \mathbb{L}_M$ defining the perfect obstruction theory). Similarly the above Theorem implies that $\mathcal{O}_M^{\mathrm{vir}}$ also depends only on M and $[E^\bullet]$. One aspect of the analogy we refer to is that — with some work — (3) follows from (4) by the virtual Riemann-Roch theorem of [5, 6]. There is further discussion of the analogy between cohomological and K-theoretic Chern classes and Fulton classes in Section 5.

To understand why Λ_M might be polynomial in t, consider what happens in the case that M is a zero dimensional Artinian scheme. (The general case is a family version of this.) By [2, Section 11] the Hilbert series of the graded $\mathcal{O}_M$-module $\bigoplus_{i\geq 0} I^i/I^{i+1}$ is $p(t)/(1-t)^d$, where $d = \dim A$ and p is polynomial in t. But $\Omega_A|_M \cong \mathcal{O}_M^{\oplus d} \otimes t$ so tensoring by $\Lambda^\bullet\Omega_A|_M = (1-t)^d\mathcal{O}_M$ gives the result.

Acknowledgements. It is an honour to dedicate this paper to Bill Fulton on the occasion of his 80th birthday. Much of my career has depended crucially on the masterpiece [7]. You should never meet your heroes, but I did and it was a genuine pleasure.

Thanks to Ben Antieau for suggesting Proposition 3.1 and Dave Anderson for lengthy discussions on the link to K-theoretic Chern and Segre classes (see Section 5). I am especially grateful to an alert referee, Jason Starr and Will Sawin for correcting my false assumptions about embedding dimension [15]. I acknowledge support from EPSRC grant EP/R013349/1.

2 K-theoretic analogue of Fulton's Chern class

Throughout this paper we fix a quasi-projective scheme M over $\mathbb{C}$, endow it with the *trivial* $\mathbb{C}^*$ action, and work with the $\mathbb{C}^*$-equivariant K-theory of coherent sheaves on M. This is

$$K_0(M)^{\mathbb{C}^*} \;\cong\; K_0(M)[t,t^{-1}]$$

as a module over $K(\mathrm{pt})^{\mathbb{C}^*} = \mathbb{Z}[t,t^{-1}]$, where t is the class of the weight one irreducible representation of $\mathbb{C}^*$. In fact we only ever use only the subgroup $K_0(M)[t]$ generated by coherent sheaves with $\mathbb{C}^*$ actions *with nonnegative weights*, and its completion

$$K_0^{\mathbb{C}^*}(M)_{\geq 0} \otimes_{\mathbb{Z}[t]} \mathbb{Z}[\![t]\!] \;=\; K_0(M)[\![t]\!].$$

For E locally free we let $\Lambda^\bullet E$ denote the K-theory class $\sum_{i=0}^{\operatorname{rank} E}(-1)^i\Lambda^i E$. If E carries a $\mathbb{C}^*$ action then on the total space of $\pi\colon E^* \to M$ the tautological section of π^*E^* is $\mathbb{C}^*$-equivariant, giving an equivariant Koszul resolution $\Lambda^\bullet\pi^*E \xrightarrow{\sim} \mathcal{O}_M$ and so the identity $\Lambda^\bullet\pi^*E = \mathcal{O}_M$ in $K_0(E^*)^{\mathbb{C}^*}$. Applying $\pi_*\colon K_0(E^*)^{\mathbb{C}^*} \to K_0(M)[\![t]\!]$ to this when E has only strictly positive weights gives $\Lambda^\bullet E \otimes \operatorname{Sym}^\bullet E = \mathcal{O}_M$. That is, $\Lambda^\bullet E = 1/\operatorname{Sym}^\bullet E$ in $K_0(M)[\![t]\!]$.

Remark. This paper commutes with operations such as localisation with respect to a nontrivial $\mathbb{C}^*$ action on M (as used in [17], for instance) since any such $\mathbb{C}^*$ action commutes with the trivial $\mathbb{C}^*$ action used here.

Let $\mathbb{C}^*$ act on $\Omega_A|_M$ with weight 1 and on I^i/I^{i+1} with weight i. Consider

$$\Lambda_M \;:=\; \Lambda^\bullet\Omega_A\big|_M \otimes \left(\bigoplus_{i\geq 0} I^i/I^{i+1}\right) \;\in\; K_0(M)[\![t]\!]. \tag{5}$$

Theorem 2.1 *This* Λ_M *is independent of the smooth ambient space* $A \supset M$ *and is* polynomial in t, *defining the K-theoretic Fulton classes*

$$\Lambda_M = \sum_{i=0}^{d} (-1)^i \Lambda_M^i t^i \in K_0(M)[t].$$

Here d *is the embedding dimension of* M. *When* M *is smooth,* $\Lambda_M^i = \Omega_M^i$.

Remark. By "embedding dimension" we mean the smallest dimension of a smooth variety A containing M. It is natural to wonder if we can replace this in the theorem by the maximum of the dimensions of the Zariski tangent spaces $T_x M$ over all closed points $x \in M$. This number is in general smaller than the embedding dimension [15].

Proof of Theorem Fix two embeddings $M \subset A_i$, $i = 1,2$ with ideal sheaves I_i and cones $C_i := C_M A_i$. We get the induced diagonal inclusion

$$M \subset A_1 \times A_2$$

with ideal I_{12} and cone $C_{12} := C_M(A_1 \times A_2)$. This gives the exact sequence of cones [7, Example 4.2.6],

$$0 \longrightarrow T_{A_2}\big|_M \longrightarrow C_{12} \longrightarrow C_1 \longrightarrow 0. \tag{6}$$

That is, I_{12}^i / I_{12}^{i+1} has an increasing filtration beginning with I_1^i / I_1^{i+1} and with graded pieces $\mathrm{Sym}^j \, \Omega_{A_2}|_M \otimes I_1^{i-j} / I_1^{i-j+1}$. Therefore, in completed equivariant K-theory,

$$\begin{aligned}
\Lambda^\bullet \Omega_{A_1 \times A_2}\big|_M &\otimes \left(\bigoplus_{i \geq 0} I_{12}^i / I_{12}^{i+1} \right) \\
&= \Lambda^\bullet \Omega_{A_1 \times A_2}\big|_M \otimes \left(\bigoplus_{i,j \geq 0} \mathrm{Sym}^j \, \Omega_{A_2}\big|_M \otimes I_1^{i-j} / I_1^{i-j+1} \right) \\
&= \Lambda^\bullet \Omega_{A_1}\big|_M \otimes \Lambda^\bullet \Omega_{A_2}\big|_M \otimes \mathrm{Sym}^\bullet \, \Omega_{A_2}\big|_M \otimes \bigoplus_{i \geq 0} I_1^i / I_1^{i+1} \\
&= \Lambda^\bullet \Omega_{A_1}\big|_M \otimes \bigoplus\nolimits_{i \geq 0} I_1^i / I_1^{i+1}.
\end{aligned}$$

This gives the independence from A. We now show (5) is polynomial in t of degree $\leq d := \dim A$.

Writing (5) as $\pi_* \pi^* \Lambda^\bullet \Omega_A|_M$, where $\pi \colon C_M A \to M$ denotes the projection, the power series in t comes from the fact that π is not proper. So we only need to show that $\pi^* \Lambda^\bullet \Omega_A|_M$ is equivalent in $K_0^{\mathbb{C}^*}(C_A M)$ to a class pushed forward from M with $\mathbb{C}^*$ weights in $[0, d]$.

The basic idea of the proof is the following. Suppose we could pick a $\mathbb{C}^*$-invariant section s of $\pi^*T_A|_M$ which — on the complement $C_MA\backslash M$ of M — has vanishing locus of the expected dimension 0. By $\mathbb{C}^*$ invariance this locus is empty, so the vanishing locus of s is just M (on which s must be zero since $T_A|_M$ has weight -1). Therefore the Koszul complex $\pi^*\Lambda^\bullet\Omega_A|_M$ is exact on $C_MA\backslash M$. It is equivalent in K-theory to its cohomology sheaves which are supported on $M \subset C_MA$ and have $\mathbb{C}^*$ weights in $[0,d]$. This gives the result required.

In general there only exist sections of $\pi^*T_A|_M$ *with poles* so we have to twist by a divisor D in the proof below, complicating matters. The reader is encouraged to read it first in the no-poles case — where $i = 0$, there are no $\mathcal{O}_D$ terms in (7), and everything reduces to the Koszul complex (10). The general case involves the generalised Koszul complexes (11).

So pick a line bundle $L \gg 0$ on M and $D \in |L|$. We work with

$$E := T_A\big|_M \otimes L$$

since it has sections which we can use to cut down to lower dimensions. We give L the trivial $\mathbb{C}^*$ action of weight 0 so that E has weight -1. We have

$$\Lambda^k\Omega_A\big|_M = \Lambda^k E^* \otimes L^k = \Lambda^k E^* \otimes (\mathcal{O} - \mathcal{O}_D)^{-k}.$$

Using $\otimes$ for the *derived* tensor product, in K-theory we deduce[1]

$$\begin{aligned} \Lambda^\bullet\Omega_A\big|_M &= \sum_{k=0}^{d}(-1)^k\Lambda^k E^* \otimes \sum_{i=0}^{d}\binom{k+i-1}{i}\mathcal{O}_D^{\otimes i} \\ &= \sum_{i=0}^{d}\mathcal{O}_D^{\otimes i} \otimes \sum_{k=0}^{d}(-1)^k\Lambda^k E^* \otimes \operatorname{Sym}^{k-1}\mathcal{O}^{i+1}. \end{aligned} \tag{7}$$

Here we have used $[\mathcal{O}_D^{\otimes i}] = 0$ for $i > d \geq \dim M$. For $L \gg 0$ basepoint free this follows from taking divisors $D_1, \dots, D_{d+1} \in |L|$ with empty intersection.

More generally, fix any $i \geq 0$. Since the sections $\pi^*H^0(L)$ of π^*L are basepoint free, we may pick generic divisors $D_1, \dots, D_i$ such that their intersection $D^{(i)} := D_1 \cap \cdots \cap D_i$, and the intersection $\pi^*D^{(i)}$ of their pullbacks π^*D_i, are *transverse*. That is, we have an equality of *derived* tensor products

[1] When $k = 0 = i$ we have to work with the standard negative binomial convention that $\binom{k+i-1}{i} = \binom{-1}{0} = 1$. Therefore we also set $\operatorname{Sym}^{-1}\mathcal{O}^{i+1}$ to be $\mathcal{O}$ for $i = 0$ and 0 for $i > 0$.

$$\mathcal{O}_{D_1} \otimes \cdots \otimes \mathcal{O}_{D_i} = \mathcal{O}_{D^{(i)}} \quad \text{and} \quad \mathcal{O}_{\pi^* D_1} \otimes \cdots \otimes \mathcal{O}_{\pi^* D_i} = \mathcal{O}_{\pi^* D^{(i)}}.$$

Thus by (7) we can write $\pi^* \Lambda^\bullet \Omega_A|_M$ as the sum over $i = 0, \ldots, d$ of

$$\sum_{k=0}^{d} (-1)^k \pi^* \Lambda^k E^*\big|_{D^{(i)}} \otimes \mathrm{Sym}^{k-1} \mathcal{O}^{i+1}_{\pi^* D^{(i)}}. \tag{8}$$

So we are left with showing (8) is equal, in equivariant K-theory, to a class pushed forward from $D^{(i)} \subset \pi^* D^{(i)}$ with $\mathbb{C}^*$ weights in $[0, d]$.

Consider the projectivised cone $\mathbb{P}(C_M A) = \mathrm{Proj} \bigoplus_{j \geq 0} I^j / I^{j+1}$ and its fibre product

$$\mathbb{P}(C_M A) \times_M D^{(i)} \xrightarrow{q} D^{(i)}. \tag{9}$$

It has dimension $d - i - 1$, while $q^* E(1)$ has rank d and is globally generated. Therefore $q^* E(1)$ has $i + 1$ linearly independent sections on (9) by a standard argument (see [13, page 148], for instance).

Equivalently, $\pi^*\big(E|_{D^{(i)}}\big)$ has $i + 1$ $\mathbb{C}^*$-equivariant sections which are linearly independent away from the zero section $D^{(i)} \subset \pi^* D^{(i)}$ and whose scheme-theoretic degeneracy locus is precisely $D^{(i)} \subset \pi^* D^{(i)}$ (i.e. its 0th Fitting ideal is the irrelevant ideal $\bigoplus_{j>0} I^j / I^{j+1} \subset \mathcal{O}_{\pi^* D^{(i)}}$).

Therefore, taking $i = 0$ to begin with, we get one equivariant section of $\pi^* E$ and a Koszul complex

$$\pi^* \Lambda^d E^* \longrightarrow \pi^* \Lambda^{d-1} E^* \longrightarrow \cdots \longrightarrow \pi^* E^* \longrightarrow \mathcal{O}_{C_M A} \tag{10}$$

with cokernel $h^0 \cong \mathcal{O}_M$. Considering it as a sheaf of dgas, its other cohomology sheaves are modules over h^0, so they are also supported scheme-theoretically on $M \subset C_M A$. Since M is $\mathbb{C}^*$-fixed, and $\mathbb{C}^*$ acts on the $\Lambda^k E^*$ term of (10) with weight $k \in [0, d]$, this shows the (pushdown to M of) the cohomology sheaves of (10) have $\mathbb{C}^*$ weights in $[0, d]$. Therefore, for $i = 0$, its K-theory class (8) has $\mathbb{C}^*$ weights in $[0, d]$.

For $i \geq 1$ we use the generalised Koszul complex

$$\begin{aligned} &\pi^* \Lambda^d E^*|_{D^{(i)}} \otimes \mathrm{Sym}^{d-1} \mathcal{O}^{i+1}_{\pi^* D^{(i)}} \longrightarrow \pi^* \Lambda^{d-1} E^*|_{D^{(i)}} \otimes \mathrm{Sym}^{d-2} \mathcal{O}^{i+1}_{\pi^* D^{(i)}} \\ &\longrightarrow \cdots \longrightarrow \pi^* \Lambda^2 E^*|_{D^{(i)}} \otimes \mathcal{O}^{i+1}_{\pi^* D^{(i)}} \longrightarrow \pi^* E^*|_{D^{(i)}}. \end{aligned} \tag{11}$$

It is exact away from $D^{(i)} \subset \pi^* D^{(i)}$, since its twist by $\pi^* \Lambda^d E\big|_{D^{(i)}}$ is the obvious resolution of $\Lambda^{d-1} Q$, where Q is the locally free quotient $0 \to \mathcal{O}^{i+1} \to \pi^* E|_{D^{(i)}} \to Q \to 0$; since $\operatorname{rank} Q = d - i - 1 < d - 1$, this is zero. Furthermore, its cohomology sheaves are well-known[2] to be supported scheme-theoretically on $D^{(i)}$.

Since $D^{(i)}$ is $\mathbb{C}^*$-fixed, and $\mathbb{C}^*$ acts on the $\Lambda^k E^*$ term of (11) with weight k for $k = 0, 1, \ldots, d$, this shows the (pushdown to $D^{(i)}$ of) the cohomology sheaves of (11) have $\mathbb{C}^*$ weights in $[0, d]$. Therefore the K-theory class (8) of (11) has $\mathbb{C}^*$ weights in $[0, d]$. □

If we pick a locally free resolution $F^* \xrightarrow{s} I$ on A (i.e. a vector bundle $F \to A$ with a section s cutting out $s^{-1}(0) = M \subset A$) then we can express Λ_M differently as follows. Give F the $\mathbb{C}^*$ action of weight -1, so the embedding

$$C_M A \hookrightarrow F|_M \tag{12}$$

induced by s is equivariant. Let $\iota\colon M \hookrightarrow F|_M$ and $\pi : F|_M \to M$ denote the zero section and projection respectively.

Lemma 2.2 *The K-theoretic Fulton class equals*

$$\Lambda_M \;=\; L\iota^* \mathcal{O}_{C_M A} \otimes \left.\frac{\Lambda^\bullet \Omega_A}{\Lambda^\bullet F^*}\right|_M .$$

Proof Applying $\pi_*\iota_* = \mathrm{id}$ to the right hand side gives

$$\pi_*\left(\iota_*\mathcal{O}_M \overset{L}{\otimes} \mathcal{O}_{C_M A}\right) \otimes \left.\frac{\Lambda^\bullet \Omega_A}{\Lambda^\bullet F^*}\right|_M .$$

Then $\iota_*\mathcal{O}_M = \pi^*\Lambda^\bullet F^*|_M$ by the Koszul resolution of the zero section $M \hookrightarrow F|_M$, so by (1) we get

$$\Lambda_M \;=\; \left(\bigoplus_{i\geq 0} I^i/I^{i+1}\right) \otimes \Lambda^\bullet\Omega_A|_M . \qquad \square$$

3 de Rham cohomology

Consider the pushforward of Λ_M to the formal completion $\widehat{A}$ of A along M. Its K-theory class looks remarkably similar to that of Hartshorne's algebraic de Rham complex [8]

$$\left(\Lambda^\bullet \Omega_{\widehat{A}}, d\right). \tag{13}$$

[2] Nonetheless, I couldn't find a decent reference. On $\rho\colon \mathbb{P}^i \times \pi^* D^{(i)} \to \pi^* D^{(i)}$ the $i+1$ sections of $\pi^* E$ give a section of $\rho^*\pi^* E(1)$ with scheme-theoretic zero locus $\mathbb{P}^i \times D^{(i)} \subset \mathbb{P}^i \times \pi^* D^{(i)}$. Hence the Koszul complex $\Lambda^\bullet(\rho^*\pi^* E^*(-1))$ of this section has cohomology sheaves supported on $\mathbb{P}^i \times D^{(i)}$. To this complex we apply $R\rho_*(\,\cdot\, \otimes \mathcal{O}(-i))[i]$ to give (11) and the result claimed.

If we discard the de Rham differential in (13) and filter by order of vanishing along M (with nth filtered piece $I^n \otimes \Lambda^\bullet \Omega_{\widehat{A}}$) then the associated graded is

$$\Lambda^\bullet \Omega_A\big|_M \otimes \Big(\bigoplus_{i\geq 0} I^i/I^{i+1}\Big). \tag{14}$$

This is just (the push forward from M to $\widehat{A}$ of) Λ_M with its $\mathbb{C}^*$ action forgotten. However convergence issues stop us from equating (14) with (13) in K-theory. Putting the $\mathbb{C}^*$ action back into (14) we get convergence to Λ_M in completed equivariant K-theory, but in general there is no $\mathbb{C}^*$ action on (13).

The algebraic de Rham complex (13) — with the de Rham differential — has been shown by Illusie [10, Corollary VIII.2.2.8] (and more generally Bhatt [4]) to be quasi-isomorphic to the pushforward of the *derived de Rham complex* $\Lambda^\bullet \mathbb{L}_M$ of M. (Here $\Lambda^\bullet$ denotes the alternating sum of *derived* exterior powers.) And, as suggested to us by Ben Antieau, we can prove that the K-theory class of $\Lambda^\bullet \mathbb{L}_M$ (again *without* its de Rham differential) can be identified with Λ_M when M is a local complete intersection.[3]

Proposition 3.1 *Let $\mathbb{C}^*$ act on $\mathbb{L}_M$ with weight 1, and suppose M is lci. Then $\Lambda_M = \Lambda^\bullet \mathbb{L}_M$ in $K_0(M)[\![t]\!]$.*

Proof For M lci we have $\mathbb{L}_{M/A} = I/I^2[1]$ so the exact triangle $\mathbb{L}_A|_M \to \mathbb{L}_M \to \mathbb{L}_{M/A}$ gives, in K-theory,

$$\mathbb{L}_M \;=\; \Omega_A\big|_M - I/I^2.$$

Using the weight one $\mathbb{C}^*$ action on $\mathbb{L}_M$ this gives

$$\Lambda^\bullet \mathbb{L}_M \;=\; \Lambda^\bullet \Omega_A\big|_M \otimes \operatorname{Sym}^\bullet I/I^2 \;\in\; K_0(M)[\![t]\!].$$

Furthermore I/I^2 is locally free so

$$\operatorname{Sym}^i I/I^2 \longrightarrow I^i/I^{i+1}$$

is a surjection from a locally free sheaf to a sheaf of the same rank. It is therefore an isomorphism and we have

$$\Lambda^\bullet \mathbb{L}_M \;=\; \Lambda^\bullet \Omega_A\big|_M \otimes \Big(\bigoplus_{i\geq 0} I^i/I^{i+1}\Big). \qquad \square$$

By Theorem 2.1 this means $\Lambda^\bullet \mathbb{L}_M$ is in fact polynomial in t, so we can set $t = 1$ to get a class in non-equivariant K-theory.

[3] When M is not lci $\mathbb{L}_M$ will have homology in infinitely many degrees in general (even before we take exterior powers), so it is not clear how to define a K-theory class for $\Lambda^\bullet \mathbb{L}_M$.

However it does *not* follow (and indeed is not in general true) that the push forward of Λ_M can be equated with the algebraic de Rham complex (13) when M is lci. Firstly, the de Rham differential does not preserve the $\mathbb{C}^*$ action we have used, so we cannot lift Illusie's theorem to *equivariant* K-theory. Secondly, this therefore gives us convergence issues; Illusie and Bhatt use the "Hodge completion" of the derived de Rham complex to get their quasi-isomorphism, and this differs from our completion.

4 A formula for the virtual structure sheaf

The foundations of cohomological virtual cycles are laid down in [3, 12]; we use the notation from [3]. The foundations for K-theoretic virtual cycles (or "*virtual structure sheaves*") are laid down in [5, 6]; we use the notation from [6].

Again let M be a quasi-projective scheme over $\mathbb{C}$. A perfect obstruction theory $E^\bullet \to \mathbb{L}_M$ is a 2-term complex of vector bundles $E^\bullet = \{E^{-1} \to E^0\}$ with a map in $D(\operatorname{Coh} M)$ to the cotangent complex[4] $\mathbb{L}_M$ which is an isomorphism on h^0 and a surjection on h^{-1}.

We sometimes call $E^\bullet$ the *virtual cotangent bundle* $\mathbb{L}_M^{\mathrm{vir}}$ of M. Its rank is the *virtual dimension* $\mathrm{vd} := \operatorname{rank} E^0 - \operatorname{rank} E^{-1}$.

By [3] this data defines a cone $C \subset E_1 := (E^{-1})^*$ from which we may define M's *virtual cycle*

$$\big[M\big]^{\mathrm{vir}} := \iota^![C] \in A_{\mathrm{vd}}(M),$$

where $\iota\colon M \to E_1$ is the zero section. Siebert [16] proved the alternative formula

$$[M]^{\mathrm{vir}} = \big[s\big((E^\bullet)^\vee\big) \cap c_F(M)\big]_{\mathrm{vd}}.$$

The construction of Section 2 allows us to give a K-theoretic analogue. Firstly, the K-theoretic analogue of $[M]^{\mathrm{vir}}$ is the *virtual structure sheaf* [6]

$$\mathcal{O}_M^{\mathrm{vir}} := \big[L\iota^*\,\mathcal{O}_C\big] \in K_0(M), \tag{15}$$

where $L\iota^*\,\mathcal{O}_C$ is a bounded complex because ι is a regular embedding.

Theorem 4.1 *The virtual structure sheaf* (15) *can be calculated in terms of the K-theoretic Fulton class* Λ_M (5) *as*

[4] Or its $\tau^{[-1,0]}$ truncation.

$$\mathcal{O}_M^{\mathrm{vir}} = \left[\frac{\Lambda_M}{\Lambda^\bullet(E^\bullet)}\right]_{t=1} = \left[\frac{\Lambda_M}{\Lambda^\bullet \mathbb{L}_M^{\mathrm{vir}}}\right]_{t=1}. \tag{16}$$

In particular, when M is smooth $\Lambda_M = \Lambda^\bullet \Omega_M$ and $[E^\bullet] = \Omega_M - \mathrm{ob}_M^*$, so (16) recovers $\mathcal{O}^{\mathrm{vir}} = \Lambda^\bullet \mathrm{ob}_M^*$.

Proof By [3], the perfect obstruction theory $E^\bullet \to \mathbb{L}_M$ induces a cone $C \subset E_1$ which Siebert [16, proof of Proposition 4.4] shows[5] sits inside an exact sequence of cones

$$0 \longrightarrow T_A\big|_M \longrightarrow C_M A \oplus E_0 \longrightarrow C \longrightarrow 0. \tag{17}$$

Here A is any smooth ambient space containing M with ideal I, so that $C_M A = \mathrm{Spec} \bigoplus_{i\geq 0} I^i/I^{i+1}$.

As before we give the E_i and $T_A|_M$ weight -1 (so E^i and $\Omega_A|_M$ have weight 1) and let $\iota\colon M \hookrightarrow E_1$ and $\pi : E_1 \to M$ denote the zero section and projection respectively. Then

$$\mathcal{O}^{\mathrm{vir}} = L\iota^* \mathcal{O}_C = \pi_* \iota_* L\iota^* \mathcal{O}_C = \pi_*\big(\mathcal{O}_C \overset{L}{\otimes} \iota_* \mathcal{O}_M\big) = \pi_*\big(\mathcal{O}_C \otimes \Lambda^\bullet E^{-1}\big)$$

evaluated at $t = 1$, by (15) and the Koszul resolution of $\iota_* \mathcal{O}_M$. By (17),

$$\pi_* \mathcal{O}_C = \mathrm{Sym}^\bullet E^0 \otimes \Big(\bigoplus_{i\geq 0} I^i/I^{i+1}\Big) \otimes \Lambda^\bullet \Omega_A\big|_M,$$

so by (5),

$$\mathcal{O}^{\mathrm{vir}} = \left[\Lambda_M \otimes \mathrm{Sym}^\bullet E^0 \otimes \Lambda^\bullet E^{-1}\right]_{t=1} = \left[\Lambda_M / \Lambda^\bullet(E^\bullet)\right]_{t=1}. \qquad \square$$

Corollary 4.2 *$\mathcal{O}_M^{\mathrm{vir}}$ depends only on M and the K-theory class of $E^\bullet$.*

Of more interest in enumerative K-theory is the *twisted* virtual structure sheaf

$$\widehat{\mathcal{O}}_M^{\mathrm{vir}} = \mathcal{O}_M^{\mathrm{vir}} \otimes \det(E^\bullet)^{1/2}$$

of Nekrasov–Okounkov [14]. Here we twist by a choice of square root of the virtual canonical bundle $\det(E^\bullet) = \det E^0 \otimes (\det E^{-1})^*$ when one exists (or use $\mathbb{Z}[\frac{1}{2}]$ coefficients to ensure that it does). The above shows it depends only on M, the K-theory class of $E^\bullet$, and the choice of square root (or "*orientation data*").

[5] After possibly replacing $E^\bullet$ by a quasi-isomorphic 2-term complex of vector bundles.

5 K-theoretic Chern classes

The K-theory of complex vector bundles is an oriented cohomology theory, so admits a notion of Chern classes. In particular the rth Chern class of a rank r bundle E — the K-theoretic Euler class of E — is

$$c_r^K(E) := \Lambda^\bullet E^* \in K^0(M). \tag{18}$$

Whenever there exists a transverse section of E this is the class $[\mathcal{O}_Z]$ of the structure sheaf of its zero locus $Z \subset M$.

Dave Anderson asked if the K-theoretic Fulton class Λ_M of (5) could be written in terms of K-theoretic Chern classes and Segre classes.[6] I do not think there is a simple formula along these lines, essentially because the formal group law for K-theory is nontrivial, complicating the K-theoretic analogue of the formula (21) below. For instance if we simply substitute c^K and s^K into the right hand side of the definition (2)

$$c_F(M) := c\big(T_A|_M\big) \cap s(C_{M/A}) \in A_\bullet(M) \tag{19}$$

we do *not* get Λ_M.

However if we first note the usual Fulton class $c_F(M)$ can be re-expressed in terms of a $\mathbb{C}^*$-equivariant Euler class, then there is a nice analogue in $\mathbb{C}^*$-equivariant K-theory. Begin by rewriting (19) as

$$c_F(M) = \Big[t^d c_{\frac{1}{t}}\big(T_A|_M\big) \cdot t^{m-d} s_{\frac{1}{t}}(C_{M/A})\Big]_{t=1} \tag{20}$$

where $m = \dim M$, $d = \dim A$ and $c_t := \sum t^i c_i$, $s_t := \sum t^i s_i$. Now let $\mathfrak{t}$ denote the standard weight one irreducible representation of $\mathbb{C}^*$, so we can recycle the notation t as $t := c_1(\mathfrak{t}) \in H^2(B\mathbb{C}^*)$. This allows us to rewrite the first term in the brackets of (20) as the equivariant Euler class

$$t^d c_{\frac{1}{t}}\big(T_A|_M\big) = e^{\mathbb{C}^*}(T_A|_M \otimes \mathfrak{t}) \tag{21}$$

in the $\mathbb{C}^*$-equivariant cohomology,[7] or Chow cohomology, of M (with the trivial $\mathbb{C}^*$ action). By (18) its K-theoretic analogue is therefore

$$t^d c_{\frac{1}{t}}\big(T_A|_M\big) \longleftrightarrow \Lambda^\bullet(T_A^*|_M \otimes \mathfrak{t}^{-1}). \tag{22}$$

For the same reason I would like to think of the second term in the brackets of (20) as an *inverse* equivariant Euler class of $C_{M/A}$,

$$t^{m-d} s_{\frac{1}{t}}(C_{M/A}) \text{ “ = ” } \frac{1}{e^{\mathbb{C}^*}(C_{M/A} \otimes \mathfrak{t})}. \tag{23}$$

[6] See [9, 1] for one definition of K-theoretic Segre classes.
[7] As usual we localise by inverting t.

When $C_{M/A}$ is a bundle this makes perfect sense, at least. In that case the K-theoretic analogue of (23) is, by (18),

$$\frac{1}{\Lambda^\bullet(C^*_{M/A}\otimes \mathfrak{t}^{-1})} = \mathrm{Sym}^\bullet(C^*_{M/A}\otimes \mathfrak{t}^{-1}) = \bigoplus_{i\geq 0}\mathfrak{t}^{-i}(I^i/I^{i+1}),$$

after completing with respect to $\mathfrak{t}$ or localising appropriately. So it is natural to think of this as giving the K-theoretic analogue

$$t^{m-d}s_{\frac{1}{t}}(C_{M/A}) \longleftrightarrow \bigoplus_{i\geq 0}\mathfrak{t}^{-i}(I^i/I^{i+1}) \tag{24}$$

even when $C_{M/A}$ is not a bundle. Substituting (22, 24) into (20) gives

$$\Lambda^\bullet(\Omega_A|_M\otimes \mathfrak{t}^{-1})\otimes\bigoplus_{i\geq 0}\mathfrak{t}^{-i}(I^i/I^{i+1}).$$

On replacing $\mathfrak{t}$ by $\mathfrak{t}^{-1}$ this is precisely our K-theoretic Fulton class (5).

References

[1] D. Anderson, *K-theoretic Chern class formulas for vexillary degeneracy loci*, Adv. Math. **350** (2019), 440–485. arXiv:1701.00126.

[2] M. F. Atiyah and I. G. Macdonald, *Introduction to commutative algebra*, Addison-Wesley, 1969.

[3] K. Behrend and B. Fantechi, *The intrinsic normal cone*, Invent. Math. **128** (1997), 45–88. alg-geom/9601010.

[4] B. Bhatt, *Completions and derived de Rham cohomology*, arXiv:1207.6193.

[5] I. Ciocan-Fontanine and M. Kapranov, *Virtual fundamental classes via dg-manifolds*, Geom & Top **13** (2009), 1779–1804. math.AG/0703214.

[6] B. Fantechi and L. Göttsche, *Riemann-Roch theorems and elliptic genus for virtually smooth schemes*, Geom & Top **14** (2010), 83–115. arXiv:0706.0988.

[7] W. Fulton, *Intersection theory*, Springer-Verlag (1998).

[8] R. Hartshorne, *On the de Rham cohomology of algebraic varieties*, Publ. Math. IHES **45** (1976), 5–99.

[9] T. Hudson, T. Ikeda, T. Matsumura and H. Naruse, *Degeneracy loci classes in K-theory – determinantal and Pfaffian formula*, Adv. in Math. **320** (2017), 115–156. arXiv:1504.02828.

[10] L. Illusie, *Complexe cotangent et déformations, II*, Lecture Notes Math. **283**, Springer (1972).

[11] M. Levine and F. Morel, *Algebraic cobordism*, Springer Monographs in Math., 2007.

[12] J. Li and G. Tian, *Virtual moduli cycles and Gromov-Witten invariants of algebraic varieties*, J. Amer. Math. Soc. **11** (1998), 119–174. alg-geom/9602007.

[13] D. Mumford, *Lectures on curves on an algebraic surface*, Princeton University Press (1966).

[14] N. Nekrasov and A. Okounkov, *Membranes and Sheaves*, Algebraic Geometry **3** (2016), 320–369. arXiv:1404.2323.
[15] W. Sawin and J. Starr, https://mathoverflow.net/questions/351280, (2020).
[16] B. Siebert, *Virtual fundamental classes, global normal cones and Fulton's canonical classes*, in: Frobenius manifolds, ed. K. Hertling and M. Marcolli, Aspects Math. **36** (2004), 341–358, Vieweg. math.AG/0509076.
[17] R. P. Thomas, *Equivariant K-theory and refined Vafa-Witten invariants*, Comm. Math. Phys. **378** (2020), 1451–1500. arXiv:1810.00078.

Printed in the United States
by Baker & Taylor Publisher Services